Understanding Analog Side Channels Using Cryptography Algorithms

Alenka Zajić • Milos Prvulovic

Understanding Analog Side Channels Using Cryptography Algorithms

Alenka Zajić
Atlanta, GA, USA

Milos Prvulovic
Atlanta, GA, USA

ISBN 978-3-031-38581-0   ISBN 978-3-031-38579-7   (eBook)
https://doi.org/10.1007/978-3-031-38579-7

© The Editor(s) (if applicable) and The Author(s), under exclusive license to Springer Nature Switzerland AG 2023

This work is subject to copyright. All rights are solely and exclusively licensed by the Publisher, whether the whole or part of the material is concerned, specifically the rights of translation, reprinting, reuse of illustrations, recitation, broadcasting, reproduction on microfilms or in any other physical way, and transmission or information storage and retrieval, electronic adaptation, computer software, or by similar or dissimilar methodology now known or hereafter developed.

The use of general descriptive names, registered names, trademarks, service marks, etc. in this publication does not imply, even in the absence of a specific statement, that such names are exempt from the relevant protective laws and regulations and therefore free for general use.

The publisher, the authors, and the editors are safe to assume that the advice and information in this book are believed to be true and accurate at the date of publication. Neither the publisher nor the authors or the editors give a warranty, expressed or implied, with respect to the material contained herein or for any errors or omissions that may have been made. The publisher remains neutral with regard to jurisdictional claims in published maps and institutional affiliations.

This Springer imprint is published by the registered company Springer Nature Switzerland AG
The registered company address is: Gewerbestrasse 11, 6330 Cham, Switzerland

Paper in this product is recyclable.

*To
our daughter Mia
for her patience with us.*

Preface

Side channels are one of the methods for obtaining information about program execution. Traditionally, they are used in computer science to extract information about a secret key in cryptographic algorithms. What makes side channels different from other ways of extracting information about program execution is that they rely on how a system implements program execution, rather than what the program's algorithm specifies. Analog side channels are particularly powerful because they are not easy to suppress, and it is also difficult to detect that someone is collecting information from the system. Although side channels are very powerful tool, they are poorly understood.

The aim of this book is to explain analog side channels and demonstrate new use cases for them. In the first part of the book, we discuss how analog side channels are generated, the physics behind them, how to model and measure them, and their analogies to wireless communication systems. In the second part of the book, we introduce new applications that benefit from leveraging side channels. In addition to breaking cryptographic algorithms, we show how analog side channels can be used for malware detection, program profiling, hardware profiling, hardware/software attestation, hardware identification, and hardware Trojan detection.

This book has a wide intended audience; this includes electrical engineers who might be interested in how analog side channels work, how to model them, measure them, improve signal to noise ratio, and invent new signal processing techniques. But our intended audience also includes computer scientists, who might be interested to learn about new applications of side channels to improve system security, new techniques for breaking cryptography keys, new techniques for attestation, and new techniques for hardware Trojan detection.

Atlanta, GA, USA
Alenka Zajić
Milos Prvulovic

Contents

1	**Introduction**		1
	1.1 A Brief History of Analog Side Channels		2
	1.2 Overview of Remaining Chapters		3
	References		5
2	**What Is an Analog Side Channel?**		7
	2.1 Classification of Analog Side Channels		7
	References		11
3	**Analog Side Channels**		13
	3.1 Inverter: A Representative Building Block of Digital Circuits		13
	3.2 Instruments to Measure Analog Side Channel Signals		17
	3.3 Digital Circuits as Sources of Power Side Channels		18
	3.4 Digital Circuits as Sources of Electromagnetic Side Channels		21
	3.5 Digital Circuits as Sources of Temperature Side Channels		22
	3.6 Digital Circuits as Sources of Acoustic Side Channels		23
	3.7 Digital Circuits as Sources of Backscattering Side Channels		24
		3.7.1 Equivalent Circuit Model for a Toggling Circuit	27
	3.8 Digital Circuits as Sources of Photonic Side Channels		31
	References		34
4	**Unintentionally Modulated Side Channels**		37
	4.1 Direct vs. Unintentionally Modulated Side Channels		37
	4.2 How Are Side-Channel Signals Modulated?		38
	4.3 Unintentional Side Channel Modulations vs. Traditional Communication Modulations		43
	4.4 How to Identify Unintentionally Modulated Side Channel Signals Created by Computer Activity?		45
		4.4.1 An Algorithm for Finding AM and FM Unintentional Carriers in Computer Systems	46
		4.4.2 Identifying Carrier Frequencies	48
		4.4.3 Identifying Modulation	51

		4.4.4	Illustration of How the Algorithm Works	52
	References			56

5 Relationship Between Modulated Side Channels and Program Activity ... 57
- 5.1 Mathematical Relationship Between Electromagnetic Side Channel Energy of Individual Instructions and the Measured Pairwise Side Channel Signal Power ... 57
  - 5.1.1 Definition of Individual Instructions' ESE ... 57
  - 5.1.2 Derivation of Single-Instruction ESE from Measured Spectral Power ... 59
- 5.2 Mathematical Relationship Between Electromagnetic Side Channel Energy of Sequence of Instructions and the Measured Pairwise Side Channel Signal Power ... 67
  - 5.2.1 Definition for Emanated Signal Power (ESP) of Individual Instructions as They Pass Through Pipeline . 68
  - 5.2.2 Estimating ESP from the Total Emanated EM Signal Power Created by a Program ... 68
- 5.3 Practical Use for the Relationship Between Emanated Energy and Program Execution ... 74
- References ... 75

6 Parameters that Affect Analog Side Channels ... 77
- 6.1 Defining the Parameters that Impact Side Channels ... 77
  - 6.1.1 Signal Bandwidth and Sampling Rate ... 77
  - 6.1.2 Definition of Center Frequency, Span, Resolution Bandwidth, and Sweep Time ... 81
  - 6.1.3 Thermal and Coupling Noise in Electrical Measurements ... 83
- 6.2 How Sampling Rate Impacts Side Channel Measurements? ... 85
- 6.3 How Bandwidth Impacts Side Channel Measurements? ... 87
- 6.4 An Example of How Bandwidth Impacts Side Channel Signals ... 87
  - 6.4.1 Modular Exponentiation in OpenSSL's RSA ... 88
  - 6.4.2 Impact of Signal Bandwidth on Branch Decision Recognition ... 92
- 6.5 How Distance Affects EM Side Channel Results? ... 93
  - 6.5.1 Path Loss Measurements for EM Side Channel Signals ... 93
  - 6.5.2 Indoor and Outdoor Models for Propagation of Side Channel Signals ... 96
  - 6.5.3 How to Select Antennas and Probes to Improve Signal to Noise Ratio? ... 104
- 6.6 How Carrier Power, Frequency, and Distance Affect the Backscattering Side Channel? ... 108
- References ... 109

7	**Modeling Analog Side Channels as Communication Systems**		113
	7.1	Introduction	113
	7.2	Software-Activity-Created Signals and Communication Model	115
	7.3	Communication Model for Covert Channel Software-Activity-Created Signals	117
	7.4	Quantifying the Information Leakage of Covert Channel Software-Activity-Created Signals	122
		7.4.1 Bit Error Rate and Power Spectral Density	122
		7.4.2 Practical Considerations and Examples of Cover Side Channels	125
		7.4.3 Demonstration of the Analog Covert Channel on More Complex Systems	128
	7.5	Capacity of Side Channel Created by Execution of Series of Instructions in a Computer Processor	131
	7.6	Modeling Information Leakage from a Computer Program as a Markov Source Over a Noisy Channel	132
		7.6.1 Brief Overview of Markov Model Capacity Over Noisy Channels	132
		7.6.2 Markov Source Model for Modeling Information Leakage from a Sequence of Instructions	133
	7.7	Introducing Information Leakage Capacity for the Proposed Markov Source Model	135
		7.7.1 Reducing the Size of the Markov Source Model	136
		7.7.2 An Empirical Algorithm to Evaluate the Leakage Capacity	137
	7.8	Estimating Channel Input Power in the Proposed Markov Model	139
	7.9	Practical Use of Information Leakage Capacity	140
		7.9.1 Leakage Capacity for an FPGA-Based Processor	142
		7.9.2 Experimental Results and Leakage Capacity for AMD Turion X2 Laptop	143
		7.9.3 Experimental Results and Leakage Capacity for Core 2 DUO Laptop	144
		7.9.4 Experimental Results and Leakage Capacity for Core I7 Laptop	145
		7.9.5 Utilizing the Capacity Framework for Security Assessment	146
	References		149
8	**Using Analog Side-Channels for Malware Detection**		151
	8.1	Introduction	151
	8.2	Frequency-Domain Analysis of Analog Side Channel Signals for Malware Detection	152
		8.2.1 Periodic Activities and Their Spectral Components	152
		8.2.2 Tracking Code in Frequency Domain Over Time	153

		8.2.3	Tracking Modulated Code in Frequency Domain Over Time and its Application to Malware Detection	154
	8.3	Illustration of Malware Detection Using Analog Side Channel Analysis in Frequency Domain		155
		8.3.1	Spectral Samples (SS)	156
		8.3.2	Distance Metric for Comparing SSs	156
		8.3.3	Black-Box Training	158
		8.3.4	Monitoring	159
		8.3.5	Experimental Results	160
		8.3.6	Measurement Setup	161
		8.3.7	File-Less Attacks on Cyber-Physical-Systems	162
		8.3.8	Shellcode Attack on IoTs	166
		8.3.9	APT Attack on Commercial CPS	168
		8.3.10	Further Evaluation of Robustness	169
	8.4	Time-Domain Analysis of Analog Side Channel Signals for Malware Detection		173
	8.5	Time Domain Analog Side Channel Analysis for Malware Detection		174
		8.5.1	Training Phase: Dictionary Learning	175
		8.5.2	Learning Words	175
		8.5.3	Word Normalization	176
		8.5.4	Dictionary Reduction Through Clustering	176
		8.5.5	Monitoring Phase: Intrusion Detection	177
		8.5.6	Matching and Reconstruction	177
		8.5.7	Detection	178
		8.5.8	System Parameters	178
		8.5.9	Experiments with Different Malware Behaviors	183
		8.5.10	Experiments with Cyber-Physical Systems	188
		8.5.11	Experiments with IoT Devices	189
	8.6	MarCNNet: A Markovian Convolutional Neural Network for Malware Detection and Monitoring Multi-Core Systems		190
		8.6.1	Emanated EM Signals During Program Execution	191
		8.6.2	Markov-Model-Based Program Profiling: MarCNNet	193
		8.6.3	Input Signal and Training Phase	195
		8.6.4	Testing While Multiple Cores Are Active	197
		8.6.5	Experimental Setup and Results	200
	References			207
9	**Using Analog Side Channels for Program Profiling**			**211**
	9.1	Introduction		211
	9.2	Spectral Profiling		212
		9.2.1	Training Phase	213
		9.2.2	Profiling Phase	215

	9.2.3	Sequence-Based Matching	216
	9.2.4	Illustration of Program Profiling Results	218
	9.2.5	Loops with Input-Independent Spectra	220
	9.2.6	Accuracy for Loop Exit/Entry Time Profiling	221
	9.2.7	Runtime Behavior of Loops	222
	9.2.8	Effects of Changing Architecture	223
	9.2.9	Effects of Window Size	223
9.3	Time-Domain Analysis of Analog Side Channel Signals for Software Profiling		224
9.4	ZoP: Zero-Overhead Profiling via EM Emanations		225
	9.4.1	Training 1	227
	9.4.2	Training 2	230
	9.4.3	Inferring Timing for the Uninstrumented Code Using Time Warping	231
	9.4.4	Profiling	232
	9.4.5	Empirical Evaluation	236
	9.4.6	Illustration of Program Profiling Results Using ZOP	238
9.5	P-TESLA: Basic Block Program Tracking Using Electromagnetic Side Channels		240
	9.5.1	Signal Preprocessing: Amplitude Demodulation	242
	9.5.2	Program Execution Monitoring	248
	9.5.3	IoT Device Monitoring	254
9.6	PITEM: Permutations-Based Instruction Tracking via Electromagnetic Side-Channel Signal Analysis		255
	9.6.1	Determining Instruction Types by Using EM Side Channel	256
	9.6.2	Generating a List of Instructions Under Investigation	257
	9.6.3	Generating Microbenchmarks	257
	9.6.4	Recording EM Emanations for Microbenchmarks	258
	9.6.5	Data Processing to Obtain EM Signatures	258
	9.6.6	Generating Correlation Matrix	259
	9.6.7	Identifying Instruction Types	260
	9.6.8	Detecting Permutations of Instruction Types	260
	9.6.9	Picking an Instruction to Represent Each Instruction Type and Generating Microbenchmarks for Permutations	262
	9.6.10	Training: Generating Templates for Each Permutation	263
	9.6.11	Testing: Predicting Testing Measurements Using Templates	263
	9.6.12	Results	264
	9.6.13	Further Evaluation of Robustness	273
	9.6.14	Performance for Different SNR Levels	273
References			276

10	**Using Analog Side Channels for Hardware Event Profiling**		279
	10.1 Introduction		279
	10.2 Memory Profiling Using Analog Side Channels		280
		10.2.1 Side-Channel Signal During a Processor Stall	281
		10.2.2 Memory Access in Simulated Side-Channel Signal	283
		10.2.3 Memory Access in Measured EM Side-Channel Signals	284
		10.2.4 Prototype Implementation of EMProf	285
		10.2.5 Validation of EMProf	287
		10.2.6 Validation by Cycle-Accurate Architectural Simulation	289
		10.2.7 Validation by Monitoring Main Memory's Signals	290
		10.2.8 Experimental Results	293
		10.2.9 Effect of Varying Measurement Bandwidth	294
		10.2.10 Code Attribution	296
	10.3 Profiling Interrupts Using Analog Emanations		297
		10.3.1 Overview of PRIMER	298
		10.3.2 Training PRIMER	300
		10.3.3 Validation	303
		10.3.4 Results	305
		10.3.5 Applications	309
		10.3.6 Profiling Exceptions: Page Faults	310
		10.3.7 Profiling Interrupts: Network Interrupts	314
		10.3.8 Potential Other Uses of PRIMER	318
	References		319
11	**Using Analog Side Channels for Hardware/Software Attestation**		323
	11.1 Introduction		323
	11.2 Software Attestation		325
	11.3 EMMA: EM-Monitoring Attestation		326
	11.4 Experimental Evaluation		334
	11.5 Sensitivity Analysis		340
	References		343
12	**Using Analog Side Channels for Hardware Identification**		345
	12.1 Introduction		345
	12.2 Side Channel Signals Carrying Information About Hardware Properties		346
		12.2.1 Spectral Features	347
		12.2.2 Device Excitation	347
	12.3 Signal Compression and Processing		349
	12.4 Training and Testing Process		351
	12.5 Experimental Validation		352
		12.5.1 Measurement Setup	352
		12.5.2 Test Devices	354
		12.5.3 Measurement Projection	356

		12.5.4	Recognition of Memory Components	359
		12.5.5	Recognition of Processor Components	361
		12.5.6	Recognition of Ethernet Transceiver Components	364
	References			366

13 Using Analog Side Channels to Attack Cryptographic Implementations ... 369

	13.1	Introduction		369
	13.2	Historical Overview of Analog Side-Channel Attacks on RSA		371
	13.3	Historical Overview of Analog Side-Channel Attacks on Elliptic Curve		376
		13.3.1	Point-by-Scalar Multiplication	377
	13.4	One and Done: An EM-Based Attack on RSA		381
		13.4.1	Threat Model	381
		13.4.2	Attack Method	382
		13.4.3	Receiving the Signal	382
		13.4.4	Identifying Relevant Parts of the Signal	382
		13.4.5	Recovering Exponent Bits in the Fixed-Window Implementation	384
		13.4.6	Recovering Exponent Bits in the Sliding-Window Implementation	387
		13.4.7	Full Recovery of RSA Private Key Using Recovered Exponent Bits	389
		13.4.8	Measurement Setup	392
		13.4.9	Results for OpenSSL's Constant-Time Fixed-Window Implementation	393
		13.4.10	Results for the Sliding-Window Implementation	396
		13.4.11	Mitigation	397
	13.5	The Noncc@Once Attack		399
		13.5.1	Attack Setting	399
		13.5.2	Attack Overview	399
		13.5.3	Leveraging Constant-Time Swap	400
		13.5.4	Attack Overview for Constant-Time Swap Operation	400
		13.5.5	Leakage Amplification	401
		13.5.6	A Physical Side Channel Attack	401
		13.5.7	EM Signal Acquisition and Signature Location	401
		13.5.8	Identifying Conditional Swap Signals	403
		13.5.9	Recovering the Value of the Swap Condition from a Snippet	403
		13.5.10	Nonce and Key Recovery	405
		13.5.11	Experimental Evaluation	405
		13.5.12	Experimental Setup	406
		13.5.13	Attack Results	406
		13.5.14	Other Elliptic Curves and Primitives	410
		13.5.15	Mitigation	410
	References			415

14	**Using Analog Side Channels for Hardware Trojan Detection**		**419**
	14.1	Introduction	419
	14.2	What Is a Hardware Trojan?	419
		14.2.1 Hardware Trojans: Characteristics and Taxonomy	420
		14.2.2 Overview of Methods for HT Detection	424
	14.3	Hardware Trojan Detection Using the Backscattering Side Channel	425
	14.4	HT Detection Algorithm	428
		14.4.1 Training	428
		14.4.2 Detection	429
	14.5	Experimental Setup	430
		14.5.1 Training and Testing Subject Circuit Designs	431
	14.6	Evaluation	433
		14.6.1 Dormant-HT Detection with Cross-Training Using the Backscattering Side Channel Signal	434
		14.6.2 HT Detection of Dormant vs. Active HTs Using the Backscattering Side Channel	435
		14.6.3 Comparison to EM-Based HT Detection	436
		14.6.4 Impact of Hardware Trojan Trigger and Payload Size	437
		14.6.5 Impact of HT Trigger and Payload Position	439
		14.6.6 Further Evaluation of HT Detection Using More ICs and HTs	440
	14.7	Detection of Counterfeit IC Using Backscattering Side Channel	445
		14.7.1 Counterfeit ICs with Different Layout	445
		14.7.2 Counterfeit ICs with Changed Position	446
	14.8	A Golden-Chip-Free Clustering for Hardware Trojan Detection Using the Backscattering Side Channel	446
		14.8.1 The Impact of HTs on Backscattering Side-Channel Signal	448
		14.8.2 Graph Model for Clustering Results	448
		14.8.3 Experimental Setup and Testing Scheme Formulation	452
		14.8.4 Hardware Trojan Benchmark Implementation	454
		14.8.5 Testing Scheme Formulation	454
		14.8.6 Evaluation of Existing HT Benchmarks	455
		14.8.7 Evaluation of Changing Size of Hardware Trojan Triggers	458
	References		460

Index 465

Chapter 1
Introduction

Side channels are methods for obtaining information about computation by observing the side-effects of that computation, rather than its inputs/outputs [1, 5, 6]. Traditionally, computation is an implementation of a cryptographic algorithm, and the information extracted using side channels is about the cryptographic key that is being used. However, recent uses of side channels go well beyond such *side channel attacks*, and many of these new uses of side channel are *benevolent*, i.e., the side channels are used in a way that benefits the user that is running the program. What makes side channels different from other ways of extracting information about program execution is that side channels rely on *how* a system implements program execution, rather than *what* the program's algorithm specifies.

To illustrate this concept, consider the following two examples:

Example 1.1 We would like to know if it is raining outside. There are many ways to determine that. For example, one can watch a weather report on a local news channel. One can also check if a precipitation meter reads non-zero rainfall, or use an improvised method like sticking an empty cup out through a window for a few seconds. However, none of these methods are considered to be side channels, because information that is obtained is about the desired fact itself. On the other hand, if one observes that people are passing by in a hallway with umbrellas in their hands, or observes wet floors in hallways, these would be side channel observations about whether it is raining or not. These side channel methods do not obtain information directly, but rather deduce the information from some other information and, as a result, this deduced information has some degree of uncertainty. For example, wet umbrellas in people's hands may have been caused by rain that, at the time of the wet-umbrella observation, has already stopped or, plausibly, by an umbrella-testing event going on farther down the hallway. Similarly, wet floors may be caused by rain outside, but also by floors having been cleaned recently.

Example 1.2 We would like to find a password for another user on a computer system. If we can see the password when it is entered by that user, if that password was written on post-it note or, when we try to log in as that user, if the system responds with a message that reveals some password information, we can obtain the information we are looking for. However, none of these are side-channels. Side channels, in contrast, would involve noting that different keys on a keyboard click differently when typing, or that the system displays an incorrect-password message less quickly when the entered password is a partial match for the correct password, and then using the sounds or system response times to infer the password.

A very useful way of categorizing side channels considers what kind of property is being observed directly, which results in three general categories of side channels:

- Input/Output-Observable side channels,
- Software-Observable side channels,
- Physically-Observable (analog) side channels.

In Input/Output-observable side-channels, information is obtained by observing properties of the target's input/output (I/O) behavior [2]. In Software-observable side channels, information is obtained by a program that executes on the target system [4]. Finally, in physically-observable side channels, information is obtained by measuring physical properties of the target system [3, 7]. In this book, we focus on physically-observable side channels, which are commonly called *analog* side channels.

Note that a system is (usually) not designed to leak information through side channels. Instead, side channels are preset because they are side-effects of how the computer system implements program execution, and it is typically hard to eradicate the information leakage through side channels. In fact, the cryptographic community has spent, and continues to spend, a lot of effort to design algorithms, or to implement existing algorithms, in ways that prevent (or at least reduce) information leakage through side channels.

The aim of this book is to demystify analog side channels and demonstrate new use cases for them. In the first part of the book (Chaps. 1–7), we will discuss how analog side channels are generated, the physics behind them, the modeling and measurements of analog side channels, and the analogies between analog side channels and communication systems. In the second part of the book (Chaps. 8–14), we will introduce new applications that leverage analog side channels. In addition to "breaking" cryptographic algorithms, we will show how analog side channels can be used for beneficial purposes, such as malware detection, program profiling, hardware profiling, hardware/software attestation, hardware identification, and hardware Trojan detection.

1.1 A Brief History of Analog Side Channels

During World War II, Bell Telephone supplied the U.S. military with a mixer device that encrypted teleprinter signals by XOR'ing them with key material from single-use tapes—an implementation of one-time-pad encryption that is theoretically very secure. However, the mixer device used electromechanical relays in its operation and, while testing the device, Bell Telephone engineers discovered that they are able to detect electromagnetic (EM) spikes at a distance from the mixer and, from these spikes, recover the unencrypted (*plaintext*) content of the messages. They informed the Signal Corps, but were met with skepticism over whether the phenomenon they discovered in the laboratory could really be dangerous in practice. To demonstrate the importance of their discovery, the engineers proceeded to recover plaintext messages from a Signal Corps' crypto center on Varick Street in Lower Manhattan, by receiving EM signals form a neighboring building. Now alarmed, the Signal Corps asked Bell Telephone to investigate further. The engineers identified three problem areas: radiated signals, signals conducted on wires extending from the facility, and magnetic fields. As possible solutions, they suggested shielding, filtering, and masking. After this event, existence of side channel attacks and proposed defensive measures became classified, and activities and methods to prevent side channel attacks were referred as TEMPEST. In the 1950s and 1960s, side channel attacks have been used by various spying agencies as a way of obtaining information with little risk of getting caught.

Side channel attacks in the commercial space have emerged in the 1980s, when it was discovered that EM leakage from the local oscillators in domestic television sets can be used to detect, from outside a home, whether that household owns a TV set and, consequently, should be paying an annual license fee to support public broadcast services. In 1985, Wim van Eck, a Dutch researcher, demonstrated that computer data can also leak, by showing how a modified TV set can, at a distance, reconstruct the picture that is being shown on a VDU (Video Display Unit) of a computer. In 1990s, side channel attacks are mainly studied by cryptographers, who were trying to write a code in a way that minimizes leakage from micro controllers embedded in credit cards. In 2010, the authors of this book showed that even more advanced computers, such as laptops and desktops, leak information, at a distance, about loops and other software constructs as they execute. Since then, our objective was to understand physics behind side channels, relationship between software, hardware, and side channels and help reduce guesswork not only in preventing side channel attacks but also in building defenses against malware intrusions and hardware Trojans, and in establishing new applications that greatly benefit from side channel leakage.

1.2 Overview of Remaining Chapters

The remainder of the book is organized as follows:
- Chapter 2 briefly defines analog side channels and how they can be categorized.
- Chapter 3 explains how analog side channels are created in electronics, and how measurements of different physical properties of the system can lead to

extracting similar or different information about program execution. This chapter also explains the relationship between physical properties of electronic circuits, such as current flows, reflection coefficients, and impedances, and the observed analog side channel signals.
- Chapter 4 introduces unintentionally modulated side channels. Here we discuss differences between direct and unintentionally modulated side channels, explain how digital logic ends up creating modulated side channel signals, demonstrate existence of unintentionally modulated side channels and their relationship to performing computation, and discuss how to identify unintentionally modulated side channel signals in the spectrum.
- Chapter 5 discusses the relationship between modulated side channels and computation that produces them. This relationship is important to understand because it tells us what to look for in the signal when analyzing program activity or looking for malware, or any other side channel activity.
- Chapter 6 discusses which parameters impact both direct and unintentionally modulated side channel measurements and analysis. Here we define parameters such as signal bandwidth, sampling rate, capacity, etc. and discuss how these parameters impact side channels. We also discuss how measurement equipment impacts side channel measurements, e.g., different antennas, probes, distance, noise.
- Chapter 7 discusses how side channels can be modeled as communication systems, and also how side channels differ from traditional communication systems. In particular, we focus on covert channels, where a program is intentionally designed or modified to transmit information through side channel signals, because this scenario in many ways resembles a communication system.
- Chapter 8 discusses various signal processing approaches and techniques for malware detection using analog side channels. Analog side channels allow a device to be monitored from outside, with no access to hardware or software of the monitored device. This approach also allows for detecting zero-day malware, which is not possible with many other malware detection methods.
- Chapter 9 discusses various signal processing approaches and techniques for program profiling using analog side-channels. We first describe spectral profiling, which monitors electromagnetic (EM) emanations unintentionally produced by the profiled system, looking for spectral "lines" produced by periodic program activity (e.g., loops). This approach allows us to determine which parts of the program have executed at what time and, by analyzing the frequency and shape of each spectral "line", obtain additional information such as the per-iteration execution time of a loop. If more detail is needed, such as which basic blocks or individual instructions have been executed, time-domain signal analysis is more advantageous because it allows us to look not only at timing-related aspects of the signal, but also the shape of the signal over time.
- Chapter 10 discusses how analog side channels can be used for profiling of hardware events, such as cache misses and interrupts. These events impact performance of software but are difficult to correctly measure in detail because they occur relatively frequently, so traditional methods that record these events

on the profiled device itself often end up causing more performance degradation than the hardware events they are intended to measure.
- Chapter 11 shows how analog side channels can be used for attestation of embedded devices. This approach leverages the side-effects of the device's hardware behavior, but without requiring any specific support from that hardware.
- Chapter 12 discusses how EM side-channels can be used to recognize/authenticate components integrated onto a motherboard. By focusing on components on a motherboard, this method provides an opportunity for designers and manufacturers to authenticate devices assembled by third parties. The purpose is to detect counterfeit ICs based on changes in the EM emanations that comprise the EM side-channel. This method is intended for detecting types of counterfeiting where the physical design of the component is altered.
- Chapter 13 overviews analog side-channel attacks on the well known cryptographic algorithms, RSA and ECC, and demonstrates how they are vulnerable to analog side-channels, even when implemented as constant-time code. We also demonstrate how these vulnerabilities can be mitigated.
- Finally, Chap. 14 defines hardware Trojans (HTs) and reviews their characteristics, and then discusses backscattering side channel can be used to detect HTs and why it outperforms other analog side channels.

References

1. Dakshi Agrawal, Bruce Archambeault, Josyula R. Rao, and Pankaj Rohatgi. The EM side-channel(s). In *Cryptographic Hardware and Embedded Systems - CHES 2002, 4th International Workshop, Redwood Shores, CA, USA, August 13–15, 2002, Revised Papers*, pages 29–45, 2002.
2. Michael Backes, Markus Dürmuth, Sebastian Gerling, Manfred Pinkal, and Caroline Sporleder. Acoustic side-channel attacks on printers. In *19th USENIX Security Symposium, Washington, DC, USA, August 11–13, 2010, Proceedings*, pages 307–322, 2010.
3. Daniel Genkin, Adi Shamir, and Eran Tromer. RSA key extraction via low-bandwidth acoustic cryptanalysis. In *Advances in Cryptology - CRYPTO 2014 - 34th Annual Cryptology Conference, Santa Barbara, CA, USA, August 17–21, 2014, Proceedings, Part I*, pages 444–461, 2014.
4. David Gullasch, Endre Bangerter, and Stephan Krenn. Cache games–bringing access-based cache attacks on aes to practice. In *Security and Privacy (SP), 2011 IEEE Symposium on*, pages 490–505. IEEE, 2011.
5. Paul Kocher. Timing attacks on implementations of diffie-hellman, rsa, dss, and other systems. In *Annual International Cryptology Conference*, pages 104–113. Springer, 1996.
6. Paul Kocher, Joshua Jaffe, and Benjamin Jun. Differential power analysis: leaking secrets. In *Proceedings of CRYPTO'99, Springer, Lecture notes in computer science*, pages 388–397, 1999.
7. Alenka Zajic and Milos Prvulovic. Experimental demonstration of electromagnetic information leakage from modern processor-memory systems. *IEEE Transactions on Electromagnetic Compatibility*, 56:885–893, Aug 2014.

Chapter 2
What Is an Analog Side Channel?

An analog side channel is a method for obtaining information about execution of a program by measuring physical properties of the system that executes the program.

2.1 Classification of Analog Side Channels

A very instructive way of categorizing analog side channels is according to which physical property of the system is being measured:

- **Temperature side channel**; Information from the temperature side channel is obtained by observing how temperature changes (fluctuates) over time in a computing system or its components as the program is executed [3]. This is a slow-changing side channel and can usually only provide information about transitions between longer-lasting (each containing millions of instructions) phases in program execution (Fig. 2.1).
- **Acoustic side channel**; Information from the acoustic side channel is obtained by observing vibrations within, or sounds coming from, the computing system or its components [6]. This side channel can be measured from some distance. It provides a low-pass-filtered (i.e., with rapid changes smoothed over hundreds or thousands of processor cycles) version of what is happening in program execution (Fig. 2.2).
- **Power side channel**; Information from the power side channels is obtained by observing fluctuations in the power consumption of the system (or its components) as the program is executed [9]. This side channel captures a low-pass-filtered version, typically smoothing out rapid changes across tens or hundreds of processor cycles, of what is happening in program execution. To use this side channel, one typically needs physical contact with the system itself (Fig. 2.3).

Fig. 2.1 Temperature side channel

Fig. 2.2 Acoustic side channel

Fig. 2.3 Power side channel

- **Electromagnetic (EM) side channel**; Information from the electromagnetic side channel is obtained by observing fluctuations in the electromagnetic field around, or by receiving electromagnetic waves that emanate from, a computing system or its components [14]. This side channel can be observed from some distance, and it typically provides information with high temporal resolution (i.e., without removing fast-changing activity) about program execution (Fig. 2.4).
- **Photonic side channel**; Information from the photonic side channel is obtained by observing optical emanations in a computing system or its components [4]. This side channel can be measured from some distance, but that is usually achieved by sacrificing fast-changing activity. In contrast, if obtained by scanning across the chip (or imaging the chip), this side channel can yield very detailed information with high temporal resolution (Fig. 2.5).

2.1 Classification of Analog Side Channels

 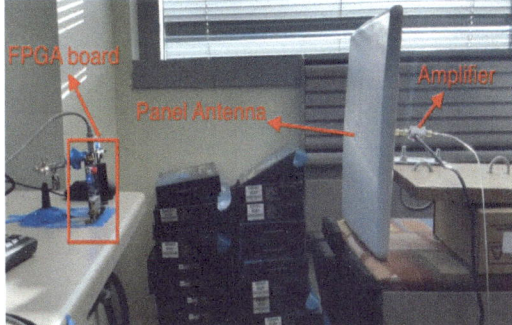

Fig. 2.4 EM side channel

Fig. 2.5 Photonic side channel

- **Backscattering side channel**; Information from the backscattering side channel is obtained by sending an EM wave toward the system and observing the EM signal that returns (is back-scattered) from the system. This side channel actually observes impedance variation in the computing system or its components [11], can be measured from some distance. It can have very high temporal resolution, providing information not only about what happens from one processor cycle to another, but also what happens during a clock cycle, as the program is executed (Fig. 2.6).
- **Fault injection**; Information from faults is obtained by physically introducing errors in computation as the program executed, and then observing the outcome [2]. Although the property that is actually observed is typically the output of the system, fault injections are often considered to be physically observable (analog) side channels because they use a physical method of injecting an error. While error injection can yield very detailed information about some aspect of the computation, this method is very intrusive (can cause a crash or even permanent damage to the system), so it is not suitable for continuous observation of program execution (Fig. 2.7).

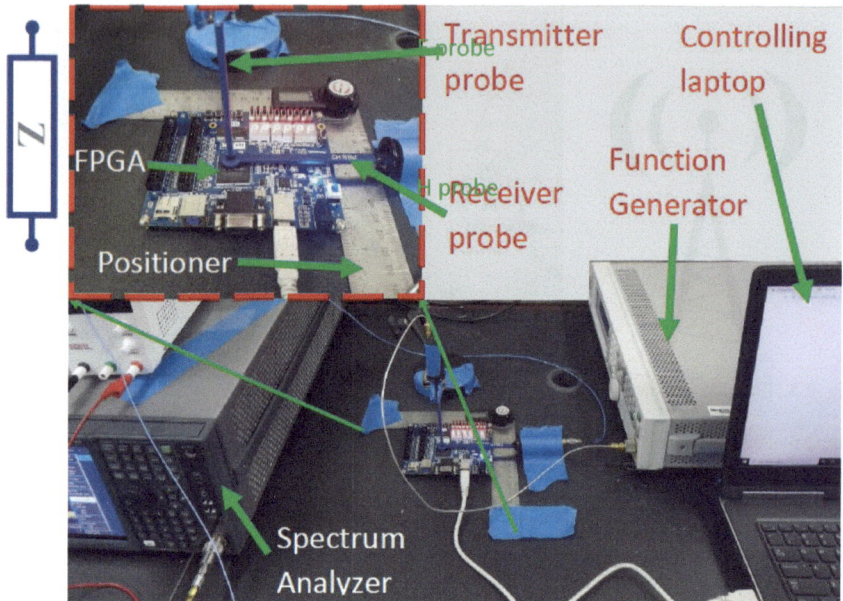

Fig. 2.6 Backscattering side channel

Fig. 2.7 Fault injection side channel

Analog side-channels can also be classified according to whether they involve only passive observation of existing physical properties, or they involve active application of some physical phenomenon to the system.

- **Passive analog side channels**, which include power, temperature, electromagnetic, and acoustic side channels, are all consequences of current flows, and the resulting voltage fluctuations, as transistors switch from one state to another within the electronics of the computer system. More details on how this switching creates each of the passive side channel signals will be provided in Chap. 3.

- **Active analog side channels**, which include fault injection, backscattering, and most photonic side channels, involve application of some signal to the system to *create* the behavior or analog signal that is observed to extract information.

In addition to the physical property that is observed to extract information, uses of analog side channels can also differ in how the analog signal is analyzed to extract information about program execution. These signal analysis methods are typically categorized into:

- Simple Analysis, where information can be obtained by identifying specific shapes in the signal traces, e.g., by visually recognizing parts of the signal trace that correspond to specific program activity. Depending on which analog property the signal corresponds to, they are often referred to in the literature as Simple Power Analysis (SPA) and Simple EM Analysis (SEMA).
- Differential Analysis, where statistical techniques are applied to the signal, e.g., difference of means, correlation, error correction, etc. In the literature, this type of analysis is typically called Differential Power Analysis (DPA) when applied to power signals, or Differential EM Analysis (DEMA) when applied to EM signals.
- High Order Differential Analysis considers data from multiple sources simultaneously.

We note that this categorization of analysis techniques was devised for side channel *attacks* that attempt to extract cryptographic keys [1], but can also be applied to analysis techniques for other uses of analog side channels, e.g. software profiling [12, 13], malware detection [7, 8, 10], hardware Trojan detection [11], profiling of hardware events [5], etc.

In the next chapter we will go into more detail on how analog side channels are physically created by a processor as it executes a program, and how these side channels can be modeled.

References

1. Monjur Alam, Haider A. Khan, Moumita Dey, Nishith Sinha, Robert Locke Callan, Alenka Zajic, and Milos Prvulovic. One&done: A single-decryption em-based attack on openssl's constant-time blinded RSA. In *27th USENIX Security Symposium, USENIX Security 2018, Baltimore, MD, USA, August 15–17, 2018.*, pages 585–602, 2018.
2. Josep Balasch, Benedikt Gierlichs, and Ingrid Verbauwhede. An in-depth and black-box characterization of the effects of clock glitches on 8-bit mcus. In *2011 Workshop on Fault Diagnosis and Tolerance in Cryptography*, pages 105–114, 2011.
3. Julien Brouchier, Tom Kean, Carol Marsh, and David Naccache. Temperature attacks. *Security Privacy, IEEE*, 7(2):79–82, March 2009.
4. Elad Carmon, Jean-Pierre Seifert, and Avishai Wool. Photonic side channel attacks against rsa. In *2017 IEEE International Symposium on Hardware Oriented Security and Trust (HOST)*, pages 74–78, 2017.
5. Moumita Dey, Alireza Nazari, Alenka Zajic, and Milos Prvulovic. Emprof: Memory profiling via em-emanation in iot and hand-held devices. In *2018 51st Annual IEEE/ACM International Symposium on Microarchitecture (MICRO)*, pages 881–893. IEEE, 2018.

6. Daniel Genkin, Adi Shamir, and Eran Tromer. RSA key extraction via low-bandwidth acoustic cryptanalysis. In *Advances in Cryptology - CRYPTO 2014 - 34th Annual Cryptology Conference, Santa Barbara, CA, USA, August 17–21, 2014, Proceedings, Part I*, pages 444–461, 2014.
7. Haider Khan, Monjur Alam, Alenka Zajic, and Milos Prvulovic. Detailed tracking of program control flow using analog side-channel signals: a promise for iot malware detection and a threat for many cryptographic implementations. In *Cyber Sensing 2018*, volume 10630, page 1063005. International Society for Optics and Photonics, 2018.
8. Haider Khan, Nader Sehatbakhsh, Luong Nguyen, Milos Prvulovic, and Alenka Zajic. Malware detection in embedded systems using neural network model for electromagnetic side-channel signals. *J Hardw Syst Security*, 08 2019.
9. Paul Kocher, Joshua Jaffe, and Benjamin Jun. Differential power analysis: leaking secrets. In *Proceedings of CRYPTO'99, Springer, Lecture notes in computer science*, pages 388–397, 1999.
10. Alireza Nazari, Nader Sehatbakhsh, Monjur Alam, Alenka Zajic, and Milos Prvulovic. Eddie: Em-based detection of deviations in program execution. In *2017 ACM/IEEE 44th Annual International Symposium on Computer Architecture (ISCA)*, pages 333–346. IEEE, 2017.
11. Luong Nguyen, Chia-Lin Cheng, Milos Prvulovic, and Alenka Zajic. Creating a backscattering side channel to enable detection of dormant hardware trojans. *IEEE Transactions on Very Large Scale Integration (VLSI) Systems*, 27(7):1561–1574, 2019.
12. Richard Rutledge, Sunjae Park, Haider Khan, Alessandro Orso, Milos Prvulovic, and Alenka Zajic. Zero-overhead path prediction with progressive symbolic execution. In *Proceedings of the 41st International Conference on Software Engineering*, pages 234–245. IEEE Press, 2019.
13. Nader Sehatbakhsh, Alireza Nazari, Alenka Zajic, and Milos Prvulovic. Spectral profiling: Observer-effect-free profiling by monitoring em emanations. In *The 49th Annual IEEE/ACM International Symposium on Microarchitecture*, page 59. IEEE Press, 2016.
14. Alenka Zajic and Milos Prvulovic. Experimental demonstration of electromagnetic information leakage from modern processor-memory systems. *IEEE Transactions on Electromagnetic Compatibility*, 56(4):885–893, Aug 2014.

Chapter 3
Analog Side Channels

Passive analog side channels are a consequence of flows of current to, from, or within electronic circuitry. These side channels tend to be named after the quantity that is actually measured, and so we have power side channels (power consumption), EM side channels (various EM emanations and fields), acoustic side channels (sound), photonic side channels (transistor dopants), and temperature side channels. Active analog side channels, which involve applying some external influence to the system and then measuring the effects of that influence, tend to be named after the physical interaction between the external physical signal process and the system itself, e.g., backscattering side channels or fault injection side channels, but in some cases they are also named after the signal that is applied or measured, e.g., photonic side channels. In the rest of this chapter we will explain the relationship between physical properties, such as current flows, reflection coefficients, and impedances in electronic circuits, and the various analog side channels.

3.1 Inverter: A Representative Building Block of Digital Circuits

Digital circuits are mostly built out of *logic gates*, which take one or more digital inputs and produce a digital output, and *state elements*, which store digital information (bits). Most digital circuits are designed in CMOS technology, which is a way to implement each logic gate and state element out of basic circuit elements, such as transistors, resistors, capacitors, etc. Because CMOS has the dominant position in modern digital electronics, our examples will use CMOS, but we note that similar reasoning can be applied to other technologies.

In CMOS, both gates and state elements are implemented using two types of transistors, without the need for resistors, capacitors, and other circuit elements. There are many types of logic gates that a circuit may use, such as not gates (inverters), and gates, or gates, etc., and also a variety of state elements (various kinds of latches and flip-flops). However, the basic principles of operation for all these types of gates and state elements are similar, so we will use the simplest gate—an inverter—to explain how the internal operation of a logic gate (or state element) relates to various side channels. A CMOS inverter consists of two transistors, as shown in Fig. 3.1, and is connected to a digital input A and produces a digital output that is connected to inputs of other gates, which in this figure are represented by capacitance C_{out}. A transistor acts as a digital switch, where the voltage on the transistor's gate connection controls whether the transistor is *on* or *off*, i.e., whether it allows or prevents current flow between the transistor's other two connections. In Fig. 3.1, the gates of both transistors are connected to input A. The gate of a transistor controls whether the transistor acts as a high-resistance path or a low-resistance path between its other two connections, i.e., whether the transistor acts as switch that is *off* or a switch that is *on*. The two transistors in the inverter are of *complementary* types. The N-type transistor (the bottom one in Fig. 3.1) is *on* when a voltage that is close to V_{dd} is applied to its gate, and i.e., input A represents a value of 1, otherwise this transistor is *off*. The P-type transistor (the top one in Fig. 3.1) has the opposite reaction to its gate (and input A)—*off* when A is 1 and *on* when A is 0.

The gate shown in Fig. 3.1 then behaves as follows. When input A is at logic 0 (close to V_{ss}), the upper transistor is *on*, i.e., it has a relatively low resistance R_{on}, so it connects the inverter's output to V_{dd}, the lower transistor is *off*, so the output's connection to V_{ss} has a very large resistance, i.e., it is practically disconnected from V_{ss}. Thus the output of the inverter has a digital value of 1 (close to V_{dd}). On the other hand, when input A is at logic 1 (close to V_{dd}), the upper transistor is *off* and the lower transistor is *on*, so the inverter's output is 0 (close to V_{ss}). Because the output of this logic gate is 1 when its input is 0, and the output is 0 when the input is 1, this gate correctly implements its logical-not (inverter) functionality.

Note that input A is connected only to gate connections of the two transistors. In CMOS, all gates and state elements are constructed using P-type and N-type

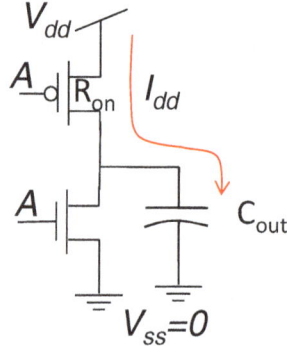

Fig. 3.1 Inverter—example of a building block for digital circuits

3.1 Inverter: A Representative Building Block of Digital Circuits

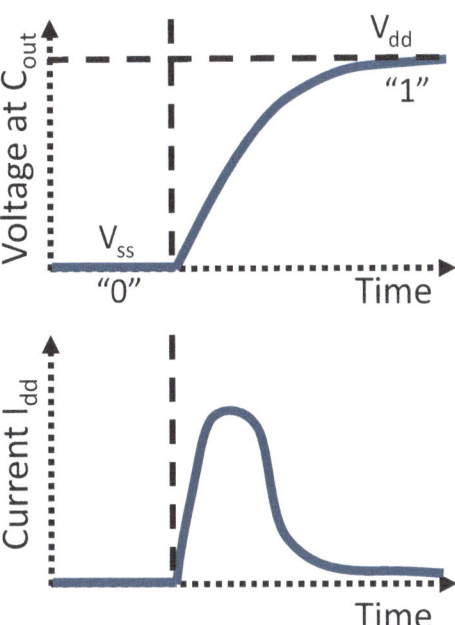

Fig. 3.2 Voltage and current consumption as a function of time on capacitor C_{out} while charging

transistors, with outputs of all gates and state elements connected only to gate connections of other transistors. These gate connections have very high resistance to both V_{ss} and V_{dd}. Therefore, as long as the inverter's output is held at 1, there is almost no current flow from V_{dd} to the output. Similarly, as long as the inverter's output is held as 0, there is almost no current flow from the inverter's output to V_{ss}. This means that the inverter has almost no current—and consumes almost no power—to maintain its output value. This is also true for other logic gates and state elements in CMOS.

However, the transistor gate connections to which the inverter's output is connected do have some capacitance, and the total capacitance of these connections is represented in Fig. 3.1 by a capacitor C_{out}. Because of this capacitance, when the value of the inverter's input changes from 0 to 1, the inverter's output does not instantaneously change from 1 to 0. Instead, there is a short period of time during which current I_{dd} flows from V_{dd} into C_{out}, which charges the capacitance C_{out} to V_{dd}) in order to actually bring the output value to 1. This change in voltage of C_{out}, and in the current I_{dd}, over time is illustrated in Fig. 3.2. Similarly, when input A is 1, C_{out} is discharged by a current flow from C_{out} into V_{ss}. Here we note that power consumption is $P_{dd} = V_{dd}I_{dd}$, and is close to zero except when charging the capacitor C_{out}. Therefore, power is consumed in a short burst whenever the output of the inverter (or any other logic gate) changes from 1 to 0. This is a very important observation because, in any given clock cycle, which logic gates do switch depends on data values and, in a microprocessor, on which instructions the program is currently executing. For most analog side channels, it is the combined currents and power consumption of his switching that "embeds" program- and data-related

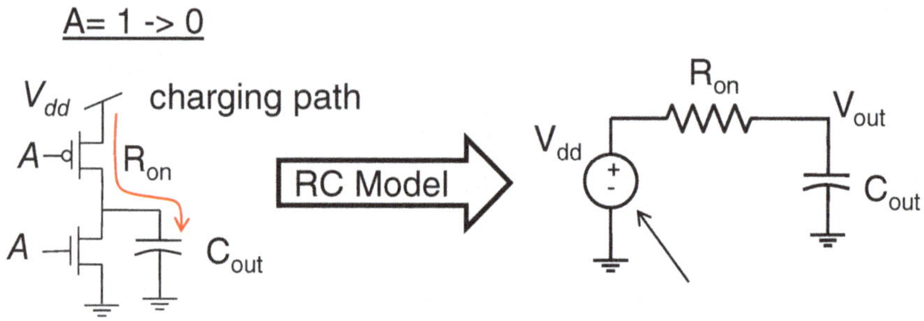

Fig. 3.3 Model of an inverter when switching from input 1 to 0

information into the side channel signal. The power side channel is obtained by measuring fluctuations of V_{dd} or I_{dd}. Using a simple RC model, other analog side channels can be explained as well.

Figure 3.3 illustrates how an inverter can be modeled as a charge based model ($R_{on} C_{on}$ circuit) when the inverter's input is switching from 1 to 0. The capacitor C_{out} starts out with no charge, i.e., the output voltage $V_{out}(t = 0)$ is zero. To calculate the output voltage $V_{out}(t)$ and the current through the circuit $I(t)$, we need to solve the following first order differential equation

$$V_{dd} = V_{out}(t) + RI(t) = V_{out}(t) + RC\frac{dV_{out}}{dt}. \tag{3.1}$$

First order differential equations have solution in the form

$$V_{out}(t) = V_{homogeneous}(t) + V_{particular}, \tag{3.2}$$

where $V_{homogeneous}(t)$ is $A_1 e^{\frac{-t}{RC}}$ and $V_{particular}$ is a constant A_2. To solve the differential equation for A_1 we apply the initial condition (uncharged capacitor) to a general solution, i.e.,

$$V_{out}(t = 0) = A_1 e^{\frac{-t}{RC}} + A_2, \tag{3.3}$$

which leads to $A_1 = -V_{dd}$. To obtain A_2, we use initial condition for particular solution and substitute into differential equation, i.e.,

$$R_{on} C_{out}(0) + A_2 = V_{dd}, \tag{3.4}$$

which leads to $A_2 = V_{dd}$. Finally, we can write the expression for output voltage on the capacitor as $V_{out} = V_{dd}\left(1 - e^{\frac{-t}{RC}}\right)$. The current in this circuit can be calculated as

$$I = \frac{V_{dd} - V_{out}}{R} = \frac{V_{dd}}{R} e^{\frac{-t}{RC}}. \tag{3.5}$$

Using these functions, we can plot how output voltage and current through capacitor change over time as shown in Fig. 3.2. This plot illustrates the power side channel of an inverter. In later sections, we will also use this model to illustrate how other analog side channels are created.

3.2 Instruments to Measure Analog Side Channel Signals

To measure analog side channel signals, we are typically interested in measuring instantaneous signal power over time. For that, we can use real-time oscilloscopes, real-time spectrum analyzers, or software defined radios. Here we will compare these tree measurement instruments and how to use them for side channel measurements.

Oscilloscopes fall into two groups:

- Real-time oscilloscopes—needed for side channel measurements.
- Sampling oscilloscopes—can be used for side channel measurements when averaged samples are beneficial, e.g., hardware Trojan detection.

Real-time oscilloscopes digitize a signal in a real-time. They act like a video camera, taking a series of "frames" of the signal over time. These "frames" are called samples. The amount of samples the real-time oscilloscope captures in each second is called the sampling rate. On the other hand, a sampling oscilloscope requires that the signal be repeated many times. In one repetition, the sampling oscilloscope takes only one sample, but in each repetition this sample is taken at a different point in the signal's timeline, so eventually it gets a highly accurate digitized version of the signal. However, if "repetitions" of the signal are not all the same, e.g., in they vary in timing, one can collect several measurements for each point in the signal's timeline, yielding a recorded signal in which the value of each point corresponds to the average among the versions of the actual signal. Still, this averages out short-duration differences that we are usually looking for in side-channel traces—not very helpful!

Spectrum analyzers typically work in sampling mode, where samples are taken at different frequencies over many repetitions of the signal, but many spectrum analyzers have a real-time option that can be additionally purchased. An important parameter for a real-time spectrum analyzer is its real-time bandwidth, the span of frequencies that can be recorded at the same time, which needs to be selected to capture the bandwidth that is needed for monitoring the desired side-channel signal. A spectrum analyzer can move its real-time bandwidth from one frequency band to another, while an oscilloscope's bandwidth is always a band from zero to its maximum frequency. Signals that are located around some high frequency in the spectrum thus require very high-bandwidth (and thus very expensive) real-time oscilloscopes, or a much more modestly priced real-time spectrum analyzer. On the other hand, an oscilloscope can have a higher sampling rate, which is sometimes more important than noise floor or real-time bandwidth. Traditionally, an RF spectrum analyzer measures a signal with respect to frequency, i.e., in the frequency domain, while an oscilloscope measures a signal with respect to time, i.e., in the time domain. With additional purchased software, both can display signal in both time and frequency domain. An RF spectrum analyzer usually presents a terminated input to the signal to be measured at a defined impedance—usually 50 Ω, while an oscilloscope usually presents a high impedance input to the signal being measured (usually 1 MΩ) but can also be set to 50 Ω on some instruments. Finally, spectrum analyzers are typically designed for weak signals, while most oscilloscopes are not, so the noise floor (how much noise is recorded even in the absence of any signal) is typically lower (better) in a spectrum analyzer than in an oscilloscope.

A software-defined radio (SDR) is essentially a more compact version of an oscilloscope or spectrum analyzer. The quality of the side-channel signal captured by an SDR depends on the SDR's real-time bandwidth, noise floor, sampling rate, memory depth, and how many bits its analog-to-digital converter has. While SDRs can be used for many side channel applications, they are typically less powerful measurement tools than lab-grade spectrum analyzers and oscilloscopes.

3.3 Digital Circuits as Sources of Power Side Channels

Digital circuits have many gates, each consuming power when it "toggles" (changes value), and the overall power side channel signal is a sum of the individual contributions of all the gates that switch during a given interval of time (e.g., a clock cycle). Figure 3.4 shows the measured power signal when a chip alternates between active (lots of toggles) and idle (few toggles) behaviors.

Here we can note that, even when the chip is continuously active, the power signal still leaks information. Specifically, in this case it leaks how many bits have

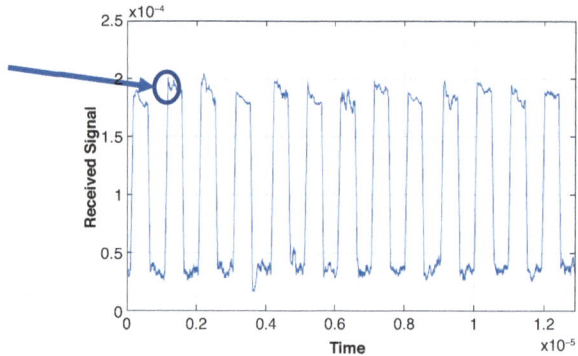

Fig. 3.4 Power signal when chip alternates between active (lots of toggles) and idle (few toggles)

3.3 Digital Circuits as Sources of Power Side Channels

changed each cycle (Hamming distance). Specifically, in our simplified RC model in Fig. 3.3, the energy stored in the capacitor can be computed as

$$E_{\text{cap}} = \int_0^\infty P_{\text{cap}}(t)dt, \quad (3.6)$$

where $P_{\text{cap}}(t)$ is the power stored in the capacitor C when inverter input is switching from 1 to 0. This can be calculated as

$$\begin{aligned}
E_{\text{cap}} &= \int_0^\infty I(t)V_{\text{out}}(t)dt = \int_0^\infty \left(\frac{V_{\text{dd}}^2}{R}\right) e^{-\frac{t}{RC}} \left(1 - e^{-\frac{t}{RC}}\right) dt \\
&= \left(\frac{V_{\text{dd}}^2}{R}\right) \int_0^\infty \left(e^{-\frac{t}{RC}} - e^{-\frac{2t}{RC}}\right) dt = \\
&= \left(\frac{V_{\text{dd}}^2}{R}\right) \left(-RCe^{-\frac{t}{RC}} + \frac{RC}{2}e^{-\frac{2t}{RC}}\right) \Big|_0^\infty \\
&= \left(\frac{V_{\text{dd}}^2}{R}\right) RC(-0 + 0 + 1 - 0.5) \\
&= \frac{1}{2}CV_{\text{dd}}^2. \quad (3.7)
\end{aligned}$$

and this energy is expended when the capacitor is charged and later emptied.

Typically, when measuring power consumption of an electronic device, e.g., to estimate its battery life or requirements for its power supply unit, we are interested in average power consumption over time, and also peak power consumption. In contrast, when measuring the power side channel, we are interested in how instantaneous power changes over time. To collect the signal, we also need appropriate oscilloscope probes, such as those shown in Fig. 3.5, that need to be attached at the appropriate points on the measured system's power supply or circuit board.

Power Rail Probe

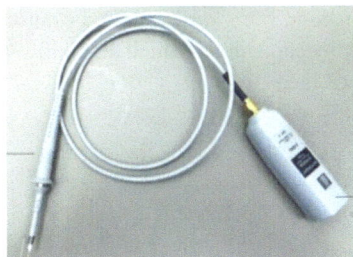

Micro SMD Clip

Fig. 3.5 Oscilloscope probes

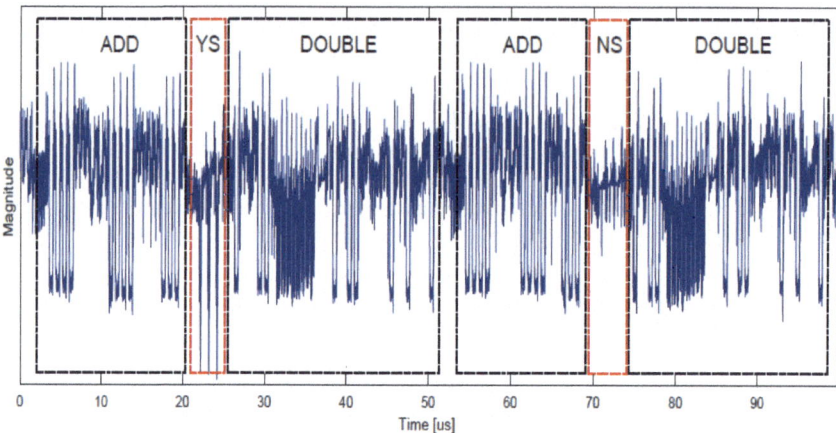

Fig. 3.6 An example of a power trace in Ecliptic curve code

Finally, Fig. 3.6 shows the power trace collected in a system whose processor is executing a program—an Elliptic curve cryptography (ECC) operation that consists of three types of program activity, two that are labeled ADD and DOUBLE, and a third one that is labeled as YS or NS, depending on the data values it uses. From this power trace we can observe several properties of the code execution:

- Repetitive patterns: typically coarse, structure of algorithm, and implementation (e.g., loops),
- Time: what happens when, program flow,
- Amplitude: what happens at a given moment in time, data flow,
- The same operation (YS/NS), executed with different operand values, can have different power consumption.

Power side channels can be observed at lower frequencies at which code executes, or at higher frequencies as a modulated replica of the code execution. More discussion on modulated side-channels will be presented in Chap. 4. Note that power side channel information is limited by where power probes can be positioned in the circuit. Typically, it is hard to attach probes at the source of activity (processor, memory, etc.). Instead, probes are typically positioned either on power cables or near power-regulating circuits. At these locations, we can observe only low-pass filtered versions of the original side channel source (i.e., logic gates), because the currents drawn by digital logic pass through many capacitors (which act as low-pass filters) between the digital logic and the point of measurement.

3.4 Digital Circuits as Sources of Electromagnetic Side Channels

During a change from output value 0 to output value 1, there is a short burst of current I_{dd} to fill the output capacitance from V_{dd}. Similarly, when output value changes from 1 to 0, there is a short burst of current to drain the output capacitance to V_{ss}. Each such current burst causes a sudden variation of the EM field (Faraday's law) around the device, which can be monitored by inductive probes, as illustrated in Fig. 3.7. A probe measures the local field. The output signal can be proportional to electric field (collected with electric probe) or magnetic field (collected with magnetic probe). They are typically collected with similar instruments as power side-channels, i.e., real-time oscilloscopes, spectrum analyzers, and software-defined radios.

Whenever a bit flips, the resulting signal exhibits a high frequency damped oscillation. This is illustrated in Fig. 3.8, where an EM signal trace is shown and high frequency oscillations are marked with red circles.

Integrated circuits (IC) emit EM radiation as:

Fig. 3.7 Measuring EM side-channel with an inductive probe

Fig. 3.8 EM signal trace, with high frequency oscillations marked with red circles

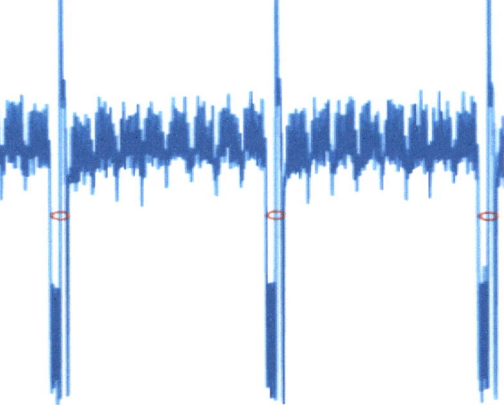

Fig. 3.9 EM radiation from ICs: conductive emissions and EM near-field emissions

Conductive emissions

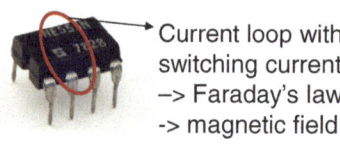
Current loop with switching current
–> Faraday's law
-> magnetic field

- Conductive emissions—signal is radiated from the integrated circuit pins and connecting traces, which are connected to the integrated circuit (IC),
- Electric and magnetic near–eld emissions—EM -field is generated due to current loops in the IC,
- EM far field emissions—EM modulated signals.

The first two EM radiation effects are illustrated in Fig. 3.9. Modulated EM side-channels are going to be discussed in detail in Chap. 4.

From the EM trace we can observe several properties of the code execution:

- Repetitive patterns: typically coarse, structure of algorithm and implementation (e.g., loops),
- Time: what happens when, program flow,
- Amplitude: what happens at a given moment in time, data flow,
- The same operation, executed with different operand values, creates EM signals with different amplitudes.

Note that EM side channels do not suffer from the same low-pass filtering effects that limit power side channels, because probes can be positioned close to the sources of emanations, so fast changes in the signal are not filtered out through the power's supply stabilization system.

3.5 Digital Circuits as Sources of Temperature Side Channels

Figure 3.3 in Chap. 3.3 illustrates how an inverter can be modeled as a charge based model (R_{on} C_{on} circuit) when inverter input is switching from 1 to 0. Thermal dissipation on an inverter is the source of thermal side-channels. In the simplified model, thermal energy is dissipated on the resistor. Energy dissipation as a heat from the resistance can be computed as:

$$E_{\text{heat}} = \int_0^\infty P(t)dt, \qquad (3.8)$$

3.6 Digital Circuits as Sources of Acoustic Side Channels

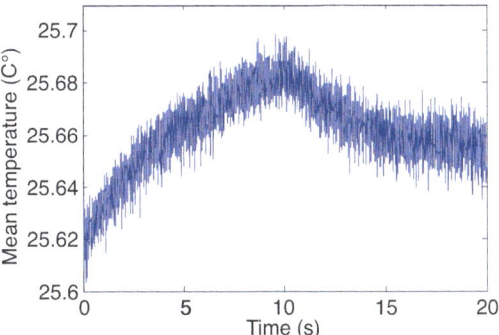

Fig. 3.10 Mean temperature fluctuations as a function of time

where $P(t)$ is the power consumed over the resistance R. This can be calculated as

$$E_{\text{heat}} = \int_0^\infty I(t)^2 R \, dt = \int_0^\infty \left(\frac{V_{dd}}{R}\right)^2 e^{-\frac{2t}{RC}} R \, dt$$

$$= \left(\frac{V_{dd}^2}{R}\right) \int_0^\infty e^{-\frac{2t}{RC}} dt = \left(\frac{V_{dd}^2}{R}\right) \left(-\frac{RC}{2}\right) e^{-\frac{2t}{RC}} \Big|_0^\infty$$

$$= \left(\frac{V_{dd}^2}{R}\right) \left(-\frac{RC}{2}\right) (0 - 1)$$

$$= \frac{1}{2} C V_{dd}^2. \tag{3.9}$$

Figure 3.10 illustrates how mean temperature of an IC changes over time. We can observe that the signal is noisy and changes slowly with time. Note that the time scale in Fig. 3.10 is in seconds, whereas the time-scale in Figs. 3.4 and 3.6 was in microseconds. This means that the fast fluctuations that are present in the EM side-channel are filtered out in the temperature side-channel, and even the slower changes that are present in the power side channel are too fast for the temperature side channel, so the temperature side-channel can capture only large changes in program execution, e.g., changes from one long-lasting phase in program execution to another, or a change from one program to another.

3.6 Digital Circuits as Sources of Acoustic Side Channels

We continue to use Fig. 3.3 as a charge based model (R_{on} C_{on} circuit). As the capacitor is discharged, and the current flows through the resistor, small mechanical movements can be created, e.g., because the opposite charges on the capacitor are removed, causing the attraction between them to decline, and thus allowing a

Fig. 3.11 Microphone collects acoustic vibrations from electronics

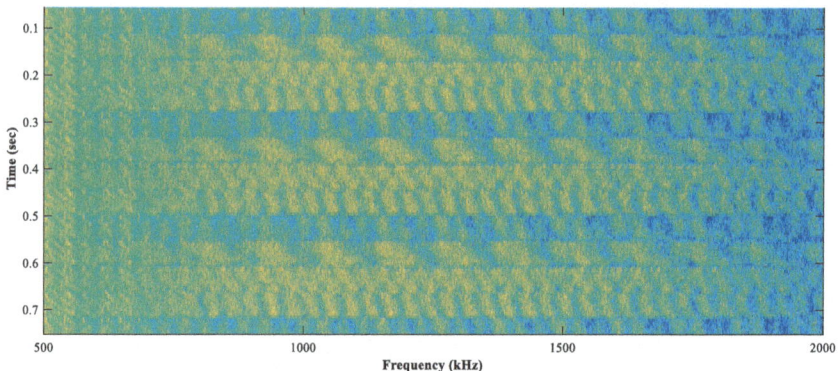

Fig. 3.12 Spectrogram of acoustic measurements when plotter motor is on and off

tiny expansion of the insulator material between them, or because the material of the resistor slightly expands as heat is generated by the current flowing through it. Similarly, charging the capacitor also causes slight movements. These small mechanical movements create vibrations can be measured by observing acoustic fluctuations and/or vibrations in a computing system as illustrated in Fig. 3.11, i.e., they are a source of acoustic side channels.

Figure 3.12 illustrates spectral features in the acoustic side channel signal. We can observe that there are spectral lines when motor is off. They are spaced 25 kHz apart and represent clock of the voltage regulator of the board that is controlling the motor. When the motor is on, microcontroller's program passes through 4 loops and, depending on the speed of the motor, the length of the loop varies, i.e., the speed of the motor can be estimated from the signal.

3.7 Digital Circuits as Sources of Backscattering Side Channels

So far, we have discussed passive analog side-channels, such as power or EM, which are a consequence of current-flow changes that are dependent on the activity inside the electronic circuit. Recently, we have introduced a new class of side-channels that

3.7 Digital Circuits as Sources of Backscattering Side Channels

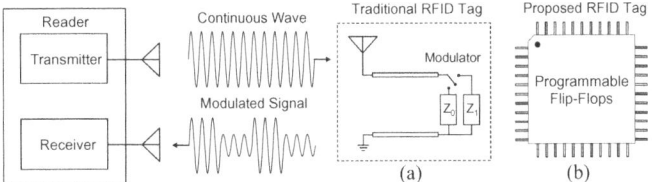

Fig. 3.13 Comparison between (**a**) traditional RFID tag and (**b**) proposed RFID tag [2]

is a consequence of impedance changes in switching circuits, and we refer to it as a backscattering side-channel [7].

Traditional backscattering communication, illustrated in Fig. 3.13a, refers to a radio channel where a reader sends a continuous carrier wave (CW) signal and retrieves information from a modulated wave scattered back from a tag. During backscatter operation, the input impedance of a tag antenna is intentionally mismatched by two-state RF loads (Z_0 and Z_1) to vary the tag's reflection coefficient and to modulate the incoming CW [4, 11].

In [7], our hypothesis was that logic gates and state elements in digital electronics, which have two output states (0 and 1), also have different RF loads in those states, as illustrated in Fig. 3.14 for a CMOS NAND gate, and can thus reflect a modulated signal. For example, when the gate's output is 1, its NMOS transistors are off and its PMOS transistors are on, creating a path between V_{out} and V_{DD} through resistance R_1 that corresponds to the on-resistance of the PMOS transistors which are, in a NAND gate, connected in series. This path produces the high output voltage that represents a digital value of 1. On the other hand, output value of 0 involves creating a path between V_{out} and the ground, through resistance R_2 that corresponds to the on-resistance of the NMOS transistors which are, in a NAND gate, connected in parallel.

The values of R_0 and R_1 are different [9], so the switching between digital output values of 1 (R_1) and 0 (R_0) creates impedance variation, which is analogous to the impedance switching in typical RFID tags. That impedance switching, in turn, creates a variation in the circuit's radar cross section (RCS), which modulates the backscatter signals. The same reasoning can be applied to other CMOS gates and state elements.

Because most digital circuits are almost entirely built out of logic gates and state elements gates (NAND, NOR, or OR gates), the behavior described can be generalized to any digital circuit. Therefore, if we send a continuous carrier wave (CW) into any digital circuit, the impedance variation will create differences in the circuit's radar cross section (RCS), which modulates the carrier wave and sends it back as a backscattered signal. Different circuits have different impedance variations when their outputs switch between digital values (0 and 1), so they should create different backscattered signals. By analyzing those signals, we can get useful information about the circuit and its switching activity. Since these

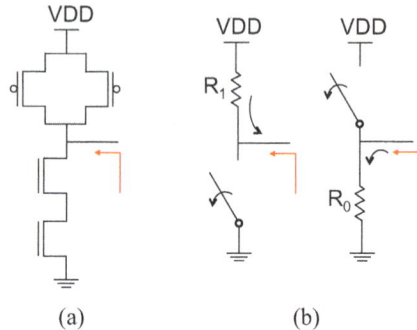

Fig. 3.14 (**a**) CMOS NAND gate. (**b**) Equivalent impedance circuits [2]

signals are generated by the impedance switching of the circuit, we refer to them as backscattering side-channel.

To test this hypothesis, we have used a Field-Programmable Gate Array (FPGA) to create a toggling circuit that consists of a cyclic shift register out of flip-flops. In that circuit, the flip-flops continuously switch their values between high and low at a certain frequency f_m, as illustrated in Fig. 3.15. The clock frequency of this circuit is $f_c = 50$ MHz. A near-field probe was used to send a CW of frequency $f_{carrier}$ to the FPGA, and another probe was used to receive the backscattered signals. The Altera DE0-Cyclone V FPGA development board was used as the test device.

In the test circuit, the flip-flops repeatedly toggle between values of 0 and 1, with a frequency of f_m. This toggling modulates the CW. The switching frequency (also called modulation frequency) f_m directly relates to the modulated signal bandwidth, i.e., the first harmonic of the modulated backscatter signal will be located at $f_{carrier} \pm f_m$, where $f_{carrier}$ is the frequency of the (externally transmitted) carrier signal. By changing $f_{carrier}$, we can easily up-shift or down-shift the modulated signals. Note that to avoid mixing with pure EM signals, the carrier frequency $f_{carrier}$ should not be a harmonic of the circuit's clock frequency f_c.

Figure 3.16 shows that side-bands in the backscattered signal are actually created by changing switching frequency of the FPGA circuits. The switching frequencies of the flip-flops in the FPGA board are varied from $f_m=900$ kHz to 1.2 MHz and 1.6 MHz. The carrier frequency is $f_{carrier} = 3.031$ GHz and the carrier signal power is $P_t = 15$ dB. A frequency of 3.031 GHz was selected because it is not a harmonic of f_c. The standby curve (violet) is the measured backscatter signal when the FPGA board is turned on but not toggling. For the other 3 curves, distinct modulated sidebands are observed at 3.031 GHz $\pm$ 900 kHz, 3.031 GHz $\pm$ 1.2 MHz, and 3.031 GHz $\pm$ 1.6 MHz. The signal strength of all the sidebands is at least 40 dB above the noise floor. Therefore, the results in Fig. 3.16 confirm that switching electronics can establish a backscatter channel.

To further explain this backscattering mechanism, the following subsection will discuss this circuit impedance model that includes transistors' equivalent resistance and parasitic capacitance.

3.7 Digital Circuits as Sources of Backscattering Side Channels

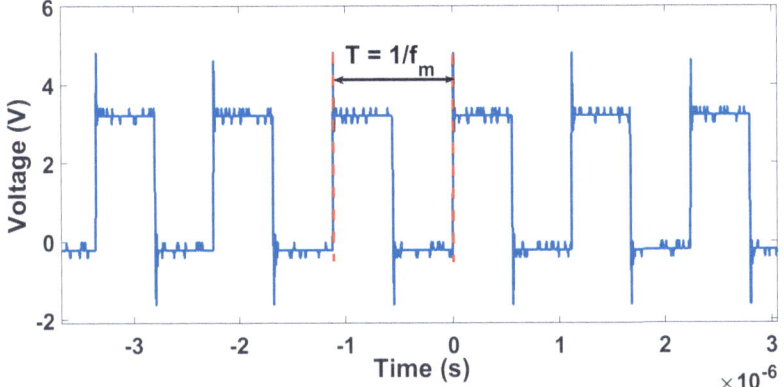

Fig. 3.15 Measured voltage at the output of flip-flops switching at f_m=900 kHz [7]

Fig. 3.16 Backscattered signals at $f_{carrier}$=3.031 GHz [7]

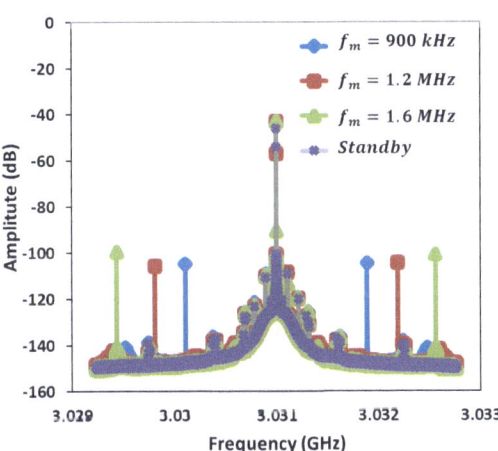

3.7.1 Equivalent Circuit Model for a Toggling Circuit

The equivalent impedance of the digital logic (transistors) changes as the toggling pattern and the number of toggled flip-flops change in a circuit. Such impedance variation in the digital logic modulates the incoming CW in a form of amplitude modulation. Hence, by controlling the activity of the chip's logic and flip-flops, we can transmit information through backscatter modulation. For illustration purposes, the equivalent circuit impedance model shown here is developed based on an FPGA due to its field-programmability. The "field-programmable" term in FPGAs refers to their ability to implement a new function on the chip after its fabrication is complete.

Link budget is a commonly used metric to evaluate the coverage of a radio system, and the link budget for traditional backscatter radio can be expressed as follows [4]

$$P_{\text{rx\_backscattered}} = \frac{P_{\text{tx}} G_{\text{tx}} G_{\text{rx}} L_{\text{refl}}^2 M \lambda^4}{(4\pi d)^4}, \tag{3.10}$$

where $P_{\text{rx\_backscattered}}$ is the received backscattered power, P_{tx} is the transmit power, $G_{\text{tx}}/G_{\text{rx}}$ is the Tx/Rx antenna gain, L_{refl} is the reflection loss of the tag, M is the modulation loss factor, λ is the wavelength, and d is the distance between the Tx/Rx and the tag. The modulation loss factor M, however, is computed from the impedances that are used in the design of a traditional backscatter radio (e.g., RFID tag), so it does not readily accommodate modulation that is created by switching in digital circuits whose impedances (with respect to the reflected radio wave) are not available. To fill this gap, we have proposed a modified M that interprets the tag impedance as a function of logic utilization of digital circuits. For simplicity, we assume that the "antenna" for backscatter radio effect in a digital circuit mostly consists of that circuit's power delivery network, whose wiring spans the entire circuit and is connected to all the individual gates.

From the power delivery network's point of view, the impedance of a digital circuit is the parallel combination of output impedances of individual elements (simplified as flip-flops in our work) [12]. The more elements (flip-flops) are connected, the more NMOS/PMOS transistors are connected in parallel, which reduces impedance. In other words, the total impedance of the circuit is inversely related to utilization of the logic. Given this relationship between logic utilization and input impedance, a modified M can be expressed as [2, 3]

$$M(x\%) = \frac{1}{4} \left| \frac{Z_1(x\%) - 377^*}{Z_1(x\%) + 377} - \frac{Z_0(x\%) - 377^*}{Z_0(x\%) + 377} \right|^2, \tag{3.11}$$

where $Z_1(x\%)$ and $Z_0(x\%)$ are the estimated high state (1) impedance and low state (0) impedance of the tag, and parameter x represents the percentage of total logic resources being configured. $Z_1(x\%)$ and $Z_0(x\%)$ are defined as

$$Z_1(x\%) = \frac{Z_1(10\%)}{\frac{x\%}{10\%}}, \tag{3.12}$$

and

$$Z_0(x\%) = \frac{Z_0(10\%)}{\frac{x\%}{10\%}}. \tag{3.13}$$

$Z_1(10\%)$ and $Z_0(10\%)$ are the estimated high state (1) and low state (0) impedance of an FPGA chip with 10% of total resources utilized. Note that we select 10% since it is less than the smallest logic utilization being configured in this work, which is 15%. The parameter x is in the denominator since the input impedance of the digital circuits is inversely related to the logic utilization, x. The input impedance of the

3.7 Digital Circuits as Sources of Backscattering Side Channels

tag is equal to free space impedance, 377 Ω, since there is no antenna but only air at the interface between the carrier signal and FPGA chip.

Therefore, to model the backscatter radio effect for a toggling digital circuit implemented in an FPGA chip, we need to estimate the equivalent impedances Z_1 and Z_0 (which we collectively refer to as ($Z_{1/0}$). Transistors' equivalent impedance ($Z_{1/0}$) consists of resistance ($R_{1/0}$) and reactance ($X_{1/0}$), where $R_{1/0}$ is the real part of the one/zero state impedance and $X_{1/0}$ is the imaginary part of the one/zero state impedance, i.e., $Z_{1/0} = R_{1/0} + jX_{1/0}$. Resistance and reactance together determine the magnitude and phase of the impedance. According to [9], the equivalent reactance ($X_{1/0}$) of transistors is dominated by the parasitic capacitance of the transistors, i.e., $X_{1/0} \approx -1/(\omega C_{1/0})$, whereas transistors' parasitic inductance can be overlooked due to the relatively short channel length (i.e., each technology reduces it). As a result, the proposed circuit model focuses on the equivalent R and C of the transistors.

A simplified internal structure of an FPGA is shown in Fig. 3.17a, where logic blocks are connected by programmable-routing interconnects and arranged in a two-dimensional grid. This symmetrical grid is connected to I/O blocks which make off-chip connections. Logic blocks can be simplified as lookup tables, each consisting of multiple latches or flip-flops which, in turn, consist of logic gates. To simplify discussion, we will again use an inverter, shown in Fig. 3.17b, as the simplest representative of CMOS logic gates. As shown in Fig. 3.17c, the equivalent output models for the inverter are very similar to those shown in Fig. 3.14 for a NAND gate.

The impedances that determine the RCS of the circuit, however, also include the various capacitances. These capacitances include parasitic capacitances of the inverter itself, and also the capacitances of elements connected to the output of the inverter. In Fig. 3.18, we show a cascaded inverter, i.e., an inverter whose output is connected to inputs of N other inverters. The inverter itself has a gate-drain

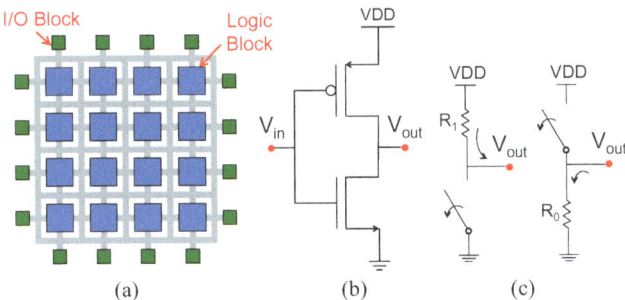

Fig. 3.17 (a) Simplified internal structure of an FPGA; (b) a CMOS inverter; (c) equivalent output model of the two output values of the inverter, with resistances R_1 (PMOS on-resistance) and R_0 (NMOS on-resistance) [3]

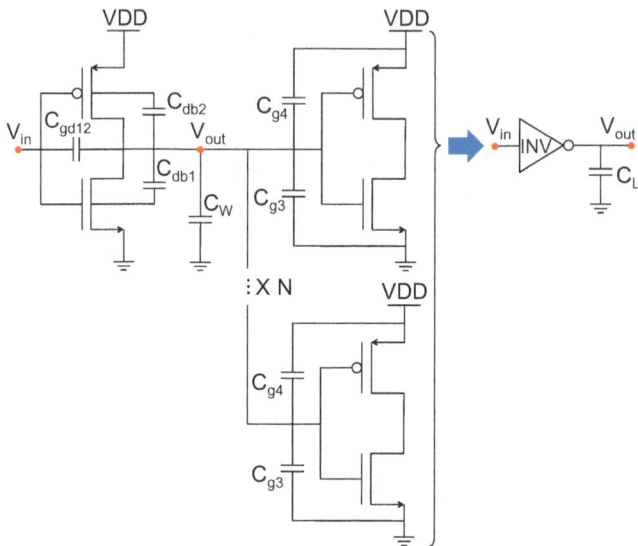

Fig. 3.18 Parasitic capacitance observed at the output of the cascaded inverter during transient state [3]

capacitance (C_{gd12}) and the diffusion capacitances of the two transistors (C_{db1} and C_{db2}). The wiring (interconnect) between the inverter's output and the inputs of other gates has capacitance C_w, and at the inputs of each of the cascaded gates are transistor gates, each with a gate capacitance (C_{g3} and C_{g4} for the two transistors in each cascaded inverter). Fortunately, in our model all these capacitances can be lumped together into one single capacitor C_L that is connected to V_{out} and the ground.

In order to estimate $Z_0(10\%)$ and $Z_1(10\%)$, we first calculate the values of $R_{1/0}$ and $X_{1/0}$ using transistor's SPICE model parameters (22nm PTM LP model in [1], which shares the same technology node as the Altera Cyclone V FPGA we use) to determine a range for $R_{1/0}$, $X_{1/0}$, and then perform curve fitting between the measured backscattered power and the modeled backscattered power to estimate the optimal value of $Z_{1/0}$. According to [13], R_n (R_0) and R_p (R_1) can be estimated by $R_{n/p} = (R_{lin,n/p} + R_{sat,n/p})/2$, where $R_{lin,n/p}$ and $R_{sat,n/p}$ are the on-resistance of the NMOS/PMOS in the linear and saturation regions, respectively, and can be calculated as [3]

$$R_{lin,n/p} = \frac{V_{lin}}{I_{lin}} = \frac{(V_{DD} - V_{SS} - V_{th,n/p})/2}{\frac{3}{8}k'_{n/p}(\frac{W}{L})_{n/p}(V_{DD} - V_{SS} - V_{th,n/p})^2}, \quad (3.14)$$

and

3.8 Digital Circuits as Sources of Photonic Side Channels

Table 3.1 Expressions and estimated values for the parasitic capacitances [3]

Capacitor	Expression	High to low value [fF]	Low to high value [fF]
C_{gd1}	$2 \cdot C_{on} \cdot W_n$	0.0143	0.0143
C_{gd2}	$2 \cdot C_{op} \cdot W_p$	0.0143	0.0143
C_{db1}	$K_{eqbpn} \cdot A \cdot D_n \cdot C \cdot J_n + K_{eqswn} \cdot P \cdot D_n \cdot C \cdot J \cdot S \cdot W_n$	0.0524	0.0662
C_{db2}	$K_{eqbpp} \cdot A \cdot D_p \cdot C \cdot J_p + K_{eqswp} \cdot P \cdot D_p \cdot C \cdot J \cdot S \cdot W_p$	0.0692	0.0584
C_{g3}	$C_{ox} \cdot W_n \cdot L_n + 2 \cdot C_{on} \cdot W_n$	0.074	0.074
C_{g4}	$C_{ox} \cdot W_p \cdot L_p + 2 \cdot C_{op} \cdot W_p$	0.074	0.074
C_w	From extraction	0.085	0.085
C_L	$C_{gd1} + C_{gd2} + C_{db1} + C_{db2} + (C_{g3} + C_{g4}) \cdot N + C_w$	0.383	0.386

$$R_{sat,n/p} = \frac{V_{sat}}{I_{sat}}$$
$$= \frac{V_{DD} - V_{SS}}{\frac{1}{2}k'_{n/p}(\frac{W}{L})_{n/p}(V_{DD} - V_{SS} - V_{th,n/p})^2}. \quad (3.15)$$

Parameter $k'_{n/p}$ represents $u_{n/p}C_{ox}$, where $u_{n/p}$ is the mobility of the NMOS/PMOS and C_{ox} is the gate capacitance. W and L are the width and length of the transistor, respectively. V_{SS} and V_{DD} are the source and drain voltage, respectively. $V_{th,n/p}$ is the threshold voltage of the NMOS/PMOS. Detailed values of the above parameters can be found in [1, 9, 10]. The expressions and estimated values of the parasitic capacitances C_{gd}, C_{db}, C_g, C_w, and C_L are presented in Table 3.1. The total reactance ($X_{1/0}$) is then estimated by $X_{1/0} \approx 1/(\omega C_{L1/L0})$, whereas transistors' parasitic inductance can be overlooked due to the relatively short channel length as technology node further miniaturizes [9]. The definitions of the corresponding parameters are presented in Table 3.2 and detailed parameter values are provided in [1, 9, 10].

3.8 Digital Circuits as Sources of Photonic Side Channels

Most photonic side channels are active side channels, using a laser beam to illuminate the device under test in order to observe side channel effects. In contrast to passive side channels and the EM backscattering side channel, each of which collects information about one physical property of the digital device, the photonic side channel (indirectly) can collect information about temperature change, electromagnetic field change, and free carrier change, depending on the

Table 3.2 Definitions of the parameters for the parasitic capacitances [3]

Parameter	Definition
$C_{ox_{n/p}}$ (fF/um^2)	Gate CAP (CAP per unit area by gate oxide)
$C_{on/p}$ (fF/um)	Overlap CAP
$CJ_{n/p}$ (fF/um^2)	Bottom junction CAP
$CJSW_{n/p}$ (fF/um)	Sidewall junction CAP
$AD_{n/p}$ (um^2)	Drain area
$PD_{n/p}$ (um)	Drain perimeter
$K_{eqbpn/p}$	Bottom plate capacitor linearization factor
$K_{eqswn/p}$	Sidewall capacitor linearization factor

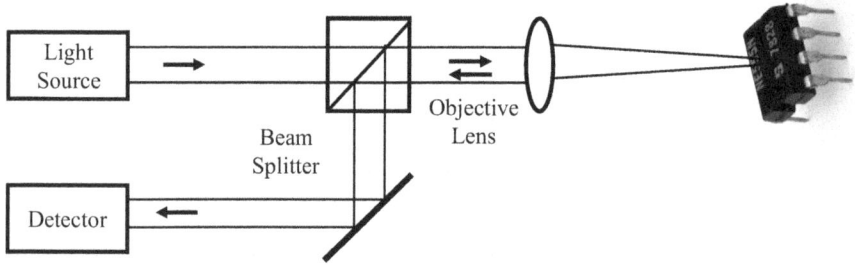

Fig. 3.19 Typical measurement setup for collection of photonic side channels

measurement system wavelength. The typical measurement setup for photonic side channel collection is illustrated in Fig. 3.19.

There are many different techniques in optics that can be used to collect photonic side channel signals [6]. All of them measure refractive index of the target, which varies with temperature, electric field and free-carrier concentration. Normal operation of a transistor involves frequent switching between on and off states, i.e., from inversion to saturation and back. The resistance of, the electric field across, and the power dissipation within the transistor's channel all change significantly between these two states, resulting in a significant change in the refractive index. Thus the photonic side channel combines several analog side-channels into one, and careful selection of the wavelength can allow a particular aspect to be enhanced or diminished in how much they affect the overall signal. For example, when the temperature increases, the material undergoes thermal expansion and the density decreases. This causes the refractive index to decrease. On the other hand, polarization increases with temperature due to a limited degree of freedom in a crystal lattice arrangement, causing the refractive index to increase. Depending on the incident photon energy, either thermal expansion resulting in density change or temperature-dependent polarization variation may dominate.

3.8 Digital Circuits as Sources of Photonic Side Channels

An empirical model of the refractive index of silicon $n(E, T)$ as a function of temperature is as follows [5]:

$$n(E, T) = \sqrt{\left(4.386 - 0.00343 * T + \frac{99.14 + 0.062 * T}{E_g^2 - E^2}\right)}, \quad (3.16)$$

where T is the temperature of the material in degree Celsius, E is the incident photon energy in eV and $E_g = 3.652$ eV is a constant. The second term represents the thermal expansion effect and the last term represents the polarization effect. Since temperature changes slowly with program execution, for temperature variation the photonic side channel provides results similar to direct measurement of temperature.

Electro-refraction is often used to describe the change in refractive index when an electric field is applied. Two electro-absorption maximums typically are observed at 1055 and 1170 nm. At these two wavelengths, electro-refraction is observed to be negligible. When the incident wavelength is less than 1055 nm, the refractive index is observed to modulate negatively with the applied electric field. At longer wavelengths, the refractive index modulation is positive. Peak negative refractive index modulation is observed at 1045 nm, and peak positive refractive index modulation is observed at 1065 nm. For all applied electric field strengths, the variation of the refractive index is in the range from 10^{-6} to 10^{-5}.

On the other hand, carrier refraction is dominant at higher wavelengths, e.g., 1300–1340 nm has been used. Carrier refraction is used to describe the variation of the refractive index with free carrier concentration. The change in refractive index due to changes in free electrons and free holes is as follows [8]:

$$\Delta n = -\left(\frac{q^2 \lambda^2}{8\pi^2 c^2 \epsilon_0 n}\right)\left(\frac{\Delta N_e}{m_e^*} + \frac{\Delta N_h}{m_h^*}\right), \quad (3.17)$$

where q is the electronic charge, ϵ_0 is the permittivity of free space, n is the refractive index of unperturbed c-Si, ΔN_e is the change in free electron density, m_e^* is the effective mass of electrons, ΔN_h is the change in free hole density, and m_h^* is the effective mass of holes.

The refractive index increases when carriers are depleted and decreases when carriers are injected. Each decade increase in free carrier density results in a decade increase in the magnitude of the refractive index modulation.

The reflectance modulation is the ratio of the change in the reflected intensity at the higher temperature to the reflected intensity at the reference temperature of 300 K.

The reflectance modulation index can be related to change in the impedance of the circuit. The reflectance modulation index is defined as

$$n = \frac{n_0 + \Delta n}{n_0} = 1 + \frac{\Delta n}{n_0}, \quad (3.18)$$

where n_0 is the reflectance index at 300 K and Δn is difference in modulation index when circuit changes states from low voltage to high and vice versa.

The reflectance nodulation index n can also be defined as a function of circuit impedance and impedance in air, i.e.,

$$n = \left| \frac{Z_1 - Z_0^*}{Z_1 + Z_0^*} \right| = \left| \frac{Z_1 - 377}{Z_1 + 377} \right|, \tag{3.19}$$

and the difference in reflectance index when circuit changes states and can be defined as

$$\Delta n = |n_1 - n_2| = \left| \frac{Z_1 - 377}{Z_1 + 377} - \frac{Z_2 - 377}{Z_2 + 377} \right| = \sqrt{4M}, \tag{3.20}$$

where M is the reflectance modulation index defined in (3.11). By selecting a wavelength that is high enough, so that the free carrier effect dominates, this side channel is very similar to the backscattering side channel. The difficulties in collecting this side channel arise when chip is packaged because signal gets significantly attenuated before it interacts with transistors. Typically, plastic packaging has to be removed to collect photonic side channel signal.

References

1. 22nm ptm low power model. http://ptm.asu.edu/latest.html.
2. C-L. Cheng, L. N. Nguyen, M. Prvulovic, and A. Zajic. Exploiting switching of transistors in digital electronics for rfid tag design. *IEEE Journal of Radio Frequency Identification*, 3(2):67–76, 2019.
3. C-L Cheng, S. Sangodoyin, L. N. Nguyen, M. Prvulovic, and A. Zajic. Digital electronics as rfid tags: Impedance estimation and propagation characterization at 26.5 ghz and 300 ghz. *IEEE Journal of Radio Frequency Identification*, 5(1):29–39, 2021.
4. J. D. Griffin and G. D. Durgin. Complete link budgets for backscatter-radio and rfid systems. *IEEE Antennas and Propagation Magazine*, 51(2):11–25, 2009.
5. G. E. Jellison, Jr, D. H. Lowndes, and R. F. Wood. The optical functions of silicon at elevated temperatures and their application to pulsed laser annealing. *J. Appl. Physics*, 76(4):3758, 1994.
6. U. Kindereit. Fundamentals and future applications of laser voltage probing, 2014.
7. L. Nguyen, C-L. Cheng, M. Prvulovic, and A. Zajic. Creating a backscattering side channel to enable detection of dormant hardware trojans. *IEEE Transactions on Very Large Scale Integration (VLSI) Systems*, 2019.
8. Soref R.A. and Bennett B.R. Electro-optical effects in silicon. *IEEE J. Quantum Electronics*, 23(1):123–129, 1987.
9. J. M. Rabaey, Chandrakasan A. P., and B. Nikolic. In *Digital integrated circuits*. Prentice hall Englewood Cliffs, 2002, 2002.
10. A. B. Sachid, P. Paliwal, S. Joshi, M. Shojaei, D. Sharma, and V. Rao. Circuit optimization at 22nm technology node. In *2012 25th International Conference on VLSI Design*, pages 322–327, 2012.

References

11. H. Stockman. Communication by means of reflected power. *Proceedings of the IRE*, 36(10):1196–1204, 1948.
12. I. Vaisband and E. G. Friedman. Stability of distributed power delivery systems with multiple parallel on-chip ldo regulators. *IEEE Transactions on Power Electronics*, 31(8):5625–5634, 2016.
13. W. Wolf. In *FPGA-Based System Design*. Prentice Hall PTR, 2004, ser. Prentice Hall Modern Semicondutor, 2004.

Chapter 4
Unintentionally Modulated Side Channels

4.1 Direct vs. Unintentionally Modulated Side Channels

Direct side channels are created as a direct consequence of intentional current flows, i.e., current flows that are necessary to implement switching between values in digital logic, as discussed in Chap. 3. These direct side channels include power, EM, acoustic, and thermal side channel signals discussed in Chap. 3. These side channel signals are created by brief bursts of current, with sharp rising edges, resulting in signals that span a wide frequency band for those side channels that do not experience low-pass filtering, i.e., EM and close-to-the-circuit power side channels. Recall that signals with a wide frequency band carry information about finer-granularity program activity, so they are more useful for side channel analysis. However, even for direct side channels that do span a wide frequency band, that band starts at zero frequency, and so it includes noise and interference that is prevalent in lower frequency bands. To isolate side channel signals from this noise and interference, the signals must be collected very close to the signal source. Alternatively, the signal can be filtered to suppress frequencies that contain interference, at the cost of also suppressing parts of the side channel signal. Finally, sophisticated signal processing techniques, such as interference cancellation, can be used to eliminate some interference, but this cancellation only affects predictable interference signals, i.e., the noise and some part of interference still remains.

In contrast, unintentionally modulated side channels are created when computational activity directly or indirectly modulates periodic carrier signals that are already present, are generated, or are "introduced" within the device. One strong source of carrier signals is the ubiquitous, harmonic rich, "square-wave" clock signal. Various communication signals, when unintentionally modulated by program activity, are another typical source of unintentionally modulated side channel signals.

Typically, carrier signals are unintentionally modulated by electronic switching activity through:

- Amplitude Modulation: Periodic square wave (e.g., clock) pulse width and height variation results in the generation and emanation of an Amplitude Modulated (AM) signal. The data signal can be extracted via AM demodulation using a receiver tuned to the carrier frequency.
- Angle Modulation: Periodic square wave (e.g., clock) pulse width variation also results in Angle Modulated Signals (FM or Phase modulation). For example, if the circuits draw upon a limited energy source the generated signal will often be angle modulated by the data signal. The data signal is recoverable by frequency and/or phase demodulation of the emanated signal.

In traditional wireless communication, modulation is intentional and the type of modulation is tightly controlled, e.g., AM radio signals are amplitude modulated and have (practically) no angle modulation. In contrast, unintentionally modulated side channels can be both amplitude- and angle-modulated, i.e., the same side channel signal typically contains both types of modulation—even when one type of modulation is dominant, the other type is typically present to some degree.

4.2 How Are Side-Channel Signals Modulated?

To understand how program activity modulates periodic signals, what these modulated side channel signals look like, and how information about program activity can be obtained from such side channel signals, we first describe how program activity can be used to *create* a periodic carrier signal and modulate that carrier signal. We then show how pre-existing carrier signals are unintentionally modulated in a way that is similar to how program-generated carriers can be modulated.

A carrier should be a periodic (repetitive) signal that varies during a period of time, and this variation is then repeated, creating a repeating pattern. To create such a carrier signal using software activity, we can use program code [6] shown in Fig. 4.1. This code consists of a `while` loop that implements repetition, and each iteration of this loop creates a single period of this repetition. Each iteration of this loop creates a variation in the side channel signal by performing some activity A for part of the time period, then a different activity B for the rest of the time period. Because execution time of an individual processor operation (instruction) cannot be individually controlled, activity A and activity B each consist of repeating the same operation `n_inst` times. As illustrated in Fig. 4.2, because the signals produced by activity A and activity B are not identical, repetition of this A-then-B pattern will create an oscillation with period T, and the repetition of this A-then-B pattern creates a carrier signal at frequency $f = 1/T$. Note that period T is equal to the execution time of one iteration of the `while` loop, so we can select the frequency f of the signal by appropriately adjusting the value of `n_inst`. For example, to produce a carrier signal with frequency f=1 MHz, we should set `n_inst` such that one iteration of the `while` loop executes in $T = 1\,\mu s$.

4.2 How Are Side-Channel Signals Modulated?

```
1    while(1){
2    // Do some instances of activity A
3      for(i=0;i<n_inst;i++){
4        ptr1=(ptr1&~mask1);
5        // Activity A, e.g. a load
6        value=*ptr1;
7      }
8    // Do some instances of activity B
9      for(i=0;i<n_inst;i++){
10       ptr2=(ptr2&~mask2);
11       // Activity B, e.g. a store
12       *ptr2=value;
13     }
14   }
```

Fig. 4.1 The A/B alternation pseudo-code [6]

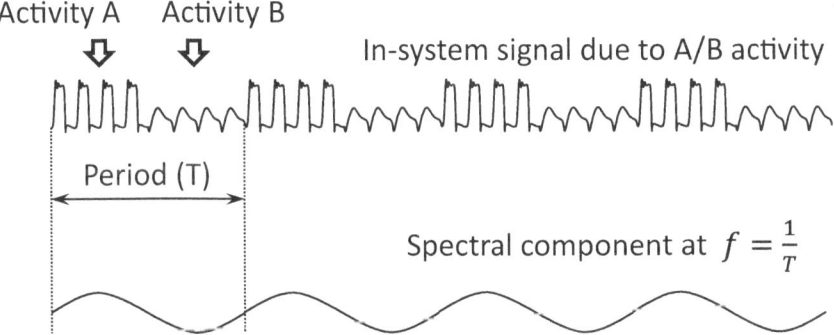

Fig. 4.2 Illustration of how program code induces emanations at a specific radio frequency by alternating half-periods of A and B activity [6]

Next, this carrier is *amplitude modulated* by inserting intervals during which only activity B is performed in both half-periods. This makes the signal have similar values during the entire period of the carrier, which means that the signal will *not* have a strong periodic behavior at the frequency of the carrier. The effect of this is the simplest form of AM modulation (on-off keying), where the carrier is either present or absent. On-off keying is traditionally used to transmit a binary value (on for 1, off for 0). However, on-off keying can also be used to transmit a sequence of pulses (on, off, on, etc.), as illustrated in the left half of Fig. 4.3. Note that this sequence of pulses is itself a periodic signal (with a period that corresponds to one on-then-off period). To simplify discussion and avoid confusion with the carrier

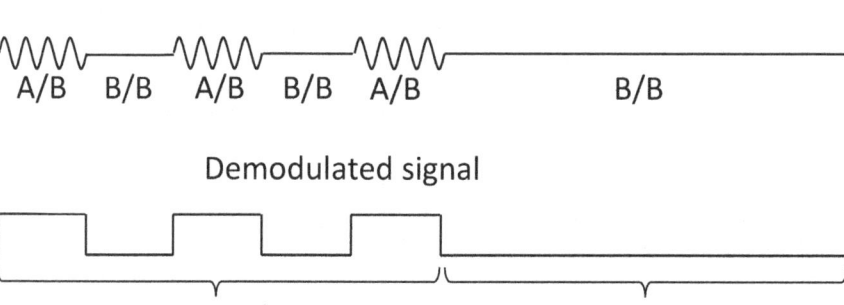

Fig. 4.3 Illustration of how the program code modulates the signal into the carrier using on-off keying (bottom) [6]

(also a periodic signal), we use "tone" to refer to this sequence of pulses.[1] Thus, we have used program activity to AM-modulate one periodic signal (the tone) onto another (the carrier) and transmit the resulting modulated signal as the EM side channel signal. If the carrier frequency is chosen such that it is in the standard AM radio band (between 535 kHz and 1.7 MHz), and if the tone frequency is chosen such that it is within the human hearing range (between 20 Hz to 20 kHz), this EM side channel signal can be received an AM-demodulated by an ordinary AM radio receiver, producing an audible tone from the AM radio's speaker.

Next, this tone itself can be on-off keyed, by producing periods of silence as shown in the right half of Fig. 4.3, and this presence or absence of a tone can be used to transmit information. To confirm this, we have performed two experiments. First, we modulated our transmitted signal with the A5 note (880 Hz) as our tone, and then created a sequence of tones and silences to transmit Morse code for "All your data is belong to us" [6]. Second, we added our transmitter code to the keyboard driver, using tone/silence on-off keying to wirelessly transmit the information about what is being typed on a laptop computer [3]. In both cases, we were able to correctly receive and demodulate transmitted signals.

Figure 4.4 illustrates how intentionally and unintentionally modulated signals using Morse code look like. Figure 4.4a shows the spectrum of a carrier that is modulated to produce Morse code (A/B Morse). This spectrum closely matches what would be expected from a signal whose carrier is modulated using an 880 Hz tone. Note that the actual carrier frequency is 198.5 kHz, not the 200 kHz that our

---

[1] The reasons for using the word "tone" will become apparent soon.

4.2 How Are Side-Channel Signals Modulated?

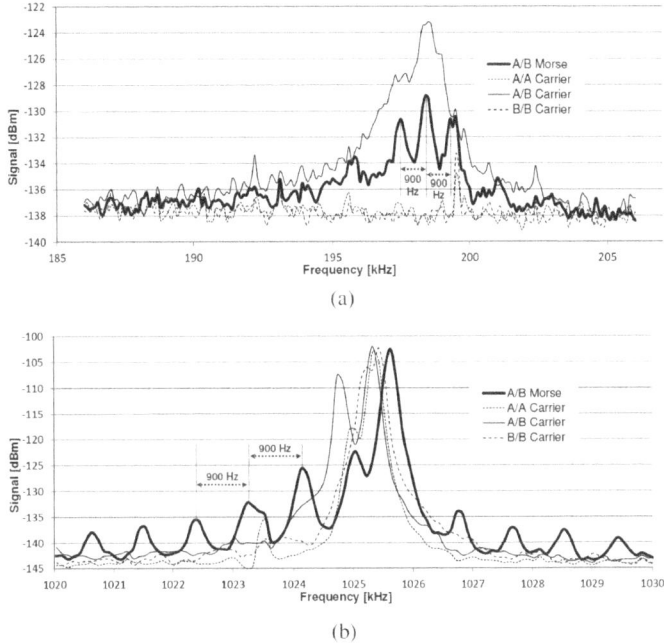

Fig. 4.4 Received spectra for the i7-based laptop when the program code is trying to create emanations at 200 kHz. (**a**) Spectrum at 200 kHz (Intended), (**b**) Spectrum at 1025 kHz (Unintended) [6]

program code is trying to achieve. This is typical when working with program-activity-modulated signals—the duration of each half-period must be a whole number of iterations in our A or B activity loop, so the program cannot precisely match the target frequency. We also show the spectrum for just the "A/B Carrier" signal, without any attempt at modulation, which shows a strong carrier-like signal at 198.5 kHz. Finally, we also show the spectra for our "carrier generation" code when the same activity is used in both half-periods ("A/A Carrier" and "B/B Carrier"). As expected, without a change in activity during each period of the carrier, this A/A or B/B activity does not result in transmitting a carrier signal.

Figure 4.4b shows a spectrum for the exact same experiments, i.e., "carrier generation" at 200 kHz), but instead of showing the frequency range around the intended carrier frequency, we show the frequency range around 1025 kHz. We observe a strong signal around 1025.5 kHz, regardless of program activity, i.e., this strong signal is an *existing* periodic signal that is not generated by our program activity. Using additional experiments, we have found that this signal is generated by memory refresh activity, which is performed automatically in hardware to ensure

Fig. 4.5 Illustration of how coupling between digital and analog circuitry may occur

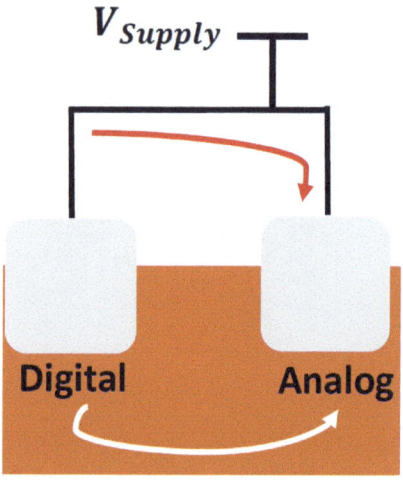

that the main memory does not "forget" its contents. However, this pre-existing carrier signal is *unintentionally modulated* by our program activity. The "A/B Carrier" activity results in an additional signal about 600 Hz below the "main" one, while the "A/B Morse" activity additionally results in numerous "ripple" signals at 900 Hz intervals. This does not directly match the spectrum for traditional AM modulation, but the "main" frequency and the first "ripple" on each side, when AM-demodulated, do result in audible Morse-code signals.

Any periodic activity in electronic circuitry can act as a carrier that may be modulated by program (or hardware) activity, and computer systems are filled with such periodic activity, such as: (i) processor clock, (ii) memory clock, (iii) voltage regulator, (iv) power supply, (v) memory refresh, etc.

Here we note that even more complex unintentional modulations may occur. For example, if digital and analog circuitry share the power supply, the analog signal's clock can pick up information from digital circuitry through substrate coupling or power supply coupling, as shown in Fig. 4.5. To illustrate this process, Fig. 4.6a shows a spectrogram, where the red part of the spectrogram denotes program activity in the baseband and the green part of the spectrogram denotes the processor clock at 64 MHz. Typically, the unintentionally modulated side channels will appear as program activity modulated onto processor clock, as shown in Fig. 4.6b. If analog and digital electronics share power plane, both processor clock and program activity may get modulated onto communication clock (e.g., WiFi), as shown in Fig. 4.6c. This type of side channel is called *screaming side-channel* [2] because the analog signals are stronger and are assumed to travel further (e.g., 10 m) than digital side-channels. This assumption should be taken with a grain of salt, however, because demonstrations exist of digital side channels being received even at distances of 200 m [5].

Fig. 4.6 Illustration of screaming side channels: (**a**) spectrum of baseband program activity and processor clock, (**b**) spectrum of unintentionally modulated program activity onto processor clock, (**c**) unintentionally modulated program activity and processor clock onto communication carrier

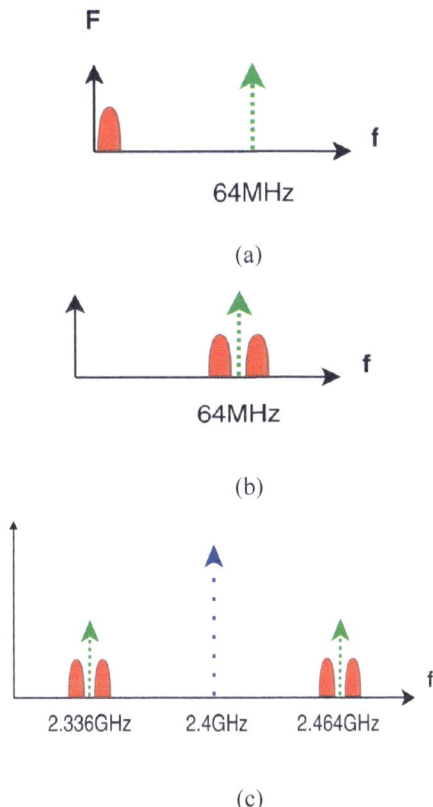

4.3 Unintentional Side Channel Modulations vs. Traditional Communication Modulations

Now that we have confirmed that unintentionally modulated side channel signals exist, a question arises—can we leverage knowledge from digital communications to receive and demodulate these signals? It is important to note that this modulation is not happening as it does in traditional communications, where the message signal is created from the digital (or analog) message, and then the carrier signal is mixed (using a mixer circuit) with this message signal to produce the transmitted signal. In digital electronics, however, no mixer circuit is (intentionally) placed in the design to perform modulation. Instead, the amplitude, width, and shape of the carrier signal's pulses are changing due to program-dependent electronic activity. Because a sequence of pulses can be represented as a sum of sinusoidal (single-frequency) signals, changes to the amplitude, width, and shape of the pulses correspond to changes in amplitude and frequency of its sinusoidal components, and these modulated sinusoidal components can then be received and analyzed as AM- or FM- modulated signals.

Furthermore, traditional communications rely on carefully designed transmit and receive signaling (i.e., carrier and baseband signals), with a thoroughly regulated allocation of the frequency spectrum. This results in carrier signals with very sinusoidal shapes and very stable frequencies, and these carriers are modulated with baseband signals where information is encoded in equal-duration periods with very precise timing. In contrast, unintentionally modulated signals in computer systems are generated by many possible "transmitters." Note that many periodic carrier signals in computer systems are generated by digital circuits and clocks, and therefore have sharp transitions that are best approximated by rectangular pulses instead of the sinusoidal waves, and the frequency of many of these carrier signals is not very stable because it was not necessary for the circuit design or, sometimes, because variations in frequency of these signals are actually preferable for the design. The duty cycle of these "carriers", (i.e., for what part of the period do they stay "high") is also often not stable, and even when it tends to be stable it may not be a traditional 50% (same duration of *on* and *off*) duty cycle. The spectrum of a pulse train with an arbitrary duty cycle is equivalent via Fourier analysis to a set of sinusoids with various amplitudes at f_c and its multiples (harmonics). In other words, for each carrier signal generated by a digital circuit or clock, additional carrier signals will also be present at $2f_c$, $3f_c$, $4f_c$, $5f_c$, etc. As the duty cycle of a signal approaches 50%, the amplitudes of the odd-numbered harmonics (f_c, $3f_c$, $5f_c$, etc.) reach their maximum, while amplitudes of the even harmonics ($2f_c$, $4f_c$, etc.) trend toward zero. For a small duty cycle (i.e., < 10%) the magnitudes of the first few harmonics (both even and odd) decay approximately linearly. Finally, note that these observations imply that amplitudes of all the harmonics are a function of the duty cycle. If program activity modulates the duty cycle of a periodic signal while keeping its period constant (i.e., causes pulse width modulation), all of the signal's harmonics will be amplitude-modulated. Whether the signal is AM or FM modulated can be determined by tracking the carrier signal as the duty cycle of the baseband signal changes. For baseband signals whose highest frequency component is much lower than the carrier frequency, the AM and FM spectra look very similar, but FM shifts the frequency of the carrier when duty cycle changes, while AM carrier does not.

The reception of unintentional modulation "signals" differs from traditional communication receivers in several ways. Since unintentional signals occur at the frequency of the unintentional carrier, they are combined with all the other noise generated by the computer system (other clocks and switching noise) and with any communications signals that happen to be present nearby (including the wireless communications of the computer system itself). Unintentional signals are treated as noise as far as regulations are concerned, so they are subject to EMC restrictions which place limits on noise power that may be emanated by an electronic device. From the side-channel point of view, this limits signal power in the EM side channel, and sometimes mitigation methods to bring EM emanations into compliance also affect signals in other side analog channels. Therefore, unintentionally modulated EM signals are typically weaker than communications signals. Some of the periodic signals, such as processor clocks, result in strong emanations that cannot be suppressed easily. Because regulators typically limit overall power emanated at any given frequency, a popular method for achieving regulatory compliance involves varying the clock frequency such that emanations are diffused across a range of frequencies rather than concentrated at one specific frequency. This also acts as motivation for not introducing frequency stability into other periodic activity—correct designs that have more frequency variation in their periodic activity, e.g., switching in voltage regulators, are actually preferable (from the regulatory compliance perspective) to equally correct designs that have less frequency variation. Additionally, since the side-channel carriers are typically generated by non-sinusoidal sources, these carrier signals typically have harmonics. As a result of this, one type of periodic activity in a circuit (e.g., clock), when modulated, results in unintentionally modulated side-channel signals at several (sometimes many) frequencies, and these frequencies corresponds to clock frequencies, switching frequencies, memory refresh periods, etc. in computers and other digital circuits. This is in contrast to traditional wireless communications, where a specific signal is typically allowed (by regulations and standards) to choose among specific frequency bands (e.g., WiFi channels) and confine its transmissions to one of those bands at any given time. Finally, communication signals have direct and obvious control of the baseband (modulation) signal, while unintentionally modulated signals from computer systems do not. We may be interested in several different system activities (baseband signals). For example, a baseband signal may be caused by processor activity and another baseband signal may be caused by memory activity. In some cases, multiple baseband signals may even modulate the same carrier.

These effects complicate the detection of unintentionally modulated signals. The presence of noise generated by the system makes it difficult to determine which signals are AM or FM carriers. Some of the unintentional AM or FM carriers are generated by spread spectrum clocked signals, making them harder to recognize. Existing methods to find AM and FM modulation based on its spectral properties (i.e., without knowing the baseband signals) are not designed to deal with these issues, and are not able to identify which carriers are modulated by a specific system activity. In the following section we describe a method for identifying unintentionally modulated side channels.

4.4 How to Identify Unintentionally Modulated Side Channel Signals Created by Computer Activity?

One possible approach to finding unintentionally modulated signals is to create a simple identifiable baseband signal. These baseband signals are generated by system activity, such as execution of specific instructions, memory accesses, etc. While we may not know, for a particular program activity the exact effect it will have on a particular carrier's baseband signal, we can create variations in a particular activity with a relatively long period (i.e. a low-frequency, which we label f_{alt}), and then expect that (in aggregate) these variations will generate a component (at frequency f_{alt}) in the baseband signal. To accomplish that, we can use the A/B program code shown in Fig. 4.1. It is important to emphasize that, while the effect of a single event (i.e., execution of a single memory access or processor instruction) on the baseband signal is unknown, as long as there is some difference between the A and B activities, there will be a signal generated at the frequency f_{alt} and also at some of the harmonics of f_{alt} ($2f_{alt}, 3f_{alt}, \ldots$). Furthermore, we can change the duty cycle of the A/B activity by using separate n_inst for A and B activity and then increasing/reducing time spent on activity A while adjusting (in the opposite direction) the time spend on activity B, such that the overall period of the activity (time of A plus time of B) remains the same.

4.4.1 An Algorithm for Finding AM and FM Unintentional Carriers in Computer Systems

Using the A/B program code as described in the previous section, we can create predictable spectral patterns in the sideband of any carrier modulated by the benchmark activity. The code is run at several different A/B alternation frequencies $f_{alt_1}, f_{alt_2}, \ldots, f_{alt_N}$, for several duty cycles $d_1, d_2, \ldots, d_m$, and each combination of alternation frequencies and duty cycles is recorded K times. The frequency spectrum for each run is recorded, the repeated runs are averaged, and the result we denote as $S(f, f_{alt_i}, d_j)$, where f is the frequency range at which the spectrum is recorded, f_{alt_i} denotes the chosen alternation frequency, and d_j denotes the chosen duty cycle. Here we chose alternation frequencies such that $f_{\Delta_i} = f_{alt_{i+1}} - f_{alt_i}$ is not constant [1, 4]. This is an important step to allow robust automated detection of both AM and FM modulations.

To illustrate what measured $S(f, f_{alt_i}, d_j)$ looks like, Fig. 4.7 plots a part of one spectrum around a carrier frequency at 382 kHz. This spectrum was recorded with $f_{alt} = 23$ kHz, so it shows a lower and upper sidebands around 359 and 405 kHz, respectively.

It is not surprising that the real alternation frequency differs from what was intended when we chose the parameters for our program code—execution time of a program varies from run to run, and the time for each instance of A and B cannot be subdivided further, so adjustment of n_inst by one results in a step from one frequency to another, while frequencies in-between these steps are not achievable by this program code.

Because the actual f_{alt} in a recorded spectrum may not be exactly as intended, before we proceed to identify the type of modulation, we first need to estimate the actual value of f_{alt} in each recorded spectrum.

To do that, for every duty cycle, we average spectra with different alternation frequencies, i.e.,

4.4 How to Identify Unintentionally Modulated Side Channel Signals Created...

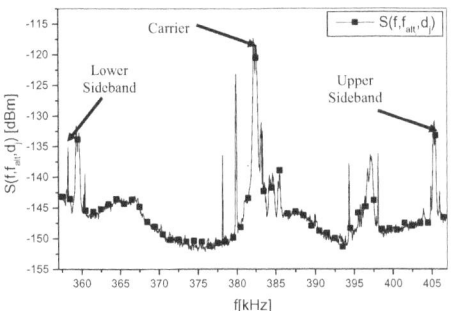

Fig. 4.7 A measured spectrum $S(f, f_{alt_i}, d_j)$ at a carrier frequency at 382 kHz and a lower and upper sidebands around 359 and 405 kHz, respectively [4]

$$S_{avg}(f, d_j) = \text{mean}_{f_{alt_i}} S(f, f_{alt_i}, d_j), \quad (4.1)$$

and create new spectra as a difference between the original and averaged spectra, i.e.,

$$S_{new}(f, f_{alt_i}, d_j) = S(f, f_{alt_i}, d_j) - S_{avg}(f, d_j). \quad (4.2)$$

This attenuates most spectral features that are not related to modulated signals we are looking for, while preserving most of those that are activity-modulated.

To find the true alternation frequency, we shift all points in this spectrum ($S_{new}(f, f_{alt_i}, d_j)$) by $\pm f_{alt_i}$, and take the pointwise minimum between two shifts i.e. we compute

$$M(f, f_{alt_i}, d_j) = \min \Big[S_{new}(f + f_{alt_i}, f_{alt_i}, d_j),$$

$$S_{new}(f - f_{alt_i}, f_{alt_i}, d_j) \Big]. \quad (4.3)$$

Figure 4.8 plots the spectrum $S_{new}(f, f_{alt_i}, d_j)$ shifted up by $f_{alt_i} = 23$ kHz (black square curve) and shifted down by $f_{alt_i} = 23$ kHz (red circle curve), their pointwise minimum $M(f, f_{alt_i}, d_j)$ (blue triangle curve). Also shown (magenta diamond curve) is the pointwise minimum computed in the same way (shifting by 23 kHz) for another spectrum whose alternation frequency is different (e.g., 29 kHz). We observe that, when the spectrum contains sidebands that correspond to f_{alt_i}, the shift in frequency aligns these sidebands at the frequency that corresponds, in the original spectrum, to the carrier that produced the sidebands (382 kHz in this case). At points that do not correspond to the modulated carrier or its sidebands, the pointwise minimum will only have a peak if two prominent spectral features (e.g., two radio unrelated signals) happen to be separated by exactly $2 f_{alt_i}$. Finally, when the spectrum is shifted by an amount that does not match the alternation frequency, the sidebands do not align and the pointwise minimum is unlikely to have a peak even at the carrier's frequency.

Fig. 4.8 A spectrum $S_{new}(f, f_{alt_i}, d_j)$ shifted up and down for 23 kHz, the pointwise minimum between these two spectra, and the pointwise minimum between two spectra with shift different from $f_{alt_i} = 23$ kHz [4]

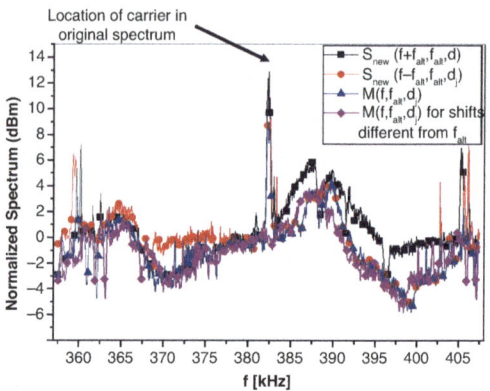

To find that actual alternation frequency, we compute this minimum-of-shifted-spectra operation with all frequency shifts that are within 25% of the intended one, in 50 Hz increments. For each of these $M(f, f_{alt_i}, d_j)$ we compute the average across f, and the shift that produced the largest average is taken as the actual alternation frequency. The intuition behind this is that shifts that correspond to the true alternation frequency will produce stronger peaks at frequencies that correspond to modulated carriers, and will possibly have other peaks that come from aligning unrelated signals. In contrast, incorrect shifts will only have the peaks that come from aligning unrelated signals, but their sideband-induced peaks will be attenuated (or completely eliminated). Thus the shift that corresponds to the actual alternation frequency tends to produce more (and stronger) peaks, which increases its average-over-f relative to other shifts.

In our experiments we have found that the actual alternation frequency is often 150 to 300 Hz away from the intended one. This difference may seem small, but some sidebands are sharply defined, e.g., the peak is only 100 to 200 Hz wide, so use of the intended rather than true alternation frequency may cause our approach to completely miss the actual sideband signals and thus not report the corresponding modulated carrier signals.

4.4.2 Identifying Carrier Frequencies

Once the actual alternation frequencies are identified, we can find the frequencies of carriers that are unintentionally modulated by program activity. To do that, we perform the following steps for each duty cycle d_j. First, for every alternation frequency f_{alt_i}, where $0 < i < N$, the spectrum $S(f, f_{alt_i}, d_j)$ (that corresponds to that alternation frequency) is shifted by $-f_{alt_i}$ to the left and by f_{alt_i} to the right. This creates $2N$ spectra that all correspond to the same duty cycle and whose sideband signals are shifted to the frequency of the carrier that produced

4.4 How to Identify Unintentionally Modulated Side Channel Signals Created...

that sideband signal. Then, the pointwise minimum among all these shifted spectra is found, i.e.,

$$M_{\text{true}}(f, d_j) =$$
$$\min \Big[S(f + f_{alt_1}, f_{alt_1}, d_j), S(f - f_{alt_1}, f_{alt_1}, d_j),$$
$$S(f + f_{alt_2}, f_{alt_2}, d_j), S(f - f_{alt_2}, f_{alt_2}, d_j),$$
$$\vdots$$
$$S(f + f_{alt_N}, f_{alt_N}, d_j), S(f - f_{alt_N}, f_{alt_N}, d_j) \Big]. \quad (4.4)$$

Intuitively, at a frequency that corresponds to a modulated carrier, the sidebands that correspond to different f_{alt} will all align and the minimum will have a peak. At other frequencies, the minimum will have a peak only if other stronger-than-usual signals happen to be present in the original spectra at *every* one of the $2N$ positions, which becomes increasingly unlikely as we increase N.

However, it is still possible that other signals happen to align and create peaks in $M_{\text{true}}(f, d_j)$. To suppress these peaks, for every alternation frequency, we also compute $M_{\text{false}}(f, k, d_j)$ by taking each spectrum (collected with f_{alt_i}) and shifting it by $\pm f_{alt_{i+k}, k \neq 0}$, then taking the point-wise minimum among such spectra:

$$M_{\text{false}}(f, k, d_j) =$$
$$\min \Big[S(f + f_{alt_{1+k}}, f_{alt_1}, d_j), S(f - f_{alt_{1+k}}, f_{alt_1}, d_j),$$
$$S(f + f_{alt_{2+k}}, f_{alt_2}, d_j), S(f - f_{alt_{2+k}}, f_{alt_2}, d_j),$$
$$\vdots$$
$$S(f + f_{alt_k}, f_{alt_N}, d_j), S(f - f_{alt_k}, f_{alt_N}, d_j) \Big]. \quad (4.5)$$

The key property of $M_{\text{false}}(f, k, d_j)$ is that it is computed in exactly the same way as $M_{\text{true}}(f, d_j)$, but the use of incorrect f_{alt} causes none of the sideband signals to be aligned with each other. This is repeated for different non-zero values of k and compute the permutations of $f_{alt_{i+k}}$, and we compute $M_{\text{false}}(f, d_j)$ as the point-wise average among $M_{\text{false}}(f, k, d_j)$ across all non-zero values of k.

Figure 4.9 plots $M_{\text{true}}(f, d_j)$ and $M_{\text{false}}(f, d_j)$ for the experiment where there is an activity-modulated carrier at 382 kHz. We can observe that the $M_{\text{true}}(f, d_j)$ has a distinctive peak at the carrier frequency, while $M_{\text{false}}(f, d_j)$ does not. However, accidental alignment of other (non-sideband) signals would produce similar peaks in $M_{\text{true}}(f, d_j)$ and $M_{\text{false}}(f, d_j)$. Thus we compute a "modulated carrier score" $MCS(f)$ as the point-wise ratio between $M_{\text{true}}(f, d_j)$ and $M_{\text{false}}(f, d_j)$:

Fig. 4.9 Minimums of shifted spectra, i.e., $M_{\text{true}}(f, d_j)$ and $M_{\text{false}}(f, d_j)$, with the carrier frequency at 382 kHz and the alternation frequency of 23 kHz [4]

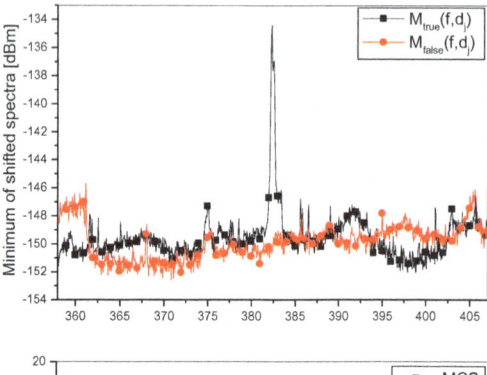

Fig. 4.10 Modulated carrier score as a function of frequency for a spectrum with the carrier frequency at 382 kHz and the alternation frequency of 23 kHz [4]

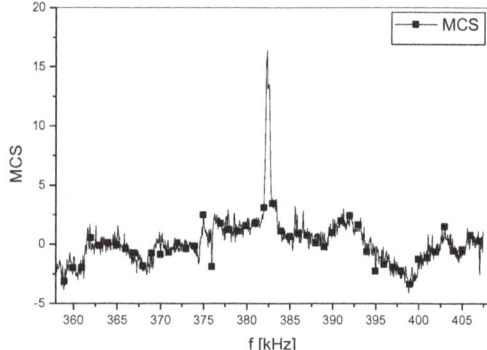

$$MCS(f) = 10 * \log_{10}\left(\frac{M_{\text{true}}(f, d_j)}{M_{\text{false}}(f, d_j)}\right). \quad (4.6)$$

Intuitively, the value of the MCS at each frequency corresponds to how much stronger (in dB) is the signal that corresponds to the sidebands of that (potential) carrier, relative to the signal that would be computed for that frequency even if no sideband were present. To illustrate this, Fig. 4.10 shows the $MCS(f)$ that corresponds to $M_{\text{true}}(f, d_j)$ and $M_{\text{false}}(f, d_j)$ from Fig. 4.9.

The $MCS(f)$ shown in Fig. 4.10 has a strong peak that strongly suggests that a modulated carrier is present at 382 kHz. However, $MCS(f)$ also has many other, smaller, peaks, so it is not easy to determine what value of MCS should be treated as the threshold for reporting modulated carriers. If the MCS threshold is set to some manually selected value, it will need to be adjusted for each evaluated computer system, environment in which the experiment is carried out, antenna position, etc.

Instead, it is highly desirable to set a threshold in terms of the probability that a reported carrier is a false positive, and then automatically determine the corresponding threshold for MCS. To accomplish this, we note that $M_{\text{true}}(f, d_j)$ and $M_{\text{false}}(f, d_j)$ should be statistically equivalent for frequencies that are *not* modulated carriers, so for those frequencies the values of $MCS(f)$ should have a zero mean and a CDF that is symmetric around that mean. In contrast, for

Fig. 4.11 Empirical joint and baseline cumulative distribution functions for MCS score [4]

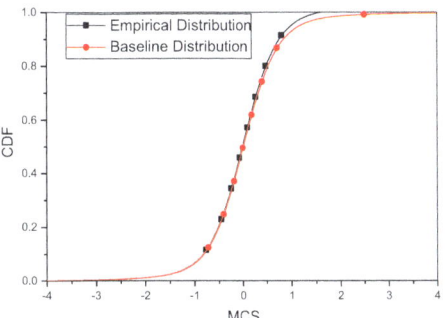

frequencies that correspond to modulated carriers, the $MCS(f)$ will have a bias toward positive values, and the magnitude of that bias increases as the power of sideband signals increases. Thus the problem of deciding how likely it is that a particular frequency has a modulated carrier becomes the problem of determining how likely it is that the $MCS(f)$ value for that frequency belongs to the positive-biased "modulated carrier" distribution rather than the symmetric "baseline" (no modulated carrier) distribution.

Although empirical data for the baseline distribution is not available (the $MCS(f)$ contains points from both distributions), the baseline distribution can be closely approximated by noting that 1) the baseline distribution is symmetric around zero and 2) negative values of $MCS(f)$ are very likely to belong to that distribution. The negative-values part of the baseline distribution is thus approximated by simply using the negative-values part of the empirical joint distribution, while the positive side of the baseline distribution is approximated by using the "mirror image" of the empirical joint distribution. Figure 4.11 shows the empirical joint distribution and the approximated baseline distribution.

It can be observed that the empirical joint distribution has more high-magnitude points than the approximated baseline distribution. Thus we can now set the probability-of-false-positive threshold (p_{fp}) to a desired value, e.g., $p_{fp} \leq 0.02$, look up the MCS value that corresponds to $1 - p_{fp}$, and report carriers whose MCS is no less than that value. For reported MCSs, we than read the actual CDF value and report it as the confidence level. For example, for $p_{fp} \leq 0.02$, we find all MCSs that have value larger than MSC that corresponds to CDF=0.98. Then, for each MCS that satisfies this criteria, we read their actual CDF value. All values should be larger than 0.98.

4.4.3 *Identifying Modulation*

Once frequencies of modulated carrier are identified, we can also identify the type and some other properties of the modulation. Specifically, to identify if the carrier has a mostly-AM-modulated or mostly-FM-modulated signal, we observe how the carrier's frequency and sideband power change as the duty cycle changes. Note that an amplitude-modulated carrier should have the same frequency for all duty cycles (although the magnitude of the carrier and baseband signals will vary as the duty cycle changes). For a frequency-modulated carrier, however, the change in the duty cycle changes the average (DC) value of the baseband signal, which results in shifting the frequency *of the carrier* (and its sidebands) in proportion to the duty cycle. Intuitively, if we plot the modulated carrier's frequency on the Y-axis and the duty cycle on the X-axis, a horizontal line corresponds to AM, while a line with a non-zero slope corresponds to FM whose Δf corresponds to the line's slope.

To reduce the number of spectra that must be collected, however, we only get a few discrete points on this line that correspond to duty cycles used in the experiments. Furthermore, the AM/FM identification (and the estimate of Δf for FM) relies on estimating the slope of the frequency-vs.-duty cycle line, so the duty cycles used in the experiments should not be too close to each other. Finally, the linear fit is imperfect—the actual duty cycle may differ from the intended one, the empirically determined frequency of the modulated carrier may contain some error, etc. Thus the key problem in identifying modulation is the problem of grouping together likely-carrier points from different duty cycles. In other words, for a likely-modulated-carrier point found for a given duty cycle, we need to determine which likely-carrier points from other duty cycles belong to the same modulated carrier. Unfortunately, simply using the points that produce the best goodness-of-fit (e.g., squared-sum-or-errors) for the frequency-vs-duty-cycle produces poor results when a frequency range contains several different modulated carriers, especially if these carriers do not have very sharply defined central frequencies. To overcome this, we note that the sideband power produced by a carrier is also a function of the duty cycle—points that have different duty cycles but belong to the same carrier should each have the sideband power $P_j = P_{max} \sin(d_j \pi)/\pi$, so their $M_{\text{true}}(f, d_j)$ should also be proportional to $\sin(d_j \pi)/\pi$. Thus our modulation-finding consists of finding, for each likely-carrier point, the linear fit (that uses one point from each duty cycle) that produces the smallest *product* of the squared sum of error for the frequency fit and the squared sum of errors for the (M_{true}) fit.

Because the slope of the linear fit is estimated, it is highly unlikely to be exactly zero. Thus we also determine the 95% confidence interval for the estimated slope, and report the carrier as AM if this confidence interval includes the zero value. Intuitively, we report a carrier as FM-modulated only if there is a high enough (95%) confidence that its frequency change is duty-cycle-induced rather than caused by other (duty-cycle-unrelated) variation in estimated frequencies of modulated carriers.

4.4.4 Illustration of How the Algorithm Works

To illustrate how this algorithm works, we show the results of finding unintentionally modulated signals on several devices—a desktop, a laptop, and a smartphone system described in Table 4.1. The signals are recorded using a spectrum analyzer

4.4 How to Identify Unintentionally Modulated Side Channel Signals Created...

Table 4.1 Description of measured devices [4]

Type	Device	Processor
Laptop	Lenovo	Intel Core 2 Duo
Phone	LG P705	Snapdragon S1
Desktop	Dell	Intel i7

Fig. 4.12 Measurement setup for laptop or desktop (left) and measurement setup for cell-phone (right) [4]

(Agilent MXA N9020A). The desktop and laptop measurements are collected with a magnetic loop antenna (AOR LA400) at a distance of 30 cm as shown on the left of Fig. 4.12. To receive weaker signals from smartphones, EM emanations were recorded using a small loop probe, with 20 turns and a 4 mm radius, that was positioned on top of the cellphone as shown on the right of Fig. 4.12. The spectra were measured from 0 to 4 MHz with a resolution bandwidth of 10 Hz.

The benchmarks are run at several different alternation frequencies $f_{alt} = \{23000, 29000, 37000, 53000\}$ Hz with duty cycles $d = \{20, 40, 50, 60, 80\}\%$. The alternation frequencies were chosen to ensure sufficient separation between sidebands of modulated signals, i.e. such that separation between f_{alt_1}, f_{alt_2}, etc. (and their harmonics) is sufficient to prevent overlap. For example, if $f_{alt_1} = 23$ kHz is chosen, frequencies in the vicinity of the harmonics of f_{alt_1} (46, 69 kHz, etc.) should be avoided when choosing other alternation frequencies. Aside from this consideration, the choice of f_{alt} is arbitrary. We have found that four alternation frequencies are sufficient for the algorithm to identify carrier frequencies. To identify if the modulation is AM or FM, we need all five duty cycles.

The benchmarks were run on the laptop and desktop systems as single-threaded Windows 7 32-bit user-mode console applications, and on the smartphone as ordinary Android applications. When possible, all unrelated programs and activities were disabled, the CPU frequency scaling was disabled, and screens/displays were turned off. We measured two sets of A/B alternation activities. The first set alternated between a load from DRAM memory and a load from the on-chip L1 cache, which we abbreviate as LDM/LDL1. This alternation is useful for exposing modulated carriers related to memory activity, in addition to some that are related to on-chip activity. The second activity alternated between loads from the on-chip L2 and L1 caches, which we abbreviate as LDL2/LDL1. This activity was chosen to help identify which carriers are modulated by on-chip activity. We also tried other

Table 4.2 Carrier frequencies found in a laptop [4]

Carrier frequency [Hz]	Harmonic no.	SNR [dB]	Type of modulation	Confidence level
383010	1	16	FM (Δf=2275 Hz)	99.8%
765949	2	12	FM (Δf=4700 Hz)	99.9%
1148959	3	10	FM (Δf=7225 Hz)	99.8%
448071	1	4	AM	99.1%

Table 4.3 Carrier frequencies found in a cell phone [4]

Carrier frequency [Hz]	Harmonic no.	SNR [dB]	Type of modulation	Confidence level
110543	1	1	FM (Δf=3100 Hz)	98.6%
1599990	1	30	AM	100%
3200000	2	22	AM	99.98%
3257391	1	1	FM (Δf=96002 Hz)	99.98%

instruction pairs (e.g., arithmetic, memory stores, etc.) and found that they lead to identification of the same modulated carriers already discovered by LDM/LDL1 and LDL2/LDL1.

We tested three devices described in Table 4.1, with two measurement sets per device (one for LDM/LDL1 and one for LDL2/LDL1). Table 4.2 summarizes the carrier frequencies found in emanations from a laptop using our algorithm, along with the type of modulation, signal to noise ratio (SNR), and confidence level for each carrier. Here, we define SNR as a difference in decibels between $M_{\text{true}}(f, d_j)$ and $M_{\text{false}}(f, d_j)$, as defined in equation (4.6). Our algorithm has found one FM carrier and its two harmonics with confidence levels above 99%. We can also observe that SNR for all three FM modulated frequencies is above 10 dB which indicates that these carriers are strong and will carry the signal to some distance away from the laptop. Our algorithm has also found one AM modulated carrier but the observed SNR is only 4 dB, which indicates that this is a weak carrier. Please note that our algorithm finds all carrier frequencies first, and then checks for possible harmonic relationships among the carriers it found.

Table 4.3 summarizes carrier frequencies, and their parameters, found for a cell phone. Here, our algorithm has found one AM carrier and its second harmonic with confidence level above 99%. The SNR for these two frequencies is above 20 dB, i.e., they are excellent candidates to carry signal outside of the cell phone. Our algorithm has also found two FM modulated carriers, but the observed SNR is only 1 dB, which indicates that these are weak carriers.

Finally, Table 4.4 summarizes carrier frequencies, and their parameters, found for a desktop. Here, our algorithm has found one AM carrier and its 11 harmonics

Table 4.4 Carrier frequencies found in a cell phone [4]

Carrier frequency [Hz]	Harmonic no.	SNR [dB]	Type of modulation	Confidence level
315488	1	28	AM	99.8 %
631006	2	28	AM	99.99 %
946654	3	22	AM	99.7%
1262312	4	21	AM	99.8%
1566849	5	19	AM	99.9%
1893447	6	18	AM	99.8%
2209415	7	13	AM	99.9%
2840661	9	5	AM	99.8%
3156239	10	6	AM	99.8%
3471917	11	8	AM	99.9%
3787705	12	6	AM	99.8%
451581	1	5	FM ($\Delta f = 550$ Hz)	99.92%
511653	1	17	AM	99.96%
1023306	2	13	AM	99.97%
1534938	3	24	AM	100%
2046601	4	25	AM	100%
2558214	5	23	AM	100%
3069877	6	20	AM	100%
3581530	7	11	AM	99.99%

with confidence level above 99%. The SNR for the first seven harmonics is above 10 dB, while SNR for the other five harmonics is above 5 dB. Furthermore, it found one more AM carrier and its seven harmonics, all with SNRs above 10 dB. Finally, it found one FM carrier with a SNR of 5 dB.

To verify the accuracy of the algorithm, we have visually inspected all spectra and confirmed that carriers found by the algorithm exist in the spectrum. From the results, it can be observed that there are only 2 or 3 fundamental frequencies in this frequency range (0 to 4 MHz), while the others are their harmonics. The fundamental frequencies that were reported are all attributable to voltage regulator and memory refresh activity on the measured system. For example in Fig. 4.4 we can observe that the two strongest sources are the voltage regulator (315 kHz) and memory refresh (software activity in the system at 511 kHz). The voltage regulator emanations can be reduced by better shielding of coils, and the memory refresh can be eliminated by creating different scheduling pattern for memory refresh. Alternatively, program code can be changed to avoid power-fluctuations and memory activity that depends on sensitive information. Please note that carrier frequencies can also be found at higher frequencies—they are typically above 500 MHz and belong to processor or memory clock. While our algorithm can find these frequencies as well, information about processor and memory clocks is usually readily available. Finding carrier frequencies in lower frequency ranges is more challenging because there is much more noise-like activity in the spectrum and it is difficult to identify information caring signals.

Automatic identification of potential carriers in the system has several benefits. From the security prospective, it allows us to quickly identify frequencies that are of interest for observing RF emanations, as well as prediction of distances from which we can expect to receive good quality signal (based on observed SNR), and the type of demodulation needed to correctly receive signals. From the system designer prospective, finding carrier frequencies helps us identify leaky circuits. For example, the unintentional FM and AM carriers found for a desktop and laptop were caused by voltage regulators and memory refresh activity. For a cell phone, several carriers were found to be caused by voltage regulators. The other carriers found on the cell phone were traced to particular IC packages or modules and were likely caused by either voltage regulators or possibly additional periodic memory activity. However, smartphones integrate many system components into System on Chip (SoC) modules and often use Package on Package (PoP) technology to integrate both the processor and memory into the same package and little information is publicly available describing these components. More information would be needed to definitively determine the circuits and mechanisms modulating these carriers—but such information is typically available to system designers of the cellphone.

References

1. Robert Callan, Alenka Zajic, and Milos Prvulovic. Fase: Finding amplitude-modulated side-channel emanations. In *2015 ACM/IEEE 42nd Annual International Symposium on Computer Architecture (ISCA)*, pages 592–603, 2015.
2. Giovanni Camurati, Sebastian Poeplau, Marius Muench, Tom Hayes, and AurAlien Francillon. Screaming channels: When electromagnetic side channels meet radio transceivers. In *Proceedings of the 25th ACM conference on Computer and communications security (CCS)*, CCS '18. ACM, October 2018.
3. Milos Prvulovic and Alenka Zajic. Rf emanations from a laptop, 2012. http://youtu.be/ldXHd3xJWw8.
4. Milos Prvulovic, Alenka Zajic, Robert L. Callan, and Christopher J. Wang. A method for finding frequency-modulated and amplitude-modulated electromagnetic emanations in computer systems. *IEEE Transactions on Electromagnetic Compatibility*, 59(1):34–42, 2017.
5. Seun Sangodoyin, Frank T. Werner, Baki B. Yilmaz, Chia-Lin Cheng, Elvan M. Ugurlu, Nader Sehatbakhsh, Milos Prvulovic, and Alenka Zajic. Side-channel propagation measurements and modeling for hardware security in iot devices. *IEEE Transactions on Antennas and Propagation*, 69(6):3470–3484, 2021.
6. A. Zajic and M. Prvulovic. Experimental demonstration of electromagnetic information leakage from modern processor-memory systems. *Electromagnetic Compatibility, IEEE Transactions on*, 56(4):885–893, Aug 2014.

Chapter 5
Relationship Between Modulated Side Channels and Program Activity

5.1 Mathematical Relationship Between Electromagnetic Side Channel Energy of Individual Instructions and the Measured Pairwise Side Channel Signal Power

Establishing a mathematical relationship between electromagnetic side channel energy (ESE) of individual instructions and the measured pairwise side-channel signal power available to an attacker can not only help programmers and computer hardware designers to anticipate the vulnerability of the system, but can also help us relate instruction probabilities with energy levels of particular instructions [1, 5, 6]. More importantly, it helps us understand the relationship between program code and the resulting analog side channel signals, and it helps us know what to expect to see in signal when particular kinds of software activities are present. Overall, this mathematical relationship removes most of the guesswork out of side channel analysis, which was not possible in the past.

5.1.1 Definition of Individual Instructions' ESE

To achieve this goal, we define a voltage signal $s_1(t)$ that corresponds to execution of one sequence of instructions and a voltage signal $s_2(t)$ corresponding to another sequence of instructions. Then, the total EM side-channel energy available to the attacker (for the purpose of deciding which of the two sequences was executed) can be defined as [5]:

$$\mathscr{P}_A(s_1(t), s_2(t)) \equiv \frac{1}{R} \int_0^{T_s} (s_1(t) - s_2(t))^2 dt, \tag{5.1}$$

where $s_1(t)$ and $s_2(t)$ are voltages measured across a resistance R (typically instruments have $R = 50\,\Omega$), and $t = (0, T_s)$ is the time interval over which the difference in program execution occurs. This difference can be, for example, execution of one path or the other in an if-then-else statement, occurrence or absence of a cache miss when a memory access is executed, etc. Because instruction sequences are composed of instructions, we are especially interested in the case where we consider two single-instruction sequences, i.e., where we compute the EM side-channel energy (ESE) available to the attacker for the purpose of deciding between two instruction sequences that are the same except for only one instruction.

Assume $x_1(t)$ and $x_2(t)$ are the characteristic signals that correspond to execution of instructions X_1 and X_2. Then, ESE is defined as [5]

$$\text{ESE}(X_1, X_2) \equiv \frac{1}{R} \int_0^{T_x} (\mathbb{E}[x_1(t) - x_2(t)])^2 \, dt, \tag{5.2}$$

where T_x is the maximum of the execution times of the instructions and $\mathbb{E}$ is the expectation operation.

Note that this single-instruction ESE corresponds to executing an instruction (one or the other) only one time. An naive way of measuring this ESE would be using the above definition directly—one would collect the side channel signal when one instruction is executed, then when the other instruction is executed, and then compute the difference between the two signals. However, execution of a single instruction is very brief and cannot be accomplished in isolation—other instructions are needed to set up execution of a program and clean-up after it, resulting in thousands of instructions being executed in addition to the one for which the signal is being collected. Thus the signal is collected over some time, with all the noise and interference over that interval of time, whereas the ESE we are trying to measure has only a small and short-lived contribution to the collected signals. Additionally, because the contribution of the ESE to the signal is so short-lived, in a naive measurement the signal must be recorded with very high real-time bandwidth, i.e., using very expensive measurement equipment. Therefore, the naive method of measuring the single-instruction ESE is likely to be very expensive and yet have very high measurement error.

To achieve a less expensive and less noisy measurement of this single-instruction ESE for specific instructions X_1 and X_2, our measurement method will be to execute the A/B alternation pseudo-code in Chap. 4, where activity A is instruction X_1 and activity B is instruction X_2, perform measurement of the EM signal power at the alternation frequency—a simple measurement for most instruments, even inexpensive ones—and then derive the single-instruction ESE from that measurement [1]. Specifically, under some simplifying assumptions that are needed to make our derivations mathematically tractable, this measurement allows the value of ESE to be calculated as follows [5]:

Theorem 5.1 (ESE) *ESE*$[X_1, X_2]$ *defined in (5.2) can be estimated from the spectral power* $P(f_{\text{alt}})$ *measured at* f_{alt} *while running the* X_1/X_2 *alternation code*

5.1 Mathematical Relationship Between Electromagnetic Side Channel... 59

with alternation frequency f_{alt}, by repeatedly executing instruction X_1 n_{inst} times, followed by n_{inst} executions of instruction X_2, as follows:

$$\text{ESE}[X_1, X_2] \approx \left(\frac{\pi}{2}\right)^2 \frac{P(f_{\text{alt}}) \cdot N}{n_X \cdot n_{\text{inst}} \cdot f_{\text{alt}}} + C(X_1, X_2), \quad (5.3)$$

where N is the number of samples taken during only one inner for-loop, n_X is the maximum of the number of samples taken during the execution of instructions X_1 and X_2 and $C(X_1, X_2)$ is a constant term that depends on the instruction pair, i.e.,

$$C(X_1, X_2) = -\frac{\mathscr{P}_A\left(s_1^{(X_1)}, s_2^{(X_1)}\right) + \mathscr{P}_A\left(s_1^{(X_2)}, s_2^{(X_2)}\right)}{2 n_X n_{\text{inst}}}. \quad (5.4)$$

In the rest of the Sect. we present the notations and assumptions under which Theorem 5.1 can be derived, and then we present the derivations for this theorem.

5.1.2 Derivation of Single-Instruction ESE from Measured Spectral Power

To make our derivations mathematically tractable, we introduce following notations and assumptions.

1. The for-loops of the A/B alternation pseudo-code in Chap. 4 generate $s_1(t)$ and $s_2(t)$ respectively. Because the emitted signal in each loop depends on the type of instruction used in the loop, to discard the ambiguities regarding the inserted instruction and the taken branch, we denote produced signals as $s_i^{(X_i)}(t)$ which is generated by the execution of the ith inner-for-loop and X_i represents any instruction inserted into that loop.
2. $s_1^{(X_1)}(t)$ and $s_2^{(X_2)}(t)$ are voltages sampled at frequency $1/T_I$ to create the sequences $s_1^{(X_1)}[n]$ and $s_2^{(X_2)}[n]$ of length $N_s = T_s/T_I$.
3. The frequency content of $s_1^{(X_1)}(t)$ and $s_2^{(X_2)}(t)$ above $1/(2T_I)$ is negligible (i.e., $s_1^{(X_1)}(t)$ and $s_2^{(X_2)}(t)$ have bandwidth $1/(2T_I)$).
4. $s_1^{(X_1)}(t)$ and $s_2^{(X_2)}(t)$ are voltages measured across a resistance R.
5. The discrete energy available to the attacker $\mathscr{P}_A\left[s_1^{(X_1)}, s_2^{(X_2)}\right]$ is then defined as

$$\mathscr{P}_A\left[s_1^{(X_1)}, s_2^{(X_2)}\right] \equiv T_I \sum_{l=0}^{N_s-1} \frac{\left(s_1^{(X_1)}[l] - s_2^{(X_2)}[l]\right)^2}{R}. \quad (5.5)$$

6. Sampled voltages can be represented as $s_1^{(X_1)}[l] = \mu_1[l] + w_1[l]$ and $s_2^{(X_2)}[l] = \mu_2[l] + w_2[l]$ where $\mu_1[l]$ and $\mu_2[l]$ are the mean voltage values, $w_1[l]$ and $w_2[l]$

are the additive noises which are i.i.d., $\mathcal{N}(0, \sigma_l^2)$. Here, σ_l^2, $\mu_1[l]$, and $\mu_2[l]$ are depended on the instruction. For example, if the sample is taken during the execution of the first-inner-loop and the embedded instruction is X_1, the sample can be written as

$$s_1[l] = \mu_1[l] + w_1[l] = X_1^v + w_1[l], \tag{5.6}$$

where $w_1[l] \sim \mathcal{N}(0, \sigma_{X_1}^2)$ and X_1^v is the average power instruction X_1 emits. We assume that additive noise describes all the variation in the signal and that noise power is dependent on the executed instruction since the electromagnetic emissions can vary according to the execution location.

7. Finally, the discrete ESE is defined as

$$\text{ESE}[X_1, X_2] \equiv \frac{T_I}{R} n_X \left(X_1^v - X_2^v\right)^2. \tag{5.7}$$

where $n_X = T_x/T_I$ is the number of samples taken during only the execution of the instructions.

To show that $\text{ESE}[X_1, X_2]$ defined in (5.2) is related to the alternation power $P(f_{\text{alt}})$ as in (5.3) where $s_m^{(X_m)}[n]$ denotes the sampled signal when instruction X_m is inserted into the mth loop given in the A/B alternation pseudo-code.

We start by noting that the signals generated by the ESE benchmarks can be represented as a specific mixture of two *periodic* signals with period N. For $n = 0, \ldots, N-1$, the first signal obtained from the first iteration of the first inner loop is denoted as [5]

$$\hat{s}_1^{(X_1)} = [o_1[0], o_1[1], \cdots, o_1[n_O - 1], x_1[0], x_1[1], \cdots, x_1[n_X - 1]],$$

such that $N = n_O + n_X$. Note that $\mathbb{E}\left[s_1^{(X_1)}[n+N]\right] = \mathbb{E}\left[s_1^{(X_1)}[n]\right]$ because $s_1^{(X_1)}[n]$ is periodic, which implies that we also assume the additive noise is also periodic. Following the same procedure for the second inner loop, we have [5]

$$\hat{s}_2^{(X_2)} = [o_2[0], o_2[1], \cdots, o_2[n_O - 1], x_2[0], x_2[1], \cdots, x_2[n_X - 1]].$$

We denote the sampled voltage at the time points where instruction X_1 is active as $x_1[i]$ and the sampled voltage at the time points where instruction X_2 is active as $x_2[i]$ where $i \in \{0, 1, \cdots, n_X - 1\}$. Similarly $o_m[n]$ represents the other instructions in the benchmark necessary to make the benchmark practical (e.g., to create a loop around instruction X_1 or instruction X_2) for the mth loop.

First, we derive the ESE between two instructions from the measurement given in (5.5) as follows [5]:

$$\mathscr{P}_A\left[s_1^{(X_1)}, s_2^{(X_2)}\right] \equiv T_I \sum_{l=0}^{N_s-1} \frac{\left(s_1^{(X_1)}[l] - s_2^{(X_2)}[l]\right)^2}{R}$$

5.1 Mathematical Relationship Between Electromagnetic Side Channel...

$$= \frac{T_1}{R} \sum_{k=0}^{Nn_{\text{inst}}-1} \left(s_1^{(X_1)}[k] - s_2^{(X_2)}[k] \right)^2$$

$$\approx \frac{\kappa}{2} \left[\sum_{k=0}^{N-1} \left(\hat{s}_1^{(X_1)}[k] - \hat{s}_2^{(X_2)}[k] \right)^2 \right], \quad (5.8)$$

where $\kappa = (2T_1 n_{\text{inst}})/(R)$ and (5.8) follows the periodicity of $s_1^{(X_1)}[n]$ and $s_2^{(X_2)}[n]$. Using assumptions made above, we can write $x_1[i] = X_1^v + w_1^{(X_1)}[i]$, $x_2[i] = X_2^v + w_2^{(X_2)}[i]$ and $o_m[i] = O^v + w_m^{(O)}[i]$ where $w_1^{(X_1)}[i] \sim \mathcal{N}(0, \sigma_{X_1}^2)$, $w_2^{(X_2)}[i] \sim \mathcal{N}(0, \sigma_{X_2}^2)$, $w_m^{(O)}[i] \sim \mathcal{N}(0, \sigma_O^2)$ and $m \in \{1, 2\}$. Therefore,

$$\mathcal{P}_A\left[s_1^{(X_1)}, s_2^{(X_2)} \right] \approx$$

$$\approx \frac{\kappa}{2} \sum_{k=0}^{n_X-1} \left(X_1^v - X_2^v + w_1^{(X_1)}[k] - w_2^{(X_2)}[k] \right)^2$$

$$- \frac{\kappa}{2} \sum_{k=0}^{n_O-1} \left(w_1^{(O)}[k] - w_2^{(O)}[k] \right)^2$$

$$= \frac{\kappa n_X}{2} \frac{1}{n_X} \sum_{k=0}^{n_X-1} \left(X_1^v - X_2^v + w_1^{(X_1)}[k] - w_2^{(X_2)}[k] \right)^2$$

$$- \frac{\kappa n_O}{2} \frac{1}{n_O} \sum_{k=0}^{n_O-1} \left(w_1^{(O)}[k] - w_2^{(O)}[k] \right)^2. \quad (5.9)$$

Assuming that a large-enough number of samples is taken during executions of instructions, the sum operations given in (5.9) can be replaced with the expectation operation. Therefore, we obtain

$$\mathcal{P}_A\left[s_1^{(X_1)}, s_2^{(X_2)} \right] \approx$$

$$\approx \frac{\kappa n_X}{2} \mathbb{E}\left[\left(X_1^v - X_2^v + w_1^{(X_1)}(k) - w_2^{(X_2)}(k) \right)^2 \right]$$

$$- \frac{\kappa n_O}{2} \mathbb{E}\left[\left(w_1^{(O)}[k] - w_2^{(O)}[k] \right)^2 \right]$$

$$= \frac{\kappa}{2} \left(n_X (X_1^v - X_2^v)^2 + n_X \sigma_{X_1}^2 \right.$$

$$\left. + n_X \sigma_{X_2}^2 + 2n_O \sigma_O^2 \right). \quad (5.10)$$

Furthermore, observe that (5.10) can be modified in terms of ESE as follows [5]:

$$\mathscr{P}_A\left[s_1^{(X_1)}, s_2^{(X_2)}\right] \equiv$$

$$\equiv n_{\text{inst}}\text{ESE}[X_1, X_2] + \frac{1}{2}\left(\mathscr{P}_A\left[s_1^{(X_1)}, s_2^{(X_1)}\right]\right.$$
$$\left.+ \mathscr{P}_A\left[s_1^{(X_2)}, s_2^{(X_2)}\right]\right). \quad (5.11)$$

If we define

$$\hat{C}(X_1, X_2) = \mathscr{P}_A\left[s_1^{(X_1)}, s_2^{(X_1)}\right] + \mathscr{P}_A\left[s_1^{(X_2)}, s_2^{(X_2)}\right], \quad (5.12)$$

the ESE of two instructions in time domain can be written as

$$\text{ESE}[X_1, X_2] = \frac{2\mathscr{P}_A\left[s_1^{(X_1)}, s_2^{(X_2)}\right] - \hat{C}(X_1, X_2)}{n_{\text{inst}}}. \quad (5.13)$$

This ESE is expressed in the time domain, where measuring ESE of two sequences can be cumbersome because the number of samples required can be huge. With that in mind, we now derive the equation given in (5.3), which allows ESE to be computed by measuring power at f_{alt}—a relatively simple measurement for any spectrum analyzer or software defined radio.

To relate $s_1^{(X_1)}[n]$ and $s_2^{(X_2)}[n]$ to our benchmarks, we define a square wave $p[n]$ with a 50% duty cycle as [5]

$$p[0 \le n < Nn_{\text{inst}}] = 1, \quad (5.14)$$

$$p[Nn_{\text{inst}} \le n < 2Nn_{\text{inst}}] = 0, \quad (5.15)$$

where $p[n]$, $s_1^{(X_1)}[n]$, and $s_2^{(X_2)}[n]$ are periodic with period $2Nn_{\text{inst}}$, since we assume the additive noise is also periodic. Then, we can take the discrete Fourier series of these signals over $2Nn_{\text{inst}}$ samples. We refer to $S_1^{(X_1)}[k]$, $S_2^{(X_2)}[k]$, and $P[k]$ as the discrete Fourier series (DFS) of $s_1^{(X_1)}[n]$, $s_2^{(X_2)}[n]$ and $p[n]$ respectively, defined for $0 \le k < 2Nn_{\text{inst}}$.

We next define

$$v[n] = p[n]s_1^{(X_1)}[n] + (1 - p[n])s_2^{(X_2)}[n], \quad (5.16)$$

which represents the signal created by the sequence of instructions executed by the A/B alternation code.

We start the derivation of relationship between ESE and measured spectral power by observing that the DFS of $v[n]$ defined in (5.16) can be written as [5]

5.1 Mathematical Relationship Between Electromagnetic Side Channel...

$$\begin{aligned} V[k] &= P[k] * S_1^{(X_1)}[k] + (1 - P[k]) * S_2^{(X_2)}[k] \\ &= S_2^{(X_2)}[k] + P[k] * \left(S_1^{(X_1)}[k] - S_2^{(X_2)}[k] \right), \end{aligned} \quad (5.17)$$

where $*$ denotes periodic convolution, defined as [3]

$$x_1[n]x_2[n] \xrightarrow{\text{DFS}} X_1[k] * X_2[k] = \frac{1}{M} \sum_{l=0}^{M-1} X_1[l] X_2[k-l]. \quad (5.18)$$

The discrete Fourier series of a periodic signal $x[n]$ taken *over a period of* ML *samples* is equal to [5]

$$\begin{aligned} X[k] &= \sum_{n=0}^{ML-1} x[n] e^{-j\frac{2\pi}{ML}kn} \\ &= \sum_{m=0}^{M-1} \left(x[m] \sum_{l=0}^{L-1} e^{-j\frac{2\pi}{ML}k(m+Ml)} \right) \\ &= \sum_{m=0}^{M-1} \left(x[m] e^{-j\frac{2\pi}{ML}km} \sum_{l=0}^{L-1} e^{-j2\pi k \frac{l}{L}} \right) \\ &= \sum_{m=0}^{M-1} \left(x[m] e^{-j\frac{2\pi}{ML}km} L \sum_{l=-\infty}^{\infty} \delta[k - lL] \right) \\ &= \begin{cases} L \sum_{m=0}^{M-1} x[m] e^{-j\frac{2\pi}{ML}km} & \text{for } k = Ll. \\ 0 & \text{for } k \neq Ll \end{cases} \end{aligned} \quad (5.19)$$

The discrete time periodic impulse train with period M can be written as [3]

$$\sum_{l=-\infty}^{\infty} \delta[k - Ll] = \frac{1}{L} \sum_{l=0}^{L-1} e^{-j2\pi k \frac{l}{L}}. \quad (5.20)$$

Now we consider $V[1]$, the 2nd Fourier coefficient (the first harmonic) of the $v[n]$ sequence:

$$V[1] = S_2^{(X_2)}[1] + \frac{\sum_{m=0}^{2Nn_{\text{inst}}-1} P[1-m] \left(S_1^{(X_1)}[m] - S_2^{(X_2)}[m] \right)}{2Nn_{\text{inst}}}. \quad (5.21)$$

Using the Eq. (5.19), we can observe that $S_1^{(X_1)}[k]$ and $S_2^{(X_2)}[k]$ are non-zero only for $k = 2n_{\text{inst}} l$ for $l = 0, 1, \ldots, N-1$. Then $V[1]$ simplifies to [5]

$$V[1] = \frac{\sum_{l=0}^{N-1} P[1 - 2n_{\text{inst}}l] \left(S_1^{(X_1)}[2n_{\text{inst}}l] - S_2^{(X_2)}[2n_{\text{inst}}l]\right)}{2Nn_{\text{inst}}}. \quad (5.22)$$

Then, $V[1]$ can be further expanded as follows [5]

$$V[1] = \frac{P[1]\left(S_1^{(X_1)}[0] - S_2^{(X_2)}[0]\right)}{2Nn_{\text{inst}}}$$
$$+ \frac{P[1 - 2n_{\text{inst}}]\left(S_1^{(X_1)}[2n_{\text{inst}}] - S_2^{(X_2)}[2n_{\text{inst}}]\right)}{2Nn_{\text{inst}}} \quad (5.23)$$
$$+ \ldots$$

Here we note that next few higher order odd harmonics can be similarly expanded while the even harmonics are zero. Additionally, we note that $P[k]$ is the kth coefficient of the discrete Fourier series for a square wave with period $2Nn_{\text{inst}}$ and can be written as ([3], Example 8.3)

$$\frac{|P[k]|}{2Nn_{\text{inst}}} = \frac{\sin(\pi k/2)}{2Nn_{\text{inst}} \cdot \sin\left(\frac{\pi k}{2Nn_{\text{inst}}}\right)}$$
$$\Rightarrow \frac{|P[k]|}{2Nn_{\text{inst}}} \approx \frac{\sin(\pi k/2)}{\pi k}$$
$$\Rightarrow \frac{|P[1]|}{2Nn_{\text{inst}}} \approx \frac{1}{\pi}. \quad (5.24)$$

The last two steps follow by recognizing that [5]

$$2Nn_{\text{inst}} \cdot \sin\left(\frac{\pi k}{2Nn_{\text{inst}}}\right) = \pi k \cdot \frac{\sin\left(\frac{\pi k}{2Nn_{\text{inst}}}\right)}{\frac{\pi k}{2Nn_{\text{inst}}}}, \quad (5.25)$$

and noting that $\sin(x)/x \to 1$ as $x \to 0$ (i.e., large n_{inst}). Since n_{inst} is typically > 100, this approximation is valid. For $n_{\text{inst}} > 100$, $|P[1]| > 100|P[1 - n_{\text{inst}}]|$, so the higher order terms in (5.23) can be neglected, giving

$$|V[1]| \approx \frac{|P[1]|}{2Nn_{\text{inst}}} \cdot \left|S_1^{(X_1)}[0] - S_2^{(X_2)}[0]\right|$$
$$\Rightarrow \pi |V[1]| \approx \left|S_1^{(X_1)}[0] - S_2^{(X_2)}[0]\right|. \quad (5.26)$$

5.1 Mathematical Relationship Between Electromagnetic Side Channel...

The next step is to calculate the difference between $S_1^{(X_1)}[0]$ and $S_2^{(X_2)}[0]$. Here we note that, we decompose $s_1^{(X_1)}[n] = \hat{o}_1[n] + i_1^{(X_1)}[n]$ where the first N samples of $\hat{o}_1[n] = [o_1[0], o_1[1], \ldots, o_1[n_O - 1], 0, \cdots, 0]$ and the first N samples of $i_1^{(X_1)}[n] = [0, \ldots, 0, x_1[0], x_1[1], \cdots, x_1[n_X - 1]]$. We can decompose $s_2^{(X_2)}[n]$ similarly. By the linearity of the Fourier transform

$$\begin{aligned} S_1^{(X_1)}[k] - S_2^{(X_2)}[k] &= I_1^{(X_1)}[k] + \hat{O}_1[k] - \left(I_2^{(X_2)}[k] + \hat{O}_2[k]\right) \\ &= I_1^{(X_1)}[k] - I_2^{(X_2)}[k] + \left(\hat{O}_1[k] - \hat{O}_2[k]\right). \end{aligned} \quad (5.27)$$

The DFS coefficient $I_1^{(X_1)}[0]$ is

$$\begin{aligned} I_1^{(X_1)}[0] &= \sum_{n=0}^{2Nn_{\text{inst}}-1} i_1^{(X_1)}[n] \\ &= \sum_{r=0}^{2n_{\text{inst}}-1} \sum_{s=0}^{n_X-1} i_1^{(X_1)}[rn_{\text{inst}} + n_O + s] \\ &= \sum_{r=0}^{2n_{\text{inst}}-1} \sum_{s=0}^{n_X-1} \left(X_1^v + w_1^{(X_1)}[rn_{\text{inst}} + n_O + s]\right) \\ &= 2n_{\text{inst}} n_X X_1^v + 2n_{\text{inst}} \sum_{n=n_O}^{N} w_1^{(X_1)}[n], \quad (5.28) \end{aligned}$$

where the last equality follows the assumption that noise is also circular. Similarly, $I_2^{(X_2)}[0] = 2n_{\text{inst}} n_X X_2^v + 2n_{\text{inst}} \sum_{n=n_O}^{N} w_2^{(X_2)}[n]$ and $O_i[0] = 2n_{\text{inst}} n_O O^v + 2n_{\text{inst}} \sum_{l=0}^{n_O-1} w_i^{(O)}$ where $i \in \{1, 2\}$. Therefore,

$$\begin{aligned} S_1^{(X_2)}[0] - S_2^{(X_2)}[0] = 2n_{\text{inst}} \bigg[& n_X(X_1^v - X_2^v) \\ &+ \sum_{l=0}^{n_O-1} \left(w_1^{(O)}[l] - w_2^{(O)}[l]\right) \\ &+ \sum_{l=0}^{n_X-1} \left(w_1^{(X_1)}[l] - w_2^{(X_2)}[l]\right) \bigg]. \quad (5.29) \end{aligned}$$

Assuming n_O and n_X are large enough and additive noises are independent, we have [5]

$$\left|S_1^{(X_2)}[0] - S_2^{(X_2)}[0]\right|^2 \approx (2n_{\text{inst}})^2 \left[\left(n_X(X_1^v - X_2^v)\right)^2\right.$$

$$\left. + 2n_O\sigma_O^2 + n_X\sigma_{X_1}^2 + n_X\sigma_{X_2}^2\right]$$

$$= 4n_{\text{inst}}^2 D_{ab} \tag{5.30}$$

where

$$D_{ab} = n_X\left(n_X(X_1^v - X_2^v)^2 + \sigma_{X_1}^2 + \sigma_{X_2}^2\right) + 2n_O\sigma_O^2. \tag{5.31}$$

As the next step, we need to relate (5.30) to the power observed with the spectrum analyzer. The power observed with the spectrum analyzer is described by Heinzel et al. [2] and Press et al. [4])

$$P(f_{\text{alt}}) = \frac{2}{R}\left(\frac{|V[1]|}{2Nn_{\text{inst}}}\right)^2, \tag{5.32}$$

where $2Nn_{\text{inst}}$ is the number of samples taken in one period T_{alt}. We also note that

$$n_{\text{inst}} f_{\text{alt}} = \frac{1}{2NT_I}. \tag{5.33}$$

So, by plugging (5.26) and (5.30) into (5.32), we have [5]

$$P(f_{\text{alt}}) = \frac{2}{R}\left(\frac{\left|S_1^{(X_1)}[0] - S_2^{(X_2)}[0]\right|}{2Nn_{\text{inst}} \cdot \pi}\right)^2$$

$$= \frac{2}{R}\left(\frac{4n_{\text{inst}}^2 D_{ab}}{4N^2 n_{\text{inst}}^2 \cdot \pi^2}\right)$$

$$= \frac{2}{R}\left(\frac{D_{ab}}{N^2\pi^2}\right). \tag{5.34}$$

As mentioned before, since working in the time domain is cumbersome, our main goal is to measure ESE in the frequency domain. Therefore, using (5.33), (5.32) and (5.34), we obtain the relationship between ESE and $P(f_{\text{alt}})$ as follows [5]:

$$P(f_{\text{alt}}) = \frac{2}{R}\left(\frac{D_{ab}}{N^2 \cdot \pi^2}\right) \tag{5.35}$$

$$\Rightarrow \quad \frac{D_{ab}}{R} = \frac{P(f_{\text{alt}}) \cdot N^2 \cdot \pi^2}{2 \cdot} \tag{5.36}$$

5.2 Mathematical Relationship Between Electromagnetic Side Channel... 67

$$\Rightarrow \frac{D_{ab}}{RNn_{\text{inst}}} = \frac{P(f_{\text{alt}}) \cdot N \cdot \pi^2}{2 \cdot n_{\text{inst}}} \tag{5.37}$$

$$\Rightarrow \frac{2 \cdot f_{\text{alt}} T_1 D_{ab}}{R} = \frac{P(f_{\text{alt}}) \cdot N \cdot \pi^2}{2 \cdot n_{\text{inst}}} \tag{5.38}$$

$$\Rightarrow \frac{2T_1 n_{\text{inst}}}{R} D_{ab} = \frac{P(f_{\text{alt}}) \cdot N \cdot \pi^2}{2 \cdot f_{\text{alt}}} \tag{5.39}$$

$$\Rightarrow \kappa D_{ab} = \pi^2 \frac{P(f_{\text{alt}}) \cdot N}{2 \cdot f_{\text{alt}}}, \tag{5.40}$$

where (5.38) follows the equality given in (5.33). Relating the equation in (5.40) with (5.10) and (5.11), we have [5]

$$\kappa D_{ab} = \kappa \left(n_X \left(n_X (X_1^v - X_2^v)^2 + \sigma_{X_1}^2 + \sigma_{X_2}^2 \right) + 2n_O \sigma_o^2 \right)$$
$$= 2n_X n_{\text{inst}} \text{ESE}[X_1, X_2]$$
$$+ \mathcal{P}_A \left[s_1^{(X_1)}, s_2^{(X_1)} \right] + \mathcal{P}_A \left[s_1^{(X_2)}, s_2^{(X_2)} \right]. \tag{5.41}$$

To simplify the notation, we define [5]

$$C(X_1, X_2) = -\frac{\mathcal{P}_A \left[s_1^{(X_1)}, s_2^{(X_1)} \right] + \mathcal{P}_A \left[s_1^{(X_2)}, s_2^{(X_2)} \right]}{2n_X n_{\text{inst}}}, \tag{5.42}$$

which can be considered as the noise term added to ESE of (X_1, X_2) instruction pair because of the measurement done in frequency domain. Therefore, we have [5]

$$\text{ESE}[X_1, X_2] = \frac{1}{2n_X n_{\text{inst}}} \pi^2 \frac{P(f_{\text{alt}}) \cdot N}{2 \cdot f_{\text{alt}}} + C(X_1, X_2) \tag{5.43}$$

$$= \left(\frac{\pi}{2}\right)^2 \frac{P(f_{\text{alt}}) \cdot N}{n_X \cdot n_{\text{inst}} \cdot f_{\text{alt}}} + C(X_1, X_2). \tag{5.44}$$

5.2 Mathematical Relationship Between Electromagnetic Side Channel Energy of Sequence of Instructions and the Measured Pairwise Side Channel Signal Power

In this section we continue to analyze the relationship between emanated signals and program execution. Specifically, we define side channel energy of instructions as they pass through pipeline and estimating total energy emanated by a program.

5.2.1 Definition for Emanated Signal Power (ESP) of Individual Instructions as They Pass Through Pipeline

In this section, we define Emanated Signal Power (ESP) which is the channel input power.

For activity $\mathscr{A}_1$, let assume $T_{\mathscr{A}_1}$ is the execution time, $T^P_{\mathscr{A}_1}$ is the total time spent in the pipeline except the execution stage, $a_{\mathscr{A}_1}(t)$ is the characteristic signal emanated only when $\mathscr{A}_1$ is executed, and $a^P_{\mathscr{A}_1}(t)$ is the signal emanated as a consequence of processing the activity throughout the pipeline excluding the execution stage. We define ESP($\mathscr{A}_1$) as [6]:

$$\text{ESP}(\mathscr{A}_1) = \frac{\int_0^{T^P_{\mathscr{A}_1}} |a^P_{\mathscr{A}_1}(t)|^2 dt + \int_0^{T_{\mathscr{A}_1}} |a_{\mathscr{A}_1}(t)|^2 dt}{R}, \qquad (5.45)$$

where we assume the activity $\mathscr{A}_1$ stays in the pipeline for the time interval $(0, T_{\mathscr{A}_1} + T^P_{\mathscr{A}_1})$ only once, R is the resistance of the measuring instrument, and the execution step is the last step of the pipeline. Here, we need to emphasize that $a_{\mathscr{A}_1}(t)$ and $a^P_{\mathscr{A}_1}(t)$ are the desired signals emanated while processing activity $\mathscr{A}_1$ through the pipeline in isolation. They do not contain any components from any other signals and interrupts. We also need to note that, although we assume that the execution of an instruction happens at the very end of the pipeline, it is only for better illustration of the equation given in (5.45), and the execution could be done at any stage of a pipeline. We need to note that ESP provides the mean available power while executing an instruction, therefore, we assume that the noise term comprises all variations in the emanated power.

Although ESP is defined in the continuous-time domain, we have to alter this equation to cope with discrete-time analysis since measurements are done on digital devices. Let assume sampling frequency of the measuring instrument is $f_s = 1/T_s$. We also assume that the number of samples taken during the execution of the instruction $\mathscr{A}_1$ is $N_I = T_{\mathscr{A}_1}/T_s$, and the number of samples taken, when the instruction $\mathscr{A}_1$ is processed in a pipeline except for execution stage, is $P_S = T^P_{\mathscr{A}_1}/T_s$. Then, ESP in discrete time can be written as [6]

$$\text{ESP}[\mathscr{A}_1] = \frac{\sum_{m=0}^{P_S-1} |a^P_{\mathscr{A}_1}[m]|^2 + \sum_{m=0}^{N_I-1} |a_{\mathscr{A}_1}[m]|^2}{R/T_s}. \qquad (5.46)$$

5.2.2 Estimating ESP from the Total Emanated EM Signal Power Created by a Program

Measuring ESP is not a trivial task. Execution of any instruction is overlapped with execution of other instruction in the code, and also with other activities in the other

5.2 Mathematical Relationship Between Electromagnetic Side Channel...

stages of the pipeline. Therefore, we need a method to separate signal components that do not belong to the considered instruction from the signals that do belong to it. In [1], a program is designed to calculate the emanated energy difference between two instructions.

Here we again use the A/B alternation pseudo-code described in Chap. 4. However, instead of inserting two different activities into for-loops of the code, we insert instruction under observation in the first for-loop of the code, and NOP instruction into the second for-loop of the code. We note here that NOP instruction keeps the processor idle for one clock cycle. Hence, if the execution time of the activity in the first for-loop takes more than one clock cycle, the number of NOPs in the second for-loop has to be chosen carefully so that both loops take equal amount of time. In other words, the number of iterations of the first for-loop, n_{inst}, has to be equal to number of iterations of the second for-loop $n_{inst2} = n_{inst}$. Here, we assume the emitted signal power at all stages of a pipeline for NOP forms the baseline that we use to normalize the power consumption of other instructions relative to NOP. Therefore, for the mathematical tractability, we assume that the signal measured while execution of NOP is a consequence of additive Gaussian white noise.

After running the modified code and measuring the power at the alternation frequency, the next step is to derive the relationship between the total emitted power and ESP. Let $s(t)$ be the emanated signal when the outer loop iterates for one time. We assume that the frequency content of $s(t)$ is negligible for the frequencies above $f_s/2$, and lasts for T_E seconds. Therefore, the total number of samples taken during the experiment is equal to $N_T = T_E/T_s$. Let T_L be the execution time of any inner for-loop only for one period. Then, the number of samples taken in a period can be written as $N_L = T_L/T_s$. Therefore, the relationship between N_T and N_L becomes $N_T = 2 \times n_{inst} \times N_L$.

Now, let $\tilde{\mathscr{P}}_{\mathscr{A}_1}(f_{\text{alt}})$ be the power measured around this frequency while executing the code under the assumptions stated above. The following theorem gives the relationship between the total emanated signal power and the instruction power.

Theorem 5.2 (ESP) *Let $\mathscr{P}_{\mathscr{A}_1}(f_{\text{alt}})$ be the normalized emanated power which is defined as*

$$\mathscr{P}_{\mathscr{A}_1}(f_{\text{alt}}) = \tilde{\mathscr{P}}_{\mathscr{A}_1}(f_{\text{alt}}) - \mathscr{P}_{\text{NOP}}(f_{\text{alt}}) \qquad (5.47)$$

where $\mathscr{P}_{\text{NOP}}(f_{\text{alt}})$ is the measured emanated power when both for-loops of the code are employed with NOP. The mathematical relationship between ESP[$\mathscr{A}_1$] and $\mathscr{P}_{\mathscr{A}_1}(f_{\text{alt}})$ while running the activity $\mathscr{A}_1$ in the first for-loop can be written as:

$$\text{ESP}[\mathscr{A}_1] = \left(\frac{\pi}{2}\right)^2 \frac{\mathscr{P}_{\mathscr{A}_1}(f_{\text{alt}}) \cdot N_L}{(N_I + P_S) \cdot f_{\text{alt}} \cdot n_{inst}}. \qquad (5.48)$$

To prove this, we assume that the sampled sequence of $s(t)$ is $s[m]$, and each sample can be written as $s[m] = i[m] + w[m]$ where $i[m]$ is the emanated signal sample and $w[m]$ is additive independent and identically distributed (i.i.d.) white noise with zero mean and variance σ_w^2. We assume that the noise term contains all disruptive signal powers and their variations.

Let $s_1^{L_1}[m]$ be the sequence corresponding to only one period of the first for-loop signal, and the length of $s_1^{L_1}[m]$ is N_L. We can decompose $s_1^{L_1}[m]$ into three different sequences. Assuming the depth of the pipeline is P_S, these sequences are:

- The samples of the considered instruction including all pipeline stages [6]:

$$a_{\mathscr{A}_1}[m] = \left[0, \cdots, 0, a_{\mathscr{A}_1}^1, a_{\mathscr{A}_1}^2, \cdots, a_{\mathscr{A}_1}^p, a_{\mathscr{A}_1}[0], \right.$$
$$\left. \cdots, a_{\mathscr{A}_1}[N_I - 1], a_{\mathscr{A}_1}^{p+1}, \cdots, a_{\mathscr{A}_1}^{P_S}\right]$$

where $a_{\mathscr{A}_1}^i$ is the ith sample of the emitted signal when $\mathscr{A}_1$ is in a pipeline stage rather than execution.

- The samples of other activities (that are not $\mathscr{A}_1$, but which were needed in the A/B alternation code to make it practical), including the pipeline effect:

$$o_{L_1}[m] = [o[0], o[1], \cdots, o[N_L - 2], o[N_L - 1]].$$

Here, we need to note that the samples taken for the first iteration of the inner for-loop will be different than the other iterations even for the ideal case due to pipeline depth. Although it looks like the periodicity is not valid for $o_{L_1}[m]$, we can able to ignore it thanks to the assumption that n_{inst} is large.

- Finally, the last sequence comprises of all other components which are assumed to be Gaussian and given as [6]

$$w_{L_1}[m] = [w[0], w[1], \cdots, w[N_L - 1]].$$

Combining all these sequences, we have [6]

$$s_1^{L_1}[m] = a_{\mathscr{A}_1}[m] + o_{L_1}[m] + w_{L_1}[m].$$

Following the same decomposition for the second for-loop signal, called $s_2^{L_2}[n]$, we have [6]

* $o_{L_2}[m] = [o[0], o[1], \cdots, o[N_L - 2], o[N_L - 1]]$,
* $w_{L_2}[m] = [w[0], w[1], \cdots, w[N_L - 1]]$,

which leads to

$$s_2^{L_2}[m] = o_{L_2}[m] + w_{L_2}[m].$$

Here, we assume that NOP consumes very little energy as it passes through the stages of a pipeline, which means it produces a signal whose power is close to zero.

5.2 Mathematical Relationship Between Electromagnetic Side Channel... 71

Observe here, that since both loops are almost identical except the part where $\mathscr{A}_1$ is inserted, we assume that $o_{L_1}[m]$ and $o_{L_2}[m]$ are identical to each other, therefore, we refer both sequences as o[m]. Let p[m] be a square wave with 50% duty cycle and period of $2N_L n_{inst}$ samples, and **s**[m] be the one period signal of the outer for-loop. Let also $\mathbf{a}_{\mathscr{A}_1}[m]$ and **o**[m] be generated by concatenating $2 \cdot n_{inst}$ instances of $a_{\mathscr{A}_1}[m]$ and $o_{L_1}[m]$, respectively. Furthermore, we can assume that noise components are i.i.d. for both for-loops. Therefore, we have [6]

$$\mathbf{s}[m] = p[m]\mathbf{a}_{\mathscr{A}_1}[m] + \mathbf{o}[m] + \mathbf{w}[m].$$

The first harmonic of **s**[m] can be written as [6]

$$\mathbf{S}[1] = \frac{\sum_{\gamma=0}^{2N_L n_{inst}-1} P[1-\gamma]\mathbf{A}_{\mathscr{A}_1}[\gamma]}{2N_L n_{inst}} + \mathbf{O}[1] + \mathbf{W}[1]. \tag{5.49}$$

We know that **O**[k] and $\mathbf{A}_{\mathscr{A}_1}[k]$ have nonzero frequency components only if $k = 2 \cdot n_{inst} \cdot l$, $\forall l \in \{0, \cdots, N_L - 1\}$, and $|P[1]| \gg |P[1 - 2n_{inst}]|$. Therefore, (5.49) can be approximately written as [6]

$$\mathbf{S}[1] \approx \frac{P[1]}{2N_L n_{inst}} \mathbf{A}_{\mathscr{A}_1}[0] + \mathbf{W}[1]. \tag{5.50}$$

If we take the magnitude square of both sides, we have [6]

$$|\mathbf{S}[1]|^2 = \left| \frac{P[1]}{2N_L n_{inst}} \mathbf{A}_{\mathscr{A}_1}[0] + \mathbf{W}[1] \right|^2$$

$$= \left| \frac{P[1]}{2N_L n_{inst}} \mathbf{A}_{\mathscr{A}_1}[0] \right|^2 + |\mathbf{W}[1]|^2$$

$$- \frac{\mathfrak{Re}\left\{ P[1]\mathbf{A}_{\mathscr{A}_1}[0]\mathbf{W}^*[1] \right\}}{N_L \cdot n_{inst}}, \tag{5.51}$$

where $(\cdot)^*$ is conjugation and $\mathfrak{Re}\{\cdot\}$ takes the real part of its argument. Assuming

$$\left| \frac{P[1]}{2N_L n_{inst}} \mathbf{A}_{\mathscr{A}_1}[0] \right| \gg \mathfrak{Re}\left\{ P[1]\mathbf{A}_{\mathscr{A}_1}[0]\mathbf{W}^*[1] \right\},$$

the first harmonic of **s**[m] can be simplified further as [6]

$$|\mathbf{S}[1]|^2 \approx \left| \frac{P[1]}{2N_L n_{inst}} \mathbf{A}_{\mathscr{A}_1}[0] \right|^2 + |\mathbf{W}[1]|^2. \tag{5.52}$$

72　　　　　　　　5 Relationship Between Modulated Side Channels and Program Activity

To proceed further, we need to have the expression for $\mathbf{A}_{\mathscr{A}_1}[0]$. Utilizing the discrete Furrier series (DFS), we have [6]

$$\mathbf{A}_{\mathscr{A}_1}[0] = \sum_{\gamma=0}^{2N_L n_{inst}} \mathbf{a}_{\mathscr{A}_1}[\gamma] \stackrel{(a)}{=} 2n_{inst} \sum_{\gamma=0}^{N_L} \mathbf{a}_{\mathscr{A}_1}[\gamma], \qquad (5.53)$$

where (a) follows the fact that $\mathbf{a}_{\mathscr{A}_1}[m]$ is periodic with N_L samples. Since, at each period, only $N_I + P_S$ of $\mathbf{a}_{\mathscr{A}_1}[m]$ have nonzero values, and assuming $N_I + P_S$ is large enough, (5.53) can be written as [6]

$$\mathbf{A}_{\mathscr{A}_1}[0] = 2(N_I + P_S)n_{inst}\mathbb{E}\left[\mathbf{a}_{\mathscr{A}_1}[m]\right]$$
$$= 2(N_I + P_S)n_{inst}\mu_{\mathscr{A}_1}. \qquad (5.54)$$

Using Eq. (5.46), ESP[$\mathscr{A}_1$] can also be written as [6]

$$\text{ESP}[\mathscr{A}_1] = \frac{T_s(N_I + P_S)}{R}\mathbb{E}\left[|\mathbf{a}_{\mathscr{A}_1}[m]|^2\right]$$
$$= \frac{T_s(N_I + P_S)}{R}\left(\mu_{\mathscr{A}_1}^2 + \sigma_{\mathscr{A}_1}^2\right)$$
$$\approx \frac{T_s(N_I + P_S)}{R}\mu_{\mathscr{A}_1}^2, \qquad (5.55)$$

where $\sigma_{\mathscr{A}_1}$ is the standard deviation of the samples while an instruction is executed, and (5.55) follows the assumption that the variation in measured signal during the execution of an instruction is much smaller than its mean value. Combining (5.54) with (5.55), we have [6]

$$\text{ESP}[\mathscr{A}_1] \approx \frac{T_s}{4R(N_I + P_S)n_{inst}^2}|\mathbf{A}_{\mathscr{A}_1}[0]|^2. \qquad (5.56)$$

The final step is to show how ESP[$\mathscr{A}_1$] and the alternation power $\mathscr{P}(f_{\text{alt}})$ are related to each other. The relation between the first harmonic of the signal and the power measure through the spectrum analyzer is given as [2]

$$\mathscr{P}(f_{\text{alt}}) = \frac{2}{R}\left(\frac{|S[1]|}{2 \cdot N_L \cdot n_{inst}}\right)^2. \qquad (5.57)$$

Let $\mathscr{P}_{\mathscr{A}_1}(f_{\text{alt}})$ be the measured alternation power when $\mathscr{A}_1$ is inserted into first for-loop, and the second loop is kept empty. On the other hand, let $\mathscr{P}_0(f_{\text{alt}})$ be the measured power when both for-loops are kept empty (here, we need to remark that keeping the loops empty means inserting as many NOP as needed to get the same total number of clock cycles as we got with $\mathscr{A}_1$). Finally, let $\bar{\mathscr{P}}_{\mathscr{A}_1}(f_{\text{alt}})$ be the normalized alternation power for the instruction $\mathscr{A}_1$ which is defined as [6]

5.2 Mathematical Relationship Between Electromagnetic Side Channel... 73

$$\mathscr{P}_{\mathscr{A}_1}(f_{\text{alt}}) = \mathscr{P}_{\mathscr{A}_1}(f_{\text{alt}}) - \mathscr{P}_0(f_{\text{alt}}).$$

The critical observation is that the term related to $\mathscr{A}_1$ in (5.52) is zero when both for-loops are kept empty. Assume $\mathbf{S}_{\mathscr{A}_1}[1]$ and $\mathbf{S}_0[1]$ denote the first harmonics of the signal when 1) $\mathscr{A}_1$ is inserted, and 2) both loops are kept empty, respectively. Considering this setup, we can write [6]

$$\left|\mathbf{S}_{\mathscr{A}_1}[1]\right|^2 - \left|\mathbf{S}_0[1]\right|^2 \approx \frac{1}{\pi^2}\left|\mathbf{A}_{\mathscr{A}_1}[0]\right|^2, \tag{5.58}$$

where we utilize the approximation that $\pi|P[1]| \approx 2N_L n_{inst}$. Exploiting the definition of normalized alternation power, and using the equations given in (5.56), (5.57), and (5.58), we can write [6]

$$\begin{aligned}
\mathscr{P}_{\mathscr{A}_1}(f_{\text{alt}}) &= \mathscr{P}_{\mathscr{A}_1}(f_{\text{alt}}) - \mathscr{P}_0(f_{\text{alt}}) \\
&= \frac{2/R}{(2N_L n_{inst})^2}\left(\left|\mathbf{S}_{\mathscr{A}_1}[1]\right|^2 - \left|\mathbf{S}_0[1]\right|^2\right) \\
&= \frac{2/R}{(2N_L n_{inst})^2}\frac{1}{\pi^2}\left|\mathbf{A}_{\mathscr{A}_1}[0]\right|^2 \\
&= \frac{\text{ESP}[\mathscr{A}_1]}{(\pi N_L)^2}\frac{2(N_I + P_S)}{T_s}.
\end{aligned} \tag{5.59}$$

To emphasize the relation between the power at the alternation frequency and ESP, we can write [6]

$$f_{\text{alt}} \cdot n_{inst} = \frac{1}{2 \cdot N_L \cdot T_s}. \tag{5.60}$$

Plugging the Eq. (5.60) into (5.59), we have [6]

$$\begin{aligned}
\mathscr{P}_{\mathscr{A}_1}(f_{\text{alt}}) &= \left(\frac{2}{\pi}\right)^2 \frac{\text{ESP}[\mathscr{A}_1]}{N_L/(N_I + P_S)}\frac{1}{2N_L T_s} \\
&= \left(\frac{2}{\pi}\right)^2 \frac{\text{ESP}[\mathscr{A}_1]}{N_L/(N_I + P_S)} f_{\text{alt}} n_{inst}.
\end{aligned} \tag{5.61}$$

To finalize our proof, we need to keep $\text{ESP}[\mathscr{A}_1]$ alone on the one side. Therefore, we have [6]

$$\text{ESP}[\mathscr{A}_1] = \left(\frac{\pi}{2}\right)^2 \frac{\mathscr{P}_{\mathscr{A}_1}(f_{\text{alt}}) \cdot N_L}{(N_I + P_S) \cdot f_{\text{alt}} \cdot n_{inst}}. \tag{5.62}$$

5.3 Practical Use for the Relationship Between Emanated Energy and Program Execution

One of the key steps in analyzing EM emanations is to establish a relationship between software activity and the resulting EM signals. Switching activity in electronic circuits draws currents which depend on data values, and these currents generate EM emanations. When the electronic circuit is a processor, the "data values" in its circuitry include not only actual data (operands of instructions) but also instructions themselves, e.g., the processor's instruction decoder is an electronic circuit that takes the instruction as an input and produces various instruction-dependent values as its outputs.

The currents in a processor typically depend on both the instructions that are executed and the data values that these instructions operate on. However, *differences in instruction execution* caused by different data values may generate much stronger EM side-channel emanations than the data values themselves, especially for highly performance-optimized processors where the instruction's actual operation is only a small part of the activity that occurs for that instruction (fetch, decoding, scheduling, etc.). To understand this, consider that a data value of 0 may draw slightly less current than a data value of 1. In contrast, accessing array element A[X] may cause a cache hit or cache miss depending on the data value X. Since a cache miss draws much more current than a cache hit, the EM emanations caused by these currents will also be very different. As another example, consider an encryption algorithm that has an if-then-else that considers a bit of the secret key, resulting in executing different program codes (the then-path or the else-path) depending on whether this bit is 0 or 1. Since different computations generate different EM emanations, attackers may be able to infer the key bit's value by observing EM emanations to determine which computations (instructions) were performed. The ESE values for specific pairs of events, e.g., ESE for cache hit vs. cache miss, or ESEs among instructions on the then-path and the else-path, help designers and programmers quantify how much each such data-dependent behavior helps an attacker deduce data values (e.g., secret keys) used by the program as it executes.

In previous sections we have derived a quantification metric which does not present or imply a specific side-channel attack, but instead provides direct quantitative feedback to programmers and hardware designers about which instructions (or combination of instructions) have the greatest potential to create side-channel vulnerabilities. Our analysis focuses on instruction-level activity because analysis at the circuit level (e.g., wires, transistors, and gates) does not account for the effects of system architecture and software. On the other hand, analysis of emanations at the program or program phase granularity does not provide direct feedback to pinpoint leakage sources on a circuit level. ESP and ESE overcome the difficulties in measuring information leakage in complex systems by generating controlled EM emanations to isolate the differences between instructions one pair at a time, and then measuring and analyzing these emanations in the frequency domain.

References

1. Robert Callan, Alenka Zajic, and Milos Prvulovic. A Practical Methodology for Measuring the Side-Channel Signal Available to the Attacker for Instruction-Level Events. In *Proceedings of the 47th International Symposium on Microarchitecture (MICRO)*, 2014.
2. Gerhard Heinzel, A. O. Rüdiger, and Roland Schilling. Spectrum and spectral density estimation by the discrete fourier transform (dft), including a comprehensive list of window functions and some new at-top windows. 2002.
3. A.V. Oppenheim, R.W. Schafer, and J.R. Buck. *Discrete-time Signal Processing*. Prentice Hall, 1999.
4. W.H. Press, S.A. Teukolsky, W.T. Vetterling, and B.P. Flannery. *Numerical recipes in C*. Cambridge University Press, 1996.
5. Baki Berkay Yilmaz, Robert Callan, Alenka Zajic, and Milos Prvulovic. Capacity of the em covert/side-channel created by the execution of instructions in a processor. *IEEE Transactions on Information Forensics and Security*, 13(3):605–620, 2018.
6. Baki Berkay Yilmaz, Milos Prvulovic, and Alenka Zajic. Electromagnetic side channel information leakage created by execution of series of instructions in a computer processor. *IEEE Transactions on Information Forensics and Security*, 15:776–789, 2020.

Chapter 6
Parameters that Affect Analog Side Channels

6.1 Defining the Parameters that Impact Side Channels

There are several signal-related parameters that impact quality of side channel signals and which information can be deduced from these signals. We start by defining signal bandwidth and sampling rate, followed by center frequency, span, resolution bandwidth, and sweep time, as parameters that need to be set on the measurement device and can impact signal quality. Then we discuss how these parameters impact side channels and what are practical considerations when collecting signals. Finally, we present results on how different antennas, probes, distance, noise, etc., impact side channel measurements and information that can be extracted from these signals.

6.1.1 Signal Bandwidth and Sampling Rate

The bandwidth (B) of a signal is defined as the difference between the highest (f_H) and lowest (f_L) frequency that is present in the signal as shown in Fig. 6.1. It is measured in terms of Hertz (Hz) i.e., the unit of frequency.

Figure 6.2 shows an amplitude modulated signal in time domain as it can be observed on an oscilloscope display.

When the signal is analyzed in terms of how its magnitude changes over time, we call that the Time Domain representation of the signal. The signal can also be represented as a sum of sinusoidal functions, each with a different frequency, and each with some magnitude and phase. This is the Frequency Domain representation of the signal, and Fig. 6.3 shows what maybe observed on the display of a spectrum analyzer for the time-domain signal from Fig. 6.2.

Note that every signal has both time-domain and frequency-domain representations. Given the time-domain representation, we can use the Fourier transform to

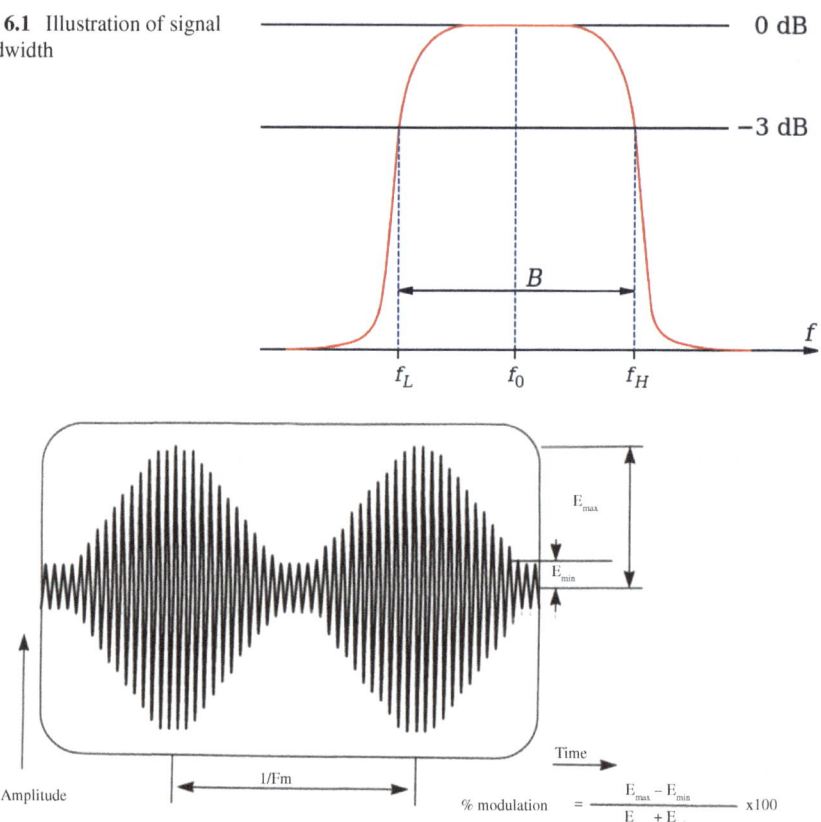

Fig. 6.1 Illustration of signal bandwidth

Fig. 6.2 Illustration of an amplitude modulated signal in the time domain

compute the frequency-domain representation, and the inverse Fourier transform converts in the opposite direction (from frequency-domain into time-domain). However, although the two representations are just different ways to represent the same signal, they are both very useful because some signal properties and signal analyses are much easier to reason about when using time-domain representation, while other signal properties and analyses are much easier to reason about when using the frequency-domain representation of the signal (Fig. 6.3).

Intuitively, the bandwidth of a signal is a measure of the range of frequencies that is used in the signal's frequency-domain representation. Fast-changing time-domain signals need high-frequency sinusoids in its frequency-domain representation. Intuitively, low-frequency sinusoids go slowly up then down and cannot represent quick up-then-down features in a signal.

The natural question is how much bandwidth we need to correctly receive a signal. We need to capture the highest frequency component of the desired signal to correctly reconstruct it's time domain. This is not always possible, e.g., spectral

6.1 Defining the Parameters that Impact Side Channels

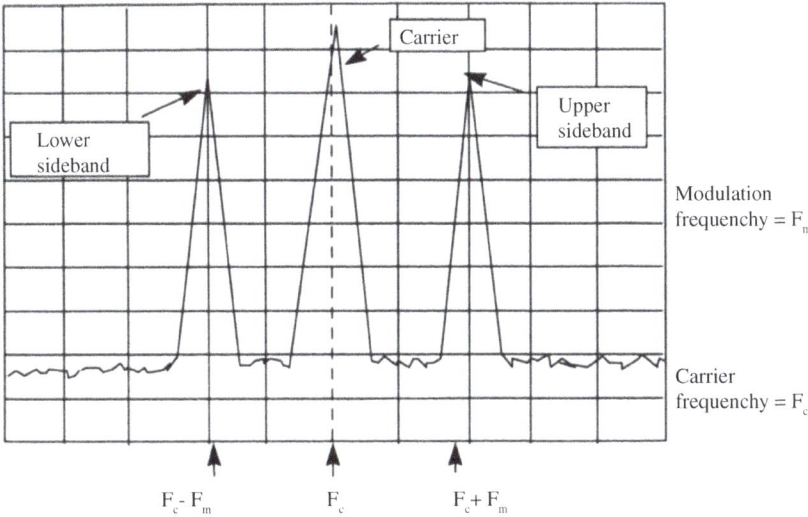

Fig. 6.3 Illustration of an amplitude modulated signal in the frequency domain

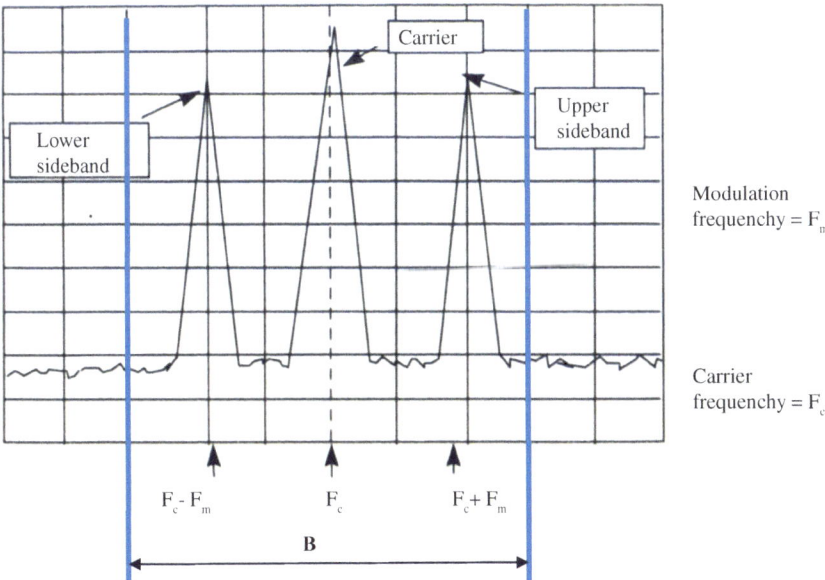

Fig. 6.4 Illustration of a signal bandwidth for an amplitude modulated signal

content of the signal may have infinite number of spectral components. For example, an infinite sequence of ideal square pulses in Fig. 6.5 has an infinite number of spectral components. Because measurement equipment always has finite bandwidth,

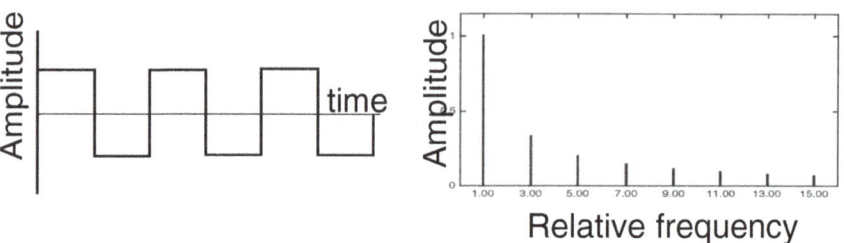

Fig. 6.5 Illustration of a square pulse train signal in time and frequency domain

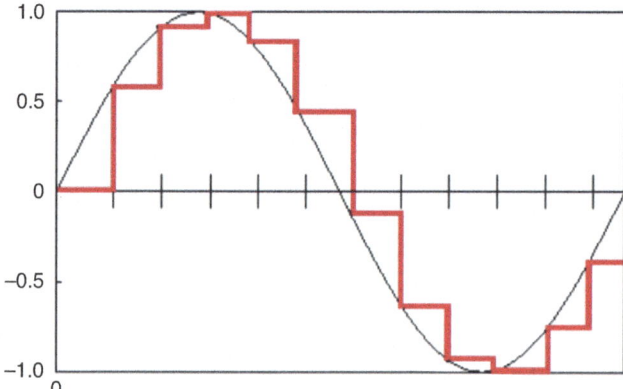

Fig. 6.6 Illustration of a sinusoidal signal and quantization process

this signal can never be perfectly captured in measurements. Should we include more signal than the highest frequency component of the signal? Based on the sampling theorem, that will be defined shortly, the receiver must have a bandwidth twice that of the sampling rate of the message to be received in order to correctly reconstruct the signal. Past that point, more bandwidth is actually harmful because as the bandwidth increases, more noise gets through, assuming that the noise is random in nature (evenly scattered across frequencies). Consequently, the signal to noise ratio becomes worse.

The sampling rate (SR) is the rate at which amplitude values are digitized from the original waveform. Figure 6.6 illustrates a sampling process of a sinusoidal signal. As a continuous signal, this sinusoid takes an infinite number of amplitude values, which is hard to store on any device. To solve this problem, we need to quantize a signal, e.g., record a limited number of amplitude values that later can be used to fully reconstruct the signal. So the questions we need to answer is how fast do we need to sample and what sampling resolution do we need.

Figure 6.7 illustrates how different sampling rates impact signal reconstruction. We can observe that higher sampling rates allow the waveform to be more accurately reconstructed.

6.1 Defining the Parameters that Impact Side Channels

Fig. 6.7 Illustration how sampling rate impacts signal reconstruction

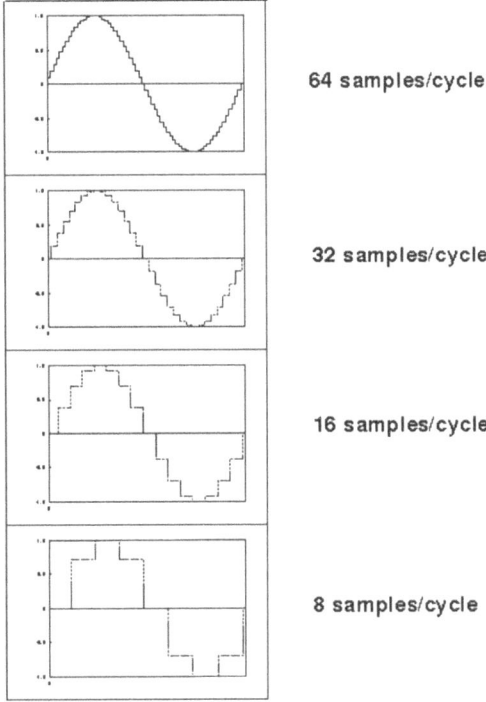

Shannon's Sampling Theorem defines the relationship between signal bandwidth and sampling rate needed for correct reconstruction of an analog signal. If a continuous, band-limited signal contains no frequency components higher than f_s, then we can recover the original signal without distortion if we sample at a rate of at least $2f_s$ samples/second. In other words, signal can be perfectly reconstructed from time-domain samples when sampling rate is twice the bandwidth of the signal. Note that $2f_s$ is called the Nyquist rate.

Quantizing Resolution While sampling rate determines how often to collect samples of the analog signal, quantizing resolution, a.k.a. ADC resolution, determines with which accuracy amplitude value can be recorded and how many digits are used to record the number. Both are very important for accurately recording side channels.

6.1.2 Definition of Center Frequency, Span, Resolution Bandwidth, and Sweep Time

Center Frequency is a halfway point between the start and stop frequencies on a spectrum analyzer display.

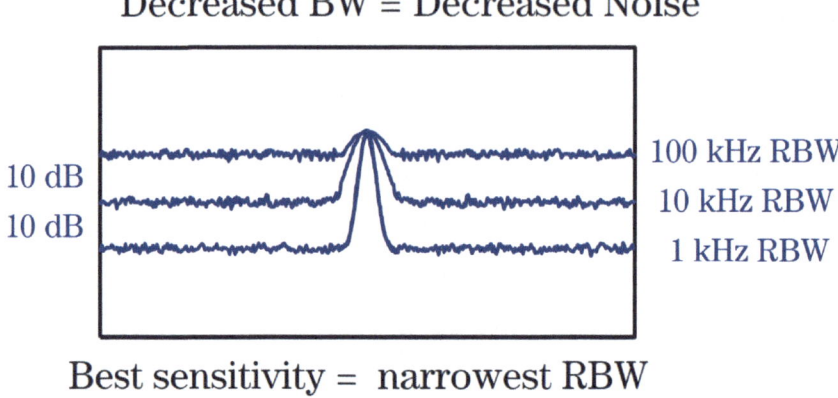

Fig. 6.8 Illustration how RBW impacts the noise floor in measurements

Span specifies the range between the start and stop frequencies. These two parameters allow for adjustment of the display within the frequency range of the instrument to enhance the visibility of the spectrum being measured.

Resolution Bandwidth or RBW filter is the bandpass filter in the intermediate frequency (IF) path. It is the bandwidth of the RF route before the detector (power measurement device). This filter determines the RF noise floor and how close two signals can be and still be resolved by the analyzer into two separate peaks. Adjusting the bandwidth of this filter allows for the discrimination of signals with closely spaced frequency components, while also changing the measured noise floor and the time needed to sweep over a range of frequencies. A wider (larger) resolution bandwidth passing more frequency components through to the envelope detector than a narrower resolution bandwidth, which results in a higher measured noise floor as illustrated in Fig. 6.8. However, on a sampling spectrum analyzer the resolution bandwidth is also the "step" with which measurements are made – with a wider resolution bandwidth the instrument needs fewer point-measurements to cover a frequency range. Even for real-time instruments, due to properties of the Discrete Fourier Transform, the amount of time during which the signal must be recorded is inversely proportional to the resolution bandwidth that is needed for a frequency-domain measurement, i.e., less measurement time is needed when the resolution bandwidth is wider.

Sweep time is defined as ratio of Span and RBW, i.e.,

$$ST = \frac{k * \text{Span}}{RBW^2}, \tag{6.1}$$

where ST is sweep time in seconds, k is proportionality constant, Span is the frequency range under consideration in hertz, and RBW is the resolution bandwidth in hertz. Note that with RBW filter set to 1 Hz resolution bandwidth, very precise

6.1 Defining the Parameters that Impact Side Channels

measurements can be collected because of the low noise floor, but the sweep time becomes very long.

6.1.3 Thermal and Coupling Noise in Electrical Measurements

One of the important parameters that impacts all measurements is thermal noise. Random motion of electrons in conducting materials cause random localized currents. Figure 6.9 shows the fluctuations of current across a wire as a function of time. Here we note that average magnitude of these fluctuations is zero, but the variance is not zero.

Similarly we can measure voltage fluctuations across a resistor. While other effects may impact these measurements, we can consider the voltage fluctuations across a resistor to be primarily due to thermal noise. Figure 6.10 shows the voltage

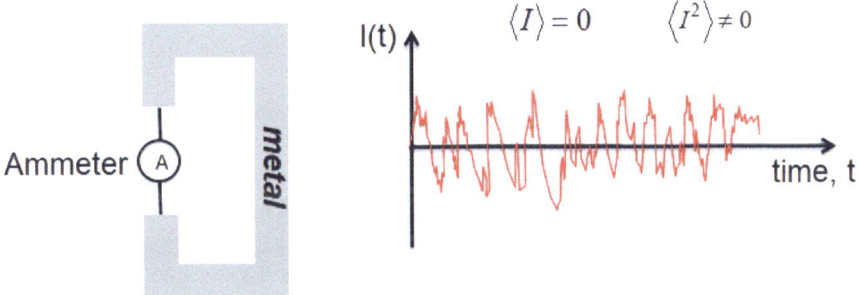

Fig. 6.9 Measurement of current thermal noise over time

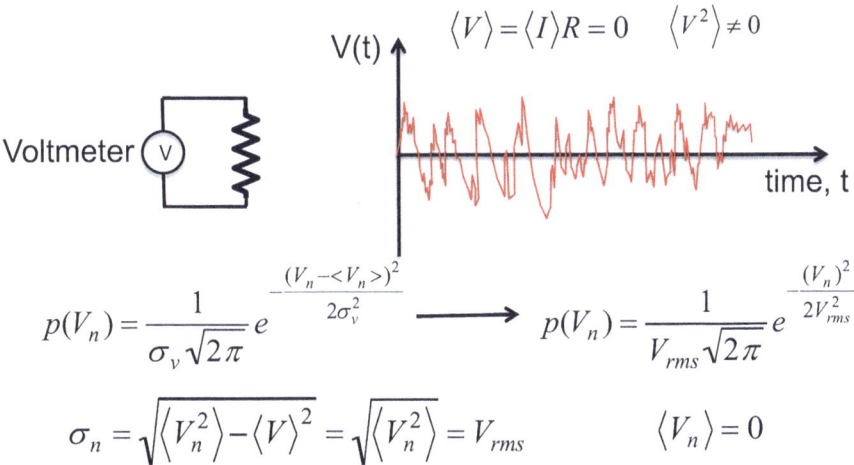

Fig. 6.10 Measurement of voltage thermal noise over time

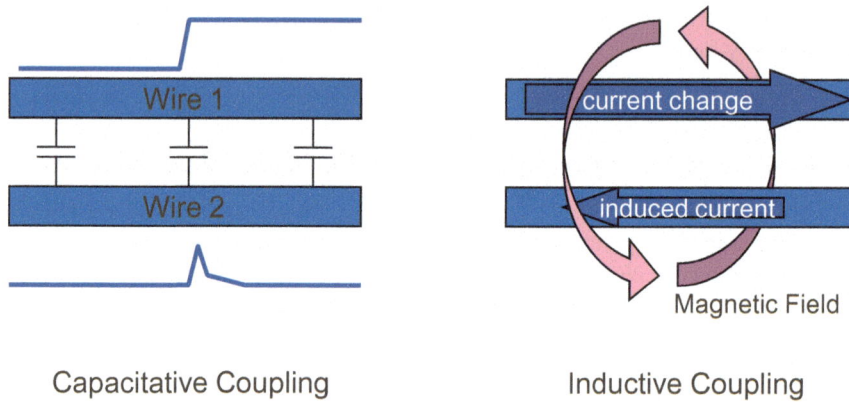

Fig. 6.11 Illustration of capacitive and inductive coupling noise

amplitude fluctuations across a resistor as a function of time. Here we note that average voltage magnitude is also zero, but the variance is not zero. Thermal noise is typically modeled as a Gaussian random variable with zero mean, and with variance that corresponds to root mean square voltage V_{rms}.

Coupling noise is another type of nose that can impact measurements. Figure 6.11 shows capacitive and inductive coupling noise. Here, a rising edge in the signal that is carried on one wire (or transmission line) can create signal spikes in the neighboring wires due to capacitive coupling. This is treated as noise in all practical applications. However, this "noise" carries information about rising and falling edges on a line, so it can also be viewed as a side channel. Similarly, strong currents in one wire or transmission line can induce currents in neighboring lines. While this is typically called "inductive coupling noise", it can also be viewed as a potential side channel. System designers have several techniques to minimize capacitive and inductive coupling, such as, decoupling capacitors or spreading wires further apart. However, it is hard to completely remove this noise—a problem that becomes harder with increasing miniaturization of electronics.

Signal to Noise Ratio, often abbreviated SNR or S/N, is an engineering term for the ratio between the magnitude of a signal (meaningful information) and the magnitude of noise that is received, recorded, or measured along with the signal. Because many signals have a very wide dynamic range, SNRs are often expressed using a logarithmic scale (i.e., in decibels).

Data Rate is the amount of digital data that is moved from one place to another in a given time. The data rate can be viewed as the speed of travel of a given amount of data from one place to another. In general, the greater the bandwidth of a given path, the higher its data rate.

Capacity is the maximum data rate that can be achieved on a given path (e.g., a communications channel).

6.2 How Sampling Rate Impacts Side Channel Measurements?

Figure 6.12 illustrates a sampling process for an analog signal. We can observe that an analog signal has a value at any point in time. To record or analyze an analog signal, however, we need to digitize it. Typically, digitization is performed by, at regular intervals, recording the value (amplitude) of the signal. Each of these measurements is called a sample, and the signal is represented as a sequence of samples.

An important parameter in digitization of the analog signal is how often we are going to collect samples. The rate at which samples are collected, e.g., how many samples are collected per second, is the sampling rate. For example, if samples collected every microsecond, the sampling rate is 1 million samples per second (1Ms/s). Intuitively, samples represent the original signal well if we can use the digitized signal (samples) to reconstruct the value of the original signal at any point in-between the samples. The reconstruction is considered to be good if time for rise/fall in the signal is equal to (or larger than) time between samples. However, if multiple rises/falls can happen between samples, the sampled signal will fail to represent that, as illustrated in Fig. 6.13. Figure 6.14 illustrates how with higher sampling rate fast-changes signals can be reconstructed well too. Side channel signals are typically fast changing signals, so they need a high sampling rate to capture all the features of the signal. However, there are practical considerations that need to be taken into account when choosing instruments to collect side channel data. All real digitizers have limited bandwidth and sampling rate. Usually the sampling rate is 2.5 times the bandwidth to give it some slack vs. the theoretical 2 times the bandwidth. Equipment prices go up very quickly as sampling rates increase. For example, an oscilloscope with 2 GHz of instantaneous bandwidth may cost $28,000, but a similar oscilloscope with for 4 GHz of instantaneous bandwidth would cost $44,000 and with 8 GHz the price would go up to $82,000. The price

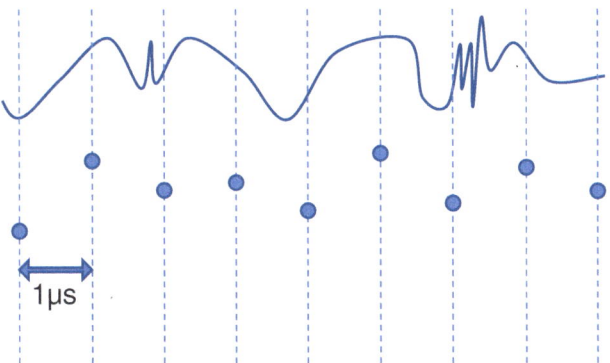

Fig. 6.12 Illustration of digitization of an analog signal

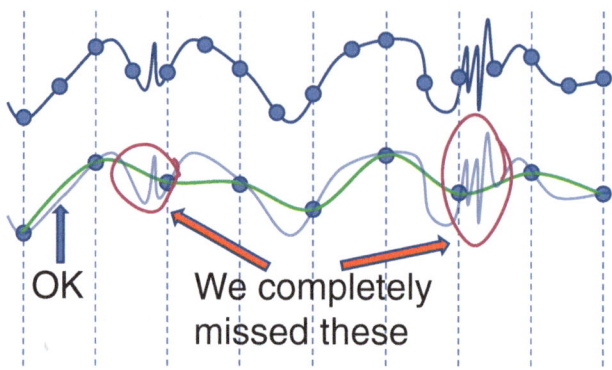

Fig. 6.13 Illustration of poorly reconstructed analog signal

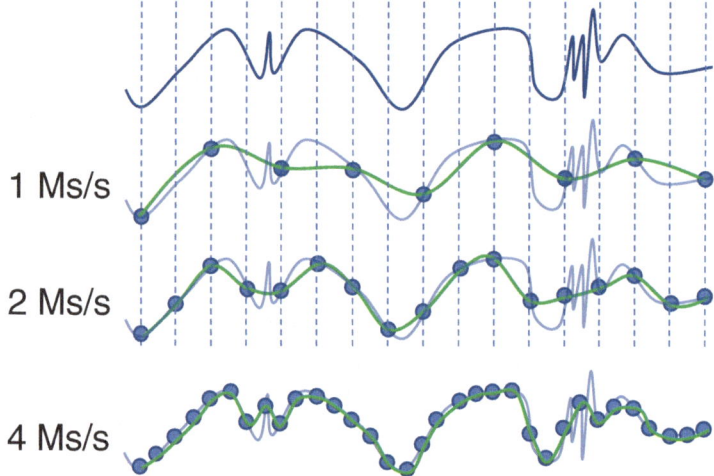

Fig. 6.14 Illustration of selection of sampling rate on signal reconstruction

can be even higher if we want to have excellent analog to digital conversion (ADC) resolution. For example, the prices above were for 10-bit ADC resolution. That is not great for side channel analysis, but a typical oscilloscope will have only an 8-bit ADC! Finally, prices increase if we want an instrument with a high sampling rate that can continuously record signals for a while—when the sampling rate doubles, it also means that twice as much sample memory is needed to record for the same amount of time, and that memory needs to be twice as fast. For example, the prices above are for only a few milliseconds of continuous recording at the instrument's full sampling rate! Larger sample memories, especially in instruments with very high sampling rates, can usually be added as (very expensive) options when purchasing the instrument.

6.3 How Bandwidth Impacts Side Channel Measurements?

Every received signal is filtered to limit its bandwidth before it is digitized. Filtering removes high-frequency oscillations but accurately preserves lower-frequency components. Intuitively, lower-frequency components are like a moving mean of the fast-changing signal. Filtering removes the fast-changing variation around that moving mean, but the moving mean itself is preserved very accurately. Unfortunately, many interesting side channel features are short in duration (need high-frequency components to capture the actual shape), but have non-zero average (not just oscillation around some slower-changing signal). After band-limiting (filtering) and sampling, feature may still be represented in the signal, but sometimes only as a small change in magnitude.

Based on previous reasoning, it seems that the more bandwidth for signal recording, the better. There is downside to requiring always more bandwidth. For example, it is more expensive to record higher-bandwidth signal. Signal also uses more memory, takes longer to analyze.

There is also a question of accuracy. Is analysis more accurate if we use higher-bandwidth signal? More bandwidth always adds noise, and may also add more of the relevant signal. If a low-bandwidth signal already contains most of the signal, using extra bandwidth to capture the rest of the signal is likely to reduce the overall SNR of the signal and can be harmful! Analysis works better when using just enough bandwidth, and gets worse with more (or less) bandwidth!

6.4 An Example of How Bandwidth Impacts Side Channel Signals

To illustrate how the quality of signal analysis is affected by bandwidth used to record the side channel signal [11], our example analysis will be an attack on two implementations of modular exponentiation. Modular exponentiation is an essential part of cryptographic encryption and signing in RSA, and our attack will attempt to determine the outcome of specific branch decisions that consider bits of the secret exponent and, from that, reconstruct the secret key that is being used. The two implementations we will use are the sliding window implementation, which was used in past versions of OpenSSL, and a slight variation of the modern fixed-window implementation. We first briefly describe the two implementations of modular exponentiation, and then we discuss how the success of an EM side channel attack depends on the bandwidth of the recorded signal.

6.4.1 Modular Exponentiation in OpenSSL's RSA

Modular exponentiation is the main part of cryptographic encryption and signing operations when using RSA. Specifically, secret-key (encryption and signing) operations in RSA compute $v^p \bmod m$, where v is the (encrypted) message, p is the (secret) exponent, and m is the modulus.

Sliding window implementation was used for RSA operations in past versions of OpenSSL, and its source code is still present in current versions of OpenSSL's source code (function BN_mod_exp_simple in crypto/bn/bn_exp.c in OpenSSL 3.0.5, the latest version as of this writing). This implementation of modular exponentiation conceptually follows the grade-school exponentiation approach, where the exponent is examined starting from the most significant bit, squaring the result for each bit of the exponent, and multiplying the result with the message when the bit of the exponent has a value of one. However, rather than process one bit of the exponent at a time, the sliding window algorithm splits the exponent into multi-bit chunks called *windows* and performs multiplication with the (appropriately pre-exponentiated) message an entire window at a time. In preparation for this, for each possible value of the window (wvalue), the value of $v^{\text{wvalue}} \bmod m$ has been pre-computed and placed in the table *val* at an index that corresponds to wvalue.

Initially, the position of the most significant bit (MSB) of the very first window is set to be the most significant bit of the exponent (line 1), and the result variable r is set to a big-number value of 1 (line 2). Note that, like the message and the exponent, the result r is a very large number (>1000 bits) that is kept in a data structure that contains an array of word-sized (e.g., 32-bit) integers, so we cannot simply use r=1 in line 2. Instead of using regular word-sized operations, big-number operations are implemented as functions whose names begin with BN_, and each such function typically involves a number of word-sized operations. For example, BN_one involves setting the word that stores the least significant part of the big-number value to 1, but it also sets all other words in this big-number data structure to zero.

The main loop of the exponentiation then begins (line 4). In each iteration of this loop, a window value is obtained from the exponent and the result is updated for that window value, until we have gone past the zero-index bit-position (i.e., the least significant bit) in the exponent—at that point the entire exponent has been processed and the final result of the modular exponentiation is in the result variable r. In the sliding-window implementation of exponentiation, the window value must begin and end with a non-zero bit, except in the special case when the window's MSB is 0. In that case, that bit alone will be a (zero-valued) window. Line 5 checks for this case. A one-bit zero-value window requires only a squaring of the result (line 6), so a new window will begin at the next bit position (line 7) and a new iteration of the main loop has begun (line 8).

If the starting bit of the window is non-zero, the code at lines 10 through 20 attempts to form a multi-bit window. Since we already know the MSB or the window (it is non-zero, i.e., its value is 1) the value of the window is initialized to 1 (line 10)

6.4 An Example of How Bandwidth Impacts Side Channel Signals

and the size of the window (in bits) is initialized to 1 (line 11). The loop at line 12 iterates over the bits that are examined for potential inclusion into the window, up to the maximum permitted size of the window (which is typically set to be 5 or 6 bits). Variable i is the number of bits that have been examined, and it starts at 1 because the LSB of the window has already been found. Line 13 checks if the would-be window would go past the last (least significant) bit of the exponent, in which case the window cannot be expanded further and the window-building loop is exited at line 14. Line 15 examines the next candidate bit. Since a window cannot end with a 0-valued bit, a 0-valued bit is simply skipped without expanding the window. However, a 1-valued bit causes the window to be expanded to include that bit, as well as all the zero-valued bits between it and the rest of the window. This is done by shifting the wvalue by enough bit positions to make room for the bits that are being included (line 16), setting the least significant bit of wvalue to 1 (line 17) because the new least significant bit of the window is 1—note that the other bits that are being included into the window are all zero-valued, otherwise they would have been included when they were encountered—and setting wend (line 18) to the new size of the window. Note that some windows are not max_window_bits in size—the window's size depends on where a one-valued bit was last included. For example, if max_window_bits is 6, when the bits examined for the window are 100000, the window only has one bit (the initial 1) because none of the loop iterations finds a 1-valued bit. Similarly, a two-bit window is built when the examined bits are 110000, a three-bit window is built when examined bits are 111000 or 101000, etc.

```
1   wstart = bits - 1;  /* Topmost bit of the exponent */
2   BN_one(r);          /* Result starts our as 1 */
3
4   while(wstart>=0) {
5           if (BN_is_bit_set(p, wstart) == 0) {
6                   BN_mod_mul(r, r, r, m, ctx);
7                   wstart--;
8                   continue;
9           }
10          wvalue = 1;
11          wsize = 1;
12          for (i = 1; i < max_window_bits; i++) {
13                  if (wstart - i < 0)
14                          break;
15                  if (BN_is_bit_set(p, wstart - i)) {
16                          wvalue <<= (i - wsize+1);
17                          wvalue |= 1;
18                          wsize = i+1;
19                  }
20          }
21          for (i = 0; i < wsize; i++)
22                  BN_mod_mul(r, r, r, m, ctx);
```

```
23              BN_mod_mul(r, r, val[wvalue >> 1], m, ctx);
24              wstart -= wsize;
25    }
```

Once the window is formed, the result is squared for each bit in the window (lines 21 and 22) and then multiplied (line 23) with the value from table val that was pre-computed by exponentiating the message with the value of the window. Because all window values are odd (the LSB of wvalue is always 1), if the entire wvalue were used to index into the table, the even-numbered table elements would never be used. To avoid wasting half of the table size, the LSB of wvalue is removed when indexing the table (wvalue » 1), allowing the table to only have $2^{\text{max\_window\_bits}-1}$ entries. Finally, wstart is updated to begin a new window (line 24). The main loop (line 4) will keep forming and processing windows of bits from the exponent until no more exponent bits remain to be processed. Once that happens, the entire exponentiation is complete and the final result of that exponentiation is in variable r.

Fixed window implementation is the state-of-the-art implementation of modular exponentiation. It is designed to avoid a number of side channel attacks to which the sliding window implementation was susceptible to. Specifically, in prior cryptanalysis work, a cache-based attack [1, 13] was used to identify the sequence of squaring and multiplications, while for analog side channels the squaring and multiplications were identified in the signal by using especially chosen messages [6, 7] that create a large difference in their side channel signals. Since each squaring corresponds to moving one bit toward the least significant bit of the exponent, and the multiplication corresponds to the end of a non-zero window, the sequence of squarings and multiplications reveals the position of each 1-valued bit that ends a window. Furthermore, when the number of squarings between two multiplications is less than the maximum allowed size of the window, that indicates the number of zero-valued bits that follow the last bit that was included in the window. By iteratively applying such exponent-reconstruction rules to the sequence of squarings and multiplications, a significant percentage of the exponent's bits were recovered—about 49% of the bits are recovered when the maximum window size is 4, and this percentage of recovered bits declines somewhat when the maximum window size is larger.

To avoid these attacks that rely on exponent-dependent sequence of squarings and multiplications, OpenSSL has switched to a *fixed-window* exponentiation, where all windows are of the same size, and all possible values of the window (including the all-zeros window) are represented in the pre-computed table *val*. This results in multiplications that are always separated by an equal number of squarings, regardless of the exponent. Furthermore, because every bit up to the window size is included in the window, there are no branch decisions that depend on the bits of the exponent—in fact, the sequence of *all* branch decisions is always the same for the entire exponentiation, regardless of the exponent. The listing below show the main part of the exponentiation code (in function BN_mod_exp_mont_consttime within the bn_exp.c file in OpenSSL's source code), again with some redaction (mainly removing error-checking) for clarity:

6.4 An Example of How Bandwidth Impacts Side Channel Signals

```
1   while (bit >= 0) {
2      wvalue = 0;
3         for (i = 0; i < window_bits; i++, bit--) {
4                 BN_mod_mul_montgomery(&tmp, &tmp, &tmp, mont, ctx);
5                 wvalue = (wvalue << 1) + BN_is_bit_set(p, bit);
6         }
7         MOD_EXP_CTIME_COPY_FROM_PREBUF(&am, top, val, wvalue,
8      window_bits);
9         BN_mod_mul_montgomery(&tmp, &tmp, &am, mont, ctx);
10  }
```

Note that this code uses the more efficient Montgomery multiplication algorithm (BN_mod_mul_montgomery). It also improves resilience to cache-based side channel attacks using a more complicated organization of the lookup table val, which requires a specialized function MOD_EXP_CTIME_COPY_FROM_PREBUF to retrieve the table entry that corresponds to wvalue and place it in the (big-number) variable am so it can be multiplied with the result.

Because the existing branch decisions in this code leak no information about the exponent, we change this code by replacing its line 5 with:

$$wvalue = (wvalue << 1);$$
$$if(BN\_is\_bit\_set(p, bit))$$
$$wvalue\, |=1;$$

This change introduces an exponent-dependent branch decision, with the branch outcome resulting in only a single-instruction difference in program execution. This will allow us to experimentally assess how accurately a single-instruction difference in execution can be identified from the side-channel signal.

These two implementations have three categories or branch decisions that we attempt to recover in our experiments:

The first category consists of branches at lines 5 and 21 in the sliding-window implementation, which result in relatively large changes in what is executed after them. The branch at line 5 results in either calling BN_mod_mul at line 6, or in executing the entire window-forming loop (line 12), followed by the result-squaring loop (line 21), before BN_mod_mul is executed (at line 23). This large difference in what is executed, when identified in the side channel signal, tells the attacker whether the next window in the exponent is a single-zero-bit window or a regular window. The loop branch at line 21 results either in calling BN_mod_mul at line 22 or in exiting the loop, calling BN_mod_mul at line 23, and then entering another iteration of the main loop. Compared to the branch at line 5, the outcome of this branch is more difficult to identify using side channel signals because a long-lasting BN_mod_mul is called almost immediately after this branch, regardless of its outcome, and the largest difference that results from its outcome occurs only *after* BN_mod_mul returns. However, if the outcomes of this branch are correctly identified from the side channel signal, they inform the attacker about the size of each regular window. Because each regular window has values of 1 in its MSB and LSB bit-positions, knowledge of the window's size tells the attacker not only where the next window begins, but also yields all the bits in a window of length 1 (wvalue is 1) or 2 (wvalue is 11), and two of the bits (the MSB and LSB) for larger windows. Although the "middle" bits in larger regular windows are still unknown to the attacker after this attack, the attacker can sometimes recover the secret key by combining the bits that are revealed in this side channel attack and the information contained in the public key.

The second category consists of the branch at line 15 in the sliding window implementation. This branch results in either executing or not executing the code at lines 16–18. This code consists of four instructions, after which the two options converge. Note that each of these branch outcomes, if successfully recovered from the side-channel signal, tells the attacker the value of each bit in each regular window. When combined with outcomes of the branch at lines 5 (which are much easier to recover from the signal), the attacker has all of the bits in the exponent!

The third category consists of the exponent-dependent branch in our modified fixed-window implementation creates only a single-instruction (a bitwise OR instruction) difference in what is executed by the processor. Since this branch tests each bit of the exponent from, recovery of each outcome of this branch amount to recovering of all bits in the exponent!

6.4.2 Impact of Signal Bandwidth on Branch Decision Recognition

We have measured the accuracy of identifying branch decisions for these three categories of branches, when the signal is received using 20, 30, 40, 80, and 160 MHz of bandwidth [11]. Note that the processor clock frequency is 1 GHz, so these bandwidth values correspond to only 2 3, 4, 8, and 16% of the processor's clock frequency. From these results, we can see that the first category of branch decisions (causing large changes in subsequent program execution) can be identified nearly perfectly, even when using limited signal bandwidth. For the other two categories of branch decisions, i.e., branches that cause 4-instruction and single-instruction changes in program execution, some identification is possible even when using bandwidth that is only 2% of the processor's clock frequency, but reconstruction accuracy significantly improves when bandwidth is increased to 4% of the clock frequency, and then modestly improves as bandwidth is expanded to 8% and then 16% of the processor's clock frequency. However, we found it surprising that identification accuracy for 1-instruction differences in our modified fixed-window implementation was consistently better than the identification accuracy for 4-instruction differences in the sliding-window implementation. Upon further investigation we have found that, in addition to signal bandwidth and the amount of execution change, the accuracy of branch decision identification also depends on how much is known about the branches that belong to the same signal "snippet" during analysis. In the fixed-window implementation, the outcomes of all other branch decisions in the signal "snippet" are already known (because the entire exponentiation follows a fixed sequence of branch decisions), so only one bit of information is extracted from the signal "snippet", i.e., the identification decision must choose between only two possibilities. In contrast, in the sliding-window implementation, five exponent-dependent branches occur in rapid succession, i.e., in the same signal "snippet". This means that five bits of information must be extracted from the signal "snippet", i.e., the identification decision is a choice among 32 possibilities.

Additional experimental results [11] show that fine-grained differences in program execution (only a few instructions, or even one instruction when the surrounding code exhibits very stable behavior) can be identified with >90% accuracy, even when receiving the signal at a rate that is only 4% of the monitored system's clock cycle rate, as long as the signal quality is relatively good (a signal-to-noise ratio of 20 dB or better). We also find that accuracy further improves as bandwidth is increased and when the signal-to-noise ratio is improved. Another finding is that larger differences (tens of instructions) in program execution can be identified with >99% accuracy even when using bandwidth that is only 2% of the processor's clock cycle rate, and even with relatively poor signal quality (a signal-to-noise ratio of only 5 dB).

6.5 How Distance Affects EM Side Channel Results?

As the distance between the source of EM emanations and the receiver's probe/antenna increases, the power of the received signal decreases. In telecommunications, this decrease in signal power with distance is called *path loss*, and in this section we show experimental results for path loss for EM side channel signals, present theoretical models for this path loss, and finally we discuss how to choose antennas and probes for receiving EM side channel signals in light of these experiments and models.

6.5.1 Path Loss Measurements for EM Side Channel Signals

We measure, at several different distances, the received signal power that is directly created by A/B alternation code from Chap. 4 [2, 19, 20]. This code executes on a NIOS II soft processor that is implemented on the Cyclone II FPGA chip of an Altera DE-1 board. The software executes on a "bare metal" processor, i.e., the processor executes only the A/B alternation code, with no operating system or other software activity. Similarly, only the processor is implemented on the FPGA chip—the remaining logic that is available on the FPGA chip is kept inactive. The instructions we use as A and B activity in A/B alternation are shown in Table 6.1. These instructions include loads and stores that go to different levels of the cache/memory hierarchy, simple (ADD and SUB) and more complex (MUL

Table 6.1 NIOS instructions for our DE1 FPGA A/B alternation measurements

	Instruction	Description
LDM	ldw r21, 0(r21)	Load from main memory
LDL1	ldw r21, 0(r21)	Load from L1 cache
ADD	addi r22,r22,173	Add imm to reg
SUB	subi r22,r22,173	Sub imm from reg
MUL	muli r22,r22,173	Integer multiplication
DIV	div r22,r22,r22	Integer division
NOI		No instruction

Fig. 6.15 Measurement setup for near-field measurements (left) and measurement setup for far-field measurements (right) [21]

and DIV) integer arithmetic, and we also include a "No Instruction" case where the appropriate line in the A/B alternation code is simply left empty.

In general, there are two types of measurements for EM signals—near-field and far-field. Near-field measurements typically use probes, whereas far-field measurements use antennas. When using a probe for EM side-channel measurements, the probe's type, position, and orientation affect the strength of the emanations it receives. A small "sniffer" probe placed a few millimeters above components picks up signals from only the components near the probe, but receives these signals very strongly. On the other hand, placing a probe with a larger effective area far away (> 2 m) will pick up signals from all the parts of the system, but is often not sensitive enough to pick up the weakest signals. To allow us to pick up emanations from all the parts of the system while at the same time being close enough to pick up the weakest signals tested, we settled on a compromise: a medium sized multiple-turn loop (16 cm^2 loop area, 20 turns) places above the processor as shown Fig. 6.15 (left) [21]. For our measurements, the loop was rotated in all three directions to collect the components of the magnetic field in all three (x, y, and z) directions. We then compute the total magnetic field. The signal power collected by the loop probe was measured using a spectrum analyzer (Agilent MXA N9020A) with a resolution bandwidth of only 1 Hz, to minimize the effects of variation in unrelated signals and noise.

We also compared our measurements to a theoretical model of the expected path loss. Because we were using a magnetic loop probe in the near-field, we had expected that the magnetic field would decay as $1/r^3$, and that the emanations source can be modeled as a magnetic dipole. However, our measurements did not match this model. The results are a much better match if we use a model where the source of the

6.5 How Distance Affects EM Side Channel Results?

emanations is modeled as a combination of an electric monopole (Hertzian dipole) and a magnetic dipole, where only the magnetic components of the resulting EM field can be received by the (magnetic) probe. For this model, the received power is expected to be [21]:

$$P_{rx}(H) \approx <|H|^2> = P_{rx0}\left(\frac{1}{(kr^2)^2} + \frac{1}{(kr^3)^2}\right), \qquad (6.2)$$

where $k = 2\pi/\lambda$ is the wavenumber, P_{rx0} is a reference received power that corresponds to power measurements at 0.25 m, and r is the distance between the antenna and the system. One of the main challenges in predicting propagation loss for EM side-channel signals is the fact that the transmit power and transmit "antenna" gain are unknown. To overcome this problem, we perform the measurements at "zero" distance to capture all losses that signal accumulates by exiting the electronics and reaching receive antenna. Often, this distance is not exactly zero because the computer casing, thickness of the motherboard, size of the probe, etc. can add significant distance between the transmitter and receiver. In our case, that distance was 0.25 m. After estimating transmit power P_{tx}', we can use this model to predict the propagation distance.

Figure 6.16 shows both the measured received power and our model's prediction for several representative A/B instruction pairs at 215 kHz. In each measurement, we run the A/B alternation code, measure the signal's spectrum in the vicinity of the alternation frequency, and record the magnitude of the maximum-power spectral component in that spectrum. We observe that the received power for on-chip pairs of instructions (e.g., LDL1/DIV and LDL1/MUL) *decays* at the same rate as the received power for on-chip/off-chip instruction pairs (e.g., LDL1/LDM), but that off-chip/on-chip signals are weaker. We also observe an excellent match between the measured results and the model's predictions. Results of additional experiments (not shown here) at 70 and 150 kHz alternation frequencies have also shown excellent agreement between theoretical and measured results.

We also compare measured and theoretical results for unintentionally modulated EM side-channel signals, where we measure the received signal power of the (unintentional) carrier, and of the upper and lower sideband signals created by A/B alternation of various instruction pair (Fig. 6.1) on the same processor (Nios II on the Altera DE-1 Cyclone II FPGA board).

These are far-field measurements, and we use a horn antenna whose frequency range is 1 to 18 GHz with 9 dBi gain. We measure the first (main) harmonic of processor clock at 1.083 GHz, and also its second harmonic at 2.583 GHz, with their side-bands created by A/B alternation. The power across the horn antenna was measured using a spectrum analyzer (Agilent MXA N9020A) as shown in Fig. 6.15 (right).

Here, our assumption was that propagation loss can be modeled using the Friis formula, and this model indeed matches the measurements. In this model, the received power is [21]

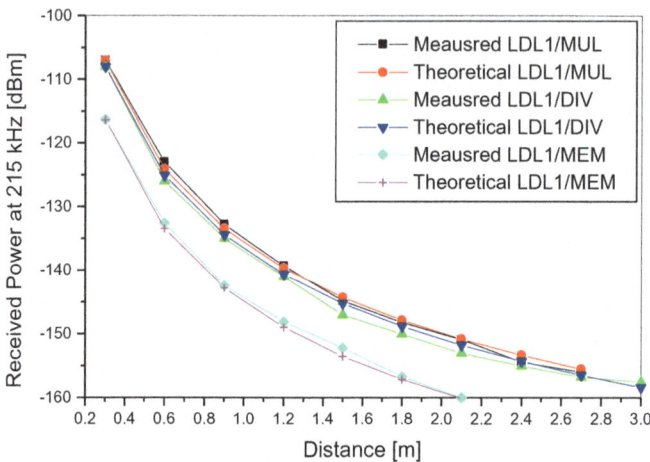

Fig. 6.16 A measured and modeled received power at 215 kHz produced by repetitive variations in a software activity [21]

$$P_{rx} = P_{rx0} \frac{1}{(kr)^2}, \tag{6.3}$$

where $k = 4\pi/\lambda$, P_{rx0} is a reference received power that corresponds to power measurements at 0.25 m, and r is the distance between the antenna and the system. One of the main challenges in predicting propagation loss for EM side-channel signals is the fact that the transmit power and transmit "antenna" gain are unknown. After estimating transmit power P_{tx}', we can use this model to predict the propagation distance.

Figures 6.17 and 6.18 compare modeled and measured received power for LDL1/LDM instructions at 1.008 GHz and 2.583 GHz. The results show that the received power on-chip/off-chip instruction pairs (e.g., LDL2/LDM) is 20–30 dB weaker then the harmonic of the processor clock signal, but still about 20–30 dB above noise floor of the measurement system. We can observe that the signals can be detected up to 3 m away.

6.5.2 Indoor and Outdoor Models for Propagation of Side Channel Signals

To test if side channels can propagate even further, we have performed several experiments in indoor and outdoor environments as described in [15] and [14]. The program activity is again the a A/B alternation of instructions from Fig. 6.1. The received power of the carrier, upper side band (USB), and lower side band (LSB) were recorded at various distances. An example received power spectrum, for the

6.5 How Distance Affects EM Side Channel Results?

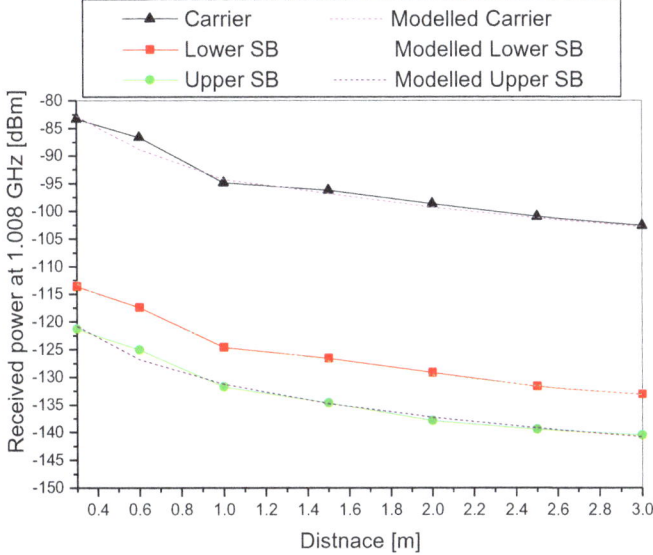

Fig. 6.17 A measured and modeled received power of processor clock harmonic at 1.083 GHz and AM modulated sidebands produced by A/B alternation [21]

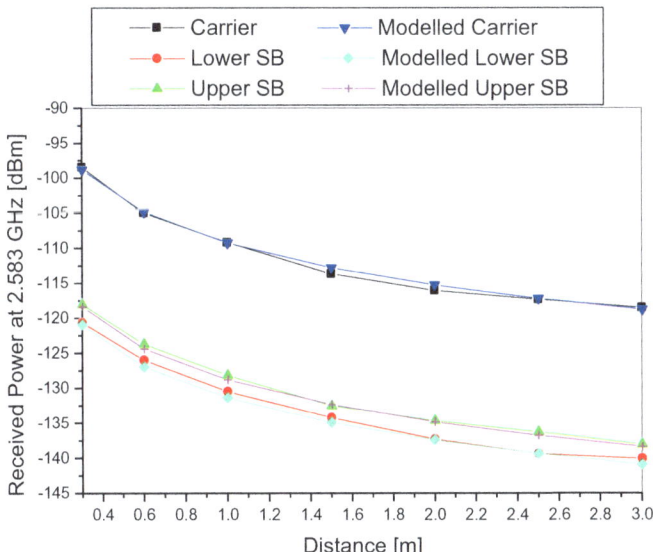

Fig. 6.18 A measured and modeled received power of processor clock harmonic at 2.583 GHz and AM modulated sidebands produced by A/B alternation [21]

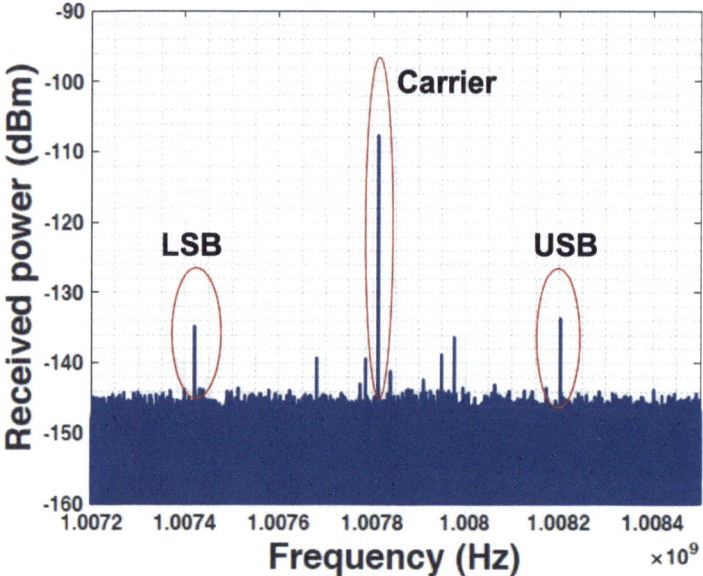

Fig. 6.19 Power spectrum of received signal in the outdoor environment for TX-to-RX distance of 100 m [15]

outdoor measurement at a distance of 100 meters, is shown in Fig. 6.19. We observe that the Carrier and its accompanying side-band signals have a significant Signal-to-Noise Ratio[1] (SNR) values at the select distances.

One of the challenges in theoretically predicting propagation loss of EM side-channel signals is that the transmit power and transmit "antenna" gain are unknown [21]. Therefore, pathloss cannot be computed using the aforementioned (traditional) power law model directly. In [15], we have modeled the received power and corresponding pathloss (at all measured distances) relative to the power received at a reference distance. We have also modified the shadowing gain model to include power variation due to both environment and board angular rotation in certain experiments. Due to differing experimental setup and data storage structures, the evaluation procedure for the various measurements are discussed separately.

For the outdoor measurements, the recorded data structure of the received power can be represented as $M^{\kappa,\psi,\alpha,p}$, where $\kappa \in [1, 2, \ldots, 7]$ denotes the indexes of the distances measured such that $d_\kappa \in [5, 10, 20, 50, 100, 130, 200]$ m, while $\psi \in [1, \ldots, \Psi = 3]$ represents the carrier and sidebands such that $\psi = 1, 2, 3$ denotes Carrier, USB, and LSB respectively.

---

[1] Note that the Signal-to-Noise Ratio of each component (i.e., Carrier, USB, and LSB) is computed by subtracting the noise level (≈ -146 dBm) from the peak of its power spectrum (in dBm) in the spectrum.

6.5 How Distance Affects EM Side Channel Results?

In all measurements we used high-gain quadrature array of nonuniform helical antennas (abbreviated as QHA) [5] and a Planar Disc Antenna (PDA) [9]. Parameter $\alpha \in [1, \ldots, A = 2]$ indicates the antenna type used with $\alpha = 1$ representing QHA while $\alpha = 2$ represents PDA. Parameter $p \in [1, \ldots, P = 2]$ indicates the Olimex board polarization where $p = 1$ corresponds to vertical polarization and $p = 2$ represents horizontal polarization. A model for the distance-dependent pathloss and shadowing gain parameters are presented in this section.

6.5.2.1 Distance-Dependent Pathloss Model for Outdoor Environments

With consideration for parameters κ, ψ, α, p, the received power from the empirical data can modeled as:

$$M^{\kappa,\psi,\alpha,p} = M_0^{\psi,\alpha,p} \cdot \left(\frac{d_\kappa}{d_0}\right)^{\eta^{\psi,\alpha,p}} \cdot \xi^{\kappa,\psi,\alpha,p} \qquad (6.4)$$

where M_0 is the power received at the reference distance d_0 (chosen as 1 m), η is the pathloss exponent, and ξ is a random variable describing shadowing gain in the environment. Figures 6.20a–d show the scatter plot of the received power at distances measured using different antennas and EM source polarization for Carrier, USB and LSB respectively. A linear regression fit for the scatter plot shows a monotonic dependence of the received power on distance. Values of parameters such as M_0 and η were extracted through the linear fit on the empirical data and have been provided in Table 6.2.

From Table 6.2 can be observed that M_0 (dBm) values from the carrier are similar in both vertical and horizontal polarizations for measurements conducted using the QHA antenna, while the sidebands (USB and LSB) slightly differed by about 3 dB between both polarizations. The closeness in value between results from both polarizations can be attributed to the fact that QHA is circularly polarized and mostly immune to any polarization changes by the EM source. For the PDA, however, a noticeable difference of about 10 dB can be observed in the M_0 (dBm) values from the Carrier, while sidebands differed by about 6–7 dB in the vertical and horizontal polarization measurements. This disparity is primarily due to the fact that the PDA is vertically polarized hence the polarization preference. It can also be observed that the pathloss exponent η values are comparable for the carrier, USB and LSB for both antennas at the respective polarizations with the exception of the PDA in the horizontal polarization.

6.5.2.2 Shadowing Gain Model for Outdoor Environments

Shadowing gain (denoted as ξ in (6.4)) is obtained by computing the deviation of the received power (M) at each measured location from the linear regression fit. The linear regression fit can be expressed as

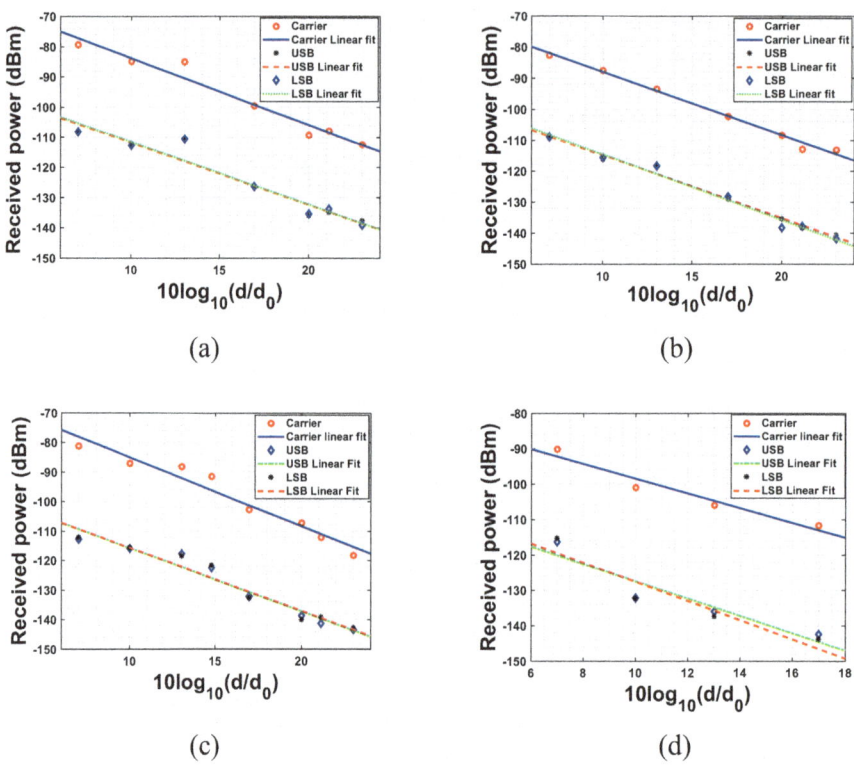

Fig. 6.20 Linear regression fit for received power over distance in LOS scenario with different antennas and polarization in the outdoor environment: (**a**) QHA V-pol, (**b**) PDA V-pol, (**c**) QHA H-pol, and (**d**) PDA H-pol [15]

Table 6.2 Extracted propagation channel parameters in the outdoor environment [15]

Vertical polarization						
	QHA			PDA		
	M_0 (dBm)	η	σ_ξ (dB)	M_0 (dBm)	η	σ_ξ
Carrier	−61.64	−2.21	3.6	−67.59	−2.03	1.1
USB	−91.09	−2.05	3.5	−94.32	−2.03	1.3
LSB	−90.68	−2.08	2.8	−93.30	−2.11	1.6
Horizontal polarization						
	QHA			PDA		
	M_0 (dBm)	η	σ_ξ (dB)	M_0 (dBm)	η	σ_ξ
Carrier	−61.70	−2.32	3.0	−77.61	−2.07	2.1
USB	−94.16	−2.14	2.7	−102.89	−2.44	3.7
LSB	−94.20	−2.13	2.8	−100.41	−2.70	4.0

6.5 How Distance Affects EM Side Channel Results?

$$\Phi^{\kappa,\psi,\alpha,p} = \Phi_0^{\psi,\alpha,p} \cdot \left(\frac{d_\kappa}{d_0}\right)^{\eta^{\psi,\alpha,p}}, \tag{6.5}$$

with the shadowing gain computed as

$$\xi^{\kappa,\psi,\alpha,p} = \frac{M^{\kappa,\psi,\alpha,p}}{\Phi_0^{\psi,\alpha,p} \cdot \left(\frac{d_\kappa}{d_0}\right)^{\eta^{\psi,\alpha,p}}}. \tag{6.6}$$

Note that Φ_0 in (6.5) is equivalent to M_0 in (6.4). We modeled the logarithmic equivalent of the extracted shadowing gain as a Gaussian distribution $\mathcal{N}(\mu_\xi(\text{dB}), \sigma_\xi(\text{dB}))$, with the first and second moment parameters computed as

$$\mu_\xi^{\psi,\alpha,p}(\text{dB}) = \mathbb{E}_\kappa \left\{ 10 \cdot \log_{10} \left(\frac{M^{\kappa,\psi,\alpha,p}}{\Phi_0^{\psi,\alpha,p} \cdot \left(\frac{d_\kappa}{d_0}\right)^{\eta^{\psi,\alpha,p}}} \right) \right\}, \tag{6.7}$$

$$\sigma_\xi^{\psi,\alpha,p}(\text{dB}) = \mathbb{D}_\kappa \left\{ 10 \cdot \log_{10} \left(\frac{M^{\kappa,\psi,\alpha,p}}{\Phi_0^{\psi,\alpha,p} \cdot \left(\frac{d_\kappa}{d_0}\right)^{\eta^{\psi,\alpha,p}}} \right) \right\}, \tag{6.8}$$

where $\mathbb{E}_{\hat{r}}\{\cdot\}$ and $\mathbb{D}_{\hat{r}}\{\cdot\}$ are the expected value and standard deviation operators over an ensemble of the parameter $\hat{r}$ respectively. The mean value of the logarithmic equivalent of the shadowing gain i.e., $(\mu_\xi(\text{dB}))$ was found to be zero in our work while the computed standard deviation values $(\sigma_\xi(\text{dB}))$ has been provided in Table 6.2. Although the values of $\sigma_\xi(\text{dB})$ are similar for measurements conducted with same antenna, it is noticeably different for measurements conducted using the different antennas.

It is important to note that the discrepancies in the results between the QHA and PDA can be attributed to the characteristics of the antennas and how they influence propagation in the channel. For example, the (approximately) 6 dB difference in M_0 (for the carrier) between QHA and PDA can be accounted for by the difference in antenna gain values at the carrier frequency of 1.008 GHz. In addition to this, the larger beamwidth of the QHA affords the aggregation of more multipath components therefore leading to a high receiver power. The shadowing gain disparity can be attributed to the variation in power of the higher number of multipath components captured by the antenna with the wider beamwidth.

6.5.2.3 Distance-Dependent Pathloss Model for Indoor Environments

A model for the distance-dependent pathloss is discussed in this section. The data structure of the received power in the indoor measurements can be represented as $Q^{\kappa_c,\psi,c,o}$, where $c \in [1,2]$ denotes the LOS and NLOS scenarios with $c = 1$

Table 6.3 Propagation channel parameters for LOS and NLOS indoor measurements [15]

LOS scenario					
	Q_0 (dBm)	$\tilde{\eta}$	$\tilde{\sigma}_{\tilde{\xi}}$ (dB)	μ_{σ_Δ} (dB)	σ_{σ_Δ} (dB)
Carrier	−73.55	−1.54	2.0	5.3	1.5
USB	−100.61	−1.57	2.5	5.1	1.9
LSB	−100.99	−1.59	1.9	5.4	1.6
NLOS scenario					
	Q_0 (dBm)	$\tilde{\eta}$	$\tilde{\sigma}_{\tilde{\xi}}$ (dB)	μ_{σ_Δ} (dB)	σ_{σ_Δ} (dB)
Carrier	−88.34	−1.69	3.9	4.6	2.1
USB	−117.11	−1.60	4.1	4.9	1.6
LSB	−116.82	−1.54	3.9	5.1	2.0

indicating LOS and $c = 2$ indicates NLOS. κ_c denotes the index of the scenario distances such that $\kappa_{c=1} \in [1, 2, \ldots, 7]$ are distance indexes for the LOS scenario while $\kappa_{c=2} \in [1, 2, \ldots, 6]$ represents distance indexes for the NLOS scenario. Note that $d_{\kappa_{c=1}} \in [1, 2, 5, 10, 15, 20 \text{ and } 25]$ m while $d_{\kappa_{c=2}} \in [1, 2, 3, 5, 8 \text{ and } 10]$ m. $\psi \in [1, \ldots, \Psi = 3]$ denotes sidebands and carrier indexes such that $\psi = 1, 2, 3$ indicates USB, Carrier and LSB respectively while $o \in [1, \ldots, O = 4]$ denotes the indexes of the rotation angles ($[0°, 90°, 180°, 270°]$) measured.

From the empirical data, we model the received power at different distances as:

$$Q^{\kappa_c,\psi,c,o} = Q_0^{\psi,c} \cdot \left(\frac{d_{\kappa_c}}{d_0}\right)^{\tilde{\eta}^{\psi,c}} \cdot \tilde{\xi}^{\kappa_c,\psi,c} \cdot \Delta^{\kappa_c,\psi,c,o}, \qquad (6.9)$$

where Q_0 is the power received at the reference distance d_0 (1 m). Parameters $\tilde{\eta}$, $\tilde{\xi}$ and Δ represent the pathloss exponent, environment-dependent shadowing gain and the board angular orientation-dependent shadowing gain ("board shadowing") in the indoor environment respectively,

$$Q_{\text{tot}}^{\kappa_c,\psi,c} = \mathbb{E}_o \left\{ Q_0^{\psi,c} \cdot \left(\frac{d_{\kappa_c}}{d_0}\right)^{\tilde{\eta}^{\psi,c}} \cdot \tilde{\xi}^{\kappa_c,\psi,c} \cdot \Delta^{\kappa_c,\psi,c,o} \right\}. \qquad (6.10)$$

To extract channel parameters Q_0 and $\tilde{\eta}$, (6.9) was averaged over an ensemble of board orientations as shown in (6.10) to create a local mean power (Q_{tot}). A linear regression fit (shown in Fig. 6.21) is then used to derive the relationship between the local mean power and distances measured. The extracted parameters Q_0 and $\tilde{\eta}$ are provided in Table 6.3. Note that in Fig. 6.21, Q_{tot} was obtained by averaging $Q^{\kappa_c,\psi,c,o}$ on the linear scale while the display is plotted in decibels. From Table 6.3, it can be observed that the extracted parameters for the sidebands are similar—irrespective of the scenario with the exception of $\tilde{\eta}$ in the NLOS case.

6.5 How Distance Affects EM Side Channel Results?

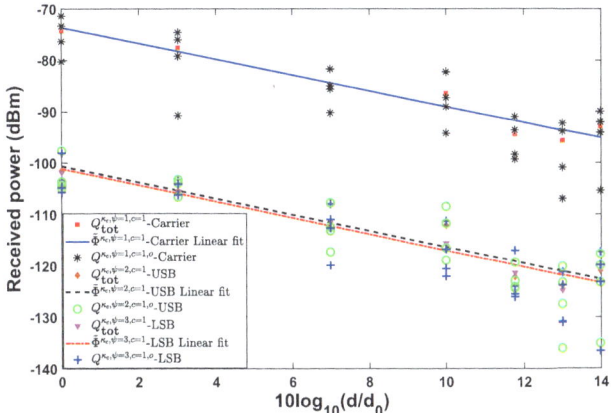

Fig. 6.21 Linear regression fit for received power over rotation and distances in an indoor LOS scenario [15]

6.5.2.4 Shadowing Gain Model for Outdoor Environments

From (6.10), the deviation of the local mean power (Q_{tot}) from the linear regression fit stems from the shadowing ($\tilde{\xi}$) caused by the environment, while the variation of Δ around Q_{tot} is caused by the board shadowing effect at different angular orientation. To extract $\tilde{\xi}$, we define the linear fit as

$$\tilde{\Phi}^{\kappa_c,\psi,c} = \tilde{\Phi}_0^{\psi,c} \cdot \left(\frac{d_{\kappa_c}}{d_0}\right)^{\tilde{\eta}^{\psi,c}}, \qquad (6.11)$$

and then compute

$$\tilde{\xi}^{\kappa_c,\psi,c} = \frac{Q_{tot}^{\kappa_c,\psi,c}}{\tilde{\Phi}_0^{\psi,c} \cdot \left(\frac{d_{\kappa_c}}{d_0}\right)^{\tilde{\eta}^{\psi,c}}}. \qquad (6.12)$$

Note that $\tilde{\Phi}_0$ is equivalent to Q_0 in (6.9).

In this work, the logarithmic equivalent of $\tilde{\xi}$ has been modeled as a Gaussian distribution $\mathcal{N}(\tilde{\mu}_{\tilde{\xi}}(\text{dB}), \tilde{\sigma}_{\tilde{\xi}}(\text{dB}))$ in the LOS and NLOS scenarios with the first and second moment parameters computed as

$$\tilde{\mu}_{\tilde{\xi}}^{\psi,c}(\text{dB}) = \mathbb{E}_{\kappa_c}\left\{10 \cdot \log_{10}\left(\frac{Q_{tot}^{\kappa_c,\psi,c}}{\tilde{\Phi}_0^{\psi,c} \cdot \left(\frac{d_{\kappa_c}}{d_0}\right)^{\tilde{\eta}^{\psi,c}}}\right)\right\}, \qquad (6.13)$$

$$\tilde{\sigma}_{\tilde{\xi}}^{\psi,c}(\text{dB}) = \mathbb{D}_{\kappa_c} \left\{ 10 \cdot \log_{10} \left(\frac{Q_{\text{tot}}^{\psi,c}}{\tilde{\Phi}_0^{\psi,c} \cdot \left(\frac{d_{\kappa_c}}{d_0}\right)^{\tilde{\eta}^{\psi,c}}} \right) \right\}. \qquad (6.14)$$

We found the logarithmic equivalent of $\tilde{\xi}$ to have a mean value of zero in both LOS and NLOS scenarios and have thus presented the extracted standard deviation ($\tilde{\sigma}_{\tilde{\xi}}$ (dB)) in Table 6.3. The board shadowing gain (Δ) was computed as

$$\Delta^{\kappa_c,\psi,c,o} = \frac{Q^{\kappa_c,\psi,c,o}}{Q_{\text{tot}}^{\kappa_c,\psi,c}}. \qquad (6.15)$$

We have modeled Δ as a zero-mean, lognormally distributed, variable at different distances in this work. We found that the standard deviation of σ_Δ differed from location to location and hence has been modeled as a random variable. The first and second order moments for the distribution of the standard deviation over an ensemble of distances are derived in (6.16) and (6.17) respectively,

$$\mu_{\sigma_\Delta}^{\psi,c}(\text{dB}) = \mathbb{E}_{\kappa_c} \left\{ \mathbb{D}_o \left\{ 10 \cdot \log_{10} \left(\frac{Q^{\kappa_c,\psi,c,o}}{\mathbb{E}_o \{Q^{\kappa_c,\psi,c,o}\}} \right) \right\} \right\}, \qquad (6.16)$$

$$\sigma_{\sigma_\Delta}^{\psi,c}(\text{dB}) = \mathbb{D}_{\kappa_c} \left\{ \mathbb{D}_o \left\{ 10 \cdot \log_{10} \left(\frac{Q^{\kappa_c,\psi,c,o}}{\mathbb{E}_o \{Q^{\kappa_c,\psi,c,o}\}} \right) \right\} \right\}. \qquad (6.17)$$

Empirical CDF plots and corresponding Gaussian fits for the logarithmic equivalent of σ_Δ for the Carrier, USB and LSB are shown in Fig. 6.22a–c. It can be observed from Table 6.3 that $\tilde{\xi}$ in the NLOS is greater than that in the LOS scenario while μ_{σ_Δ} and σ_{σ_Δ} are comparable in both scenarios. This agrees with intuition since the board shadowing effect should be similar irrespective of the scenario, while the environment shadowing gain will be more pronounced in the NLOS than in the LOS scenario.

6.5.3 How to Select Antennas and Probes to Improve Signal to Noise Ratio?

One of the challenges in studying the side-channel EM emanations is how to improve signal reception, especially at a distance and in the presence of noise. The EM side-channel propagation model in [21] suggests that the EM side-channel signal follows the Friis formula once the signal is in the far-field, and that the reception range can be extended by increasing the antenna gain.

6.5 How Distance Affects EM Side Channel Results?

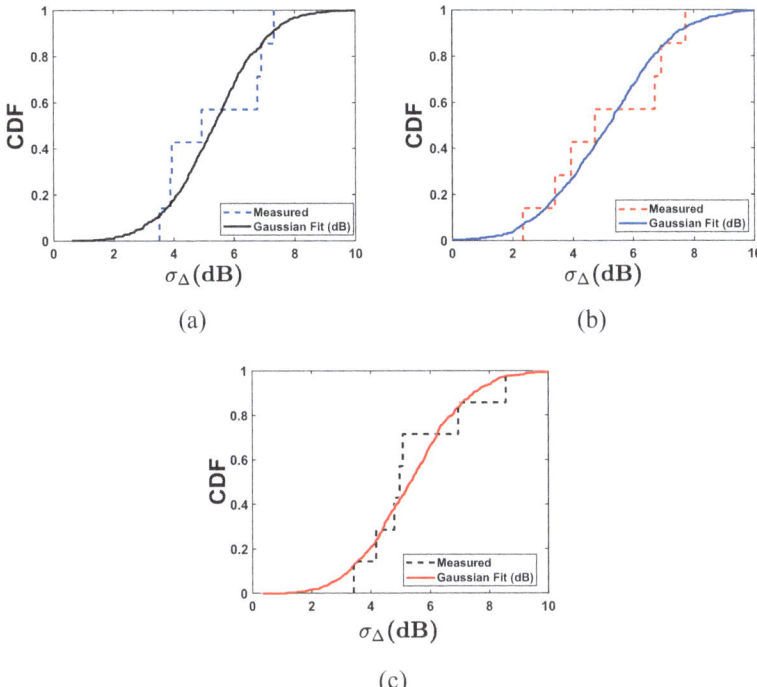

Fig. 6.22 Empirical CDF and corresponding Gaussian fit for Carrier, USB and LSB board shadowing in an indoor LOS scenario: (**a**) Carrier, (**b**) USB, (**c**) LSB [15]

Here we present one example of compact high-gain antenna and one example of a compact high-gain probe, both designed to enhance the SNR of received EM signals. In [9], our goal was to design an antenna with peak broadside gain of 19 dBi, operating around 1 GHz (a clock frequency for typical embedded processors of IoT devices), which would allow for signal reception at >4 m distance. The bandwidth requirement is around 50 MHz to capture looop-level program activity on the processor. The second requirement is that the design should be planar in nature, which can be hang on the ceilings or walls and does not occupy more than 1×1 m area, so it can be used to receive EM signals form embedded devices in a room.

Among planar radiators, microstrip arrays are widely used for high gain planar antenna applications. The elements in those arrays are generally designed to operate in their fundamental mode. For instance, in case of rectangular patch, the mode is either TM_{10} or TM_{01}. Similarly for the circular patch, the fundamental mode TM_{11} is used. For a single microstrip element that operates in the fundamental mode, the peak directivity is limited. For example, the peak directivity of a circular disc in TM_{11} mode is 9.9 dBi [4]. Substrate permittivity between 2–3 is generally used in practice, which further reduces single-element directivity to 6–7 dBi for a circular disc [10]. To achieve higher gain with planar microstrip arrays, array dimensions of N>2 are used. The array spacing is also limited—the optimum spacing for maximum directivity, with no sidelobe, is between 0.7–0.9 λ_0 [8]. For a given physical area, e.g., 1×1 m, the microstrip array on the substrate permittivity of 1, has the least number of elements and the maximum directivity. Increasing substrate permittivity reduces the element size and hence will result in increasing the number of elements. In the lower L-band design frequency range, e.g. for 1 GHz, the printed element increases the cost, due to large physical area, as substrates are expensive. Also, feed lines are generally used to excite the array elements, and these feed lines radiate on their own and are not preferable for detection of side-channel emanations.

Hence, we propose a 19 dBi gain, single feed, planar circular disc antenna design that consists of conducting metallic discs suspended on air, operates at 1 GHz, and occupies an 1.04×1.04 m area. The antenna consists of two layers of slotted conducting metal discs, which are designed to operate in the higher-order TM_{12} mode. The discs are suspended in air above the ground plane, using teflon screws at the positions of electric field nulls along each disc's radius. The upper layer is a 2×2 array of slotted circular discs electromagnetically coupled by a lower identical slotted disc, which is fed directly by a single coaxial feed. The use of a higher-order TM_{12} mode permits the use of fewer elements, due to their large electrical size when compared to the fundamental TM_{11} mode. The entire fabrication of the antenna is accomplished using aluminum metal stock sheets, using air rather than a dielectric substrate. Given the growing importance of embedded and smart systems, the proposed antenna can be used to receive EM signals from one or more embedded devices in a room. The receiving signal can then be used for software profiling [16] or security monitoring [12]. Figure 6.23 show an example of this antenna being used in a side channel measurement setup. Figure 6.24 shows measured side channel while code is executing on an IoT device 5 m away.

The challenge in researching EM cryptographic attacks is receiving a usable signal. The signals tend to be very weak (otherwise the device would not be compliant with EMC standards) and location dependent [17]. Therefore, any probe used to monitor the side-channel needs to be designed to be sensitive to weak electric

Fig. 6.23 Setup for measuring SNR for an IoT (Olimex), with the antenna (at the right edge of the image) and the IoT board (left edge of the image) [15]

6.5 How Distance Affects EM Side Channel Results?

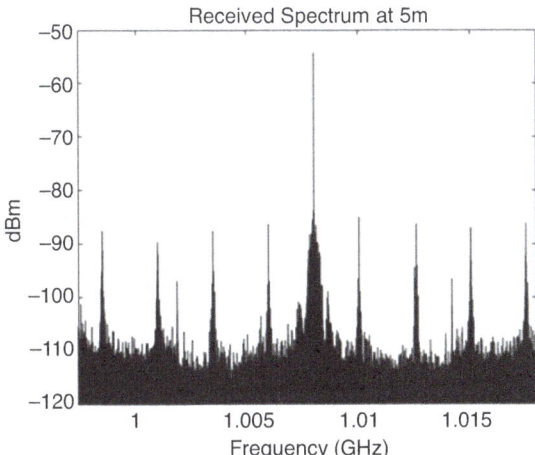

Fig. 6.24 Measured signal power at 5 m distance using the setup from Fig. 6.23 [15]

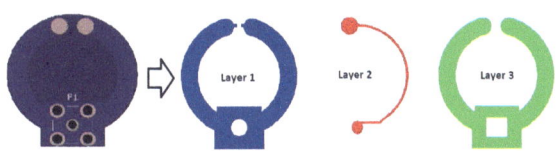

Fig. 6.25 The probe and its copper layers [18]

or magnetic fields. At the same time, the probe needs to be able to scan different parts of the device. Presented here is a shielded loop probe designed for monitoring EM side-channels by measuring the magnetic field [18].

Shielded loop probes are regularly used in EMC [3, 18]. A shielded loop composes of a loop or part of a loop with a shield enclosing it. The shield has a small gap in the middle, opposite from the output of the probe. The advantage of a shielded loop probe is that it suppresses the influence of the electric field on its measurements, ensuring that its output is the result of only the perpendicular magnetic field [3]. This makes it easier to focus on specific components on a device and locate potential emanation sources.

Figure 6.25 shows the probe and its three copper layers [18]. The first and third layers act as the shield and are connected using vias, while the second layer contains the feed line. The probe was implemented on a PCB for durability and compactness. Its small size and flatness makes it easy to scan over different parts of a devices when searching for the best location to monitor a side-channel. To obtain the required sensitivity, the probe has a diameter of 20 mm. By default, the probe has a center frequency of 0.89 GHz. The center frequency can be decreased by soldering capacitors in between the gaps in the shield or can be increased by increasing the size of the gap. Using this method, probes with center frequencies between 800 MHz and 1.8 GHz have been manufactured.

To demonstrate how well the probe suppresses the influence of the electric field, the S_{21} measured in the first part is compared to S_{21} measured when the influence of the magnetic field is minimized. For simplicity, they are referred to as S_{21H} and

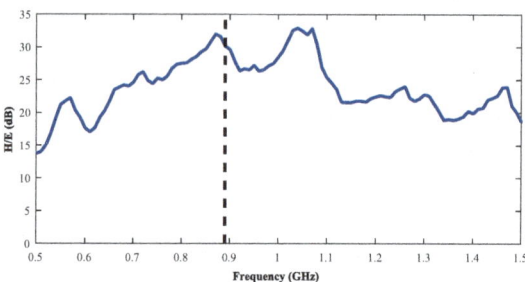

Fig. 6.26 The electric field suppression ratio for the probe. The black, dashed line indicates the probe's center frequency [18]

S_{21E}. The ratio of S_{21H} to S_{21E} is called the electric field suppression ratio (H/E). This ratio is commonly used to evaluate how strongly a probe is influenced by the electric field [3]. The suppression ratio calculated from the measurements is shown in Fig. 6.26. At 0.89 GHz, the ratio is 30.18 dB, indicating the magnetic field has a much stronger impact on the measurements than the electric field.

6.6 How Carrier Power, Frequency, and Distance Affect the Backscattering Side Channel?

For other side channels, such as EM and power, the strength of measured signal is limited by the strength of leaked signal from the device at the position of the measurement. In contrast, the backscattering side-channel does not have this limitation—it does not depend on circuits to unintentionally emanate the signal, and the received signal power can be adjusted by adjusting the power of the carrier that is transmitted toward the circuit. In this section, we analyze the impact of this carrier power on the quality of received backscattering side-channel signal. In this measurement, the toggling circuit has f_m = 1.25 MHz, and the power of the carrier is ranges from −15 to 3 dB. The results in Fig. 6.27 show that the received sideband power increases as the transmitted carrier power increases. This means that we can indeed control the strength of the backscattered side-channel signal by changing the power at the carrier transmitter. For example, when measuring far away from the device, or when there are obstacles between the antenna and the device, we can compensate for the increased path loss by transmitting a stronger carrier. Therefore, unlike the other side-channels, the backscattering side-channel is an "active" side-channel.

Here we analyze how the magnitude of backscattering side-channel signal depends on the carrier frequency. To answer this question, the toggling circuit is kept at 800 kHz and the carrier frequency is changed across the frequency range from 0.01 to 10 GHz. For each carrier frequency, we record sideband power, carrier peak power, and noise floor power level at the receiver. The results in Fig. 6.28 show that from 3 to 5 GHz, the signal-to-noise ratio of the side-channel signal is high, and a carrier frequency of 3.031 GHz produces the highest signal-to-noise ratio.

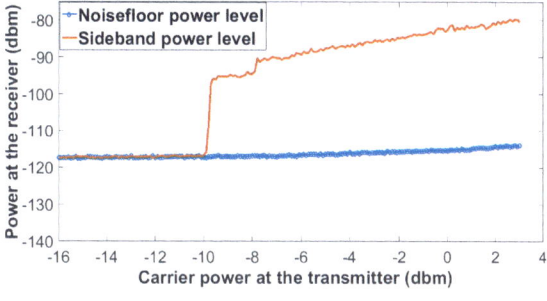

Fig. 6.27 The received power of a side-channel as a function of the carrier power at the transmitter

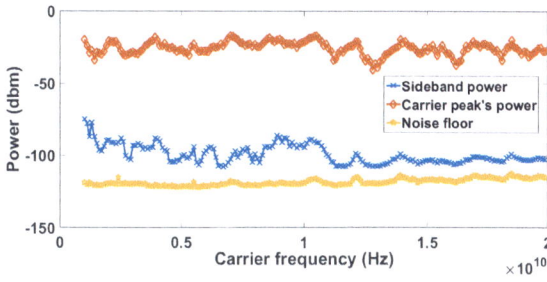

Fig. 6.28 The carrier signal strength, the side-channel signal strength, and the noise floor as functions of carrier frequency

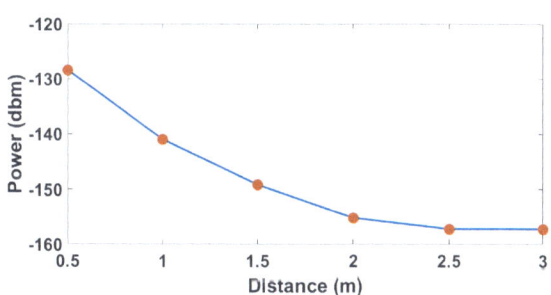

Fig. 6.29 Side-channel power as a function of the distance from the DuT

Therefore, this frequency is used to generate carrier signals for the backscattering side-channel measurements.

Finally, we analyze how the magnitude of the backscattering side-channel signals depends on the distance from the monitored device. The toggling circuit is used and the antenna-to-DuT distance is swept from 50 cm to 3 m. The power of the carrier is kept constant at 15 dB. The results in Fig. 6.29 demonstrate that the magnitude of the backscattering side-channel power (in dB) decays almost linearly with the distance up to 2 m distance. Beyond 2 m, the side-band power is "buried in the noise".

References

1. Daniel J. Bernstein, Joachim Breitner, Daniel Genkin, Leon Groot Bruinderink, Nadia Heninger, Tanja Lange, Christine van Vredendaal, and Yuval Yarom. Sliding right into disaster: Left-to-right sliding windows leak. Conference on Cryptographic Hardware and Embedded Systems (CHES) 2017, 2017.

2. Robert Callan, Alenka Zajic, and Milos Prvulovic. A Practical Methodology for Measuring the Side-Channel Signal Available to the Attacker for Instruction-Level Events. In *Proceedings of the 47th International Symposium on Microarchitecture (MICRO)*, 2014.
3. Y. Chou and H. Lu. Magnetic near-field probes with high-pass and notch filters for electric field suppression. *IEEE Transactions on Microwave Theory and Techniques*, 61(6):2460–2470, June 2013.
4. A. Derneryd. Analysis of the microstrip disk antenna element. *IEEE Transactions on Antennas and Propagation*, 27(5):660–664, 1979.
5. Dinkić, Jelena Lj and Olćan, Dragan I and Djordjević, Antonije R and Zajić, Alenka G. High-Gain Quad Array of Nonuniform Helical Antennas. *International Journal of Antennas and Propagation*, 2019, 2019.
6. Daniel Genkin, Itamar Pipman, and Eran Tromer. Get your hands off my laptop: physical side-channel key-extraction attacks on pcs. In *Conference on Cryptographic Hardware and Embedded Systems (CHES)*, 2014.
7. Daniel Genkin, Adi Shamir, and Eran Tromer. Rsa key extraction via low-bandwidth acoustic cryptanalysis. In *International Cryptology Conference (CRYPTO)*, 2014.
8. J. R. James and P. S. Hall. *Handbook of Microstrip Antennas*. London, U.K. Peregrinus, 1989.
9. P. Juyal, S. Adibelli, N. Sehatbakhsh, and A. Zajic. A Directive Antenna Based on Conducting Disks for Detecting Unintentional EM Emissions at Large Distances. *IEEE Transactions on Antennas and Propagation*, 66(12):6751–6761, Dec 2018.
10. Prateek Juyal and Lotfollah Shafai. A novel high-gain printed antenna configuration based on t_{12} mode of circular disc. *IEEE Transactions on Antennas and Propagation*, 64(2):790–796, 2016.
11. Haider Adnan Khan, Monjur Alam, Alenka Zajic, and Milos Prvulovic. Detailed tracking of program control flow using analog side-channel signals: a promise for IoT malware detection and a threat for many cryptographic implementations. In *Cyber Sensing 2018*, volume 10630 of *Society of Photo-Optical Instrumentation Engineers (SPIE) Conference Series*, page 1063005, May 2018.
12. Alireza Nazari, Nader Sehatbakhsh, Monjur Alam, Alenka Zajic, and Milos Prvulovic. Eddie: Em-based detection of deviations in program execution. In *2017 ACM/IEEE 44th Annual International Symposium on Computer Architecture (ISCA)*, pages 333–346, 2017.
13. Colin Percival. Cache missing for fun and profit. In *Proc. of BSDCan*, 2005.
14. Milos Prvulovic and Alenka Zajic. Rf emanations from a laptop, 2012. http://youtu.be/ldXHd3xJWw8.
15. Seun Sangodoyin, Frank T. Werner, Baki B. Yilmaz, Chia-Lin Cheng, Elvan M. Ugurlu, Nader Sehatbakhsh, Milos Prvulovic, and Alenka Zajic. Side-channel propagation measurements and modeling for hardware security in iot devices. *IEEE Transactions on Antennas and Propagation*, 69(6):3470–3484, 2021.
16. Nader Sehatbakhsh, Alireza Nazari, Alenka Zajic, and Milos Prvulovic. Spectral profiling: Observer-effect-free profiling by monitoring em emanations. In *2016 49th Annual IEEE/ACM International Symposium on Microarchitecture (MICRO)*, pages 1–11, 2016.
17. F. Werner, D. A. Chu, A. R. Djordjević, D. I. Olćan, M. Prvulovic, and A. Zajic. A method for efficient localization of magnetic field sources excited by execution of instructions in a processor. *IEEE Transactions on Electromagnetic Compatibility*, PP(99):1–10, 2017.
18. Frank T. Werner, Antonije R. Djordjevic, and Alenka Zajic. A compact probe for em side-channel attacks on cryptographic systems. In *2019 IEEE International Symposium on Antennas and Propagation and USNC-URSI Radio Science Meeting*, pages 613–614, 2019.
19. Baki Berkay Yilmaz, Robert Callan, Milos Prvulovic, and Alenka Zajic. Quantifying information leakage in a processor caused by the execution of instructions. In *Military Communications Conference (MILCOM), MILCOM 2017-2017 IEEE*, pages 255–260. IEEE, 2017.

References

20. A. Zajic and M. Prvulovic. Experimental demonstration of electromagnetic information leakage from modern processor-memory systems. *Electromagnetic Compatibility, IEEE Transactions on*, 56(4):885–893, Aug 2014.
21. A. Zajic, M. Prvulovic, and D. Chu. Path loss Prediction for Electromagnetic Side-Channel Signals. In *2017 11th European Conference on Antennas and Propagation (EUCAP)*, pages 3877–3881, March 2017.

Chapter 7
Modeling Analog Side Channels as Communication Systems

7.1 Introduction

In this chapter we focus on covert channels, because they resemble communication systems better, and also because they represent the worst-case scenario for information leakage through a side channel. A covert channel [15] can be used by a *spy* program to exfiltrate sensitive data from a target computer system, using legitimate software activity to secretly transmit information without using traditional (overt) communications channels (e.g., an Ethernet or WiFi connection). A typical real-world scenario involves a *privileged* but *untrusted* program, e.g., a password manager that has access to sensitive data but, because it is untrusted, has no/limited access to the outside world (i.e., this program is *sandboxed*). A malicious sandboxed program can, however, manipulate its own activity to leak sensitive information through a side channel, i.e., this program acts as a transmitter for a covert channel. If a software-observable side channel is used, the receiver is typically a *non-privileged* program (with no access to sensitive data) that is connected to the outside world, e.g., an untrusted video game that may be considered safe to execute on the computer system because it has no access to any sensitive data. If a physically-observable side channel is used, the receiver is outside of the target computer system, so information can be exfiltrated even from an target computer system computer that is *air-gapped*, i.e., physically separated from public networks [8, 9].

Like a transmitter in a traditional communications system, the spy program is designed to maximize the amount of information that is transmitted per unit time, subject to some constraints. In traditional communications, the constraints under which the transmitter must operate typically include the maximum allowed transmitter power, confining the transmitted power to a specific frequency band, etc. However, in general the designers of a traditional communication systems are free to decide how much information will be encoded in each transmitted symbol, what the transmitted signal looks like for each symbol, what is the duration of each symbol, etc. For many of these decisions, efficiency dictates that symbols should have exactly the same duration, that signals for different symbols should be as different from each other as the transmitter power limitations allows, etc. In contrast, signals generated through a spy program are far more constrained—we can only choose which instructions the program executes. Different instructions can take very different amounts of time to execute, and because some instructions produce side channel signals that are very similar to side channel signals produced by other instructions, even the best-designed spy program can only achieve a fraction of the data rate that could be achieved if the same signal power and bandwidth were to be used by a traditional communications system. This means that most theoretical analyses designed for traditional communication systems would grossly overestimate the data rates that could be achieved by spy programs that leverage physically-observable side channels.

Similarly, accidental side channel leakage from a regular program (that is not designed as a spy program) is likely to be significantly less than the leakage that would be achieved by running a spy program on the same computer system. However, some regular programs may have side-channel leakage that is almost as bad as that of a spy program, so the results that will be derived in this chapter for covert channels (with a spy program) can also serve as worst-case estimates for side channel leakage in general (using any software/program).

Covert channels are considered to be a serious security threat [18] because they can circumvent and break existing defense mechanisms (e.g., memory isolation, partitioning, access permissions, etc.) that protect data inside a computer system. When an attacker is designing the spy program, one of the primary goals is for its covert channel activity to remain secret. For example, if the spy program is a password manager, the attacker relies on it being used as a password manager, and few users would be willing to use a password manager that is known to be actively malicious. Another example involves a rogue employee that adds covert-channel transmission functionality to a reputable and widely used program, where discovery of this added "feature" would typically be traced back to that employee and likely result in significant legal consequences.

The probability of detection for covert-channel communication increases with the time it remains in active use—while "transmitting" information, the spy (part of the) program typically affects the overall performance and resource utilization of the system, and parts of the program where resources or time are spent often end up being scrutinized. For example, developers routinely profile software to find parts of the program where significant amounts of execution time are spent, as optimizations to such parts of the program tend to result in noticeable performance improvements to the entire program.

Therefore, probability of detection is reduced when the covert channel has a high data rate—it can transmit some required amount of information (e.g., a set of passwords) quickly, adding little to the overall execution time of the system or program. This means that, from the defensive point of view, the severity of any covert channel can be measured in terms of its data rate, so a very important step in analyzing any covert channel is to find the lower and upper bounds for its data rate.

7.2 Software-Activity-Created Signals and Communication Model

Millen was the first to establish a connection between Shannon's information theory and information flow models in computer systems [19], with a calculation of the capacity (maximum possible data rate) of such a covert channel. However, that model assumes a synchronous channel (where information is transmitted at a constant rate that is known to the receiver), and this is not a realistic assumption for covert channels in computer systems, where the timing of execution in the spy program varies due to a number of hardware features (e.g., pipeline stalls, dynamic scheduling of instructions, cache hits and misses, branch prediction, etc.). Additionally, while a communication system is designed to have a low error rate and can easily avoid insertion or deletion of transmitted symbols, a covert-channel transmission often includes insertion, deletion, and errors, e.g., interrupts and other system activity often inserts activity into the timeline of the spy program's execution.

There are many papers that discuss bounds on the capacity of channels corrupted by synchronization errors [5, 10, 14, 26, 28]. More recently, there have also been papers that discuss bounds on the capacity of channels that are corrupted by both synchronization and substitution errors [17, 24]. However, until recently, none of them have provided bounds for the capacity of a covert channel that has random shifts in the timing of its symbols, and where the probability of insertion depends on which symbol is being transmitted.

Here we present the first model [34] of an analog covert channel (EM emanations) that is created by program activity. We first discuss how this channel can be created and how information can be sent and received, and then we mathematically model this system and derive leakage limits by adding covert-channel specific concerns to derivations of capacity bounds proposed in [17].

In a covert channel, as in traditional wireless communications, some errors occur due to variations in the propagation environment. However, in addition to such channel errors, the software "transmitter" lacks precise synchronization, causing jitter that reduces the signal's effective bandwidth *and* increases the noise level. Also, the "transmitter" gets interrupted by other (system) activity, the transmitted signal may go through a channel obstructed by metal, plastic, etc. To capture all effects of the observed behavior, in Sect. 7.3 we model the transmitted sequence as a pulse amplitude modulated (PAM) signal with randomly varying pulse positions (jitter), in Sect. 7.4 we derive power spectral density and the bit error rate (BER) of the transmitted signal and compare these theoretical results to practical examples, and finally in Sect. 7.5 we derive the capacity (maximum achievable data rate) for a covert channel and compare these theoretical results to practical examples.

7.2 Software-Activity-Created Signals and Communication Model

As in previous chapters, here we use the same A/B alternation code as the software activity that produces analog side channel signals. This code (and how it works) was described in Chap. 4, but we will now review the key aspects of this code that are relevant to this chapter. To create a carrier, we use repetitive variations in a software activity as described in Chap. 4 [30, 31, 35]. We choose T, the period (duration) of each repetition, two types of activities (A and B), and write a small program shown in Fig. 7.1, that in each period, does activity A in the first half and B in the second half. The period T will be selected to correspond to a specific frequency, e.g., to produce a radio signal at 1 MHz, we should set $T = 1\,\mu s$. This carrier-generation approach is illustrated in Fig. 7.2.

For covert-channel transmission, the symbols are bits that are *amplitude modulated* into the signal by inserting intervals during which only activity B is performed in both half-periods which means any carrier signal produced by the differences between A and B should be absent when only B is used, resulting in the simplest form of AM modulation (on-off keying). This approach is illustrated in Fig. 7.3.

```
1   while(1){
2   // Do some instances of activity A
3       for(i=0;i<n_inst;i++){
4           ptr1=(ptr1&~mask1);
5           // Activity A, e.g. a load
6           value=*ptr1;
7       }
8   // Do some instances of activity B
9       for(i=0;i<n_inst;i++){
10          ptr2=(ptr2&~mask2);
11          // Activity B, e.g., a store
12          *ptr2=value;
13      }
14  }
```

Fig. 7.1 The A/B alternation pseudo-code [35]

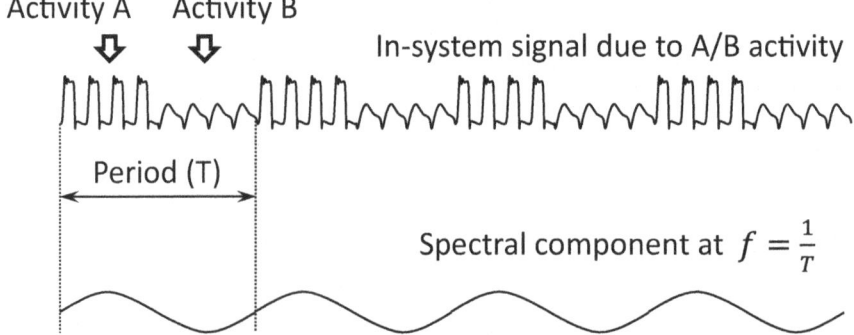

Fig. 7.2 Illustration of how a program induces emanations at a specific radio frequency by alternating half-periods of A and B activity [35]

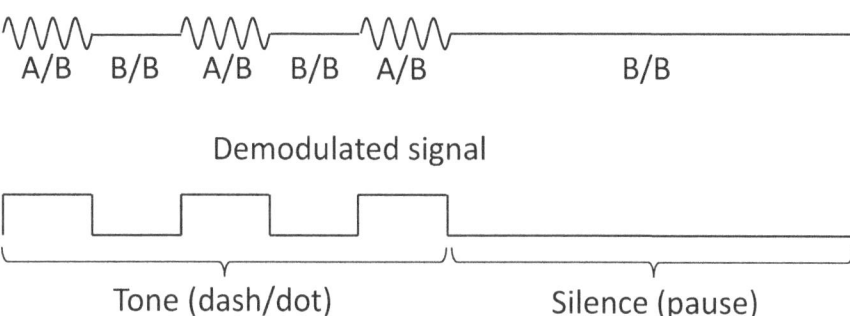

Fig. 7.3 Illustration of how A/B alternation modulates the signal into the carrier using on-off keying (bottom) [35]

To confirm that the received communication sequence is indeed the transmitted message, we performed two experiments. First, we modulated our transmitted signal with the A5 note (880 Hz), and turned this tone on/off to transmit Morse code for "All your data belong to us" [35]. Second, we placed our spy program code around the keyboard driver, which allowed us to transmit keystrokes wirelessly [23] that can be received by anyone listening to an AM radio receiver. In both cases, we were able to correctly receive and demodulate transmitted signals.

However, we observed that the timing of the instructions was not perfectly synchronized. This issue confirmed that the baseband pulses generated with on-off keying do not have equal timing, and the created carrier is spread over several kilohertz—this is in contrast to traditional communications where the carrier is well concentrated around a single frequency. This lack of synchronization in the transmitter causes significant jitter and has to be carefully modeled, as described in the following sections.

7.3 Communication Model for Covert Channel Software-Activity-Created Signals

In this section, we show a model for covert channel communication systems [34]. Before introducing the model itself, we briefly review the notations we will use, and also the properties of the baseband PAM signal that corresponds to software activity.

A PAM signal with a period of $\mathscr{T}$ can be written as [22]

$$x_p(t) = \sum_k x_k \delta(t - k\mathscr{T}) * p(t), \qquad (7.1)$$

where $\delta(\bullet)$ is Dirac delta function, $*$ is the convolution operator, $\mathbf{x}_k = (x_k, x_{k-1}, x_{k-2}, \ldots)$ is the sequence of data symbols that are chosen from a finite alphabet, and $p(t)$ is a shaping pulse. The power spectral density (PSD) of $x_p(t)$ can be written as [22]

$$S_{xp}(f) = \frac{|P(f)|^2}{\mathscr{T}} S_x(f), \tag{7.2}$$

where $P(f)$ is the Fourier transform of the shaping pulse,

$$S_x(f) = \sum_{k=-\infty}^{\infty} R_x[k] e^{-j2\pi f k \mathscr{T}}, \tag{7.3}$$

is PSD of the stationary sequence $\mathbf{x}_k$, and $R_x[k]$ is the autocorrelation function of sequence $\mathbf{x}_k$. Furthermore, if an impulse function is used as the shaping pulse (this assumption is introduced to simplify mathematical expressions), the power spectral density can be simply written as $S_{xp}(f) = S_x(f)/\mathscr{T}$.

The baseband signal from (7.1) assumes perfect symbol timing. However, transmitted signals created by program activity have synchronization problems due to variations in symbol timing. As an example, Fig. 7.4 shows the distribution of symbol duration for an EM covert channel, where bit (symbol) "0" is transmitted by executing program activity that has numerous (cache miss) memory accesses, while bit "1" is transmitted by executing program activity that only includes on-chip activity. We show distributions for "0" and "1" symbols separately because, as can be seen in Fig. 7.4, these two distributions are not the same. Their means differ

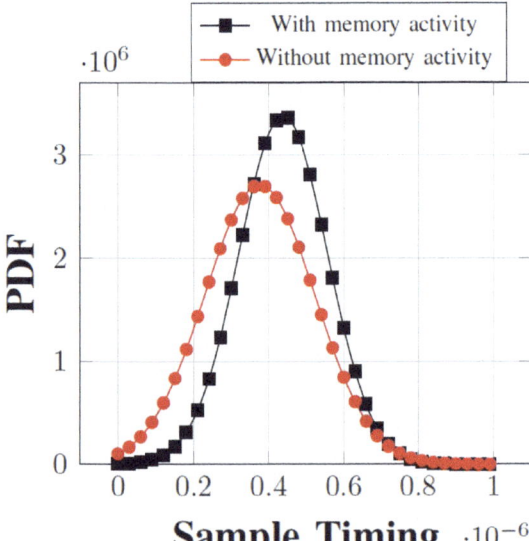

Fig. 7.4 Illustration of two timing distributions of symbols for an EM covert channel, one when memory activity is used and one with on-chip instructions is used [34]

7.3 Communication Model for Covert Channel Software-Activity-Created...

because, although the loop iteration counts for "0" and for "1" were chosen with the intention of making their durations the same, this control over the duration of the symbol is imperfect. The variance (how concentrated the distribution is) is also different for "0" and "1"—off-chip memory accesses have more variation, mainly because off-chip memory accesses take much longer to execute, so fewer of them are used to achieve similar symbol duration, and that means that variation in memory access time of individual accesses has more impact on the duration of the symbol. In contrast, many iterations of on-chip activity are executed for one symbol, so their individual variation in timing has less impact on the variation of the sum of their duration (sample variance vs. mean variance).

To deal with pulse width variations and still establish a connection using conventional communication theories, in our model the duration of the pulse is constant, but the center of the pulse varies in time. This assumption allows us to model one key aspect of varying pulse width, namely the fact that edges of pulses vary in time (jitter) due to the non-synchronous nature of the channel, while keeping the mathematics of the model tractable.

In this model, the baseband received signal can be written as

$$y_p(t) = \sum_k x_k p(t - k\mathcal{T} - \mathbf{T}_k), \tag{7.4}$$

where $\mathbf{T}_k$ is a random shift associated with a particular pulse for the transmitted symbol, x_k, whose probability density function (pdf) is denoted by $f_{\mathbf{T}_k}(t_k)$. As illustrated in Fig. 7.3, pulses are assumed to have a 50% duty cycle, and neighboring pulses are assumed to not overlap.

To ensure that neighboring pulses do not overlap, the support set of $f_{\mathbf{T}_k}(t_k)$ is set to $\{t_k \in [-\mathcal{T}/4, \mathcal{T}/4)\}$. We also assume that probability density functions $\{f_{\mathbf{T}_k}(\bullet) \,|\, \forall k \in \{-\infty, \infty\}\}$ are identical and independent distributions (i.i.d.). Figure 7.5 shows an example of a typical PAM signal with 50% duty cycle, and also its randomly-shifted version. We can observe that the time difference between neighboring pulses can increase or decrease, which mimics the variation in software activity and reflects the lack of synchronization.

To simplify Eq. (7.4) further, we will assume that $p(t)$ is an impulse function, $\delta(t)$, so the modulated baseband signal can be written as

$$y(t) = \sum_k x_k \delta(t - k\mathcal{T} - \mathbf{T}_k). \tag{7.5}$$

To evaluate the impact of the jitter (variation in symbol timing) in this system, we need to find the power spectral density (PSD) of our baseband PAM signal with a random pulse position. In [34] we have shown that the PSD of the received signal, $y(t)$, in Eq. (7.5) can be written as

$$S_y(f) = \frac{1}{\mathcal{T}} S_x(f) \Phi(f) + \frac{R_x[0]}{\mathcal{T}} (1 - \Phi(f)). \tag{7.6}$$

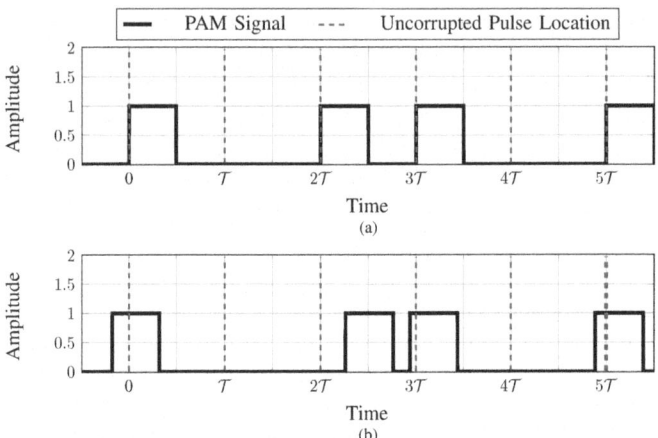

Fig. 7.5 (**a**) PAM with sequence x_k and (**b**) distribution of pulses perturbed randomly in time and modulated in amplitude when the shaping pulse is a square wave [34]

Furthermore, for an arbitrary pulse shape, the PSD of a PAM signal with a random pulse position becomes

$$S_{y_p}(f) = \frac{|P(f)|^2}{\mathcal{T}} \left(S_x(f)\Phi(f) + R_x[0]\Big(1 - \Phi(f)\Big) \right), \quad (7.7)$$

where $R_x[k]$ is the autocorrelation function and $S_x(f) = \sum_k R_x[k]e^{-j2\pi f k \mathcal{T}}$ is the power spectral density of x_k.

This result shows that transmission that uses a PAM signal with randomized pulse position (jitter) is equivalent to a traditional transmission where the PAM signal is first passed through a filter with power spectral density $\Phi(f)$, and then subjected to continuous wideband noise. The characteristics of the filter and the noise are completely determined by the probability distribution of the pulse positions (jitter). By fitting the measured spectrum of symbols into different probability distributions, we have found that the best fit for the pulse position variations is a Gaussian distribution with mean μ and standard deviation σ. Although a Gaussian distribution for pulse positions contradicts our previous assumption that pulses can be only between $-\mathcal{T}/4$ and $\mathcal{T}/4$, we can still approximate the pulse positions with a Gaussian random distribution by assuming that the tail probability beyond $-\mathcal{T}/4$ and $\mathcal{T}/4$ is almost zero. Moreover, for the tractability of the derivations, we will assume that the means of these Gaussian distributions are equal. Figure 7.6 plots the distributions of these shifts in pulse position. We can observe that the pulse shift distributions are concentrated around zero.

7.3 Communication Model for Covert Channel Software-Activity-Created...

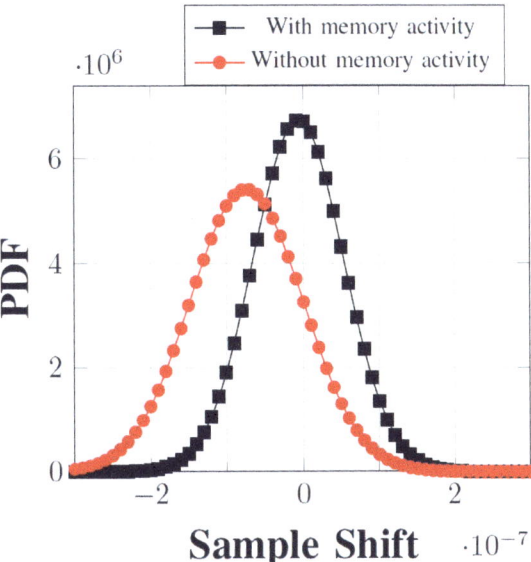

Fig. 7.6 Illustration of two distributions of pulse shift for an EM covert channel, one when memory activity is used and one with on-chip instructions is used [34]

Given that Gaussian distribution has a Fourier transform

$$\mathfrak{F}\{f(t)\} = e^{-2\pi j f \mu} e^{-2\pi^2 \sigma^2 f^2}, \quad (7.8)$$

where $\mathfrak{F}\{\bullet\}$ takes Fourier transform of its argument and $f(t)$ is any Gaussian distribution with mean μ and standard deviation σ, we can combine Eq. (7.8) with Eq. (7.7) and obtain

$$\Phi(f) = e^{-2\pi^2 f^2 (\sqrt{2}\sigma)^2}. \quad (7.9)$$

Finally, inserting (7.9) into (7.6), we calculate PSD of PAM signal with Gaussian distributed pulse positions as

$$\begin{aligned} S_y(f) &= \frac{1}{\mathcal{T}} S_x(f) e^{-2\pi^2 f^2 (\sqrt{2}\sigma)^2} \\ &\quad + \frac{R_x(0)}{\mathcal{T}} \left(1 - e^{-2\pi^2 f^2 (\sqrt{2}\sigma)^2}\right) \\ &= S_{xt}(f) + S_{nt}(f), \end{aligned} \quad (7.10)$$

where $S_{xt}(f)$ denotes the spectrum of the transmitted sequence and $S_{nt}(f)$ denotes the noise spectrum due to random pulse position.

7.4 Quantifying the Information Leakage of Covert Channel Software-Activity-Created Signals

7.4.1 Bit Error Rate and Power Spectral Density

Traditionally, performance of a communication system is evaluated by estimating its symbol error rate or its block error rate (BER). The error probability of pulse amplitude modulated signal (PAM) can be written as [22]

$$P_{PAM} = Q\left(\sqrt{\frac{P_s}{2P_n}}\right), \quad (7.11)$$

where $Q(\cdot)$ function denotes the tail probability of the standard normal distribution, P_s is the averaged transmitted power of a symbol, and P_n is the averaged noise power.

For a covert-channel attack scenario, the BER for a given data rate indicates how reliable are the data bits that are received by the attacker through the covert channel. To estimate these BER, we need the PSD of the signal, which we had already derived in (7.10). Then, we assume that the transmitted signal, $y(t)$, defined in (7.5) has been affected by channel noise, and that the received signal can be written as

$$r(t) = y(t) + n(t), \quad (7.12)$$

where $n(t)$ denotes the white Gaussian noise with zero mean and standard deviation σ_n. We also assume that the noise and the transmitted sequence are independent. Covert channels typically use low frequencies ($\sim$1 MHz) where the multi-path effect does not play a significant role, so it is reasonable to assume that inter-symbol interference (ISI) has almost negligible effect on the reliability of the covert transmission, and that the main impairment in the received signal is the noise. Here, we assume that the noise component contains both additive channel noise and all corruptive signals due to other activities in the system. Then, utilizing Eq. (7.10), the power spectral density of the received signal can be written as

$$S_r(f) = S_{xt}(f) + S_{nt}(f) + N_0/2, \quad (7.13)$$

where $N_0/2$ denotes the power spectral density of the additive white noise.

Obtaining the PSD of the transmitted symbols facilitates the calculation of BER. To utilize Eq. (7.11), we need to know the signal and noise powers. Since our transmitted signal experiences jitter due to the variations in the symbol position, we start by calculating PSD of the jitter noise and the signal.

Considering the variation in symbol positions, the PSD of the transmitted sequence for on-off keying (OOK) is given as [34]

7.4 Quantifying the Information Leakage of Covert Channel Software-...

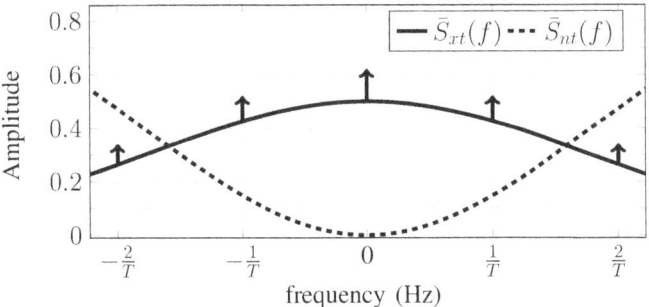

Fig. 7.7 PSD of the normalized signal and of the jitter noise due to random pulse position when $\mathcal{T} \approx 15\sigma$ [34]

$$S_y(f) = \frac{R_x[0]}{\mathcal{T}} \left(\bar{S}_{xt}(f) + \bar{S}_{nt}(f) \right)$$

$$= \frac{R_x[0]}{\mathcal{T}} \Bigg(\underbrace{\left(\frac{1}{2} + \frac{1}{2\mathcal{T}} \sum_m \delta(f - m/\mathcal{T}) \right) \Phi(f)}_{\bar{S}_{xt}(f)}$$

$$+ \underbrace{(1 - \Phi(f))}_{\bar{S}_{nt}(f)} \Bigg), \quad (7.14)$$

where $\bar{S}_{xt}(f)$ and $\bar{S}_{nt}(f)$ are the normalized power of the signal and jitter noise, respectively.

Figure 7.7 illustrates the behavior of the normalized the signal and noise power when $\mathcal{T} \approx 15\sigma$.

Since the noise power due to jitter behaves like white noise when $\Phi(f) \approx 0$, at the receiver front-end, we employ a low-pass filter whose bandwidth and magnitude are $1/2T$ and 1, respectively. The total signal power can be obtained as

$$P_{st} = \frac{R_x[0]}{\mathcal{T}} \cdot \int_{-\frac{1}{2\mathcal{T}}}^{\frac{1}{2\mathcal{T}}} \bar{S}_{xt}(f) df$$

$$= \frac{R_x[0]}{\mathcal{T}} \cdot \left(\frac{1}{2\mathcal{T}} + \frac{\sqrt{\pi}}{4\mathcal{T}} \mathrm{erf}\left(\pi\sigma/\mathcal{T}\right)/(\pi\sigma/\mathcal{T}) \right), \quad (7.15)$$

and total noise power due to jitter is equivalent to

$$P_{nt} = \frac{R_x[0]}{\mathcal{T}} \cdot \int_{-\frac{1}{2\mathcal{T}}}^{\frac{1}{2\mathcal{T}}} \bar{S}_{nt}(f) df$$

$$= \frac{R_x[0]}{\mathcal{T}} \cdot \left(\frac{1}{\mathcal{T}} - \frac{\sqrt{\pi}}{2\mathcal{T}} \text{erf}(\pi\sigma/\mathcal{T})/(\pi\sigma/\mathcal{T}) \right), \tag{7.16}$$

where erf(•) is the error function.

Now that we have the power for both jitter noise and the received signal, we can estimate the BER by using Eq. (7.11) assuming we have a channel without synchronization problems. However, this is not a realistic assumption for software-activity-based covert channels. Fortunately, after filtering the received signal on the receiver side, the jitter power becomes flat and behaves like an extra source of channel noise. Therefore, the receiver sees channel noise with power $\hat{N}_0/2 = N_0/2 + \mathcal{T}P_{nt}$. With that approximation, we can treat our communication system as a synchronized system with extra channel noise power. Hence, the BER for the system can be approximated as

$$\text{BER} = Q\left(\sqrt{\frac{P_{xt}/\mathcal{T}}{\hat{N}_0}}\right). \tag{7.17}$$

The effect of varying jitter noise on BER is illustrated in Fig. 7.8, where SNR_i is defined as

$$\text{SNR}_i = R_x[0]/(N_0/2). \tag{7.18}$$

We can observe that, as the power of additive channel noise increases, the effect of the lack of synchronization on the erroneous transfer of bits becomes negligible. Conversely, when the channel noise power is reduced, the impact of jitter noise becomes more prominent.

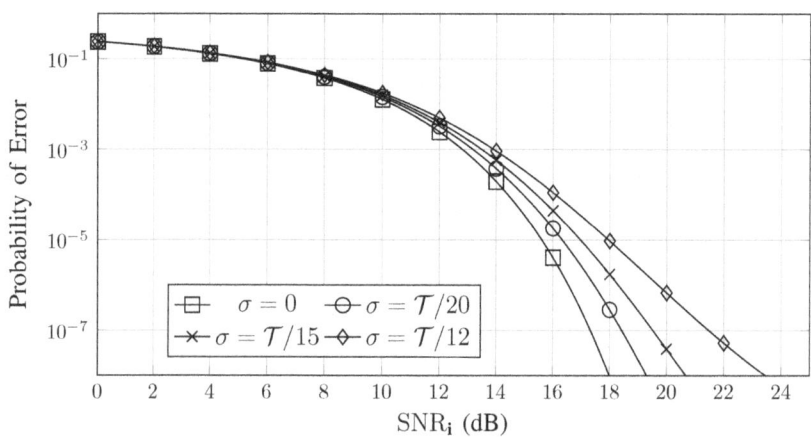

Fig. 7.8 BER for the covert wireless communication system with varying jitter noise power [34]

7.4 Quantifying the Information Leakage of Covert Channel Software-...

Finally, we investigate how the variation in symbol positions corrupts the transmitted sequence. The signal-to-jitter noise power ratio, SNR_{jitter}, at the transmitter side (caused by random pulse positioning), can be written as

$$\text{SNR}_{jitter} = \frac{P_{st}}{P_{nt}} = \frac{\frac{1}{2} + \frac{\sqrt{\pi}}{4}\text{erf}(\pi\sigma/\mathcal{T})/(\pi\sigma/\mathcal{T})}{1 - \frac{\sqrt{\pi}}{2}\text{erf}(\pi\sigma/\mathcal{T})/(\pi\sigma/\mathcal{T})}. \quad (7.19)$$

Figure 7.9 depicts how SNR_{jitter} changes as $\sigma/\mathcal{T}$ changes. Since we assume $12\sigma \leq \mathcal{T}$, we limit $\sigma/\mathcal{T}$ to be between 0 and $1/12$ (due to three-sigma rule and the assumption that the distribution of the pulse shift has non-zero probability in the region $[-\mathcal{T}/4, \mathcal{T}/4)$). As expected, as the variation in pulse position decreases, jitter noise causes less distortion in the transmitted signal.

7.4.2 Practical Considerations and Examples of Cover Side Channels

To create a practical example implementation of a covert channel, we ran a *spy* application that transmits data such that it can be received from outside the device, as described in Sect. 7.2 (and also in [30, 31]). The device is an Altera NIOS-II (soft) processor implemented on a Terasic DE1 SoC FPGA board [6]. This board is equipped with an Altera/Intel Cyclone-II FPGA chip and a variety of I/O ports (VGA, USB, Serial, etc.), and it is representative of a large class of embedded systems commonly used in the market. The spy application was written in the C programming language and was compiled using the publicly available NIOS-II compilation toolchain.

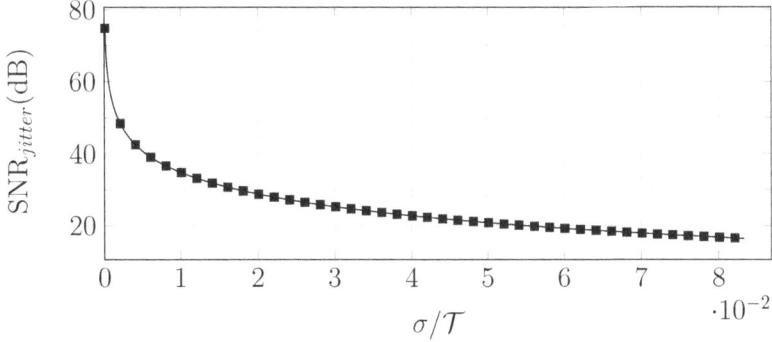

Fig. 7.9 SNR_{jitter} vs. $\sigma/\mathcal{T}$ [34]

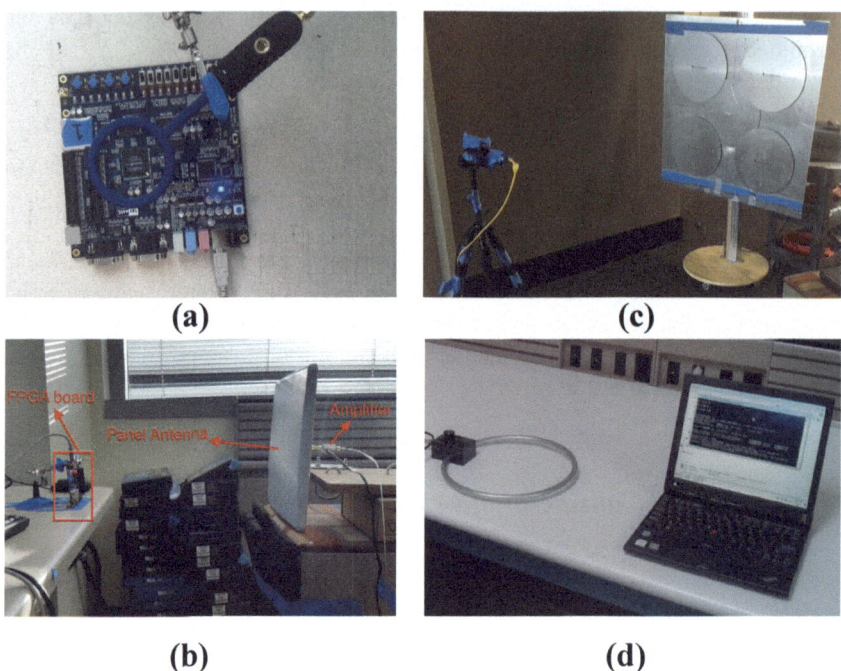

Fig. 7.10 Measurement setup for: in near-field (**a**), and at-a-distance for FPGA (**b**), OLinuXino (**c**), Laptops (**d**) [34]

To receive EM signals that are emanated by the system (and modulated by the execution of the *spy* application), we placed a magnetic probe, PBS-M [6], about 10 cm above the board so that it covers the processor area as shown in Fig. 7.10a. In these initial experiments, we placed the probe very close to the board to receive the EM signal with a high SNR. Other scenarios, e.g., where the receiver is placed away from the device, will be explored later in this chapter.

The EM signals were recorded using a spectrum analyzer (Agilent MXA N9020A). We set the sampling rate to 10 MHz, and set the spectrum analyzer's center frequency to 50 MHz (i.e., the clock frequency of the FPGA chip), and the span to 4 MHz (i.e., 2 MHz for each side-band)—our spy application creates periodic activities (i.e., A/B or A/A alternations) at 1 MHz, so the 4 MHz span allows us to pick up two harmonics in each sideband created by this periodic activity.

The received signal is shown in Fig. 7.11 when, in the A/B alternation in the spy application, activity A is a load from main memory and activity B is a load from the L1 cache. We can observe that, after AM demodulation, the received signal has the shape of a PAM signal whose pulse width fluctuates due to uncertainties in execution times of A/B activity.

To compare theoretical results with experimental results, we perform experiments where we vary A/B activity of the spy application and data rate. These

7.4 Quantifying the Information Leakage of Covert Channel Software-...

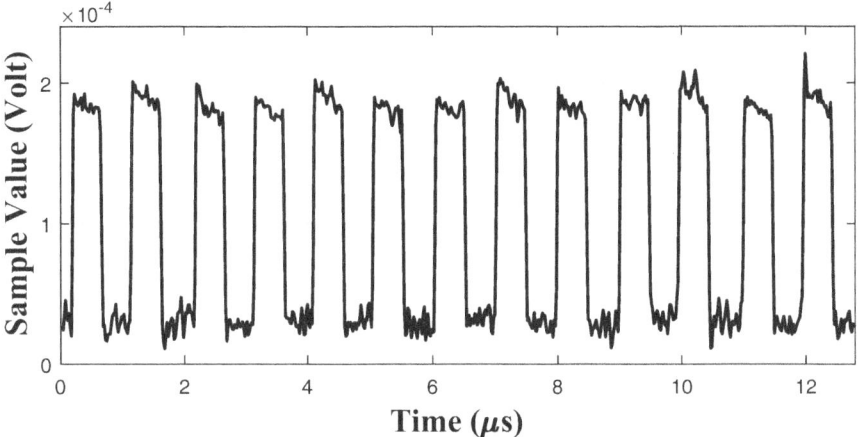

Fig. 7.11 The received baseband signal with period $\mathcal{T} = 1\,\mu s$ [34]

Table 7.1 Comparison of experimental and theoretical results in terms of BER for NIOS processor on the DE1 FPGA board [34]

Activity	Mbps	SNR (dB)	Experimental BER	Theoretical BER
ADD/MUL	4	6	0.096	0.0829
ADD/MUL	2	13	0.0013	0.0016
LDM/LDL1	4	14	0.0008	0.0009
LDM/LDL1	2	24	0	0

experiments are listed in Table 7.1, showing the A/B activity, data rate (Mbps), estimated SNRs, experimental BER, and theoretical BER for each experiment. Note that we only provide results for a few experiments here—we do not directly control the actual SNR of this communication system, so a plot that shows how the BER depends on the SNR would have only a few points and thus be unreliable (and possibly misleading). However, even these few experiments illustrate that our theoretical model is realistic for covert channels, so it can be used as a simulation tool.

Next, we discuss the jitter distributions given in Fig. 7.6, to illustrate the presence of jitter and to show why our assumption that the jitter power can be added as an extra source of white noise to calculate the BER, i.e., why our assumption in (7.17) is valid. First, we plot the PSDs of the information signal and of the jitter noise. Here, we study the scenario without memory activity in which the baseband transmitted pulses are sent with period $\mathcal{T} = 1\,\mu s$. The standard deviation of the jitter noise is calculated as $\sigma = 5.91 \times 10^{-8}$. The theoretical PSD of the transmitted signal and the jitter noise are given in Fig. 7.12a, and PSD of the filtered signal at the receiver side is given in Fig. 7.12b. We observe that the jitter noise power dominates the transmitted signal for higher data rates, so filtering the received signal helps to retrieve the information signal.

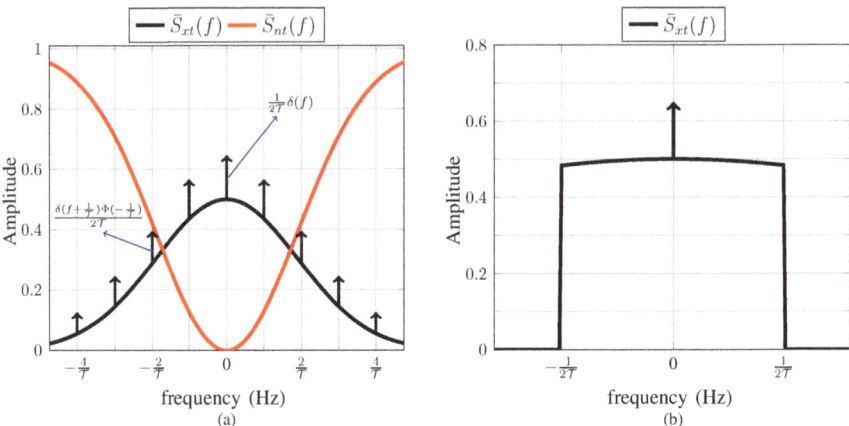

Fig. 7.12 PSD of (**a**) the transmitted signal and (**b**) its filtered version at the receiver side for the symbol without memory activity [34]

Finally, to verify our BER estimation given in (7.17), which is based on PSDs of the transmitted signal and jitter noise, we design the following experiment: We create an impulse train with period $\mathcal{T}$ and apply jitter noise by altering the location of pulses based on the normal distribution with variance σ^2. Then, we disturb the signal further by adding white noise, whose distribution can be given as $\mathcal{N}(0, N_0/2)$. The received signal is filtered with an ideal low pass filter on the receiver side and sampled with frequency $1/\mathcal{T}$. Finally, the sampled outputs are threshold to estimate the transmitted inputs. The results are shown in Fig. 7.13. Here, we plot the simulation and theoretical BER results for the cases with and without memory activity. The results confirm that simulated results agree with the theoretically derived BER, and validate the intuition that, as the jitter variance increases, BER also increases.

Please note that our A/B spy application can (in addition to PAM) also create frequency shift keying (FSK) modulation. However, the detection of FSK signals can be harder because execution time variations make it hard to disambiguate frequencies, thus increasing BER. Therefore, compared to PAM, more sophisticated receiver designs and more complex mathematical derivations are needed to achieve the same performance levels.

7.4.3 Demonstration of the Analog Covert Channel on More Complex Systems

In this section, we provide examples that show the practicality of the EM covert channel on more complex devices and with more realistic distances.

7.4 Quantifying the Information Leakage of Covert Channel Software-... 129

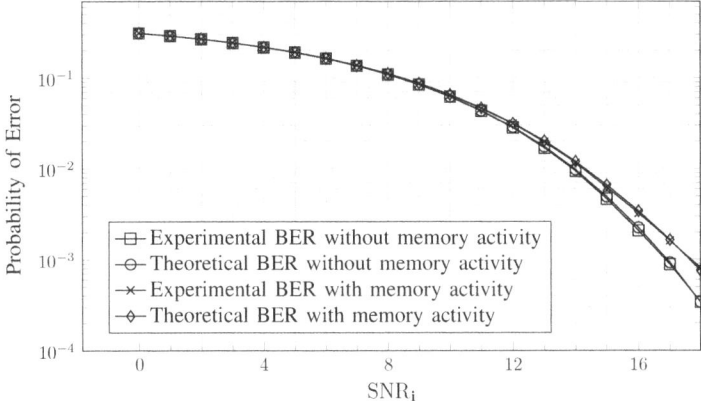

Fig. 7.13 Theoretical and experimental BER for the symbols with and without memory activity [34]

Fig. 7.14 The received signal at 1 m distance [34]

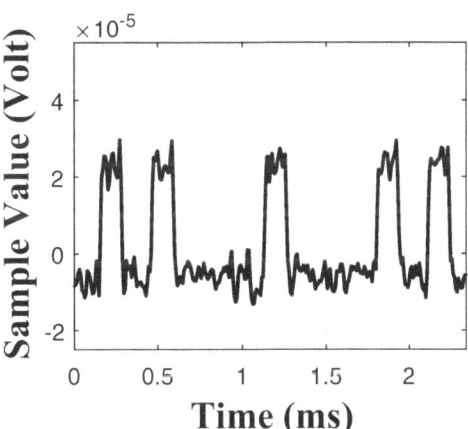

We first study the impact of distance (between the "transmitter" device and the receiver's antenna) on the bit-error-rate (BER) at the receiver. To measure the EM signals, we used a panel antenna [7]. We set the spectrum analyzer's center frequency to 2.3 GHz (i.e., the 46th harmonic of the FPGA's 50 MHz clock frequency). This frequency is chosen because it corresponds to the frequency range in which the panel antenna achieves its maximum gain. We placed the board 1 m away from the board. As shown in Fig. 7.10b. Figure 7.14 shows the received signal. This experiment confirms that the EM covert channel created by executing our spy application can be exploited at distances that make many practical scenarios feasible for an attacker.

To further study the possible range of an EM covert-channel attack and investigate its possibility on other (more complex) types of devices, we perform another experiment, this time using an embedded single-board computer called

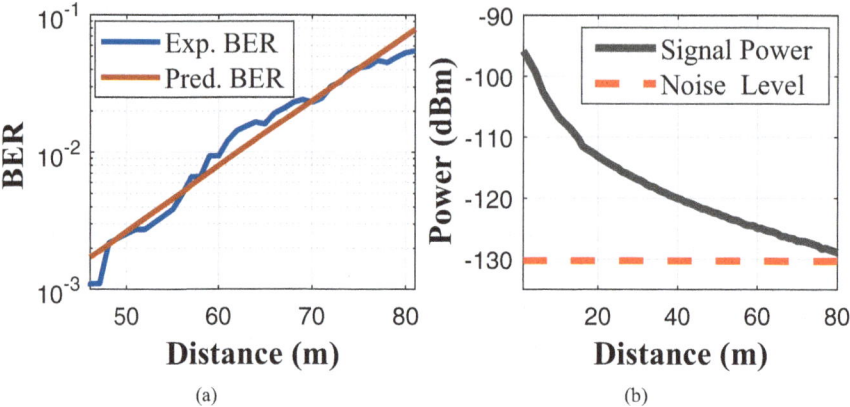

Fig. 7.15 (a) BER vs. distance, (b) The received signal power vs. distance [34]

OlinuXino [20]. This board is equipped with a more sophisticated Cortex A8 ARM core with two levels of caches, 4 MB main memory, and runs a Debian Linux operating system. OlinuXino is a representative of a popular class of single-board computers widely used in the market to control a variety of critical and commercial tasks in factory lines, hospitals, etc.

To receive the EM signals, we leveraged two different antennas: a commercially available horn antenna [1], and a high-gain custom-made disk-array antenna [12]. As before, our spy application uses A/B alternation and is written in C, but for ARM processors we compiled it with the GNU-GCC toolchain. We use the same spectrum analyzer for recording the signals, but set its center frequency set to 1 GHz (the clock frequency of the ARM core), with a span of 5 MHz, and the A/B alternation frequency of the spy application is set to 2 MHz. Please note that Fig. 7.10c shows our measurement setup, which represents a realistic indoor-environment attack scenario.

The results for BER and signal power are given in Fig. 7.15. In Fig. 7.15a, we plot the BER of measurements when the distance is more than 45 m. For closer distances, the success rate of the measurements is almost 100% (i.e., BER $< 10^{-3}$). As can be seen from this figure, the error rate (BER) linearly increases as the distance increases (as shown by the fitted line). From the results we note that covert channel can achieve a success rate more than 99.9% if the distance is less than 45 m and the signal level is at least 8 dB above the noise level of the measuring device (i.e., -130 dBm in this experiment).

Finally, to show that this EM covert channel can be created for complex computing systems such as laptops, we performed experiments on four different laptops with different processors, namely: AMD Turion X2 Ultra, Intel Core Duo T2600, Intel I7 2620M, and Intel Core 2 Extreme X9650.

In all these measurements, we use an experimental setup [3] which utilizes a magnetic loop antenna with a radius of 30 cm [11] as shown in Fig. 7.10d. The

Table 7.2 Experimental results for computer systems with distance [34]

Platform	CPU	Distance	BER
AMD turion	2.1 GHz	2.5 m	10^{-3}
Intel Core DUO	2.16 GHz	0.81 m	10^{-3}
Intel i7	2.7 GHz	1.75 m	10^{-3}
Intel Core 2	3 GHz	1.17 m	10^{-3}

center frequency for the measurements is set to 1.024 MHz and the baseband signal has 800 Hz bandwidth. The results are shown in Table 7.2. In this table, we provide the maximum distances at which we achieve a reliable communication (i.e., BER $\approx 10^{-3}$) when the transmission rate for the covert channel is 800 bits per second (bps). We observed that the signal power leaked from different platforms shows variation (depending on the packaging, board, etc.), and that affects the range of the covert channel. Compared to the previous state-of-the-art [8, 9, 16], our covert channel provides up to 5x higher data-rate and 5x lower bit-error-rate [34].

7.5 Capacity of Side Channel Created by Execution of Series of Instructions in a Computer Processor

Following the work of Millen [19], numerous papers have discussed bounds on the capacity of channels corrupted with synchronization errors [2, 4, 5, 10, 14, 26, 28], bounds on the capacity of channels corrupted with synchronization and substitution errors [17, 24, 27], or bounds on the capacity when codewords have variable length but no errors in the channel [25, 27], none of them provides the answer to how much information is "transmitted" by execution of particular sequence of instructions that do not have equal timing and are transmitted through erroneous channel.

We made the first attempts to answer this question in [29, 32], where covert channels are generated, and upper and lower leakage capacities were derived. In [31], we derived side channel leakage capacity for a discrete memoryless channel where it was assumed that each transmitted quantum of information (i.e., instruction in the code) is mutually independent but do not have equal length. Although all these papers made important steps toward assessing information leakage from side-channels, they fell short of considering the relationship among sequence of instructions, which is a result of program functionality as well as a processor pipeline depth, which impacts how much signal energy will be emanated. Recently, we derived side-channel information capacity created by execution of series of instructions (e.g., a function, a procedure, or a program) in a processor [33]. To model dependence among program instructions in a code, we used a Markov Source model, which includes the dependencies that exist in instruction sequence since each program code is written systematically to perform a specific task. The sources for channel inputs are considered as the emitted EM signals during instruction executions. To obtain the channel inputs for the proposed model, we derived a mathematical relationship between the emanated instruction signal power (ESP) as it passes through processor pipeline and total emanated signal power while running a program. This is in contrast to work in [31], where all energy emanated through side-channels is assigned to an instruction, without taking into account effect of processor pipeline depth, which significantly impacts the emanated signal. Finally, we provide experimental results to demonstrate that leakages could be severe and that a dedicated attacker could obtain important information. The presented framework considers processors as the transmitters of a communication system with multiple antennas. The antennas correspond to different pipeline stages of any processor. Moreover, inputs of the transmitter show dependency based on a Markov model which reflects the practicality of a program. Therefore, the goal here is to obtain the channel capacity of a communication system which represents the severity of the side channels.

7.6 Modeling Information Leakage from a Computer Program as a Markov Source Over a Noisy Channel

In this section, we present a Markov source model whose states are series of instructions in a pipeline [33]. We assume that channel inputs at each state are the emanated signal powers produced as combination of different instructions in a pipeline, and the channel outputs are the noise corrupted versions of the emitted signals. The reason for considering such a Markov model is that individual instructions are not independent from each other in the code, and that ordering of instructions as they pass through pipeline significantly impacts emitted signal patterns.

7.6.1 Brief Overview of Markov Model Capacity Over Noisy Channels

Channel capacity provides the limit for reliable information transmission in a communication system. Assuming Y_1^n and S_1^n represent the channel output and state sequences between $t = 1$ to $t = n$, the capacity of the Markov sources over noisy channels is defined as [13]

$$C = \max_{\substack{P_{ij} \\ (i,j) \in \mathscr{T}}} \lim_{n \to \infty} \frac{1}{n} I\left(S_1^n; Y_1^n | S_0\right), \tag{7.20}$$

where $I(\bullet)$ is the mutual information, P_{ij} is the transition probability from state i to j, and $\mathscr{T}$ is a set of valid state transitions. To maximize the overall mutual information between input and output sequences, we need to find the probability distribution of state transitions under the constraint that state transitions are only possible if $\mathcal{T}$ contains these paths. The equation given in (7.20) can be simplified further by using the chain rule, Markov, and stationary properties of the model. It has been shown in [13] that capacity can be simplified as

$$C = \max_{P_{ij}} \sum_{i,j:(i,j)\in\mathcal{T}} \mu_i P_{ij} \left[\log \frac{1}{P_{ij}} + T_{ij}\right], \quad (7.21)$$

where

$$T_{ij} = \lim_{n\to\infty} \frac{1}{n} \sum_{t=1}^{n} \left[\log \frac{P_t(i,j|Y_1^n)^{\frac{P_t(i,j|Y_1^n)}{\mu_i P_{ij}}}}{P_t(i|Y_1^n)^{\frac{P_t(i|Y_1^n)}{\mu_i}}}\right], \quad (7.22)$$

and where μ_i is the stationary probability of state i, which satisfies $\mu_i = \sum_{k\in\mathcal{S}} \mu_k P_{ki}, \forall i \in \mathcal{S}$, and $\mathcal{S}$ is the set of states. In this equation, $P_t(i|Y_1^n)$ is the probability that the state at time $t-1$ is i, and $P_t(i,j|Y_1^n)$ is the probability that the states at times $t-1$ and t are i and j respectively, given the received sequence, Y_1^n.

There is no closed form solution to the optimization problem given in (7.21) because the calculation of T_{ij} is still an open problem. However, a greedy algorithm to calculate C exists in [13]. Although that algorithm could not produce the exact results, experimental findings show that the performance gap between actual results and the algorithm's results is small.

In the following sections, we introduce our Markov Source model, obtain the channel inputs for the proposed model, and modify the expectation-maximization algorithm in [13] to quantify the side-channel information leakage.

7.6.2 Markov Source Model for Modeling Information Leakage from a Sequence of Instructions

Here, we describe a Markov source model that characterizes relationship among sequence of instructions as they pass through pipeline stages in a processor. Note that a processor pipeline is like an assembly line for instruction execution, with a sequence of activities related to computational tasks, i.e., fetching, decoding, executing, etc. [21]. We assume that channel inputs at each state are the emanated signal powers obtained as a combination of different power levels that instructions experience while passing through a pipeline, and the channel outputs are the noise-corrupted versions of the emitted signals. To include the effect of pipeline depth, states are assumed to be all possible instruction combinations because each stage performs an operation on an instruction in the queue. For example, if a pipeline has a depth of m, and the cardinality of $\mathcal{S}$ is q, the number of states will be q^m.

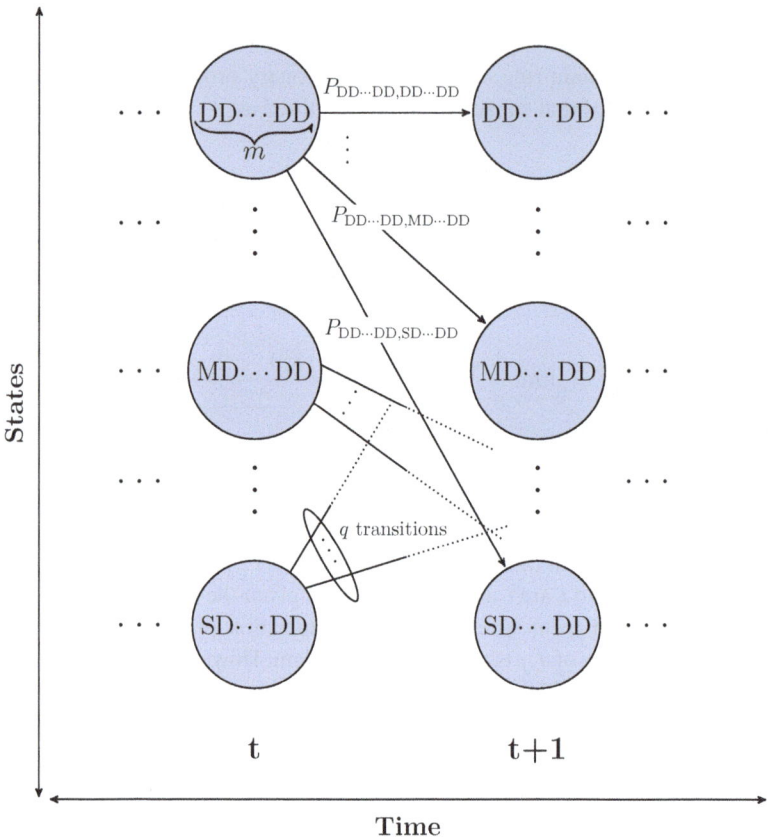

Fig. 7.16 Markov source model for the instruction execution when the pipeline depth is m, and the cardinality of the considered instruction set is three [33]

To illustrate how this Markov Model works, Fig. 7.16 shows an example of a Markov Source Model for the instruction execution when the pipeline depth is m, and the cardinality of the considered instruction set is three. In the figure, $P_{i,j}$ represents the state transition probability from state i to state j, and circles denote the states of the model. The instruction set used in the example is {D, S, M}, which corresponds to division, subtraction, and multiplication, respectively. We utilize a trellis diagram to explain the model explicitly, although transitions are time invariant, i.e., $P_{i,j}$ does not vary in time. The labels of each state is the combination of letters that represent the instructions in the pipeline. Considering these three instructions, one of the states can be labeled as "DDI$_S$DD" where $\mathbf{I}_S$ is a sequence of instructions whose length is $m - 4$. Interpretation of the state corresponding to the label is that the 1th, 2nd, ..., $m - 1$th and mth stages of the pipeline each have a D instruction.

7.7 Introducing Information Leakage Capacity for the Proposed Markov...

For each state, the number of possible paths is q, i.e., it is equal to the number of instructions in the set. For example, for the example above, there exist only three paths from each state since the instruction set contains only three elements. For example, the possible states after "DDI$_S$DD" could be "DDDI$_S$D", "MDDI$_S$D" or "SDDI$_S$D", i.e., the DDDDD instruction that was in the last stage is no longer in the pipeline, all other instructions move to the next stage in the pipeline, and a new D, M, or S instruction enters the pipeline (1st stage). Furthermore, we assume that any instruction can be followed by any other instruction. This assumption helps the model be an indecomposable channel, so that, the mutual information definition given in Eq. (7.21) is applicable.

We need to note that, by considering the Markov source model, we can successfully capture the pipeline effect because it puts constraints on the state transitions. Moreover, $P_{i,j}$ captures the frequency of the instruction orderings encountered in the program. Therefore, the capacity of the proposed model provides the worst instruction sequence distribution, i.e., the one that leaks the most information.

7.7 Introducing Information Leakage Capacity for the Proposed Markov Source Model

The capacity definition given in Eq. (7.21) is well suited for Markov source models if the states take the same amount of time. In other words, the definition is valid for the models where the transitions last equal amount of time, and the transition time is not dependent on a given state. Unfortunately, applying the same capacity definition to the proposed scheme is not appropriate because different instructions take different time to execute. Therefore, we need a capacity definition which also accounts for instruction execution times. Hence, we propose a method to quantify the information leakage, which considers both execution time of each state and the mutual information between input and output sequences.

Definition 7.1 Assuming varying execution time of instructions, maximum possible information leakage through a processor is defined as

$$C = \max_{\substack{P_{ij} \\ (i,j) \in \mathcal{T}}} \lim_{n \to \infty} \frac{I\left(S_1^n; Y_1^n \mid S_0\right)}{\sum_{i=1}^{n} \mathbf{L}(i)} \quad (7.23)$$

where $\mathbf{L}(i)$ is the length of the state executed at the ith transition.

Following the analogy between Eqs. (7.20) and (7.21), we can rearrange the equation in (7.23) as follows

$$\lim_{n\to\infty} \frac{I\left(S_1^n; Y_1^n | S_0\right)}{\sum_{i=1}^{n} \mathbf{L}(i)} = \frac{\lim_{n\to\infty} \frac{1}{n} I\left(S_1^n; Y_1^n | S_0\right)}{\lim_{n\to\infty} \frac{1}{n} \sum_{i=1}^{n} \mathbf{L}(i)} \quad (7.24)$$

$$= \frac{\sum_{i,j:(i,j)\in\mathscr{T}} \mu_i P_{ij} \left[\log \frac{1}{P_{ij}} + T_{ij}\right]}{\sum_{i\in\mathscr{S}} \mu_i L_i}, \quad (7.25)$$

where $\mathscr{S}$ is the set containing all existing states, i.e., all instruction combinations, and L_i is the execution length of the state i. Therefore, our definition can also be written as

$$C = \max_{\substack{P_{ij} \\ (i,j)\in\mathscr{T}}} \frac{\sum_{i,j:(i,j)\in\mathscr{T}} \mu_i P_{ij} \left[\log \frac{1}{P_{ij}} + T_{ij}\right]}{\sum_{i\in\mathscr{S}} \mu_i L_i}. \quad (7.26)$$

The result of this optimization provides the possible information leakage in bits per smallest number of clock cycles required to execute a state in $\mathscr{S}$ (which we call Bits/Quantum), not bits per second. The reason is that each instruction takes at least one clock cycle for any device, but clock frequencies can vary from one device to another. Since the goal is to analyze the leakage capacity at the instruction level, we provide our results in Bits/Quantum. Please note here that, even the leakage capacity of a device is small, the number of bits that a device can transmit in a second could be large. Therefore, while examining the vulnerability of any device against side channel attacks, one should combine leakage capacity with clock frequency.

7.7.1 Reducing the Size of the Markov Source Model

The main problem of the proposed Markov source model is the number of possible states and transitions. As the depth of the pipeline and the number of considered instructions increase, the number of states increases exponentially. This increase causes the iterative algorithm in [13] to be more complex. Choosing states as individual instructions will simplify the proposed scheme. For these states, the channel input signal is assigned as the emanated EM signal while executing the corresponding instruction through all pipeline stages. With this approach, the number of states increases linearly, not exponentially, as the number of instructions increases.

In Fig. 7.17, we provide an example of the state diagram when the instruction set is {D, M, S}. This model is still indecomposable based on the assumption that each instruction can follow any other instruction. Therefore, the capacity definition given in Eq. (7.25) can be used to calculate leakage capacity limits. However, this

7.7 Introducing Information Leakage Capacity for the Proposed Markov... 137

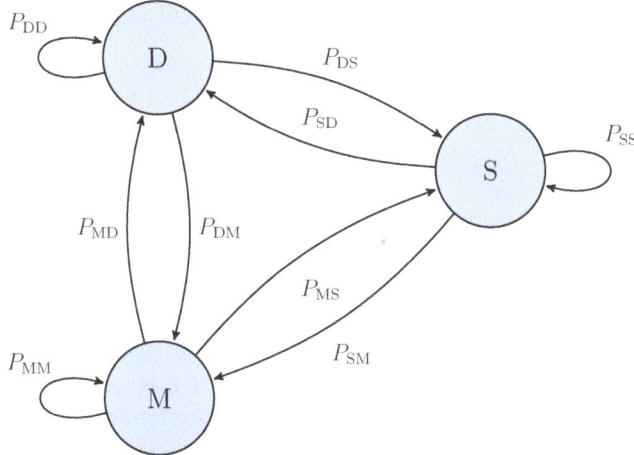

Fig. 7.17 Simplified version of Markov source model for the instruction execution when the cardinality of the considered instruction set is three [33]

definition also does not have a closed form solution, and an empirical algorithm similar to expectation-maximization (ExMa) algorithm in [13] is needed to solve the problem.

7.7.2 An Empirical Algorithm to Evaluate the Leakage Capacity

To utilize the ExMa algorithm, we have to adjust our model from the previous section to remove the execution time of the instructions from the optimization problem. To achieve this goal, we split the instructions into unit length sections, i.e., one clock cycle segments, and treat each of these segments as an individual state. To protect the overall framework and instruction sequence, we have to introduce some constraints for possible state transitions.

Let $K \in \mathscr{S}$ be a state whose length is $L_K > 1$. For the model, we divide it into L_K different states, where the states are named as K_i where $i \in \{1, \cdots, L_K\}$. Each sub-state is called:

- *Initial state* if $i = 1$, i.e., K_1,
- *Exit state* if $i = L_K$, i.e., K_{L_K},
- *Intra-state* if $i \in \{2, \cdots, L_K - 1\}$

of an instruction K. However, if the length of the instruction K equals one, we keep the instruction set unmodified. Note that the initial and exit states of K will refer to full set K for the scenario when K takes only one clock cycle. Let $\mathscr{S}_M$ and $\mathscr{T}_M$ be the set of states and state transitions, respectively, after splitting the states to have a new instruction set whose members take same amount of time. Therefore, we can rewrite (7.25) as

$$C = \max_{\substack{\mathbb{P}_{ij} \\ (i,j) \in \mathcal{T}_M}} \sum_{(i,j) \in \mathcal{T}_M} \mathfrak{u}_i \mathbb{P}_{ij} \left[\log \frac{1}{\mathbb{P}_{ij}} + \mathbb{T}_{ij} \right], \quad (7.27)$$

where $\mathbb{P}_{ij}$ refers the modified state transition probabilities, $\mathfrak{u}_i$ is the stationary distribution of the new states, and $\mathbb{T}_{ij}$ is defined as in Eq. (7.22) in the new model.

Dividing the original states into substates is not enough to protect the duality between the optimization settings given in Eqs. (7.25) and (7.27). We also have to make sure that the state transitions occur such that the instruction sequences for both settings follow the same path. For example, let L_K be equal to 2. To ensure the duality, $\mathbb{P}_{K_1 j}$ must be nonzero only if j is K_2. More formally, to guarantee the duality between the Eqs. (7.25) and (7.27), we employ constraints on transitions which only allow state transitions in the following scenarios:

R$_1$. An exit state of any instruction to an initial state of any instruction,
R$_2$. K_i to K_{i+1} of instruction K where $i \in \{1, \cdots, L_K - 1\}$.

Figure 7.18 illustrates the proposed framework. This figure is a transformed version of the Markov source model given in Fig. 7.17 based on the rules imposed by **R$_1$** and **R$_2$**. We assume that D and M take four and three times of the execution time of S, respectively. Here, M$_1$ and D$_1$ are the initial states, M$_3$ and D$_4$ are the exit states of M and D, respectively. D$_2$ and D$_3$ are the intra-states of DIV, and M$_2$

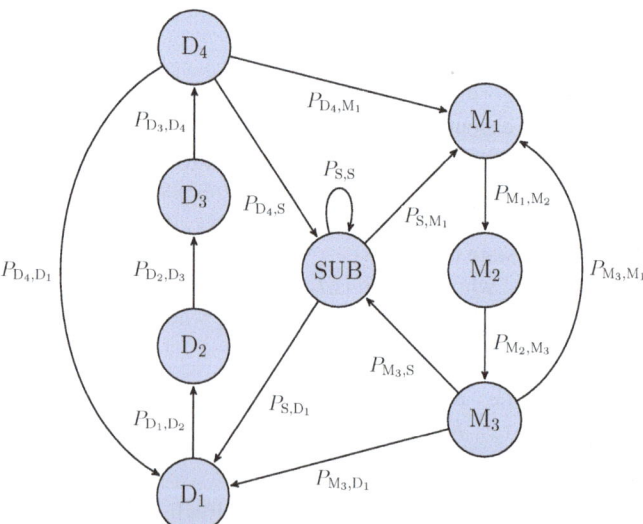

Fig. 7.18 Markov Model for the instruction execution as it goes through sub-states that take equal amount of time [33]

corresponds the intra-state of M. Note that these values are chosen arbitrarily, and only given as an illustration.

By applying the transformations introduced above, we have removed the problem of variable time of execution per instruction. The following theorem proves the models given in Sects. 7.7.1 and 7.7.2 are dual, and will lead to the same capacity results. The proof has been presented in [33].

Theorem 7.1 (Duality) *The optimization settings given in (7.25) and (7.27) are dual problems if the constraints imposed by* $\mathbf{R}_1$ *and* $\mathbf{R}_2$ *are satisfied.*

Figure 7.18 illustrates that, although we introduced some constraints on the possible state transitions, the state transition diagram is still indecomposable. Therefore, the capacity definition and corresponding iterative algorithms given in [13] can be utilized. However, to apply the algorithm, the channel inputs have to be known. In the following section, we introduce a methodology to calculate the channel input power, i.e., emitted signal power while processing an instruction through the pipeline.

7.8 Estimating Channel Input Power in the Proposed Markov Model

To obtain the channel inputs for the proposed model, we use results from Chap. 5, where a mathematical relationship between the emanated instruction power as it passes through processor pipeline and total emanated signal power while running a program is derived. Emanated signal power (ESP) in discrete time can be written as

$$\text{ESP}[\mathscr{A}_1] = \frac{\sum_{m=0}^{P_S-1} |a^P_{\mathscr{A}_1}[m]|^2 + \sum_{m=0}^{N_I-1} |a_{\mathscr{A}_1}[m]|^2}{R/T_s}. \quad (7.28)$$

and the following theorem gives the relationship between the total emanated signal power and the instruction power (See Chap. 5 for more details).

Theorem 7.2 (ESP) *Let* $\mathscr{P}_{\mathscr{A}_1}(f_{\text{alt}})$ *be the normalized emanated power which is defined as*

$$\mathscr{P}_{\mathscr{A}_1}(f_{\text{alt}}) = \mathscr{P}_{\mathscr{A}_1}(f_{\text{alt}}) - \mathscr{P}_{\text{NOP}}(f_{\text{alt}}), \quad (7.29)$$

where $\mathscr{P}_{\text{NOP}}(f_{\text{alt}})$ *is the measured emanated power when both for-loops of the code are populated with NOP. The mathematical relationship between* $\text{ESP}[\mathscr{A}_1]$ *and* $\mathscr{P}_{\mathscr{A}_1}(f_{\text{alt}})$ *while running the activity* $\mathscr{A}_1$ *in the first for-loop can be written as:*

$$\text{ESP}[\mathscr{A}_1] = \left(\frac{\pi}{2}\right)^2 \frac{\mathscr{P}_{\mathscr{A}_1}(f_{\text{alt}}) \cdot N_L}{(N_I + P_S) \cdot f_{\text{alt}} \cdot n_{inst}}. \quad (7.30)$$

Please note that the crucial assumption here is that the number of inserted NOPs must be equal to the number of clock cycles $\mathscr{A}_1$ takes to execute. The proof establishing the relationship between ESP and emanated signal power depends on this assumption. This relation provides a simple tool to obtain possible emanated signal power resulting from single instruction execution through pipeline stages.

7.9 Practical Use of Information Leakage Capacity

In this section, we provide experimental results for emanated signal power of each instruction, and evaluate leakage capacity of various computer platforms.

The experimental setup is shown in Fig. 7.19. We used a spectrum analyzer (Agilent MXA N9020A), and magnetic loop probe (Aronia H field probe PBS-H3) for FPGA board and a magnetic loop antenna (AOR LA400) for other devices. We performed our measurements by setting the alternation frequency, f_{alt}, to 80 kHz. We keep the distance as close as possible to the processor since our goal is to reveal the input powers of the transmitter, i.e., ESP. The activities used in this section correspond to x86 instructions given in Table 7.3.

Fig. 7.19 Measurement setups used in the experiments [33]

Table 7.3 x86 instructions for our setup [33]

	Instruction	Description
LDM	mov eax,[esi]	Load from main memory
STM	mov [esi],0xFFFFFFFF	Store to main memory
LDL2	mov eax,[esi]	Load from L2 cache
STL2	mov [esi],0xFFFFFFFF	Store to L2 cache
LDL1	mov eax,[esi]	Load from L1 cache
STL1	mov [esi],0xFFFFFFFF	Store to L1 cache
ADD	add eax,173	Add imm to reg
SUB	sub eax,173	Sub imm from reg
MUL	imul eax,173	Integer multiplication
DIV	idiv eax	Integer division
NOP		No operation

7.9 Practical Use of Information Leakage Capacity

To obtain the experimental results, the steps we follow are:

- Run the program given in Fig. 7.1 to measure the available total signal power at the alternation frequency.
- Calculate ESP of each instruction for all available devices based on the equation given in equation (5.48).
- Transform the Markov Chain of instructions, and define the new constraints for the new model in terms of allowable paths as in Sect. 7.7.2.
- Define the signal to noise ratio (SNR) as:

$$\text{SNR} = \frac{\sum_{i \in \mathscr{S}} (\text{ESP}[i])^2}{|\mathscr{S}| \times N_0/2}, \tag{7.31}$$

where $|\mathscr{S}|$ is the cardinality of instruction set $\mathscr{S}$.

- For a given SNR, run the ExMa algorithm [13] to obtain the stationary probabilities of each sub-state and corresponding leakage capacities.
- If the stationary probability of instructions is required, solve the following equations

$$\mu_i = \text{L} \times \text{u}_1^i, \quad \forall i \in \mathscr{S}, \tag{7.32}$$

where μ_i is the stationary probability of ith instruction for the original case, and u_1^i is the initial sub-state of ith instruction for the transformed scenario, and L is a constant that can be written as

$$\text{L} = \left(\sum_{k \in \mathscr{S}} \text{u}_1^k \right)^{-1}. \tag{7.33}$$

- Define "Quantum" as the ratio between number of required clock cycles to execute an instruction and minimum number of clock cycles to execute at least one instruction.

Please note that for some experiments, Quantum is equivalent to a clock cycle, but for some experiments, it can correspond to a couple of clock cycles. Additionally, abbreviations used in this section can be listed as follows:

- C_P: Capacity in Bits/Quantum obtained with the proposed scheme.
- C_0: Capacity in Bits/Instruction obtained by assuming execution time of all instruction takes only one clock cycle and using capacity definition given in (7.20). We also assume that the optimal stationary distribution for this capacity definition is denoted as μ_0.

- C_N: Capacity in Bits/Quantum which is calculated as

$$C_N = \frac{C_0}{\sum_{i \in \mathscr{S}} \mu_0[i] L_i}. \tag{7.34}$$

This capacity definition maps C_0 into Bits/Quantum for a fair comparison.
- C_∞: Capacity in Bits/Quantum obtained by setting SNR $= \infty$ and exploiting the proposed scheme to obtain the maximum possible leakage.

7.9.1 Leakage Capacity for an FPGA-Based Processor

This section presents the experimental results and leakage capacity for a NIOS processor on a DE1 FPGA board. The ESP and corresponding execution length of each instruction are provided in Table 7.4. Please note that length of an instruction means total execution time of each instruction in Quantums.

In Fig. 7.20, we plot the leakage capacity for FPGA as a function of SNR. We observe that C_0 exceeds C_P because C_0 considers that each instruction takes only one clock cycle. However, if we normalize C_0 to obtain C_N, we can observe that

Table 7.4 ESP values (in zJ) for DE1 FPGA board [33]

	LDM	LDL1	DIV	ADD	SUB	MUL
ESP	139.38	69.98	87.60	0.32	6.10	55.14
Length	7	4	5	1	1	4

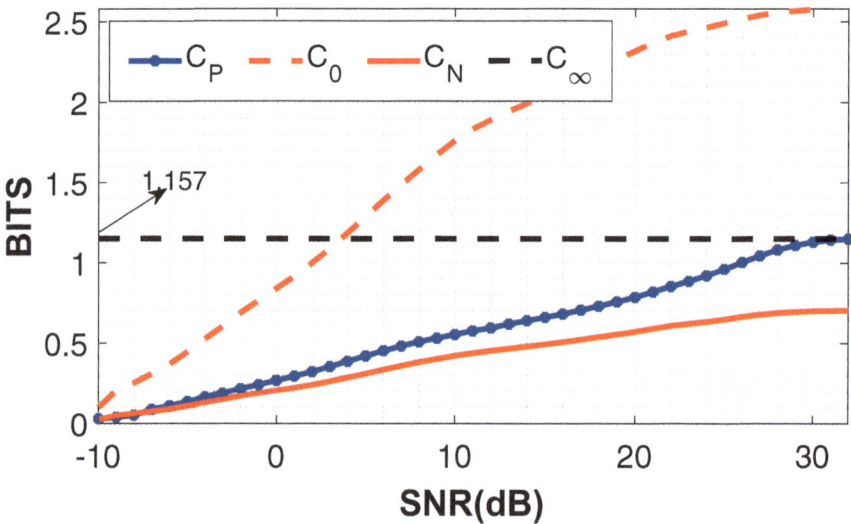

Fig. 7.20 Leakage capacity for NIOS processor on the DE1 FPGA [33]

7.9 Practical Use of Information Leakage Capacity

applying traditional Shannon theory underestimates available leakage capacity, so our leakage capacity estimation C_P is needed to establish relationship between sequence of instructions as they pass through pipeline and leakage capacity.

Additionally, we observe that leakage capacity for SNR = 59.96 dB in [31] was 1.14 Bits/Quantum. However, the method in [31] does not allow for capacity calculation as a function of SNR. On the other hand, our new method has a higher estimated leakage capacity of 1.157 Bits/Quantum when SNR is around 30 dB. This result indicates that considering the pipeline depth and the dependence between instructions, which are not included in [31], more realistically estimates leakage capacity. We also note that the leakage capacity is high even for low SNR regimes, allowing for transmission of thousands of bits per second because the clock frequencies of the current devices are high. Therefore, software and hardware designers need to consider side-channels and devise countermeasures to decrease side-channel leakages as much as possible.

7.9.2 Experimental Results and Leakage Capacity for AMD Turion X2 Laptop

This section provides the leakage capacity for a laptop with AMD Turion X2. It has 64 KB 2 way L1 Cache and 1024 KB 16 way L2 Cache. ESP values and execution lengths are given in Table 7.5.

We need to note here that LDL1, ADD, and SUB are not included into our analysis because ESP values and execution lengths of these instructions are almost equal to STL1. Therefore, including these instructions does not affect overall leakage capacity. However, if we consider STL1, LDL1, ADD, and SUB as a sub-instruction set whose members are almost identical, STL1 could be thought of as a representative of this set.

We observe that the deviation of the execution length of instructions is much larger compared to FPGA. The effect of having such a deviation can be seen from Fig. 7.21 where the gap between C_0 and C_P is significantly larger. Additionally, the leakage capacity given in [31] is 0.97 Bits/Quantum when SNR is 23.78 dB, but our new leakage capacity C_P shows that the leakage can be up to 1.36 Bits/Quantum for the same SNR region. This result indicates that all signals emanated from all stages of a pipeline carry some information and, therefore, ignoring these signals can underestimate the leakage.

Table 7.5 ESP values (in zJ) for AMD Turion X2 Laptop [33]

	LDL2	LDM	STM	STL2	STL1	MUL	DIV
ESP	150.08	84.66	64.74	188.17	0.49	0.21	7.26
Length	1	26	30	3	1	1	8

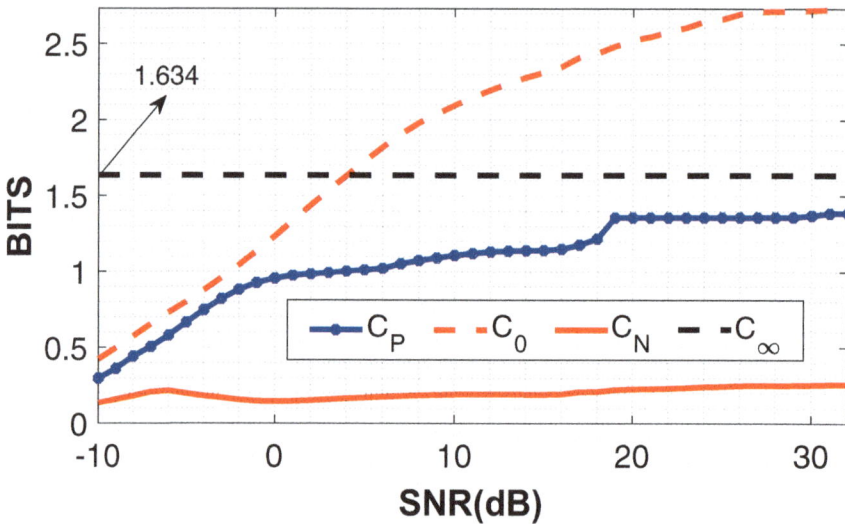

Fig. 7.21 Leakage capacity for AMD turion X2 laptop [33]

The results also show that the capacity of the laptop is moderately high even for low SNR regimes. For example, we observe that the leakage capacity of this system is approximately 1 Bits/Quantum around 0 dB SNR. Unfortunately, if the attacker is in very close proximity, and has the ideal decoder to reveal the secret information, C_P could raise up to 1.634 Bits/Quantum, which corresponds to $1.634 \cdot 10^9$ bits/second for a processor with 1 GHz clock and all instructions taking one clock cycle. We also observe that C_P could not achieve the data rate of C_∞ in the given SNR regime. For C_P to achieve maximum rate, it requires about 57 dB SNR. However, for the consistency among figures, we keep the considered SNR regime the same for each plot.

7.9.3 Experimental Results and Leakage Capacity for Core 2 DUO Laptop

In this section, we provide the results for Core 2 DUO laptop. It has 1.8 GHz CPU clock, 32 KB 8 way L1 and 4096 KB 16 way L2 caches. ESP values and lengths of instructions are given in Table 7.6. As was the case for the AMD laptop, the deviation of the instruction length is large, which causes the capacity gap between the proposed and Shannon based methods to be larger.

We do not consider the results for STL1, SUB and ADD because the lengths and ESP values of these instructions are almost same with LDL1. For this device, we assume that LDL1, STL1, SUB and ADD form a sub-instruction set, and LDL1 as the representative of this set. We observe that C_P can be up to 1.634

7.9 Practical Use of Information Leakage Capacity   145

Table 7.6 ESP values (in zJ) for Core 2 DUO Laptop [33]

	STL2	LDM	STM	LDL2	LDL1	MUL	DIV
ESP	422.16	181.58	79.94	320.48	0.75	0.06	7.02
Length	1	26	31	3	1	1	8

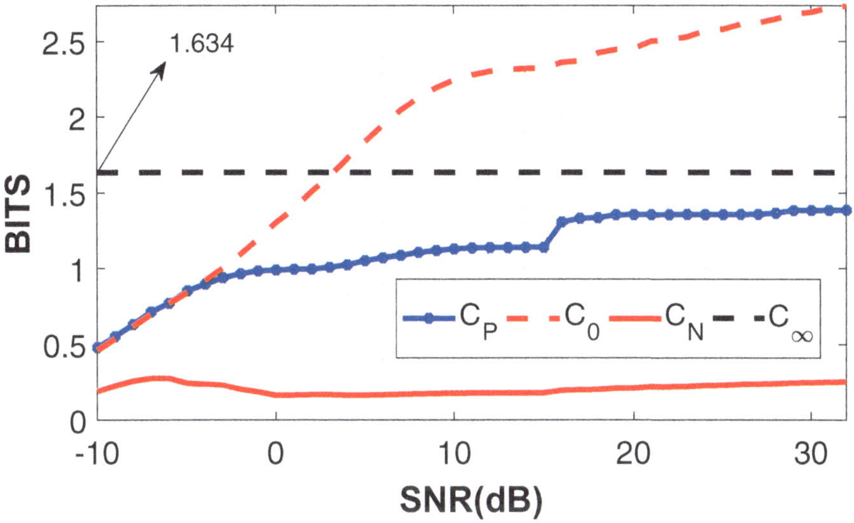

Fig. 7.22 Leakage capacity for core 2 DUO laptop [33]

Bits/Quantum if the attacker can find a way to capture emanated signals with high SNR. Furthermore, at 23.82 dB SNR, C_P is 1.36 Bits/Quantum, again higher then 1.09 Bits/Quantum capacity predicted in [31]. The difference between these results reveals the importance of considering both pipeline depth and ordering of instructions.

We also observe that the required SNR for C_P to achieve C_∞ must be at least 56 dB. However, with a moderate gain antenna and proximity to the laptop, the attacker can steal sensitive information since the leakage capacity is 0.5 Bits/Quantum even when SNR is around -10 dB. Considering the clock frequency of this processor, the side channel can have a transmission rate of thousand of bits per second under ideal circumstances (Fig. 7.22).

7.9.4 Experimental Results and Leakage Capacity for Core I7 Laptop

The last example we provide is for Core I7 laptop which has 3.4 GHz CPU clock with 64 KB 2 way L1 Data and 1024 KB 16 way L2 caches. Table 7.7 provides ESP and execution length of each instruction. The first observation here is that the

Table 7.7 ESP values (in aJ) for Core I7 Laptop [33]

	LDL2	LDM	STM	STL2	SUB	STL1	ADD	MUL	DIV
ESP	1.03	1.38	1.23	0.56	0.05	0.09	0.08	0.06	0.54
Length	1	12	15	4	1	1	1	1	8

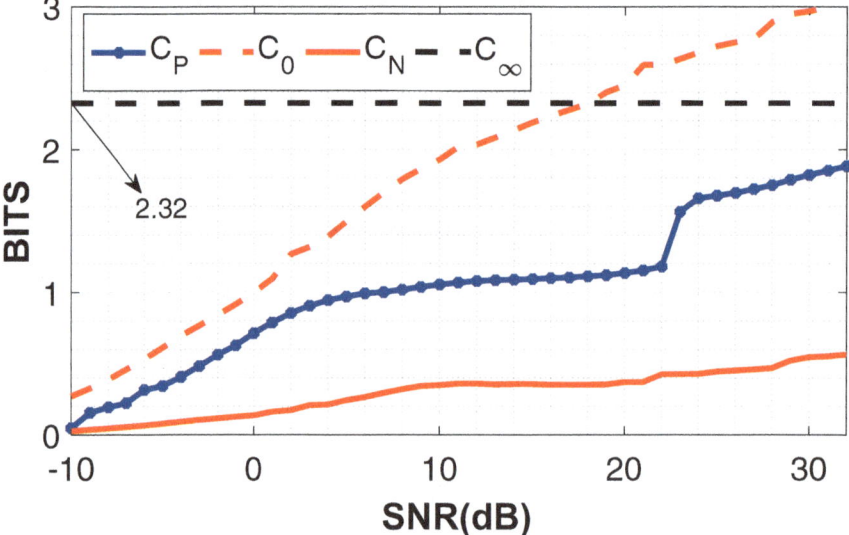

Fig. 7.23 Leakage capacity for core I7 laptop [33]

deviation of the execution length of instructions is not as large as the other laptops, which causes the gap between C_P and C_0 results to decrease as given in Fig. 7.23.

We observe that LDL1 and SUB have approximately same ESP. Therefore, SUB is considered as the representative of the group of these instructions. For ideal scenarios, C_P can go up to 2.32 Bits/Quantum. To achieve this rate, the setup must ensure at least 47 dB SNR. In addition, when SNR is 23.84 dB, the leakage capacity with the model in [31] is 0.72 Bits/Quantum, although it is obtained as 1.65 Bits/Quantum with the proposed model. Hence, including both pipeline depth and dependencies between instructions helps better quantification of leakage capacity.

Also, for the low SNR scenarios, C_P is high enough, i.e., 0.7 Bits/Quantum around 0 dB, which is relatively high given the clock frequency of the laptop.

7.9.5 Utilizing the Capacity Framework for Security Assessment

The leakage capacity definitions from this chapter provide the maximum leakage amount that any EM side/covert channel can achieve on a given device. This capacity can help designers to predict possible vulnerabilities of their products

7.9 Practical Use of Information Leakage Capacity

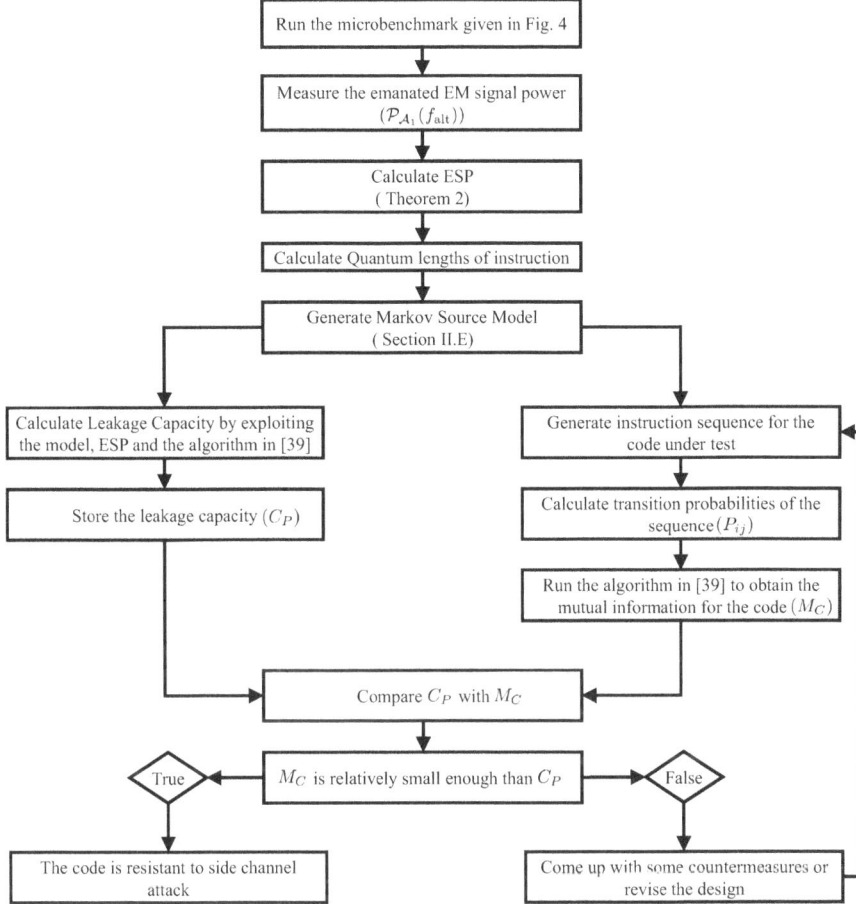

Fig. 7.24 The methodology to assess information leakage [33]

in the design stage and provide the opportunity to design countermeasures, or to redesign their systems to prevent possible side-channel attacks. Compared to the evaluation method based on success rate, which quantifies accurate retrieval rate of an attack's target (i.e., secret key bit estimation), leakage capacity defines the maximum information leakage through side channels without specifying the attack itself. Therefore, it provides a universal upper bound for EM side channels. This section provides a recipe to check whether the considered system is secure enough against side channel attacks, and explains steps to justify why they are required. The procedure for assessment is given in Fig. 7.24, and can be explained as follows:

- The first step is to collect emanated EM signal power available to an attacker while executing an instruction. Considering the clock frequency of modern computer systems, measuring single-instruction power could be problematic because of synchronization, complex pipeline structure, etc. To handle these problems, the A/B alternation code (see Fig. 7.1) is run to obtain both $\mathscr{P}_{\mathscr{A}_1}(f_{\text{alt}})$ and $\mathscr{P}_{\text{NOP}}(f_{\text{alt}})$ where $\mathscr{P}_i(f_{\text{alt}})$ is the total emanated signal power when instruction i and NOP are inserted into the first and second inner-for-loops, respectively.

- The measurements to obtain $\mathscr{P}_i(f_{\text{alt}})$ are done from near-field because the goal is to capture all emanated signal as much as possible. This approach helps to have close empirical results for signal power because actual emanated instruction power is not available. Then, ESP of each considered instruction is calculated based on the formula given in (12).
- Because of the functionality of a program, a script, etc., and the complex pipeline structure of modern computer systems, instructions shows dependency to each other. To consider the dependency among instructions, a Markov Model is created.
- The next step is to calculate the limit for information leakage using ExMa algorithm in [13]. For this algorithm, the channel inputs from each source must last the same amount of time, while instructions can take different numbers of clock cycles to execute. Therefore, the Quantum length of each instruction has to be determined as described earlier in this chapter.
- After having the execution length and utilizing the duality given in Theorem 7.1 from Chap. 7, the next step is to apply the transformation in Sect. 7.7.2. This transformation makes sure that each channel input takes same amount of time so that the ExMa algorithm in [13] can be utilized to calculate the leakage capacity for a targeted SNR regime.
- The result of the algorithm provides the leakage capacity C_P. We use this number later as the baseline to compare with the leakages of designs to understand the relative resistance of them against any possible side channel attack.
- To find the leakage for a specific code, program, design, etc., the number of transitions from ith to jth instructions is counted. These numbers are normalized to calculate P_{ij} for the inspected source code.
- With the state transition probabilities, P_{ij}, the available mutual information for the test code can be computed using the ExMa [13] once, without updating the state transition probabilities. The mutual information obtained as a result of the algorithm is denoted as M_C.
- As the last step, we compare C_P with M_C. If M_C is much smaller then C_P, and very close to zero, the designer can conclude that the code is not very leaky, at least compared to the worst-case leakage for that hardware platform. Otherwise, a new design or some countermeasures, i.e., shielding, etc., has to be considered. These steps may need to be repeated until achieving $M_C \ll C_P$.

Please note that the procedure given here does not specify the attack methodology, but provides the worst case scenario for a victim in terms of information leakage, even if the attack that can achieve the capacity limits does not exist yet. This lets designers prevent/mitigate future (still unknown at design time) attacks.

References

1. AH-118. Double Ridge Horn Antenna. https://www.com-power.com/ah118_horn_antenna.html.
2. Ross J Anderson and Fabien AP Petitcolas. On the limits of steganography. *IEEE Journal on selected areas in communications*, 16(4):474–481, 1998.
3. Robert Callan, Nina Popovic, Alenka Zajic, and Milos Prvulovic. A new approach for measuring electromagnetic side-channel energy available to the attacker in modern processor-memory systems. In *2015 9th European Conference on Antennas and Propagation (EuCAP)*, pages 1–5. IEEE, 2015.
4. Valentino Crespi, George Cybenko, and Annarita Giani. Engineering statistical behaviors for attacking and defending covert channels. *IEEE Journal of Selected Topics in Signal Processing*, 7(1):124–136, 2013.
5. M.C. Davey and D.J.C. MacKay. Reliable communication over channels with insertions, deletions, and substitutions. *Information Theory, IEEE Transactions on*, 47(2):687–698, Feb 2001.
6. DE1 FPGA on NIOS Processor. https://www.terasic.com.tw/cgi-bin/page/archive.pl?Language=English&CategoryNo=53&No=83&PartNo=2.
7. Dual Polarized Panel Antenna. http://www.l-com.com/wireless-antenna-24-ghz-16-dbi-dual-polarized-panel-antenna-n-female-connectors.
8. Mordechai Guri, Assaf Kachlon, Ofer Hasson, Gabi Kedma, Yisroel Mirsky, and Yuval Elovici. Gsmem: Data exfiltration from air-gapped computers over gsm frequencies. In *24th USENIX Security Symposium (USENIX Security 15)*, pages 849–864, Washington, D.C., August 2015. USENIX Association.
9. Mordechai Guri, Matan Monitz, and Yuval Elovici. Usbee: Air-gap covert-channel via electromagnetic emission from USB. *CoRR*, abs/1608.08397, 2016.
10. Jun Hu, T.M. Duman, M.F. Erden, and A. Kavcic. Achievable information rates for channels with insertions, deletions, and intersymbol interference with i.i.d. inputs. *Communications, IEEE Transactions on*, 58(4):1102–1111, April 2010.
11. Universal Radio Inc. Aor la390 wideband loop antenna. https://www.universal-radio.com/catalog/sw_ant/2320.html, 2014 (accessed Feb., 2019).
12. Prateek Juyal, Sinan Adibelli, Nader Sehatbakhsh, and Alenka Zajic. A directive antenna based on conducting disks for detecting unintentional em emissions at large distances. *IEEE Transactions on Antennas and Propagation*, 66(12):6751–6761, 2018.
13. A Kavcic. On the capacity of Markov sources over noisy channels. In *2009 IEEE Global Telecommunications Conference (GLOBECOM)*, volume 5, pages 2997–3001, 2001.
14. A. Kirsch and E. Drinea. Directly lower bounding the information capacity for channels with i.i.d. deletions and duplications. *Information Theory, IEEE Transactions on*, 56(1):86–102, Jan 2010.
15. Butler W. Lampson. A note on the confinement problem. *Commun. ACM*, 16(10):613–615, October 1973.
16. Ramya Jayaram Masti, Devendra Rai, Aanjhan Ranganathan, Christian Müller, Lothar Thiele, and Srdjan Capkun. Thermal covert channels on multi-core platforms. In *24th USENIX Security Symposium (USENIX Security 15)*, pages 865–880, Washington, D.C., 2015. USENIX Association.
17. H. Mercier, Vahid Tarokh, and F. Labeau. Bounds on the capacity of discrete memoryless channels corrupted by synchronization and substitution errors. *Information Theory, IEEE Transactions on*, 58(7):4306–4330, July 2012.
18. J. Millen. 20 years of covert channel modeling and analysis. In *Security and Privacy, 1999. Proceedings of the 1999 IEEE Symposium on*, pages 113–114, 1999.
19. Jonathan K. Millen. Covert channel capacity. In *Security and Privacy, 1987 IEEE Symposium on*, pages 60–60, April 1987.

20. OlinuXino. https://www.olimex.com/Products/OLinuXino/A13/A13-OLinuXino/open-source-hardware.
21. David A Patterson and John L Hennessy. *Computer Organization and Design MIPS Edition: The Hardware/Software Interface*. Newnes, 2013.
22. J.G. Proakis. *Digital Communications*. McGraw-Hill Series in Electrical and Computer Engineering. Computer Engineering. McGraw-Hill, 2001.
23. Milos Prvulovic and Alenka Zajic. Rf emanations from a laptop, 2012. http://youtu.be/ldXHd3xJWw8.
24. M. Rahmati and T.M. Duman. Bounds on the capacity of random insertion and deletion-additive noise channels. *Information Theory, IEEE Transactions on*, 59(9):5534–5546, Sept 2013.
25. Claude Elwood Shannon. A mathematical theory of communication. *Bell system technical journal*, 27(3):379–423, 1948.
26. R. Venkataramanan, S. Tatikonda, and K. Ramchandran. Achievable rates for channels with deletions and insertions. In *Information Theory Proceedings (ISIT), 2011 IEEE International Symposium on*, pages 346–350, July 2011.
27. Sergio Verdú and Shlomo Shamai. Variable-rate channel capacity. *IEEE Transactions on Information Theory*, 56(6):2651–2667, 2010.
28. Zhenghong Wang and R.B. Lee. Capacity estimation of non-synchronous covert channels. In *Distributed Computing Systems Workshops, 2005. 25th IEEE International Conference on*, pages 170–176, June 2005.
29. Baki Yilmaz, A Zajic, and Milos Prvulovic. Modelling jitter in wireless channel created by processor-memory activity. In *IEEE International Conference on Acoustics, Speech and Signal Processing, ICASSP 2018*, pages 2037–2041, 04 2018.
30. Baki Berkay Yilmaz, Robert Callan, Milos Prvulovic, and Alenka Zajic. Quantifying information leakage in a processor caused by the execution of instructions. In *Military Communications Conference (MILCOM), MILCOM 2017-2017 IEEE*, pages 255–260. IEEE, 2017.
31. Baki Berkay Yilmaz, Robert Callan, Alenka Zajic, and Milos Prvulovic. Capacity of the em covert/side-channel created by the execution of instructions in a processor. *IEEE Transactions on Information Forensics and Security*, 13(3):605–620, 2018.
32. Baki Berkay Yilmaz, Milos Prvulovic, and Alenka Zajic. Capacity of deliberate side channels created by software activities. In *Military Communications Conference (MILCOM), MILCOM 2018-2018 IEEE*. IEEE, 2018.
33. Baki Berkay Yilmaz, Milos Prvulovic, and Alenka Zajic. Electromagnetic side channel information leakage created by execution of series of instructions in a computer processor. *IEEE Transactions on Information Forensics and Security*, 15:776–789, 2020.
34. Baki Berkay Yilmaz, Nader Sehatbakhsh, Alenka Zajic, and Milos Prvulovic. Communication model and capacity limits of covert channels created by software activities. *IEEE Transactions on Information Forensics and Security*, 15:1891–1904, 2020.
35. A. Zajic and M. Prvulovic. Experimental demonstration of electromagnetic information leakage from modern processor-memory systems. *Electromagnetic Compatibility, IEEE Transactions on*, 56(4):885–893, Aug 2014.

Chapter 8
Using Analog Side-Channels for Malware Detection

8.1 Introduction

"Smart" devices, which add computational "smarts" to various devices that play a role in the physical world, are proliferating in both number and importance. These embedded systems, Cyber-Physical Systems (CPS), Internet of Things (IoT) devices, Programmable Logic Controllers (PLC), etc., are expected to be a USD 6.2 trillion market globally by 2025—this represents 8% of the entire world's GDP in 2016. Most of that is expected to be in healthcare (USD 2.5 trillion) and manufacturing (USD 2.3 trillion) [15, 20]. While "smart" devices can provide many benefits for industries and individuals, they also, unfortunately, provide new opportunities for cyber-attacks.

Computational components of cameras, cars, industrial PLCs, critical infrastructure such as the power grid (power distribution, nuclear and other power-plants, etc.), hospitals and embedded medical devices, etc. represent a rich pool of targets for cyber-attackers, and many of these targets have already been attacked (e.g., DDoS attacks [51] DNS services via Mirai malware-infected CPS, *Persiarai* [43], etc.).

Because these devices use customized and diverse hardware and software, they may not be upgraded or updated as often as general-purpose systems, and software updates are even less frequent for devices that are sensitive enough to be subject to extensive verification and/or regulatory approval requirements. This makes embedded, CPS, and IoT systems difficult to keep up-to-date with respect to the ever-evolving landscape of possible vulnerabilities and threats [52]. Furthermore, existing techniques for malware detection, such as those based on scanning for malware signatures [17], sandboxing [47], hardware-support [16, 33, 34], machine learning [45], and dynamic analysis [21], impose significant computational and cost overheads, so they are difficult to adapt to devices that often have severe performance, resource, power, and cost constraints. Moreover, many of these devices use simplified hardware and software, without virtualization, process isolation, user/system privilege level distinctions, and other features that are now common in high-performance computer systems, and that makes it easier for a successful attack to obtain full access to the device, which then allows the attacker to disable or even coopt any security-monitoring functionality that is present within the target system.

Analog side channels can be used to address the problem of detecting malware in a system that, if compromised, can no longer be trusted to truthfully implement, or even participate, in malware detection. This chapter discusses various signal processing approaches and techniques for malware detection using analog side channels. These techniques allow a device to be monitored from the outside the device, without using any hardware or software on the monitored device itself. The analog-side–channel-based methods that will be discussed in this chapter are designed to detect deviations from normal execution (as opposed to detecting specific malware signatures or behaviors), so they allow detection of zero-day (i.e., never previously seen) intrusions, which is not possible with most other methods. Our discussion begins with methods that rely on frequency-domain features of the signal, and then focuses on time-domain analysis methods and discuss the tradeoffs between time- and frequency-domain methods.

8.2 Frequency-Domain Analysis of Analog Side Channel Signals for Malware Detection

In Chapter 4, we have discussed how program activity gets modulated onto analog signals created by the processor clock or any other periodic activity within the processor, or even within the system. Then, in Chapter 5, we have discussed how these periodic activities create spectral components in the frequency domain, and also how to relate emanated electromagnetic energy to program execution. Now we will use this relationship between program activity and the frequency-domain properties of analog side-channel signals to monitor the system and detect when it begins executing malware.

8.2.1 Periodic Activities and Their Spectral Components

First, we notice that programs spend most of their time in loops! This is an important observation because loops are periodic activities, i.e., they repeat the same activity over time, as illustrated in Fig. 8.1. For example, two loops that have the same code, but that execute different numbers of iterations, would have the same repeating (per-iteration) signal in the time domain but the number of those repetitions would be different. On the other hand, two loops with different code structure, even if they have the same number of iterations, would have a

8.2 Frequency-Domain Analysis of Analog Side Channel Signals for Malware...

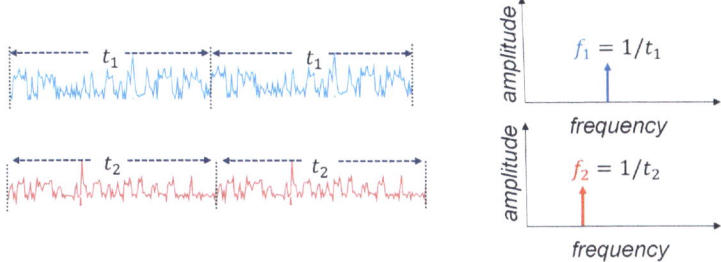

Fig. 8.1 Side channel of two periodic program activities (e.g., loops) with different code within the loop and corresponding spectral components

Fig. 8.2 Spectrogram of three loops executing one after the other

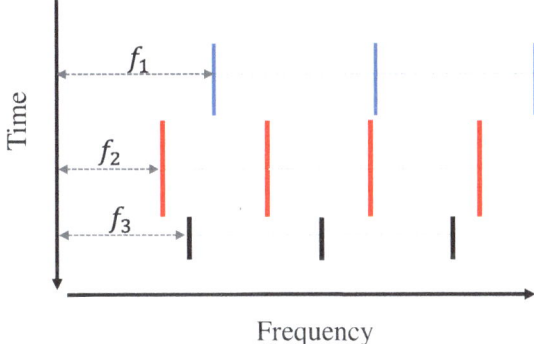

different per-iteration signal pattern, usually in both in shape and length, as shown in Fig. 8.1. The time-domain signal that corresponds to program execution can also be represented in the frequency domain. A loop is a periodic activity, i.e., can be modeled as imperfect square pulse, so in the frequency domain a loop creates spectral lines at odd multiples of the loop's fundamental frequency, and also at some even multiples of the fundamental frequency, due to imperfections of the square pulse. Figure 8.1 illustrates the first spectral component of the two loops. Note that the shape of the signal in the time domain corresponds to amplitude variations in spectral components in the frequency domain, whereas variations in per-iteration time of the loop result in variations in the frequency of the spectral components. While spectral component analysis is very effective when timing of the loops is of interest, it tends to be less effective when signal shape is also needed.

8.2.2 Tracking Code in Frequency Domain Over Time

In addition to easily recognizing program loops in the spectral domain, a spectrogram can be used to track program execution over time as illustrated in Fig. 8.2. We can observe three different spectral lines, shown as blue, red, and black, each of

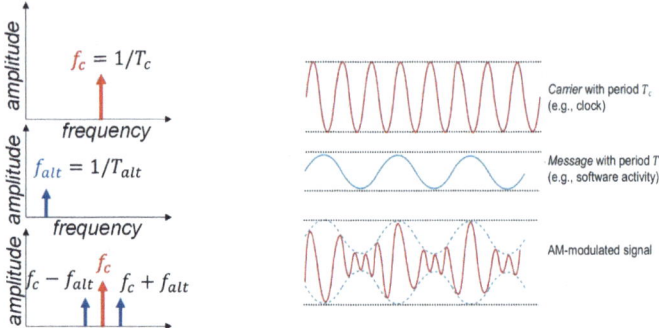

Fig. 8.3 Description of loops being modulated onto processor clock

which corresponds to a different loop in the program, and we can also observe that these three loops are executed in the program one after the other. We can also note that multiple lines or each color are shown; these correspond to multiple harmonics of the fundamental frequency of each loop. The first line is the first harmonic and that frequency is inversely proportional to execution time of one iteration of that loop. The length of the spectral line with respect to time corresponds to the duration of the entire loop, i.e., it is proportional to the number of iterations in that execution of that loop. Finally, there is some gap between when the spectral line of one loop ends and when the spectral line of the next loop begins—these gaps correspond to non-loop code that is executed while transitioning from one loop to the other.

8.2.3 Tracking Modulated Code in Frequency Domain Over Time and its Application to Malware Detection

In addition to observing loop activities at frequencies that correspond to their per-iteration time (usually low frequencies that are very noisy), loop activities also modulate higher-frequency periodic activities such as processor or memory clocks; and these modulated side channel signals have less noise. Figure 8.3 illustrates modulation process, which was described in detail in Chap. 4.

To illustrate the main idea behind using spectral components for malware detection, Fig. 8.4 shows a spectrum of a signal created by loop activity. We can observe a spike in a spectrum at $f_{clock} = 1.0079$ GHz. This signal is created by the processor clock, and it acts as a carrier that is AM-modulated by what the processor is executing (e.g., our loop). To the left and to the right of the carrier we can observe two spikes that correspond to a loop whose execution time is 77 ns ($\approx$13 MHz). This loop is from the "bitcount" benchmark from the MiBench [18] suite. Please note that these two spikes in the spectrum are a consequence of AM modulation of the carrier. Figure 8.4 indicates that, by observing the spectrogram of the signal (i.e., how the spectrum changes over time), we can deduce (1) each time when a loop in

8.3 Illustration of Malware Detection Using Analog Side Channel Analysis in...

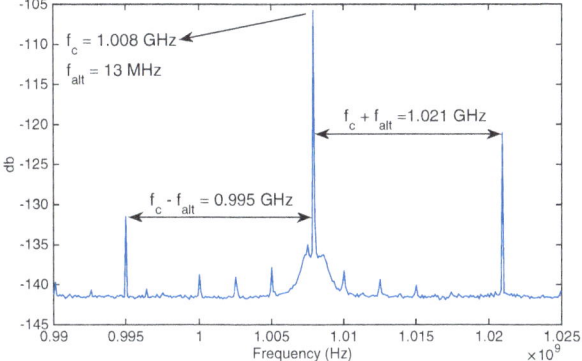

Fig. 8.4 Spectrum of an AM modulated loop activity [39]

the program is executed, (2) how much time the program has spent in each execution of each loop, (3) what is the per-iteration time of each loop, and (4) how many iterations were executed in each execution of each loop. Additionally, by collecting training signals for loops in the program when it executes without malware, we can identify deviations from this normal behavior to detect intrusions. In particular, if we observe a lower-than-expected frequency for a loop that is expected to execute at some point in the execution, we can deduce that extra code has been added to that loop's body, resulting in longer per-iteration time that reduces the loop's frequency. Conversely, a higher-than-expected frequency for a loop execution indicates that something has been removed from the loop. Furthermore, if extra code has been inserted in non-loop code that is executed in the transition between two loops, the time difference between ending of one loop and beginning of another loop can be used to detect this code modification. Finally, a modification of the loop's code that preserves the per-iteration time can be detected by analyzing the relationship among the harmonics. However, such analysis of the relationship among harmonics might be as difficult, or even more difficult, than analysis of the time-domain signal. Fortunately, program code changes that correspond to malware usually preserve original functionality (to avoid detection) and add to it, resulting in either adding program code (and thus execution time) to a loop's body or to a loop-to-loop transition.

8.3 Illustration of Malware Detection Using Analog Side Channel Analysis in Frequency Domain

In this section we discuss how spectral monitoring of program execution can be used for malware detection, discuss possible challenges in using this method. and how to overcome them using our malware detection method called REMOTE [40].

8.3.1 Spectral Samples (SS)

At a high level, REMOTE has two phases: training and monitoring. In both phases, the EM signal is first transformed into a sequence of spectral samples (SS) by using short-time Fourier transform (STFT), which divides the signal into equal-sized segments (*windows*), where consecutive segments overlap to some degree. STFT then applies the Fast Fourier Transform (FFT) to each window to obtain its spectrum. In our measurements, we use a 1 ms window size[1] with 75% overlap between consecutive windows, which provides a balance between the computation complexity and frequency/time resolution. The rest of the training and monitoring operates on this sequence of spectra, where each spectrum (i.e., the spectrum of one window) is referred to as a Spectral Sample (SS).

8.3.2 Distance Metric for Comparing SSs

In both training and monitoring, REMOTE will need to compare SSs to each other. For that, it requires a distance metric—a way to measure the "distance" between SSs in a way that corresponds how likely/unlikely they are to have been produced by execution of the same code. This distance metric should be sensitive to the aspects of the signal that change when executing different code, but insensitive to aspects of the signal that change between physical instances of the same device or over time on the same device instance. To achieve this, we create a new distance metric, ***Clock-Adjusted Energy and Peaks (CAPE)***.

Based on the insights from prior work [19, 28, 39], the frequencies of the peaks in the signal around the clock frequency are an excellent foundation for constructing a distance metric that is sensitive to which region of code is executing. Unfortunately, our experiments have shown that the clock frequency itself can vary over time and among device instances, and a change in clock frequency also changes the frequencies of loop-related peaks around it. Because the peaks' frequencies are all relative to the carrier frequency, any shift in the clock frequency also shifts the frequencies of the loop-related peaks by the same amount. Also, as the clock frequency increases, the program executes faster, leading to a lower per-iteration time T, higher frequency of the loop ($f_l = 1/T$), and thus moving the loop-related peaks away from the clock's frequency. Similarly, a reduced clock frequency moves the loop-related peaks closer to the clock frequency.

Thus the first step in computing our CAPE distance metric is to, for each frequency f that is of interest in an SS, compute the corresponding normalized frequency as $f_{norm} = (f - f_{clk})/f_{clk}$, where f_{clk} is the clock frequency for that SS. This normalized frequency is expressed as an offset from the clock frequency so that

---

[1] the window size should be determined based on sampling rate, clock frequency, and the required time resolution.

8.3 Illustration of Malware Detection Using Analog Side Channel Analysis in... 157

a shift in clock frequency does not change f_{norm} with it, and it also is normalized to the clock frequency, so it accounts for the clock frequency's first-order effect on execution time.

To make CAPE robust to weak signals and/or signals that have no well-defined peaks, we first consider the overall signal power (sum of magnitudes in the spectrum) of the signal outside the vicinity of the clock. The power of a poorly-defined peak is spread across a range of frequencies—visually it is a wide and not-very-tall "hump" rather than a narrow and tall "peak". When comparing two SSs that are different but each contain only "humps" and no sharp peaks, if we only consider whether the SSs have power concentrations at the same (clock-adjusted) frequencies, the overlap among their "humps" causes these SSs to match much better than they otherwise should, and this can prevent detection of malware-induced changes in signals. Moreover, under poor signal-to-noise conditions (e.g., when the signal is received at a distance) sharp peaks are likely to still stand out of the noise, so, due to random variation in noise, some "humps" end up below the noise level and some do not. For two SSs that should be the same (except for the noise), this causes poor matches, and this can lead to false positives. Thus to make our CAPE distance metric more robust against weak/noisy signals, we use a new insight, called the *"non-clock-energy"* test: non-clock power varies very little among SSs that do belong to the same region, while increases/decreases in a loop's overall per-iteration time concentrate less/more power toward the clock frequency in an SS. Therefore, SSs whose non-clock power differs by more than 0.5 dB are considered dissimilar by CAPE, and no further comparison between them is needed.

If the two SSs pass the "non-clock-energy" test, REMOTE compares them according to the (clock-adjusted) frequencies of their most prominent peaks. Specifically, we take N highest-magnitude frequency bins from the spectral sample (SS) that are each *(i)* not part of the *NoiseList*, and *(ii)* not within D spectral bins of a higher-amplitude spectral bin. The number N is determined differently for training and monitoring, as will be described shortly. The *NoiseList* contains frequencies of signals that are present regardless of which specific region of the application is executing. For finding the *NoiseList*, we record the EM signal several times while no program is being executed (the system is idle) and then average these recorded signals. We then choose 10 random SSs in the averaged signal, and then for each SS, sort it and find all the spikes that are at least 5 dB above the noise floor and put them in the *NoiseList*. We empirically find that choosing 10 points is sufficient to find all the strong peaks since it can accurately capture the transient behavior of the environmental noise. It is also important to point that, using this method, our detection algorithm is robust to interference from nearby devices (that are not identical to the monitored device), as their clock and other frequently-occurring peaks will end up on the *NoiseList*. The reason for ignoring D spectral bins that are too close to even-higher-magnitude ones is that a very prominent peak in the spectrum typically has "slopes" whose magnitude can exceed the magnitude of other peaks, and we found that REMOTE is more robust when its decisions are based on separate peaks rather than just a few (possibly even one) very strong peaks and a number of frequency bins that belong to their "slopes".

Finally, REMOTE combines the information about the frequencies of the peaks in the two SSs into a single value that represents the distance among the SSs. For each peak in one SS, REMOTE finds the closest-frequency peak the other. If the frequency difference is large enough, the peak votes for a mismatch, and the ratio of the mismatch votes to the number of all (mismatch and match) votes is used as the distance metric between the two SSs.

8.3.3 Black-Box Training

To train REMOTE, signals are collected as the unmodified monitored device emanates them. However, care should be taken to achieve good coverage of the software behaviors, e.g., by using the same methods that are used to test program correctness. The problem of achieving good coverage tends to be easier for many applications in the Cyber-Physical Systems (CPS) domain, especially those where correct operation is critical, because correctness concerns and the need for easy verification of correct operation motivates developers to produce code that has relatively few code regions, and with very stable patterns for how the execution transitions between them. In such cases, normal use of the device is likely to provide good coverage of the application's code regions after a while.

After signals are obtained and converted into SSs, a key part of training is to associate SSs with the code regions they correspond to. To achieve this without using instrumentation or other on-the-monitored-device infrastructure, REMOTE relies on a general observation that a given region of code tends to produce EM signals whose SSs are similar to each other, while the SSs from different regions tend to differ from each other to various degrees. This observation allows us to group SSs according to similarity, and for that we use Hierarchical Density-Based Spatial Clustering of Applications with Noise (HDBSCAN) [13], a technique that performs clustering without any a priori knowledge about which cluster (region) each sample (SS) corresponds to, and with no a priori knowledge about the number of clusters (regions). Like other clustering algorithms, HDBSCAN needs a distance metric, and in REMOTE that distance metric is the new CAPE metric defined in Sect. 8.3.2, using $N = 10$ peaks. Using this variation-robust metric allows training signals to be collected over time (e.g., over many hours), and/or on multiple device instances.

Because HDBSCAN clustering is based solely on similarity among SSs, its result may not precisely correspond to actual regions of the code, e.g., one region may produce more than one cluster if there are several distinct ways in which the region can execute, or two regions may end up in the same cluster if their execution produces very similar signals. Neither of these possibilities is a problem for REMOTE, and in fact they result in improved sensitivity and performance. If separate clusters for distinct behaviors were forced into a single cluster, the resulting unified cluster would allow a wide variety of SSs to match—all the valid SS options and *also* everything that lays in-between in the distance-space used for clustering. By creating a separate cluster for each distinct possibility, REMOTE will detect

8.3 Illustration of Malware Detection Using Analog Side Channel Analysis in... 159

anomalies that produce SSs that are not valid but lay in-between the valid ones. Conversely, when multiple regions are clustered together, they have very similar (practically indistinguishable) signals and it is more efficient and robust to treat them as one cluster. During monitoring, a Finite-State Machine (FSM) is used to keep track of the current region of the code. For each test, REMOTE compares the new SS to either the current region or any valid "next region" that has been seen during training.

8.3.4 Monitoring

During monitoring, REMOTE receives the signal and converts it to SSs in the same way it was done in training (same window size and overlap). After that, REMOTE can be viewed as a classifier that places each spectral sample (SS) into either one of the known categories (clusters identified during training) or into the "unknown" category that represents anomalous behavior, according to our CAPE distance metric (Fig. 8.5 shows the flow-chart of REMOTE's monitoring algorithm). Specifically, a candidate region is rejected if its distance metric is above 50% (fewer than half of the peaks match). If all candidates are rejected, the observed SS is categorized as "unknown," otherwise it is categorized into the candidate category with the lowest CAPE distance metric. The number of peaks used for each cluster is no longer fixed at 10—instead, it is identified for each cluster during training. We start with ten peaks, but then remove those that occur in fewer than 10% of the SS in the cluster. If this results in removing all peaks, we still retain the two most frequently occurring (among SSs from that cluster) peaks. This helps matching

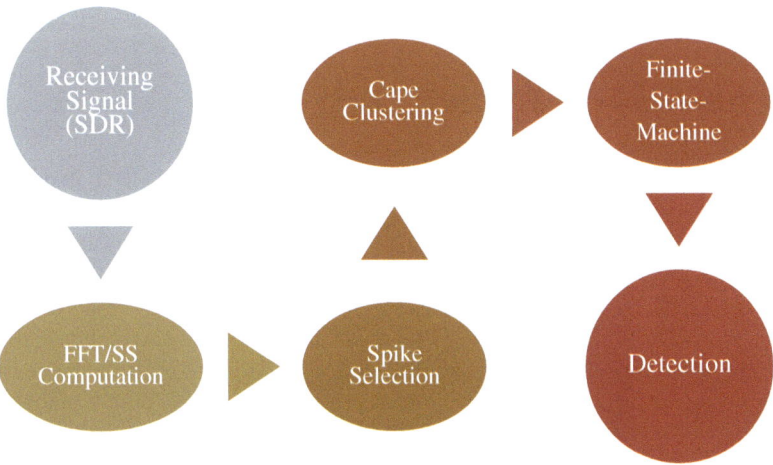

Fig. 8.5 REMOTE's monitoring flow-chart [40]

accuracy when the SSs in a cluster have few prominent peaks and a number of very weak peaks—in such cases it is more robust to use only the overall non-clock energy and the prominent peaks for matching than to use the peaks that may "disappear into the noise" due to changes in distance, antenna position, etc.

However, if the overall decision to report malware is based on only one SS, brief occurrences of strong transient noise can result in false positives. To avoid that, REMOTE only reports an anomaly if N consecutive SSs are classified as "unknown." The value of N should be chosen depending on the EM noise characteristics of the environment, but we found that N between 3 and 5 tends to work well in all our experiments. We use $N=5$ because it biases REMOTE toward avoiding false positives, while still maintaining an excellent detection latency ($N=5$ corresponds to only 1.25 ms detection latency in our setup). As mentioned in Sect. 8.3.3, an FSM is used to count N, report an anomaly, and to keep track of current valid region of code to ensure that the program follow a correct ordering of regions.

Finally, we found that in the presence of an OS, interrupts and other system activity that occurs during an SS can make that SS dissimilar to those from training. For example, an interrupt that lasts <1 ms can affect four consecutive SSs (recall that we use 1 ms windows with 75% overlap), so a naive solution would be to add 4 to N (number of consecutive "unknown" SSs that are needed to trigger anomaly reporting). Using $N = 9$ indeed eliminates interrupt-induced false positives, but also prevents detection of attacks that are brief. Unfortunately, real-world malware (e.g., the attack on *Syringe-pump* that will be described in next section) can introduce only a short burst of activity into the otherwise-normal activity of the application. Fortunately, our experiments indicate that spectral features of interrupt activity are similar to each other, so during training interrupt activity can be clustered. During monitoring, REMOTE includes these clusters as candidates, allowing it to tolerate interrupts without becoming tolerant of similar-duration deviations from expected execution.

8.3.5 Experimental Results

We use three sets of experiments to evaluate the effectiveness REMOTE in detecting different types of attack on a variety of devices.

In the first set of experiments, we use two real-world Cyber-Physical Systems (CPSs). The first CPS we use is a *Syringe-Pump*, an embedded medical device that we use as a representative of a medical cyber-physical system (Fig. 8.6). The second system is a PID controller that is used for controlling the temperature of a soldering iron. This type of system could also be used to control the temperature in other settings, such as a building or an industrial process, and thus is representative of a large class of industrial CPS/IoT systems.

For the second set of experiments, we use five applications from an embedded benchmark suite called MiBench [18] running on an IoT/embedded device, which

8.3 Illustration of Malware Detection Using Analog Side Channel Analysis in... 161

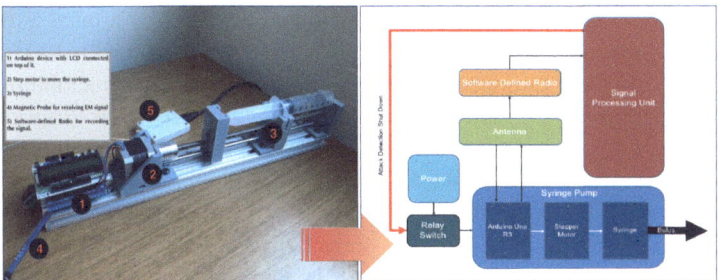

Fig. 8.6 Syringe Pump (left) and REMOTE framework (right). In our setup, the signal processing unit is implemented on a *separate* PC [40]

Fig. 8.7 The near-field setup (left) consists of a small EM probe or a hand-made magnetic probe (not shown) placed 5 cm above the system's processor. A horn antenna placed 1 m away from the board for far-field measurements (right). In all cases, a software-defined radio smaller and lighter than most portable USB hard drives, is used to record the signal [40]

are a representative of the computation that is needed in that market (e.g., automotive, industrial systems, etc.)

Finally, for the third part of our evaluations, we chose a robotic arm (LewanSoul LeArm 6DOF) [1], which is a representative of commonly-used CPS existing in the market.

8.3.6 Measurement Setup

The measurement setup is shown in Fig. 8.7. Depending on the distance, either a hand-made magnetic coil or a horn antenna is used to receive EM signals (no amplifier is used). For all measurements, we use low cost (<$30) software-defined radio (SDR) receiver (RTL-SDRv3) to record the signal. Using this radio, the entire cost for the near-field measurement setup (including the radio and a hand-made coil) is only around $35, and for the far-field measurement setup is around $100–200 (depending on the antenna). Further cost advantages can be gained if REMOTE is used in settings where multiple similar devices (with similar vulnerabilities) are used, so a single (or a few) devices can be monitored by REMOTE (especially in far-field scenarios), with random changes to which specific devices are monitored at any given time. Figure 8.6 shows the entire setup including the monitored device (Syringe-Pump in this figure) and REMOTE.

Note that all of our measurements were collected in the presence of other sources of the electromagnetic interface (EMI), including an active LCD that was intentionally placed about 15 cm behind the board. A set of TCL scripts is used to control the monitored system and the SDR (to record the signal). The entire REMOTE algorithm is implemented on a PC using Matlab2017-b.

8.3.7 File-Less Attacks on Cyber-Physical-Systems

The first part of our evaluation presents the results for two real-world CPS which are implemented on four different devices (shown in Table 8.1). To attack these devices we implement two end-to-end *file-less* attacks namely a **code-reuse** attack and an **APT** (advanced-persistent-threat) attack.

Code Reuse [11, 41] attack targets a medical CPS called Syringe Pump. A Syringe-Pump is a medical device designed to dispense or withdraw a precise amount of fluid, e.g., in hospitals for applying medication at frequent intervals [46].

The device typically consists of a syringe filled with medicine, an actuator (e.g., a stepper motor), and a control unit that takes commands (e.g., how much fluid to dispense/withdraw) and produces controls for the stepper motor. The systems must provide a high degree of reliability and assurance (typically by using a simple MAC) since imprecise or unwanted dispensing of medication, or failure to administer medication when needed, can cause significant damage to the patient's health. In our evaluation, we use the Open Source Syringe-Pump from [7].

Our code-reuse (CR) attack involves overflowing the input buffer in the serial input function, which normally reads the input, sets a flag to indicate that new input is available, and returns. Exploiting this vulnerability, the return address in the stack is overwritten by a chain of gadget's addresses to launch an attack.

Table 8.1 Boards used in this paper to evaluate REMOTE [40]

Device	Processor	Clock-rate	OS
Arduino UNO	ATMEGA-328p	16 MHz	No
DE0-CV Altera FPGA	Nios II softcore	50 MHz	No
TS-7250	ARM9	200 MHz	Debian
A13-OLinuXino	ARM Cortext A8	1 GHz	Debian

8.3 Illustration of Malware Detection Using Analog Side Channel Analysis in... 163

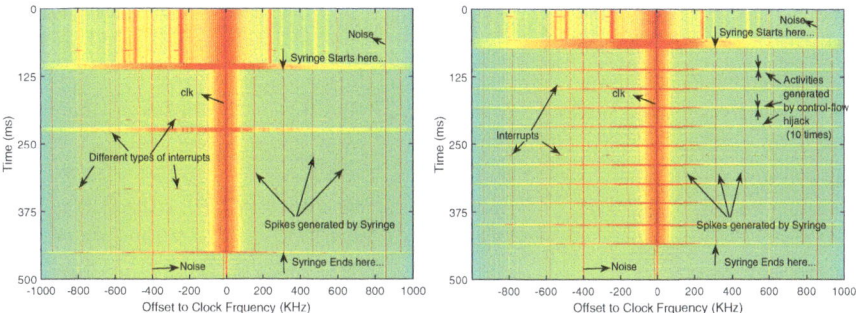

Fig. 8.8 Spectrogram of the Syringe pump application in malware-free (left) and malware-afflicted (right) runs. Note that the differences in colors between the two spectrograms correspond to differences in signal magnitude which are caused by different positioning of the antenna. Such variation is common in practice and has almost no effect on REMOTE's functionality because REMOTE was designed to be robust to such variation [40]

Since the security-critical part of this system is moving the syringe, a desirable goal for an attacker is being able to call the *MoveSyringe()* function, which is responsible for syringe movement, at an unwanted time while *skipping* the input-checking part. This checking is in the *Delay()* function, which checks the authenticity of the command (otherwise the attacker needs to hack into the C&C server to send the commands which may not be a feasible task).

We use ROPGadget [37] for finding the proper chain of gadgets to put the address of *MoveSyringe()* in a register and then go to that function directly from the *readInput()* function, skipping the checking part. After *MoveSyringe()* executes, the execution returns back to the main function, and normal behavior of the application resumes.

Figure 8.8 shows a spectrogram of the Syringe-Pump application in (left) a malware-free run, and (right) when the CR attack happens. As seen in the figure, the Syringe-Pump application has three distinct regions with clearly different EM signatures: printing debug info and reading inputs, a delay/checker function which checks the message authenticity (using a simple MAC), and actual movement of the syringe. The major difference between these two figures is the **reverse** order of "Delay" and *MoveSyringe()* parts in malicious run (bottom). In normal behavior, REMOTE expects to see $readInput \rightarrow Delay \rightarrow MoveSyirnge$ however, in a CR attack, since the return address of the $readInput$ function is overwritten by the adversary, the code immediately jumps to *MoveSyringe()* and skips the "Delay" part, thus in the spectrogram, the third region (*MoveSyringe()*) is seen before "Delay" (bottom), which violates the correct ordering of regions and will be reported as "malicious" by REMOTE.

Our evaluation of REMOTE uses one attack per run in 25 runs, with REMOTE successfully detecting each of these attacks (see Table 8.2). We then performed 25 attack-free runs and found that REMOTE produced no false positives (see Table 8.2). To further evaluate our system, we performed 1000 malware-free runs

Table 8.2 Accuracy of REMOTE for several different systems and attack scenarios using various boards and applications [40]

Applications (attacks)/accuracy		Boards		
Syringe-pump (code-reuse attack)	Arduino	Nios-II	TS-board	OLinuxino
True positive accuracy	>99.9%	>99.9%	>99.9%	>99.9%
False positive accuracy	<0.3%	<0.1%	<0.1%	<0.1%
Soldering-iron (APT attack)	Arduino	Nios-II	TS-board	OLinuxino
True positive accuracy	>99.9%	>99.9%	>99.9%	>99.9%
False positive accuracy	<0.1%	<0.1%	<0.1%	<0.1%
Robotic-arm (firmware modification)	LewanSoul LeArm 6DOF			
True positive accuracy	>98.2%			
False positive accuracy	<0.2%			
Applications (attacks)/accuracy		IoT applications		
Embedded IoT (shellcode attack)	bitcount	basicmath	qsort	fft
True positive accuracy	>99.9%	>99.9%	>99.9%	>99.9%
False positive accuracy	<0.1%	<0.1%	<0.1%	<0.1%

and 1000 malicious run on one device (Arduino)[2] for 24 hours. For these 2000 runs, REMOTE successfully found all the 1000 instances of malicious run and reported 997 out of 1000 malware-free runs as normal (i.e., only 3 out of 1000 false positive = 0.3%).

Note that, depending on the size of injection, the *MoveSyringe()* in Syringe-Pump could be very brief in time (e.g., around 3 ms as can be seen in Fig. 8.8-left), and we found that, without correctly handling the interrupts on Olimex and TS platforms (which have an operating system), we would either get very high false positives (due to interrupts), or high false negatives (by using large N to ignore short-term activity). However, as also discussed in 8.3.4, by adding training-time samples for interrupts, we can use small N, while having 0% false positives (see Sect. 8.3.10).

We then repeated our measurement for Syringe-Pump for 50 cm and then 1 m distance (using a 9 dBi horn antenna [3] connected to the SDR) and, in both cases, we also get perfect accuracy. It is also important to mention that the detection latency (i.e., the time from when the attack starts until REMOTE detects it), for all four devices is < 2 ms.

An alternative method for attacking the Syringe-Pump is by changing the *InjectionSize* (i.e., a Data-only attack). This also can be done using a CR attack. REMOTE is able to protect the Syringe-pump against such an attack because a changed *InjectionSize* will change the duration (i.e., the number of SSs) of *MoveSyringe()*. REMOTE can count the SSs which belong to *MoveSyringe()* activity and compare it to the expected number of SSs. To check how well REMOTE can

---

[2] To limit the amount of measurements and time for processing it, we picked only one of the four devices.

8.3 Illustration of Malware Detection Using Analog Side Channel Analysis in...

detect such an attack, we check the number of SSs for *MoveSyringe()* for all the 25 attack-free runs and compare it to the actual *InjectionSize*. In all the instances, REMOTE reports the correct number of SSs. Note that we are not detecting EM emanations (RF) signal produced by the motor movement; instead we are detecting the change in the code execution when this "data-only" attack is performed. i.e., we observe the signal at the clock frequency of the board and observe software execution changes, while EM emanations caused by motor movement occur at much lower frequencies.

However, if the change is less than one SS, or if the expected *InjectionSize* is unknown, REMOTE is not able to detect the change. Overall, there is a tradeoff between the size of SS and REMOTE's ability to detect small changes. Thus, to improve the effectiveness of the system, either a higher sampling-rate setup can be used (smaller SS and thus smaller detection granularity), or REMOTE can be combined with other existing methods (e.g., Data Confidentiality and Integrity (DCI) methods [14]) to protect the system against different types of data-only attacks. However, as shown in this work (for this attack and other attacks in this section), meaningful attacks typically have much larger signatures (i.e., order of milliseconds) than the current detection limit in REMOTE (200 microseconds).

The **second attack** used in this evaluation is an advanced-persistent-threat (APT) attack on an industrial CPS (called Soldering-iron). A well-known example of such attack for CPS is Stuxnet. *Soldering-iron* is an industrial CPS that allows users to specify a desired temperature for the iron, and then keeps the iron at that temperature using a proportional-integral-derivative (PID) controller. This type of controller could also be used to control temperature in other settings, such as a building or an industrial process, and thus is representative of a large class of industrial CPS. This application is significantly larger than Syringe-Pump—it has 70,000 instructions and 1,020 static control-flow edges [7].

The application starts by initializing all the components (e.g., PID controller, Iron, etc.). It then begins to control the Iron's temperature: it checks all the inputs (e.g., knob, push buttons, etc.) and then based on them decides to decrease or increase the temperature, prints new debug information on its display, etc. and then repeats this ad infinitum. The security-critical function is *keepTemp()*, where the temperature of the iron is set. This function uses an iterative process (a PID controller) to change or keep the temperature of the iron. The critical variable is *temp_hist*—it holds the last two temperatures of the iron and is used to calculate the difference between the current temperature of the iron and these two last temperatures.

To implement a Stuxnet-like malware on this application, we assume that the attacker can reprogram the device. The attacker's goal is to change a critical value under some conditions, which in turn can cause damage to the overall system. A possible modification to the code is shown in Fig. 8.9 where, based on one or several conditions (e.g., in our evaluation it checks the model of the device that is stored in memory), the temperature history can be changed. The key insight is that the added instructions will cause the spectral spikes during execution of the main loop to be

```
// The main loop
void loop() {
   int16_t old_pos = read(&rotEncoder);
   // finding the position of the control knob

   bool iron_on = isOn(&iron);
   // iron is the object for the soldering iron

   // adding malicious activity
      if (some_condition){
            iron.temp_hist[0] = maliciousVal0;
            iron.temp_hist[1] = maliciousVal1;
      } // end of malicious activity

         byte bStatus = intButtonStatus(&rotButton);
   // reading input button

      showScreen(pCurrentScreen);
      keepTemp(&iron);      }
```

Fig. 8.9 A code fragment from the main loop of the soldering iron application and a possible injected malicious activity

shifted to lower frequencies (more time per iteration) as shown in Fig. 8.10 for the A13-OLinuXino device.

To evaluate how well REMOTE can detect this type of attack, we use 7 runs in training, and use 25 runs without malware and 25 runs with malware to evaluate the monitoring algorithm. Our results show REMOTE can successfully detect all the instances of the attack (a 100% true positive rate) (see Table 8.2).

8.3.8 Shellcode Attack on IoTs

Another popular class of attacks on CPS /IoTs are *shellcode* attacks, where the adversary executes a malicious application (payload) through exploiting a software vulnerability. It is called "shellcode" because it typically starts a command shell (e.g., by executing the */bin/sh* binary) from which the attacker can control the compromised machine, but any piece of code that performs a similar task is also called shellcode. Once the attacker takes control, he/she can execute any *injected* code such as a *Denial-of-Service* attack.

In this paper, we implement this attack by invoking a shell (*/bin/sh*) via a buffer overflow exploit. We then run a malicious payload on the invoked shell: either a

8.3 Illustration of Malware Detection Using Analog Side Channel Analysis in... 167

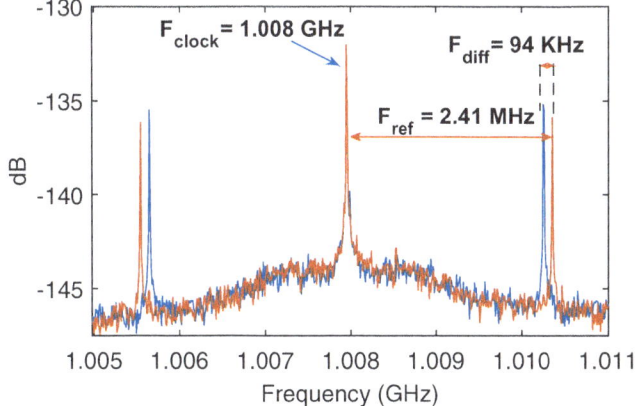

Fig. 8.10 Adding malicious activity to the main loop of the Soldering-iron application (red: without malware, blue: with malware) [40]

DDoS bot, or *Ransomwmare*. These attacks typically target devices with operating systems. In this work, we implement them on an IoT device with an ARM core (A13-OLinuXino), which is a representative of state-of-the-art IoTs.

The attacks are implemented on five representative programs from MiBench suite (*bitcount*, *basicmath*, *qsort*, *susan*, and *fft*). We chose these applications among all the MiBench applications (this benchmark is designed to represent typical behaviors of embedded system: e.g., Security, Telecom., Network, etc.) mainly because *bitcount* is a good representative of the applications that have several different distinct regions (our HDBSCAN clustering found 9 for this application) and has lots of different activities including nested-loops, recursive functions, interacting with memory, etc. *Basicmath* is chosen because it is a good representative of unstable/weak activities, since its activity in each region is very dependent on values (it is calculating different fundamental mathematics operations such as integration, square-root, etc.). We also chose *qsort* because it has lots of memory accesses, and we picked *susan* and *fft* as representatives of common and popular activities in embedded system domain (i.e., image processing and telecommunications.). In all these applications, first a buffer-overflow vulnerability is exploited and a shell with same privileges as the original application is invoked. A malicious payload (i.e., DDoS or Ransomware) is then executed in this shell.

For DDoS, we port the C&C and the bots from the Mirai open source to run on our IoT. The DDoS payload execution begins right after the shell is invoked and ends after sending 100 SYN packets. The application then resumes its normal activity. We use a computer on the local network as the target of the DDoS attack (SYN flood), and we verify on that computer that the attack is indeed taking place. As another payload, we also implement a simple Ransomware payload that uses AES-128 with CBC mode to encrypt data. This encryption represents the bulk of the execution activity created by Ransomware.

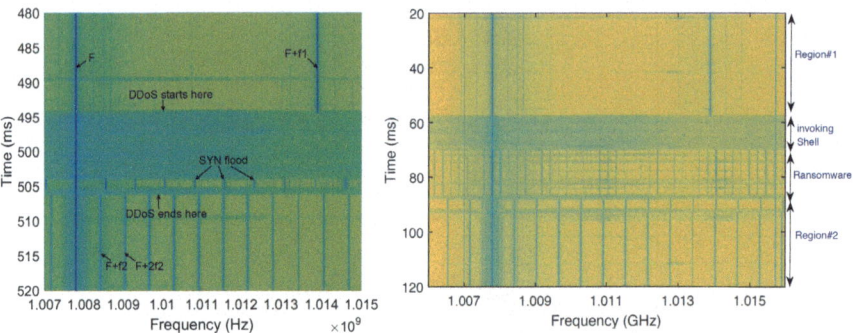

Fig. 8.11 A run (left) where exploit, shellcode, and a 100-packet payload are injected into the execution between the original loops. A run (right) where exploit, shellcode, and a Ransomware payload are injected into the execution between the original loops [40]

As in previous cases, we use 7 runs for training and then use 25 runs without malware and 25 runs with each malware (i.e., DDos and Ransomware) for all five applications. Our results (see Table 8.2) show REMOTE can successfully detect all instances of the attack (a >99.9% true positive rate), while none of the malware-free runs are incorrectly identified as malware (0% false positive rate). We have also found that just invoking a shell is visually detectable on our IoT device since it takes around 8 ms (about 32 SSs), and sending 100 SYN packets adds about 4 ms to that (see Fig. 8.11 (left) for DDoS and (right) for Ransomware).

8.3.9 APT Attack on Commercial CPS

The final system in our evaluation is a Robotic Arm, a type of device that is often used for manufacturing and is typically a critical component of any modern factory. It usually receives inputs/commands from a user and/or from sensors and moves objects based on these inputs. There is a growing concern about security of these CPSs since they are typically connected to the network and thus exposed to cyber-threats [35]. In this work, we use a commercial robotic arm (LewanSoul LeArm 6DOF [1]) which uses an Arduino board as a controller and a *Bluetooth* module to receive command. For this system, we implement an APT attack (firmware modification) where the reference libraries (e.g., library for Servo, Serial, etc.) are compromised (this can be also considered as a zero-day vulnerability). Note that, REMOTE's training is on the "unmodified" version of these library (baseline reference data). In this attack, we modify a subroutine (*writeMicroseconds()*) in Arduino's Servo library [4] by adding an extra `if/else` condition to change the speed of Servo motor randomly. We then reprogram the system with this compromised library, with the goal of causing occasional malfunctions in arm's movement in real-time.

8.3 Illustration of Malware Detection Using Analog Side Channel Analysis in... 169

We use 7 runs for training and then use 1000 runs without and 1000 runs with the firmware modification. Our results (see Table 8.2) show REMOTE can successfully detect the instances of the attack with very high accuracy($>98.2\%$ true positive rate), while only less than 0.2% of the malware-free runs incorrectly identified as malware.

8.3.10 Further Evaluation of Robustness
Interrupts and System Activity

Among the platforms we tested, the longest-duration system activity "inserted" (via an interrupt) into the application activity tends to take a few milliseconds, and it appears to be associated with display management/update because disabling lightdm [44], the display manager, eliminates these interrupts (but other kinds of interrupts still occur). In contrast, in bare-metal devices interrupts (when there are any) tend to be around a microsecond in duration. Figure 8.12 shows the (perfect) ROC curve (solid blue line) for *SyringePump* on Olimex (and Debian Linux OS) when using REMOTE as described in Sect. 8.3.1. We then prevented REMOTE from forming interrupt-activity clusters during training, and used the EDDIE's scheme, and that has resulted in a severely degraded ROC curve (red dashed line), where many false positives are detected when 4 consecutive clusters are found to be "unknown" ($N = 4$ is Sect. 8.3.4). Increasing N in this experiment reduces the false positives, but also reduces the true positives. This confirms that, our approach of addressing system activity directly in REMOTE is significantly contributing to REMOTE's ability to detect malware while not reporting false positives due to system activity.

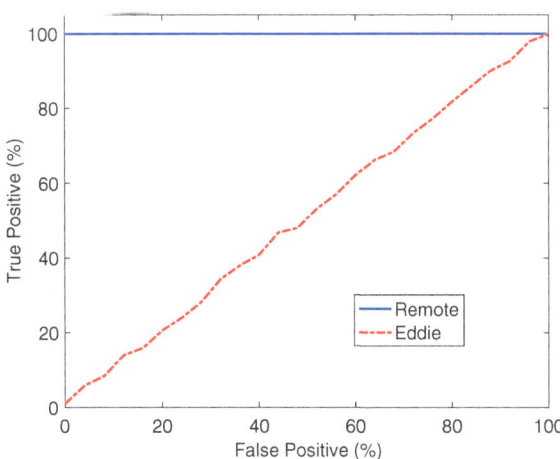

Fig. 8.12 Accuracy of REMOTE with its mechanism for addressing interrupt activity (solid blue line) and EDDIE [28] (red dashed line). The results are for the *SyringePump* software running on the Olimex board [40]

Hardware Platforms and Distance

Note that packaging and other limitations may require the EM signal to be received from some distance, which significantly weakens the signal. To evaluate the impact of distance on REMOTE, we receive the signal from distances of 5, 50 cm, and 1 m away from each of the tested devices. To limit the amount of data that is recorded, we use only two representative programs from the MiBench suite (*bitcount* and *basicmath*, described in Sect. 8.3.8), and only two representative malware behaviors—one that adds a relatively small number of instructions inside a loop (*Stuxnet*-like), and another where similar malicious activity is done all-at-once outside of loops (*DDoS*-like).

For each device and each application, we use 25 malware-free runs to obtain the false negative (malware activity not reported in a malware-affected run), and 25 runs for each of the two malware activities (75 × 3 runs for each of the platforms) to obtain false-positive rates (malware reported in a malware-free run) achieved by REMOTE. Our results show perfect accuracy (i.e., 0% false negatives and 0% false positives) for all devices and all three distances. However, if we prevent REMOTE from using total non-clock power when comparing SSs and use the scheme from EDDIE and/or Syndrome, on the TS board (which has the weakest signal among the boards tested) for 50 cm and 1 m distances we observe true positive rates of only 80% (at 50 cm) and 55% (at 1 m). Once we adjust other parameters we achieve 0% false positives (see Fig. 8.13). This confirms that, when signals are weak, comparisons based on spectral peaks alone are insufficient and other signal features (such as non-clock power used in REMOTE) must also be considered.

Manufacturing Variations

To study the effect of manufacturing variations on the EM signals and REMOTE's accuracy, i.e., to determine weather training is needed for each *type*

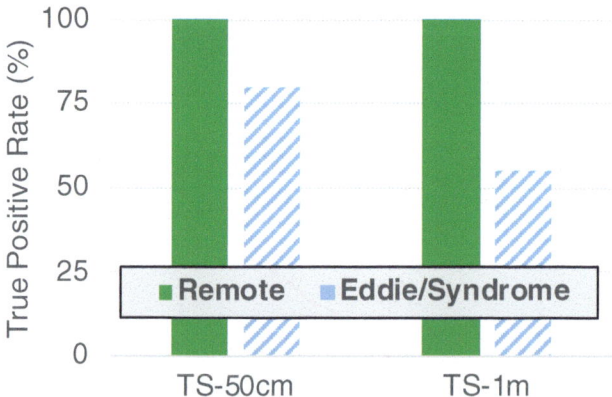

Fig. 8.13 True positive rate (with 0% false positives) of REMOTE with its non-clock-power feature when comparing SSs (dark blue) and EDDIE [28]/SYNDROME [38] (light red). The results are for *basicmath* running on the TS board [40]

of device or for each *physical instance* of a device, we use 30 physical instances of the Cyclone V DE0-CV Terrasic FPGA development board (chosen primarily because we have 30 such boards) train REMOTE on one board (randomly selected) and use that training to monitor each of the 30 instances, with 20 runs of *bitcount* on each instance, both with and without malware.

Our results show that REMOTE's accuracy remains at 100% true positives and 0% false positives throughout this experiment. However, when we prevent REMOTE from frequency-adjusting the SSs used in comparisons, we still find no degradation for 17 of the boards, but for 13 the false positive rate increases to nearly 100%. Further analysis shows that the clock frequencies of the boards vary; 17 of them (including the one trained-on) were within the frequency-tolerance (parameter D in 8.3.2) of the matching, whereas the other 13 were outside the tolerance, causing none of their peaks to vote for the cluster the signal actually should belong to. If D is then adjusted to avoid false positives, the true positive rate is severely degraded. Figure 8.14 shows one such scenario where we trained on board number 3, and test on board number 4. The figure shows the ROC curve for board number 4 when frequency-adjusting is active and inactive. We also repeated this experiment for 10 Olimex boards (we do not have 30 of those), with very similar results with and without REMOTE's frequency-adjustment. These results confirm the need for frequency-adjustment in REMOTE if training and monitoring do not use the same physical instance of a device.

Variations Over Time

In this experiment, we have recorded the signals at one-hour intervals, over a period of 24 hours, while keeping the FPGA board and the receiver active throughout the experiment, to observe how the emanated signals vary over time as device temperature (and room temperature) and external radio interference such as WiFi and cellular signals change during the day and due to the day/night transition. The set of measurements collected each hour consists of 60 *bitcount* runs, 20 without

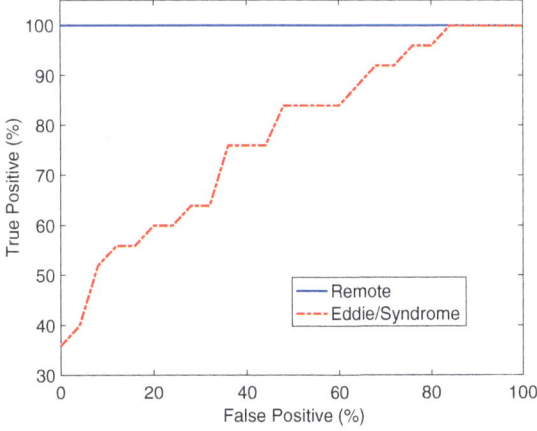

Fig. 8.14 Accuracy for REMOTE with frequency-adjusting, vs. Eddie/Syndrome for FPGA board running *bitcount* [40]

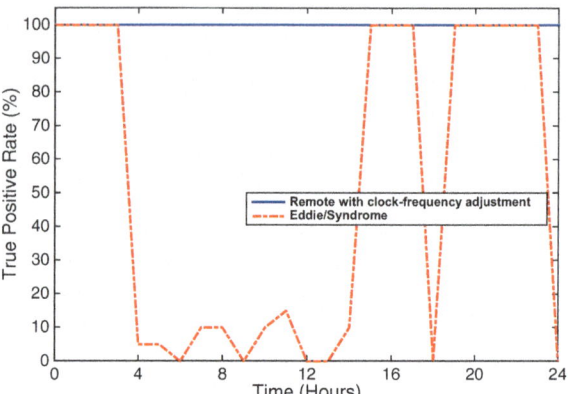

Fig. 8.15 Performance of REMOTE with its clock-frequency adjustment feature vs. Eddie/Syndrome [40]

malware and 20 for each of the two types of malware described in Sect. 8.3.10. The training data for all REMOTE analyses in this experiment was recorded just after the device (FPGA board) and the receiver (SDR) were turned on.

We observed no deviation from REMOTE's accuracy throughout this experiment (solid blue line in Fig. 8.15). We then prevent REMOTE from clock-adjusting the frequencies and repeat the experiments (on the same signal recordings), find that the detection accuracy is dramatically degraded between hours 4 through 13 and hours 23 and 24 (dashed red in Fig. 8.15). Further analysis shows that the clock frequency has shifted during these hours, coinciding with use of business-hours and off-hours thermostat setting for the room,[3] likely because the temperature had affected the board's crystal oscillator whose signal is the basis for generating the processor's clock frequency.

Multi-Tasking/Time-Sharing

In our final set of experiments, we apply REMOTE in the runs where Ransomware (see Sect. 8.3.8) is executed as a separate process, without changing the application. The OLinuXino board only has one core, so its Debian Linux OS context-switches between the two processes until the Ransomware payload completes. Figure 8.16 shows the spectrogram in one such execution. Only the application is running in the first part of the spectrogram. At some point (millisecond 812 in this spectrogram), the Ransomware process is started, and the context-switching in (approximately) 10 ms time-slices can clearly be seen beyond this point in the spectrogram. The spectrum of the malware process is clearly different from the spectrum produced by the application at this point in its execution, so we expect REMOTE to detect this malware execution scenario easily.

---

[3] The actual change in clock frequency was less than one-part-per-million of the clock frequency, well within typical design tolerances for clock signals, and with negligible impact on the processor's overall performance and power consumption.

8.4 Time-Domain Analysis of Analog Side Channel Signals for Malware... 173

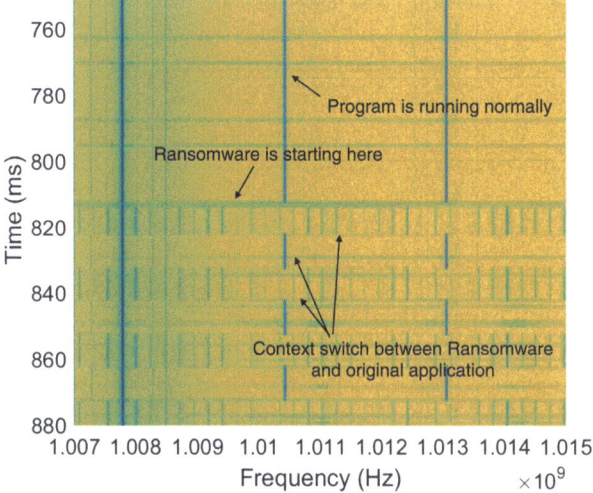

Fig. 8.16 Spectrogram of context-switching between the unmodified Bitcount application and the Ransomware process [40]

To evaluate REMOTE accuracy for this scenario, we use 25 runs, and in each run, start the Ransomware process at a different point in the run. The results of this experiment are that REMOTE successfully detects the presence of malware all these runs, even with the tolerance threshold that produces no false positives for malware-free executions. It should be noted here that in this set of runs, according to our threat model, the IoT system is running only one valid application. To successfully handle scenarios in which the system context-switches between multiple valid applications, REMOTE must be extended to identify when context switches are occurring and to keep track and validate spectral samples with the knowledge of which application(s) they might belong and where the "current" point is in each of those applications as we have demonstrated in [50].

We would like to point out here that backscattering side channels can also be used for malware detection with similar accuracy and performance [29].

8.4 Time-Domain Analysis of Analog Side Channel Signals for Malware Detection

Frequency domain analysis of analog side channel signals is efficient, and we have described how it can be effective for malware detection. However, frequency-domain analysis tends to be coarse-grained in terms of time, so it cannot detect small changes caused by stealthy malware (e.g., [28, 38]), and/or has relatively high detection latency that makes it less attractive for real-time systems (e.g., [12]), and/or does not scale well with complexity of the device (e.g., [27]). In general, when loop features are not clearly prominent in the signal or when there are no loop-like features, time domain analysis tends to be more effective. In fact, time-domain analysis is commonly used for attacks on cryptographic code, where small (in magnitude) and brief (in time) changes in the signal represent the difference between zero and one. The advantage of using time-domain is that, in addition to the timing of the execution, it can effectively use the amplitude of the signal to distinguish among different sequences of instructions executed, even when the executed time for those sequences is the same. On the other hand, time-domain analysis is time consuming—it must consider large numbers of samples for meaningful analysis. Hence, the best overall approach to the problem would involve using frequency domain signal processing continuously for coarse grain malware detection, and using time domain signal processing approach for fine grain malware detection only when frequency-domain analysis is inconclusive.

Next, we present two different approaches for time-domain analysis, both of which are illustrative of the advantages and disadvantages of time-domain analysis in general.

8.5 Time Domain Analog Side Channel Analysis for Malware Detection

In this section we describe a technique, called IDEA, that leverages concepts and methods from speech recognition to detect malware using electromagnetic side channels [23]. Note that we use IDEA here as a representative example to illustrate how fine-grain malware detection can be implemented through analog side channel signal analysis, and that other approaches can also be used. For example, machine learning techniques have been also applied to detect malware with similar success [22].

Figure 8.17 illustrates the overall workflow of IDEA. In both training and monitoring phases, the signals are demodulated, low-pass filtered, and sampled before they are subjected to the main part of the signal processing, which exploits techniques similar to template-based pattern matching to identify anomalous (potentially malicious) activity during the program execution. In the training phase, we learn a dictionary of reference EM signatures or "words" by executing trusted programs on an uncompromised reference device. Next, in the monitoring phase, we continuously monitor the target device's EM signal by matching it and reconstructing it using the dictionary. When the reconstruction error is above a predefined threshold (i.e., when there is a significant deviation from the reference EM signatures), the system reports an anomaly (intrusion).

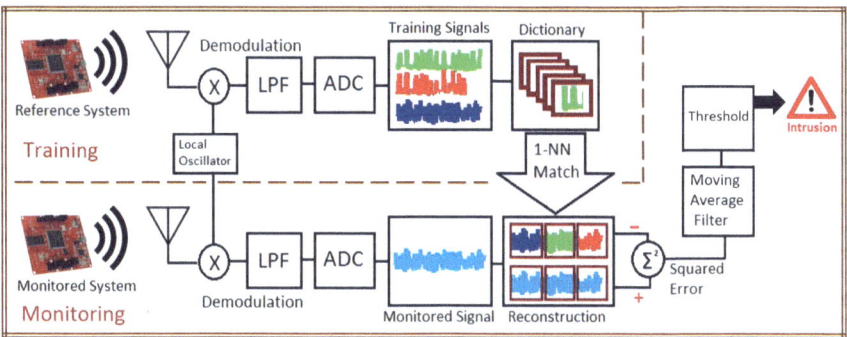

Fig. 8.17 Overview of IDEA [23]

8.5.1 Training Phase: Dictionary Learning

The training phase consists of learning a dictionary of EM signatures through the execution of trusted (i.e., OK-to-execute) programs on a reference device. We execute trusted programs on an uncompromised device, and observe and record the corresponding EM signals. We use different inputs to execute different control flow paths as described in [12]. Ideally, we would like to observe all possible control flow paths. However, in a practical scenario, this may require too many inputs (i.e., too many training examples). For example, a sequence of twenty if-then-else statements will, overall, have $2^{20} = 1048576$ different possible execution paths. However, we aim at observing most short-term control flow execution paths, and try to ensure that even if there are unobserved control flow paths, they are either highly unlikely or relatively brief. These goals are the same as those that guide program testing, so program inputs created to provide good test coverage of a program are highly likely to also satisfy the needs of our training. Once we have the training signals, we learn the dictionary words using the following process.

8.5.2 Learning Words

The AM-demodulated EM signal is split into multiple overlapping short-duration windows that are recorded as dictionary entries or "words". These words correspond to the EM signature of the underlying program execution. All dictionary words have the same word-length l. Each word is shifted by s samples from the previous one. When s is small, we end up with densely overlapping words. Consequently, we learn a dictionary with a large number of words with slight variability in time (i.e., shift). This can help to achieve shift-invariant pattern matching. Shift-invariance is necessary because of hardware events, such as cache misses, that delay (or shift) the subsequent execution (and the corresponding EM signal). A cache miss can potentially occur at many different points of the program execution. It is neither practical nor even possible to generate a training set with all possible scenarios of cache hits or misses. Therefore, creating a dictionary with densely overlapped words can help us match EM signatures better under variability due to hardware activity.

8.5.3 Word Normalization

All dictionary words are post processed by mean subtraction and scale normalization:

$$\mathbf{w} = (\mathbf{w_0} - \mu)/\sigma, \tag{8.1}$$

where $\mathbf{w}$ denotes the normalized word, μ is the mean and σ is the standard deviation of the unnormalized word $\mathbf{w_0}$.

This normalization improves the matching accuracy by ensuring that the matching is based on the pattern (i.e., the relative shape of the waveform) rather than on the actual amplitude. For instance, due to different positions of the antenna, the distance from the processor may change between the training and the monitoring phase. Hence, the training and the monitored signals can have different scales or amplifications. This normalization minimizes the impact of such issues.

8.5.4 Dictionary Reduction Through Clustering

Next, we apply clustering to reduce the number of dictionary entries. All applications have loops, which tend to generate repetitive EM patterns. Likewise, the same control flow paths are often reiterated at different points of the execution, and generate similar EM patterns. Consequently, the reference dictionary can have a large number of words or patterns that are very similar (and correspond to the same code execution). The objective of clustering is to group similar words or EM patterns into a single cluster, and then use the cluster centroid as the representative of the cluster. Using cluster centroids as dictionary words improves the computational efficiency by reducing the number of dictionary entries.

As the number of clusters k (i.e., the number of unique EM patterns) is not known a priori, popular clustering algorithms such as k-means can not be used for the dictionary reduction. Instead, we use a threshold based clustering, where threshold t is used as a parameter. Given a cluster centroid c_i, the algorithm proceeds by alternating between two steps:

Assignment Step Assign each unassigned word w_p whose Euclidean distance from the centroid c_i is less than the threshold t to the cluster S_i.

8.5 Time Domain Analog Side Channel Analysis for Malware Detection

$$\mathbf{S_i} = \{w_p : \|w_p - c_i\|^2 < t \ \wedge \ w_p \notin S_j \ \forall j, \ 1 \leq j < i\}, \tag{8.2}$$

Update Step Update the cluster centroid c_i by averaging all members of cluster S_i.

$$\mathbf{c_i} = \frac{1}{|S_i|} \sum_{w_p \in S_i} w_p, \tag{8.3}$$

Once the assignments no longer change, select a new cluster centroid randomly from the words that have not yet been assigned to any cluster. The algorithm converges when all words are assigned to a cluster.

8.5.5 Monitoring Phase: Intrusion Detection

In the monitoring phase, the EM signal is continuously monitored and matched against the dictionary, and anomalous activity is reported when the monitored signal deviates significantly from its dictionary-based reconstruction.

8.5.6 Matching and Reconstruction

The monitored EM signal is split into windows that have the same duration as the dictionary words. The signal is then reconstructed by replacing each window with its best-matching dictionary word. This is illustrated in Fig. 8.18, with windows that

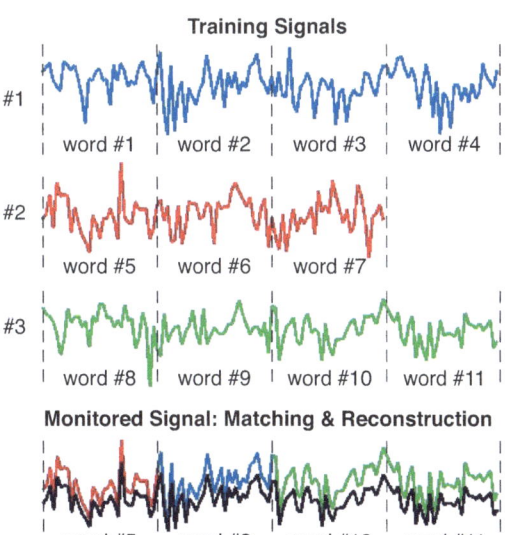

Fig. 8.18 An example of verifying the monitored signal using dictionary words. The monitored signal (shown with black lines) is verified against training (red, blue, and green) signals [23]

are each w samples in duration. After mean subtraction and scale normalization, a window is matched against the dictionary using the 1-Nearest Neighbor algorithm [9], with Euclidean distance as the distance metric. The entire signal can be reconstructed by concatenating the best-match words that correspond to the signal's sequence of windows, and this reconstructed signal is then used for anomaly (intrusion) detection.

8.5.7 Detection

The detection algorithm continuously compares the monitored signal with the reconstructed signal. Specifically, we compute the per-sample reconstruction error as the squared difference between samples of the monitored and reconstructed signal.

$$e(n) = (x(n) - y(n))^2, \tag{8.4}$$

where $x(n)$, $y(n)$, and $e(n)$ denote the normalized monitored signal, the reconstructed signal, and the squared reconstruction error signal, respectively, while n denotes the sample-index.

The detection algorithm is illustrated in Fig. 8.19. Then, we apply an L-samples long Simple Moving Average (SMA) filter to signal $e(n)$, yielding filtered signal $\tilde{e}(n)$:

$$\tilde{e}(n) = \frac{1}{L} \sum_{i=0}^{L-1} e(n-i). \tag{8.5}$$

Finally, we set a threshold on $\tilde{e}$, and whenever this threshold is breached, we report an intrusion. Figure 8.20 illustrates how reconstructed curves based on the IDEA algorithm match the original program execution vs. the execution with malware. From the plot, we can observe that reconstructed signal deviates significantly from malware-affected execution, while the deviation is small during normal execution. This allows us to detect malware.

8.5.8 System Parameters

The performance of this system depends on a number of system parameters, such as word-length l, word-shift s, and order of the SMA filter L. The rest of this section is a discussion of how these system parameters are chosen.

8.5 Time Domain Analog Side Channel Analysis for Malware Detection 179

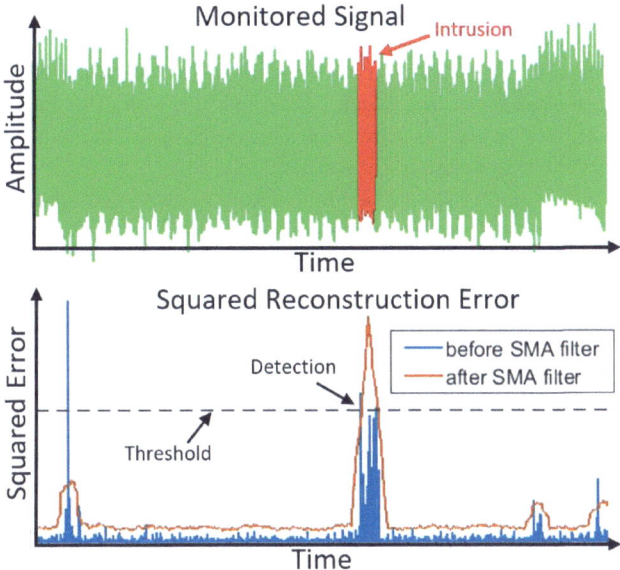

Fig. 8.19 Intrusion detection from reconstruction error: the squared reconstruction error is passed through a SMA filter to reduce false positives [23]

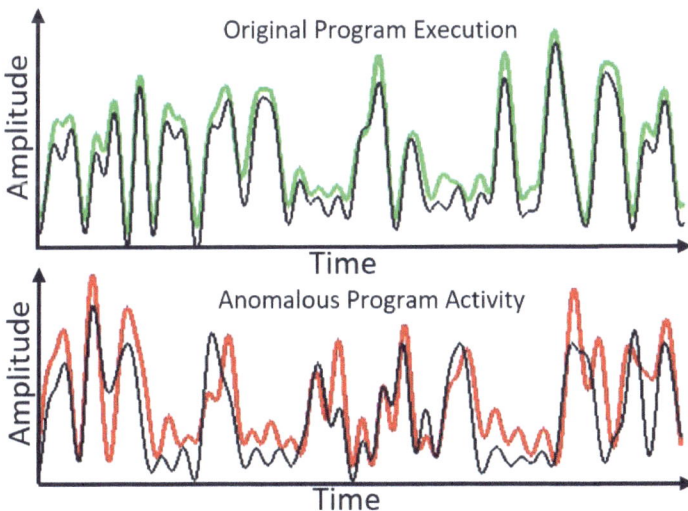

Fig. 8.20 Top green curve: original program execution; Bottom red curve: program execution with intrusion; Black curve: IDEA-reconstructed signal. Note that black curve better matches with green than with red curve [23]

8.5.8.1 Word-Length

Word-length l affects the performance of the IDEA. The optimal word-length l is a tradeoff between confidence in a match (the longer the word, the more reliable the match) and likelihood of a good match (the shorter the word, the more likely it is to find a dictionary word that matches it well).

Therefore, word-length l should be long enough to avoid good matches among a set of unrelated signals. To achieve that, a word in a dictionary should represent a relatively long sequence of processor instructions or hardware activity. This ensures that it is unlikely that a not-trained-on program will produce a sequence of executed processor instructions that is an excellent match for any dictionary entry of a trained-on program.

On the other hand, word-length l should be short enough so that random events, such as cache misses or interrupts, do not preclude good matches. For example, if we use a word that is very long, it will be difficult to find good matches in the dictionary of any reasonable size. This is because different inputs and hardware activities result in different signals when executing the same code. A reasonably-sized dictionary can contain only a small subset of the possible valid words, and a (long) window of the monitored signal will likely exhibit many input-dependent and hardware behaviors that do not match any of the dictionary words. By using a smaller word length we limit the number of word variants that can be produced, which reduces the dictionary size required for "full" coverage of these variants. Even when dictionary coverage of word variants is not complete and a window of the monitored signal has a set of input-dependent and hardware behaviors that is not represented in the dictionary, a smaller word length increases the probability that the dictionary contains a word that matches the window for most of its duration, and thus still produces a reasonably small reconstruction distance.

In order to estimate the optimal word-length l, we insert snippets of untrained-on signals into the trained or trusted reference signals. First, we record EM signals by executing a benchmark program with different inputs. Next, we follow a 10 fold cross validation to test each of these signals with and without an "untrained" insertion from a different benchmark program. Here, signals without insertion represent class 0 or "known", while signals with insertion correspond to class 1 or "intrusion". Figure 8.21 shows the histograms corresponding to "known" and "intrusion" with different word-lengths. For $w = 16$ samples, the Maximum Mean Squared Reconstruction Error (MMSRE) is low for both known (or trained) and intrusion (or untrained) signals (i.e., even an untrained signal can be matched with words in the dictionary). As a result, the two histograms overlap. However, for $w = 32$ samples, MMSRE corresponding to the known signal is significantly lower than that of the intrusion, and there exists a clear threshold between the two classes. When the word-length is much larger, i.e. $w = 128$ samples, MMSRE for both known and intrusion signals gets much higher (i.e., even a trained-on signal cannot be matched with low Euclidean distance). Hence, the two histograms cannot be separated anymore.

8.5 Time Domain Analog Side Channel Analysis for Malware Detection

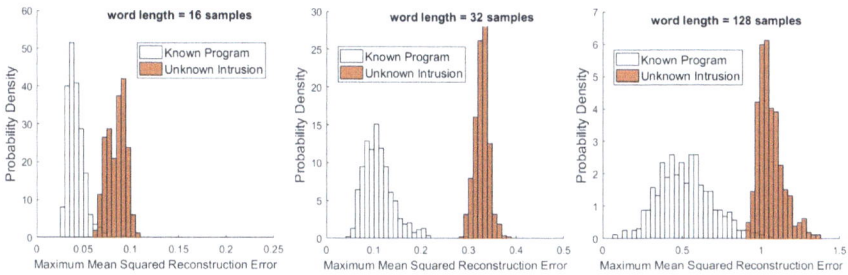

Fig. 8.21 Histogram for maximum mean squared reconstruction error with different word-lengths [23]

These experimental evaluations reveal that, for any intrusion larger than 256 samples, this system can achieve Area Under the Curve (AUC) better than 0.9995 on the Receiver Operating Characteristic (ROC) curve for any word-length between 32 to 64 samples. If not specified differently, the rest of this discussion assumes a word-length $w = 32$ of samples. This corresponds to 5 μs of execution time, which is about 250 processor clock cycles on the NIOS processor used in our experiments.

8.5.8.2 Word-Shift

Another parameter that impacts the performance of IDEA is the word-shift. Each "word" in the dictionary has to be shifted some number of samples from the previous one in order to compensate for hardware activities such as cache hit or miss. We estimate the optimal word-shift s through experimental evaluation. Again, we exploit a 10 fold cross validation in which snippets of insertions from an "untrained" program are treated as intrusion. Figure 8.22 shows the ROC curve for different word-shifts for an intrusion of 128 samples. It is clear that $s = 1$ performs the best, while larger s results in smaller AUC. This result is intuitive: $s = 1$ mimics shift-invariant signal matching most closely. However, it should be noted that, detection performance for $s = 2$ is comparable to that of $s = 1$. Hence, $s = 2$ can be used to reduce computing memory requirements, as its number of entries in the dictionary would be roughly halved compared to $s = 1$. Nevertheless, to highlight the performance of IDEA, we use $s = 1$ in our experiments.

8.5.8.3 Filter Order

In order to justify the use of the SMA filter and to determine its optimal length, consider the following detection problem. Let $\epsilon(n) \triangleq x(n) - y(n)$ denote the error signal, defined as the difference between the monitored and reconstructed signals. Observing an L-samples segment thereof $\epsilon(n-L+1), \ldots, \epsilon(n)$, we wish to decide whether:

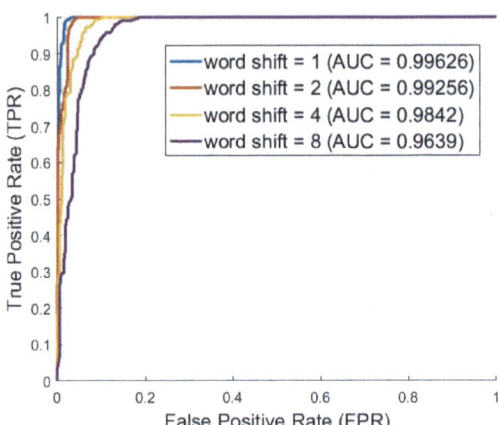

Fig. 8.22 ROC curves for intrusion detection with different word-shift [23]

- H_0 : This is a valid program execution segment; or
- H_1 : This is an intrusion code segment.

We begin by attributing two simplified statistical models to the error signal under each hypothesis: $\epsilon(n)$ is assumed to be an independent, identically distributed (i.i.d.) zero-mean Gaussian process, but with a different variance under each of the different hypotheses:

- $H_0 : \epsilon(n) \sim \mathcal{N}(0, \sigma_0^2)$,
- $H_1 : \epsilon(n) \sim \mathcal{N}(0, \sigma_1^2)$,

where $\sigma_0^2 < \sigma_1^2$ are fixed variances (assumed to be known, for now). The Likelihood Ratio Test (LRT) for deciding between the two hypotheses then takes the form:

$$\frac{f(\epsilon(n-L+1),\ldots,\epsilon(n)|H_1)}{f(\epsilon(n-L+1),\ldots,\epsilon(n)|H_0)} \underset{H_0}{\overset{H_1}{\gtrless}} \eta, \tag{8.6}$$

where $f(\cdot|H_i)$ denotes the conditional joint probability distribution function (pdf) of the observations given $H_i, i = 0, 1$, and η is a threshold value. Substituting Gaussian distributions and taking the logarithm we get

$$-\frac{L}{2}\log(\sigma_1^2) - \frac{1}{2\sigma_1^2}\sum_{i=0}^{L-1}\epsilon^2(n-i) + \frac{L}{2}\log(\sigma_0^2) + \frac{1}{2\sigma_0^2}\sum_{i=0}^{L-1}\epsilon^2(n-i) \underset{H_0}{\overset{H_1}{\gtrless}} \log(\eta), \tag{8.7}$$

which can be further rearranged as

$$\frac{1}{L}\sum_{i=0}^{L-1}\epsilon^2(n-i) \underset{H_0}{\overset{H_1}{\gtrless}} \tilde{\eta} \triangleq \frac{\kappa\sigma_0^2}{\kappa-1}\left(\log(\kappa) + \frac{2}{L}\log(\eta)\right), \tag{8.8}$$

8.5 Time Domain Analog Side Channel Analysis for Malware Detection

where $\kappa \triangleq \sigma_1^2/\sigma_0^2 > 1$ denotes the ratio between the two variances. This means that the average of squared samples of $\epsilon(n)$ over the L-samples observation interval is to be compared to some threshold $\tilde{\eta}$. This average is precisely $\tilde{e}(n)$, the output of a length-L SMA filter in (8.5), with $e(n) \triangleq \epsilon^2(n)$.

To determine the optimal value for L, note first that the mean and variance of $\tilde{e}(n)$ under H_i are (resp.) σ_i^2 and $\frac{2}{L}\sigma_i^4$, $i = 0, 1$. We note further that, if L is sufficiently large (say, larger than 10 or so), the distribution of $\tilde{e}(n)$ under each hypothesis is approximately Gaussian according to the Central Limit Theorem. Consequently, the false positive and false negative probabilities can be shown to decay monotonically in L. Therefore, to have them minimized, L should take the largest possible value that does not breech the H_1 model. Namely, L has to be chosen as the full length of the shortest possible intrusion. Since that length is not known a priori, we chose to set the filter order equal to the shortest length of insertion that we intend to detect. If not specified differently, we use $L = 256$ as the filter length in our experiments.

Note that in reality the situation is somewhat more complicated than described above: the i.i.d. Gaussian signal model for $\epsilon(n)$ is inaccurate, as its samples are expected to be correlated and not necessarily Gaussian distributed; variances σ_1^2 and σ_0^2 would rarely be known; reconstruction errors may be very high for a few brief, sporadic segments, even under H_0, resulting either from lack of full coverage in the training phase (i.e., lack of appropriate training examples that follow the same control flow path as the monitored signal) or from the variability of hardware activities between the training signal and the monitored signal. These complications certainly undermine any claim of optimality of the LRT in this case. However, the rationale behind the resulting test remains valid, justifying the use of the SMA filter and the choice of L. For example, the possibility of short occurrences of large errors under H_0 would merely increase the mean and variance of $\tilde{e}(n)$ under H_0, thereby increasing the false positive rate. Nevertheless, the dependence of this rate on L remains monotonically decreasing, still supporting our choice of the largest possible L that does not breach H_1.

In this section we present our experimental results on detecting several different types of malware on different applications and embedded systems. It is important to emphasize that this method is not limited to these specific applications and/or malware—it can be applied to any system that has observable EM emanations.

8.5.9 Experiments with Different Malware Behaviors

To show the effectiveness of the system for detecting different malware behaviors, we selected 3 applications from the SIR repository [36]: (*Replace*, *Print Tokens*, and *Schedule*) and implemented three common types of embedded system malware payloads (DDoS attacks, Ransomware attacks, and code modification) on an Altera DE-1 prototype board (Cyclone II FPGA with a NIOS II soft-processor) while executing any of these SIR applications. We selected applications from SIR repository because they are relatively compact (allowing manual checks of our results to get a deeper understanding of what affects the system accuracy), they are commonly used to evaluate performance of techniques that analyze program execution, and they have many program inputs available (each taking a different overall path through the program code) so we can use disjoint sets of inputs for training and monitoring and still have a large number of inputs for training (to improve code coverage) and monitoring (to obtain representative results).

For our implementation of a *DDoS* attack, we assume that a vulnerability (e.g., buffer-overflow) is exploited to divert control-flow (e.g., using code-reuse attack) to a code that sends DDoS packets in rapid succession, without waiting for reply from the target. After sending a burst of packets, the malware returns execution to the original application code, so the device continues to perform its primary functionality. In the case of an embedded device, e.g., Altera DE-1 (Cyclone II FPGA) board, there is no traditional network (e.g., Ethernet) port. Instead, we implement the packet-sending activity using the JTAG port.

For our implementation of a *Ransomware* attack, we implement encryption of the victim's data [10]. We assume that the attacker inserted the malicious code through firmware modification. We used *Advanced Encryption Standard*, AES-128 for encryption. Encryption of large files is time-consuming, and can be easily detected by this system. To make things more challenging, the *Ransomware* we implement encrypts only 16 bytes of data (one AES-128 encryption block). Note that more secure cyphers, e.g., AES-256, have larger encryption blocks, and are thus easier to detect.

For our implementation of *code modification*, which is the basis for an important class of malware (APT and firmware modification attacks), we assume that the malware has already successfully modified the program, and that the goal of the system is to detect when this modified part of code executes. For example, in the *Replace* benchmark, there is a function called *subline()*, which is used to search for words in an input string. In a scenario where authorities use this code to look for names in intercepted communication, the attacker's modification of this code would prevent any word that begins with specific initial letters from being reported.

Our evaluation has two parts. First, to demonstrate the effectiveness of IDEA for different kinds of original application/program activity and different devices, we port several key malware behaviors such as DDoS payload, encryption activity of a ransomware payload, and firmware modification on an IoT development board (A13-OLinuXino [30]), as well as to Intel-Altera's FPGA Nios-II softcore. In the second part, to examine practical applicability, our evaluation uses three real-world CPS applications: a medical embedded device called *SyringePump*, a proportional-integral-derivative (PID) controller for the temperature of an industrial soldering iron, and a robotic arm used in a factory's assembly line. A buffer-overflow+control-flow hijack attack and an attack that models an advanced persistent threat (APT) scenario is used to attack these applications which are implemented on a popular embedded system, Arduino UNO.

8.5 Time Domain Analog Side Channel Analysis for Malware Detection

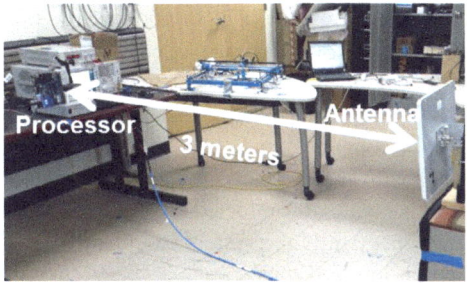

Fig. 8.23 Measurement setup used to collect EM traces from distance [23]

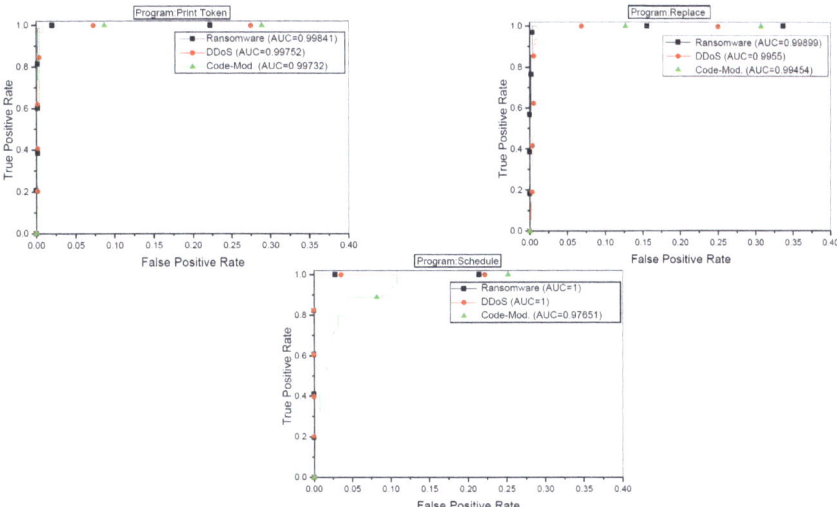

Fig. 8.24 Receiver operating characteristic curves for intrusion detection on three different benchmark programs for different variants of malware [23]

Setup

To receive EM emanations, we used a 2.4–2.5 GHz 18 dBi panel antenna that was placed 1 m, 2 m, and 3 m away from the device (see Fig. 8.23). We used a (fairly expensive) Agilent MXA N9020A spectrum analyzer to demodulate and record the EM emanations, primarily so we can have more control and flexibility in our investigations, but we will later show that a sub-$1000 software-defined radio receiver (USRP-B200mini [5]) can be used instead. Malware injections occur in 30% of the runs, randomly selected, and in each injection the injection time (relative to the beginning time of the run) is drawn randomly from a set of 10 predefined values.

Detection Performance

In order to evaluate the performance of IDEA, we use a 10-fold cross-validation. We execute each benchmark program with different inputs. Some of these executions are infected with malicious intrusions. As we do not assume any a priori knowledge about the intrusions, we do not use any of these infected executions for training the system. Likewise, we select all the system parameters through experiments with "untrained" insertions from a different benchmark program, without using any actual infected executed signal. The correct-executing signals are randomly divided into 10 roughly equal-sized subsets. In each fold, we use one of these subsets, along with the infected signals to test the accuracy of IDEA, after training with the other nine non-infected subsets. The objective of the experiment is to measure how successfully the system can identify and differentiate between the infected and the non-infected runs.

We follow this procedure with three different benchmark programs, namely *Print Tokens*, *Replace* and *Schedule*. For *Print Tokens*, a total of 637 executions were recorded, out of which 142 were infected with different variants of malware— 68 Ransomware, 66 DDoS, and 8 code-modification. In case of *Replace*, we have recorded 691 executions, including 138 infected ones (68 were Ransomware, 65 were DDoS, and 5 were code-modification). For *Schedule*, we collect 681 executions where 144 of them were infected (68 Ransomware, 67 DDoS, and 9 code-modification).

Figure 8.24 shows the ROC curves for intrusion detection on different benchmark programs with different variants of intrusions. These results show excellent performance in all three benchmark programs, with Area Under the Curve (AUC) very close to 1 for Ransomware and DDoS. Code modification creates a much smaller change in the program's execution, thus is much harder to detect. However, IDEA still detects it with an AUC > 97.5%. Specifically, detection of all code modifications is achieved by tolerating a false positive rate of no more than 1% in *Print Tokens* and *Replace*, and no more than 12% in *Schedule*.

Detection at Different Distances

Next we test IDEA at three different distances (1, 2, and 3 m) from the monitored device. The results (Fig. 8.25) show that its performance remains stable over different distances. In fact, for all three benchmark programs, the performance is quite similar at 1 m and 2 m, with AUC better than 99.5%. The performance degrades somewhat at 3 m, achieving 99% AUC for *Replace* and 98% AUC for *Print Tokens* and *Schedule*. The degradation in accuracy at the 3 m distance is mainly due to a reduced signal-to-noise ratio, as the monitored signal weakens with distance, and we believe that these results can be improved by using customized (higher-gain) antennas and low-noise amplifiers.

Detection at Different SNR

To evaluate IDEA in the presence of environmental EM noise, we introduce Additive White Gaussian Noise (AWGN) into the monitored signal, and test the system by monitoring *Replace* at three different SNRs (20, 10, and 5 dB). The results are shown in Fig. 8.26. Experimental evaluation demonstrates that the system is robust against EM noise. In fact, it achieves an excellent performance (AUC > 99.5%) at 20 dB SNR. Furthermore, at 10 dB SNR, it can still detect intrusions with

8.5 Time Domain Analog Side Channel Analysis for Malware Detection

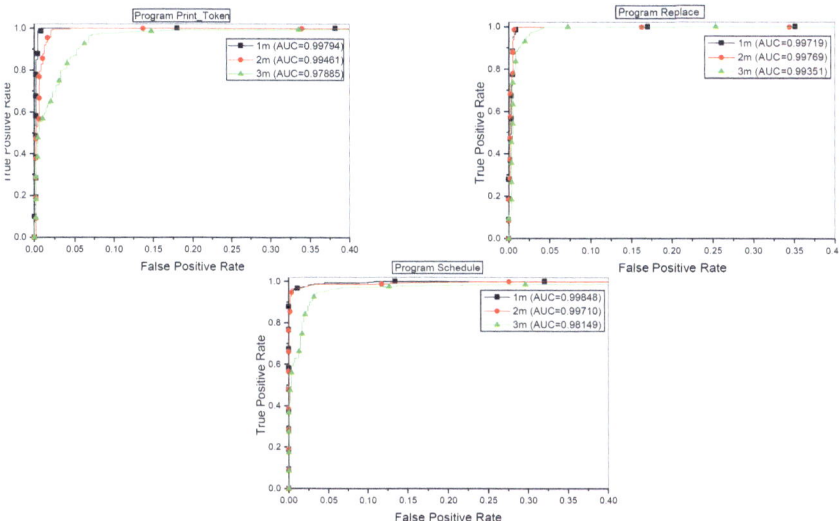

Fig. 8.25 Receiver operating characteristic curves (ROC) for intrusion detection on three different benchmark programs from different distances [23]

Fig. 8.26 ROC curves for intrusion detection at different SNR [23]

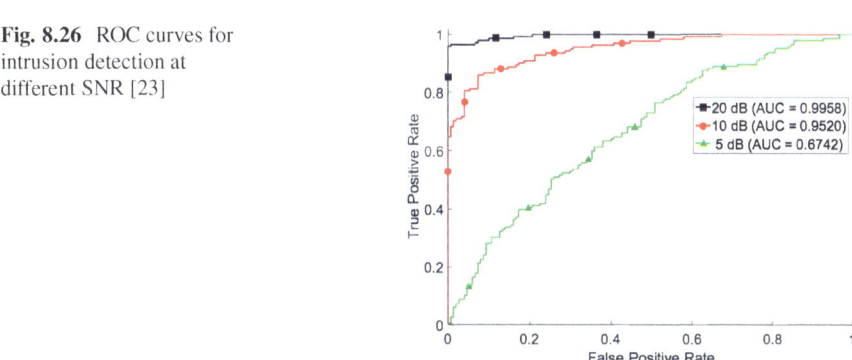

an AUC > 95%. However, at 5 dB SNR, the detection performance degrades to a 67.4% AUC.

In addition, this system is robust against EM interference from adjacent devices. It monitors the target device at a frequency band around its processor's clock frequency. Any EM interference with frequencies outside this monitored band is blocked by the anti-aliasing filter during the analog-to-digital conversion (ADC). Since each device emits EM signal at its own clock frequency, interference is limited and can be filtered out using signal processing.

8.5.10 Experiments with Cyber-Physical Systems

To assess practicality of IDEA, i.e., to demonstrate its ability to successfully detect real malware on a real CPSs, we use IDEA to monitor three industrial CPS implemented on a well-known embedded system, Arduino UNO. We use a control-flow hijacking attack, an APT, and a firmware modification attack on three different CPSs.

The first CPS we use is *Syringe Pump*, which is designed to dispense or withdraw a precise amount of fluid, e.g., in hospitals for applying medication at frequent interval [46], which is described in more detail in Section 8.3.7. We implement a *control-flow hijack* attack on this system by exploiting an existing buffer-overflow vulnerability in a subroutine that reads (*serialRead()*) the inputs. This hijack causes the program's control-flow to jump to an *injected* malicious code. We assume that the adversary is interested in disrupting the correct performance of the system by dispensing/withdrawing an unwanted amount of fluid which could cause a significant damage to the patient, thus the injected code causes the syringe to dispense a random amount of fluid. The buffer-overflow is implemented by sending a large inputs to overwrite the stack followed by the address of the "injected" malicious code which overwrites the actual return address of the *serialRead()*.

The second system is a proportional-integral-derivative (PID) controller that is used for controlling the temperature of a soldering Iron [2], which was also described in more detail in Section 8.3.7. To implement an APT on this application, we assume that the adversary's malware (like in Stuxnet) has already infiltrated the system and can reprogram the device. The adversary's goal is to change a critical value under some conditions, which in turn can cause damage to the overall physical system. In this system, a possible malicious modification is based on one or several conditions (e.g., in our evaluation it checks the model of the device that is stored in memory), the temperature history can be changed so that the system occasionally sets a wrong temperature. The injected code is only 2 lines of code (i.e., IF(X) THEN LASTTEMPHISTORY = RANDOMVALUE).

The final system in our evaluation is a robotic arm [1], which was described in more detail in Section 8.3.9. There is a growing concern in security of these CPSs since they are typically connected to the network and are exposed to cyber-threats (e.g., [35]). For this system, as in Section 8.3.9., we use the same *firmware modification* attack that we used in Section 8.3.9.

Setup

An Arduino UNO with an ATMEGA328p microprocessor clocked at 16 MHz is used to implement each CPS. An magnetic probe is used to receive EM signals from the device. Figure 8.27 shows the experimental setup for the *SyringePump*. For all measurements, we use a commercially available software-defined radio (SDR) receiver (Ettus Research B200-mini) to record the signal. This SDR costs significantly lower than a spectrum analyzer, making it a practical option for monitoring security-critical systems. For each CPS, we use 25 randomly selected

8.5 Time Domain Analog Side Channel Analysis for Malware Detection

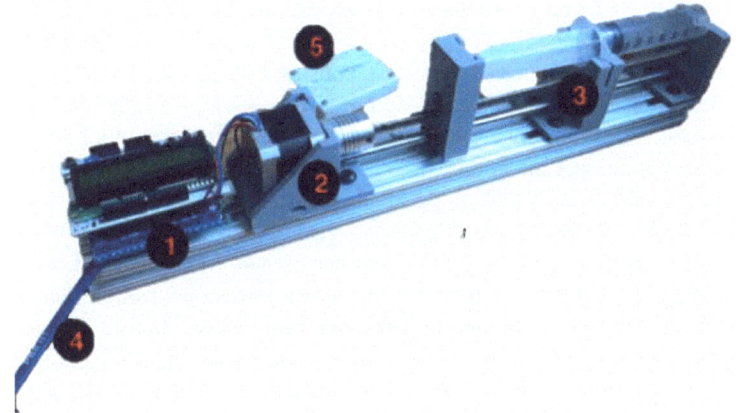

Fig. 8.27 Experimental setup for the SyringePump: (1) Arduino device with LCD, (2) stepper motor, (3) syringe, (4) magnetic probe and (5) software-defined radio [23]

Table 8.3 Experimental results for three malware on three CPS

System	AUC	Malware type
SyringePump	> 0.999	Control-flow hijack
PID controller	> 0.999	APT
Robotic arm	> 0.999	Firmware modification

signals for training, and then 25 malware-free and 25 malware-afflicted signals for testing.

Detection Performance

Table 8.3 summarizes the detection accuracy of the system on the three CPSs. As seen in the table, in all cases, IDEA has successfully detected every instance of a malware without reporting any false positive. This is promising for monitoring critical cyber-physical systems, where very small false positive rate (and a high detection accuracy) is required. Note that in, all cases, the runtime for the malicious code is significantly less (<0.01%) than the overall runtime of the application.

8.5.11 Experiments with IoT Devices

To demonstrate the robustness of the IDEA system, we also use it to monitor an A13-OLinuXino (Cortex A8 processor) IoT board. Unlike the FPGA-based system that runs the application "on bare metal," this board has a Linux-based operating system (OS). The defensive mechanisms already present in the OS make it harder to inject prototype malware activity. Instead, we model malware injection by injecting snippets of signals from a different (not-trained-on) program. For this experiment, we use *Replace* as the reference program, on to which signal-snippets from *Print Tokens* were inserted as anomalous (not-trained-on) signal. This approach also allows injections of any chosen duration, and use of different signals for different injection instances. In contrast, construction of even one short-duration actual malware instance is very challenging. For example, a single packet sent in a DDoS attack, or single-block encryption in Ransomware, lasts much longer than any of our signal-snippet injections.

To allow a direct comparison between our real-malware and signal-snippet injections, we also perform signal-snippet injection experiments on the DE-1 FPGA board. We use 10-fold cross validation, and test signals from a trained benchmark program *Replace*, with or without insertions or intrusions from *Print Tokens*. Figure 8.28 shows the experimental results. We can observe that intrusions longer than 256 samples (i.e., 200 instructions or 40 μs length) on the FPGA are detected with an AUC of 99.95%. For the IoT board, an AUC better than 99.8% is achieved for intrusions with at least 1024 samples (i.e., 800 instructions or 7.94 μs length). The difference in duration of the intrusion that is needed to achieve the same AUC on the two devices is mainly due to OS activity that is present on the IoT board and absent on the FPGA board. This OS activity introduces variation in the signals, increasing reconstruction error even for valid executions. This, in turn, raises the reconstruction error threshold for reporting an anomaly at a given confidence level, so more anomalous samples are needed to reach this increased reconstruction error threshold.

8.6 MarCNNet: A Markovian Convolutional Neural Network for Malware Detection and Monitoring Multi-Core Systems

In previous sections we have described how malware can be detected using analog side channels when program is running on a single-core processor. The tracking

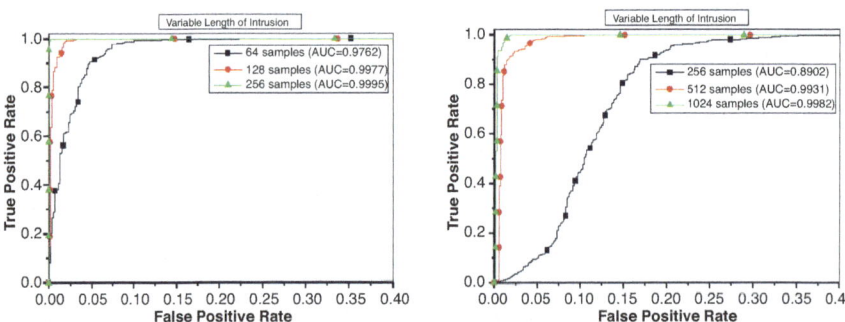

Fig. 8.28 Receiver operating characteristic curves for intrusion detection on a FPGA (top) and on an IoT Device (bottom) [23]

8.6 MarCNNet: A Markovian Convolutional Neural Network for Malware...

of program activity becomes more complex when multiple cores are active at the same time. We now present our method to track program execution on multiple cores at the same time [50]. To track multiple cores when executing simultaneously, we use a combination of Markov and convolutional neural network (CNN) models, called MarCNNet. In this framework, the CNN estimates the likelihood of being in any given state of any program by exploiting the features that are learned during the training phase, and the Markov Model monitors programs by considering the possible transitions between states.

8.6.1 Emanated EM Signals During Program Execution

As it executes, a program typically goes through a number of computational tasks, spending most of its time in each task on loops and other repetitive activity and relatively little time on transitions between them. Thus each loop or other time-consuming computational activity can be considered to be a hot-spot in terms of the time spent in that part of the code. Therefore, we can claim that execution of a program mainly consists of executing these hot spots. Additionally, because the tasks in the program typically depend on each other, i.e., one must be completed before another is begun, which hot-spots may be executed at any given time depend on which hot-spots have already been executed, i.e., after one hot-spot is executed, the very next hot-spot to be executed is not randomly chosen from among the set of all hot-spot code in the program. Instead, the very next hot-spot to execute is typically one of very few hot-spots. For example, some hot-spot A may only be followed by some hot-spot B, while another hot-spot X may only be followed by either hot-spot Y or hot-spot Z. To illustrate how execution time is dominated by hot regions, and how these regions are sequenced, we consider the `Bit_count` benchmark from MiBench suite [18]. Figure 8.29a shows the spectrogram of the received EM signal when only one core is active on the monitored device. This spectrogram was generated by first demodulating the signal, and then computing and plotting the STFT of the demodulated signal. Lighter regions in the figure represent frequency components that are relatively strong at that point in time. Here, the most powerful frequency component at the center, which is present for the entire execution of the program, corresponds to the clock frequency of the device, while the other strong frequency components that are symmetric around the clock frequency represent the modulated frequency components due to looping operations that exist in the program. The main observation is that this program contains seven dominant regions which are executed in a specific order. The frequency components observed in the spectrogram are related to execution time of a single iteration of a loop. For example, if the iteration takes T_{alt} seconds, we observe an RF signal component at the alternation frequency, $f_{alt} = 1/T_{alt}$. However, if the iteration time varies in time due to program activities, we observe a *smearing* around the expected frequency, as seen in the first loop of `Bit_count`.

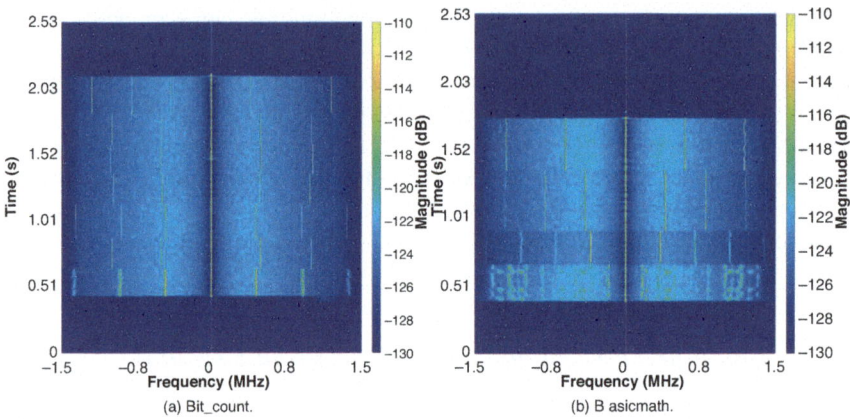

Fig. 8.29 Spectrogram when both Bitcount and Basicmath are running [50]

Similarly, the spectrogram of the received signal when a single core is running Basicmath (another benchmark from MiBench) is given in Fig. 8.29b. This time, four different regions are observed with non-overlapping frequency components. We also observe that these regions also follow a sequence.

The question here is how the emanated signal is composed when multiple cores are active. To investigate the mixtures of signals while multiple programs are executed at the same time, we run Basicmath and Bit_count in parallel on two different cores on the same device. The spectrogram of the received signal is given in Fig. 8.30. We observe that the received signal is superposition of the signals when Bit_count and Basicmath were executed on a single core, with lower signal-to-noise ratio (SNR). Therefore, it is possible to reconstruct the spectrogram of emanated signals when both cores are active if the relative initialization times of the programs are known. Note here that a perfect superposition of signals is not possible because, whenever multiple cores are active, the system draws extra power that causes a certain amount of decrease in the SNR of the received signals. However, the received signal in Fig. 8.30 illustrates that it still preserves the characteristics of both programs: the same frequency components from single-core measurements are still present. Therefore, the STFT magnitudes of the received signals for multi-core devices can be modeled as the summation of STFT outputs of single core measurements with some additive white noise. As long as one-core measurements are available for each program, any possible signal combination can be generated irrespective of the time that programs are initialized.

8.6 MarCNNet: A Markovian Convolutional Neural Network for Malware... 193

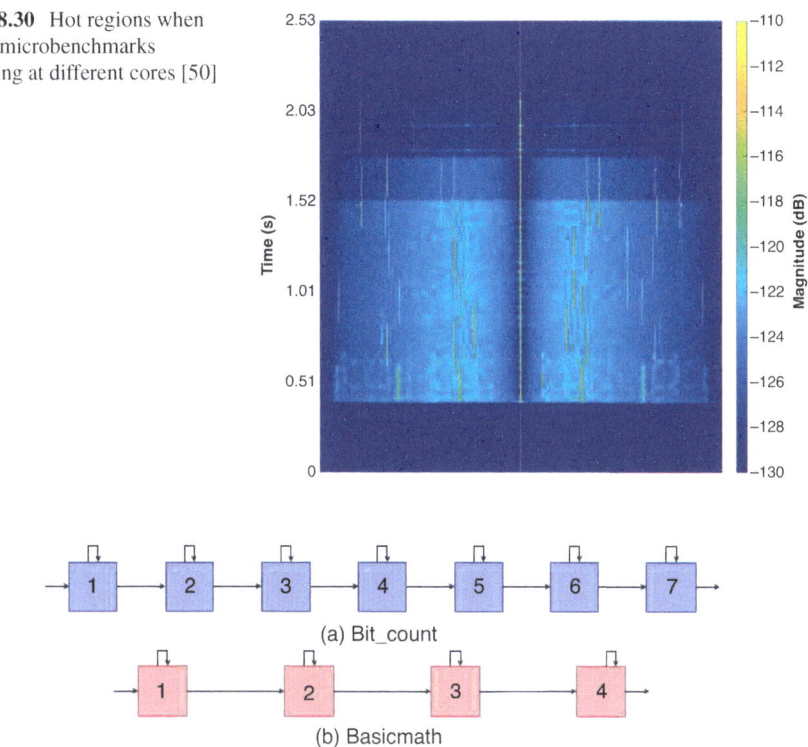

Fig. 8.30 Hot regions when both microbenchmarks running at different cores [50]

Fig. 8.31 Markov models representing the execution of benchmarks [50]

8.6.2 Markov-Model-Based Program Profiling: MarCNNet

In this section, we introduce Markovian Convolutional Neural Network Model, called MarCNNet. To simplify discussion, consider a device with two cores and two programs: Bit_count and Basicmath.

Remembering that execution of a program is a sequence of hot regions, with very few candidates for the region that is to be executed at any step in that sequence, we first investigate the spectrograms given in Fig. 8.29. For both cases, *hot paths* dominate the run-time of the programs and demonstrate a similar pattern during a loop execution. These loops can be considered to be states, of a given program as they activate distinct frequency components, which can be used to reveal the current state. Moreover, these states follow an order/path that is defined by the program. To represent this dependency among hot regions, we use Markov models, where hot regions are states and transitions among states exist only when one region can directly follow after another.

The state transition diagrams of Markov Models for both benchmarks are given in Fig. 8.31. The states of these two benchmarks depend only on the previous hot

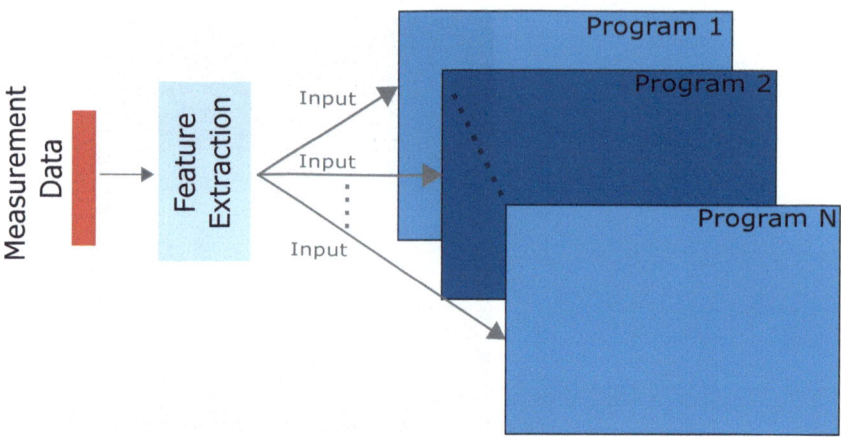

Fig. 8.32 Convolutional neural network model to track N different programs [50]

region, meaning that there is only one path from each state for these benchmarks. Please note that, although these Markov Models are very simple (each is a sequence), the proposed model can track a program with many branching operations (i.e., more complicated Markov Models), and Markov Models of typical programs are typically a mix of sequences (e.g., a loop follows another in the code) and low-degree branching (e.g., an if-then-else decides which of modules/loops to execute next).

The model is designed to have N parallel units, under the assumption that the system is running N programs. While the Markov model represents the dependency of the execution paths, the question is how to identify the current execution point of a given program. For that, we remember that *hot paths* are generally a result of looping activities within a program. Assuming that each iteration of the loop takes equal time to execute, we expect that the same frequencies are activated until the end of the loop. However, the execution times of software activities vary, which causes shifts in the frequency domain [48]. To be able to deal with problems that arise due to working on multi-core systems and the spread of the frequency components due to execution time variation, we use a convolutional neural network, which has good built-in tolerance to local variations [25]. Convolutional neural networks are generally used for image, speech and time series, and the structure of the overall model is important to achieve better true classification rates [6, 24].

The overall CNN model in *MarCNNet* is shown in Fig. 8.32. The model has N parallel units, which correspond to the N different programs that are monitored. The detailed CNN model of one CNN unit is shown in Fig. 8.33, and consists of: (1) 3 convolutional layers, (2) 3 dense layers, and (3) 1 output layer whose size changes based on the number of states of the corresponding program.

After the first and second convolutional layers, we apply max-pooling operations with a stride of 5 and a kernel size of 10. The kernel sizes of the convolutional layers are 55, 35, 15, respectively. We also utilize a dropout layer before each convolutional

8.6 MarCNNet: A Markovian Convolutional Neural Network for Malware...

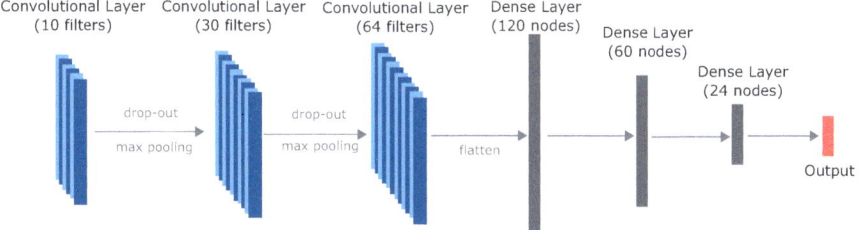

Fig. 8.33 Detailed convolutional neural network model for the branches [50]

layer with a dropout rate of 0.05 [42]. ReLU activation function is used in each layer except the output layer. We use a cross entropy loss function which is defined as

$$loss[\mathbf{o}, class\_id] = -\log\left(\frac{\exp(\mathbf{o}[class\_id])}{\sum_n \exp(\mathbf{o}[n])}\right), \quad (8.9)$$

where *class_id* is the state label of the considered training signal and **o** is a vector containing the outputs of the neurons at the output layer. We analyze the model with various hyper-parameters, i.e., # of layers, stride, etc., and come up with the parameters given above. However, we do not claim that this model is *the* optimal model because there are infinitely many options for the model selection, yet the proposed CNN model extract features that are distinctive to profile a system.

8.6.3 Input Signal and Training Phase

In this section, we first describe the input signal that is fed to the CNN, and then explain the training phase of the overall framework that extracts and learns the distinctive features of *hot paths* of each program.

Extracting Features for the Neural Network
Monitoring in real-time, with high sampling rates, results in very large input data and requires pre-processing before the data is fed to the neural network. The measured signal also has non-relevant activity and this interference should be minimized for accurate program tracking. One straightforward approach is to use time series samples with a specified window. However, because of the high sampling rates of the measuring devices (which increases the dimension of the input layer), interrupts and corruption due to environmental signals, this method can result in lower accuracy rates and generate false malware alerts [12]. Also, this approach has a larger number of parameters to optimize. To reduce the size of the data and increase the SNR level, we utilize the first phase of the *Two-Phase-Dimension-Reduction* methodology proposed in [49]. This method first calculates the STFT of

the signal and averages the magnitudes of STFT outputs. For a better explanation of the method, let $\mathscr{F}$ be the STFT window size, O_S be the number of non-overlapping samples at each STFT calculation, Ξ be the number of STFT operations to average, and $\mathbf{e}_F$ be the extracted features that are inputs to the neural network. The features can be written as

$$\mathbf{e}_F[k] = \sum_{n=1}^{\Xi} |X_n[k]|, \qquad (8.10)$$

where $k \in \{0, 1, \cdots, \mathscr{F} - 1\}$, and

$$X_n[k] = \sum_{\xi=0}^{\mathscr{F}-1} \Theta[\xi + (n-1)O_S] \exp(-j2\pi k\xi/\mathscr{F}), \qquad (8.11)$$

where Θ is the measured raw signal. To generate $\mathbf{e}_F[k]$, the number of time series samples used in the framework is Φ where

$$\Phi = \mathscr{F} + (\Xi - 1) \times O_S.$$

Therefore, the size of the input vector for the neural network is reduced from Φ to $\mathscr{F}$, which helps to improve SNR by diminishing the power of unrelated activities and reduce the number of parameters of the neural network, thus reducing its complexity.

However, we do not provide $\mathbf{e}_F$ to the neural network in the linear domain because signal power is dominated by its DC (carrier) component, while other frequency components have relatively small amplitudes, which causes computation to experience floating-point precision and other problems. Instead, we convert the input vector into the logarithmic (decibels) domain:

$$\hat{\mathbf{e}}_F = 20 \log_{10}(|\mathbf{e}_F|).$$

Training the CNN Model

The proposed neural network has N parallel units that are independent of each other. Therefore, in the training phase, the framework can be divided into N different CNN models, and treated as a number of separate training problems. However, the training signals are collected during single-core execution, but the neural network should be trained to identify hot-spots in multi-core execution. Therefore, we generate signals for network training from the available one-core measurements using Algorithm 8.1. The possible scenarios and corresponding strategies used in the algorithm are as follows:

→ *Strategy-1*: The training signal is fed to the CNN without changes. The assumption here is that, during testing, the signal may be collected when only a single core is active and in the same/similar environment that was used for training.

→ *Strategy-2*: The training signal is combined with noise and fed to the CNN, to make the output of the CNN robust under different noise scenarios. In other words, the assumption here is that testing may be in a different environment from training, with different noise figures.

→ *Strategy-3*: This strategy mimics the behavior of a multi-core signal that has the same noise figure with the collected training signals. It superimposes signals from different states of different programs by employing random weights to consider the destructive/constructive effect of multi-core signals on each other. This approach decreases the time required to collect experimental data because it does not require performing experiments with random initialization of program combinations. Another benefit of such a combination is that when more than one program is running, the neural network can still be able to monitor the system by producing a more confident estimate.

→ *Strategy-4*: This strategy reflects the testing scenario in which multiple cores are active and the test signal is measured in an environment with a different noise figure. It combines STRATEGY-2 and STRATEGY-3, i.e., the generated input is the noisy version of the weighted sum of different training signals.

In the algorithm, $\mathbf{E}_F$ represents a matrix of extracted features and $\mathbf{E}_F^{\xi,i}$ denotes the data for the state i ($i \in \{1, 2, 3, 4\}$ for Basicmath) and the program ξ, (e.g., Basicmath).

After establishing such a training procedure, the CNN model needs to be trained to learn the weights of the layers by applying backpropagation [26]. Please note that we do not consider the Markovian structure of the problem in the training phase of the CNN. In other words, the training is performed by ignoring the Markov part of the proposed framework. The Markov model is only used in the testing (monitoring) phase.

8.6.4 Testing While Multiple Cores Are Active

In this section, we introduce our monitoring procedure to detect malware. Although the main goal is to generate an alert when malware becomes active, we can extend this approach to also identify which of the programs is affected by malware. However, multi-core activity presents some problems that are not present in single-core malware detection:

- Some frequency components can be activated by several cores at the same time, and this can result in misclassification and/or false malware alert.
- Activating more than one core causes extra power consumption, thereby increasing the power of noise (emanations unrelated to program activity) in the signal. This decreases the overall SNR of the received signal.

Data: the_state_id, the_program_id, $\mathbf{E}_F$
// Generate two matrices based on the data labels.
$\mathbf{M}_0 = \mathbf{E}_F^{\text{the\_program\_id, the\_state\_id}}$
$\mathbf{M}_1 = \mathbf{E}_F^{\text{!the\_program\_id, !the\_state\_id}}$
// $\mathbf{M}_0$ contains all the data that belongs to the_state_id of the program the_program_id.
// $\mathbf{M}_{T_i}$ contains a subset of data belonging to i^{th} program.
// ttt: Generated training signals.

Randomly choose the training data generation strategy
if STRATEGY - 1 *is chosen* **then**
| ttt = $\mathbf{M}_0$
end
else if STRATEGY - 2 *is chosen* **then**
| Choose a random value for the white noise power, σ^2, such that SNR > 10 dB.
| Generate the additive noise components with mean zero and standard deviation σ.
| Assign $\mathbf{N}$ as the matrix that contains these noise components.
| ttt = $\mathbf{M}_0 + \mathbf{N}$
end
else if STRATEGY - 3 *is chosen* **then**
| Choose a random value, p, from a uniform distribution.
| Choose data from $\mathbf{M}_1$ to generate matrices $\mathbf{M}_{T_i}$ where $i \in 1, \cdots, N_C$ and N_C is the number of cores. The sizes of $\mathbf{M}_{T_i}$ are exactly equal to $\mathbf{M}_0$.
| ttt = $p_0 \mathbf{M}_0 + \sum_{i=1} p_i \mathbf{M}_{T_i}$ where $\sum_{i=0} p_i = 1$.
end
else
| // STRATEGY - 4
| Combine STRATEGY - 2 and STRATEGY - 3:
| ttt = $p_0 \mathbf{M}_0 + \sum_{i=1} p_i \mathbf{M}_{T_i} + \mathbf{N}$
end

Result: ttt

Algorithm 8.1: Overview of the training generation algorithm [50]

When the malware affects the entire spectrum, the proposed framework could not identify which program contains the malware. However, it still reports the presence of malware. One possible strategy to identify the affected program is to, upon detection of malware, temporarily switch to single-core execution that would reveal which program is affected by malware.

We summarize the monitoring procedure in Fig. 8.34. The main intuition behind the algorithm is to follow the states according to firings of CNN neurons (Fig. 8.35), and to account for the fact that false firings in the CNN can be a result of another unrelated activity distorting the actual signal, but the program tends to stay in each hot reagion for a while, and has to follow an order to execute the states (hot regions) of the program. Therefore, even if neurons other than the expected one(s) produce the strongest outputs, we only consider the expected ones among outputs that are above a threshold.

8.6 MarCNNet: A Markovian Convolutional Neural Network for Malware...

Fig. 8.34 Monitoring procedure in MarCNNet [50]

Algorithm 2: Profiling Procedure

Data: Θ, $\mathcal{M}_M$, $\mathcal{M}_N$.

// $\mathcal{M}_M$: Markov Model.
// $\mathcal{M}_N$: CNN Model after training.
// t_M: Threshold for announcing malware.
// t_S: Minumum time for execution of any state.
// t_L: Threshold for announcing a signal belongs to a cluster.
// o_F: Output of $\mathcal{M}_N$.

not_detected = # of states + 1
current_state = *not_detected*
transitions = []
while *true* **do**
 | Apply the procedure in Section IV-A to obtain $\hat{e}_F$.
 | Feed $\hat{e}_F$ to $\mathcal{M}_N$ to obtain o_F.
 | **if** *current_state != not_detected* **then**
 | | -Check whether the values of o_F for *current_state* and the state candidates of $\mathcal{M}_M$ are larger then threshold.
 | | -Apply softmax function for the states that are above the threshold.
 | | -Choose the most likely state as the next state candidate.
 | | **if** *The chosen state is current_state* **then**
 | | | Append *current_state* to transitions
 | | **end**
 | | **else if** *the time spent in current_state is larger than t_S* **then**
 | | | Update *current_state* and append to transitions
 | | **end**
 | | **else**
 | | | Append *not_detected* to transitions
 | | **end**
 | **end**
 | **else**
 | | **if** $o_F[1]$ *is larger than t_L* **then**
 | | | // The program is started!
 | | | Append 1 to transitions
 | | | *current_state* = 1
 | | **end**
 | | **else**
 | | | Append *not_detected* to transitions
 | | **end**
 | **end**
 | **if** *A program execution is started and the last entries of transitions corresponding to the last t_M seconds are equivalent to not_detected* **then**
 | | **if** *Program not ended* **then**
 | | | Alert malware!!!!
 | | **end**
 | **end**
end

Result: transitions

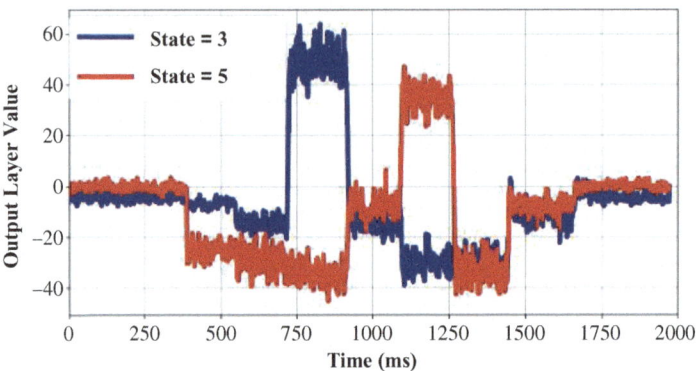

Fig. 8.35 The values of output layers for the states 3 and 5 [50]

Table 8.4 Devices used in our evaluation [50]

Device	# of cores	Clock frequency (GHz)
A13-OLinuXino-Micro [31]	1	1
A20-OLinuXino-LIME2 [32]	2	1
Alcatel Ideal [8]	4	1.1

Because other computer activity can stall the execution of monitored programs, the CNN output may not provide accurate estimates, and the Markov Model would then detect false positives (false alerts that malware is present). Therefore, the algorithm defines two parameters: t_M and t_S. t_M is the minimum time during which unknown states must be observed before a malware alert can be raised. t_S is the minimum time that each state must takes. These parameters helps avoid inaccurate transitions between states, which in turn would cause false alarms. However, these parameters also limit the sensitivity of the anomaly detection framework, so they must be selected cautiously to achieve a balance between the risk of false positives and the risk of false negatives.

8.6.5 Experimental Setup and Results

To show that the MarCNNet monitoring method is effective for different devices, we perform experiments on 3 different devices, each with a different number of cores. Table 8.4 shows for each device how many cores it has and what its clock frequency is.

The experimental setup for each device is shown in Fig. 8.36. The experiments use a lab-made magnetic probe and a spectrum analyzer to measure the emanated signals. We first present our results for a single-core, and then for multi-core runs.

8.6 MarCNNet: A Markovian Convolutional Neural Network for Malware...

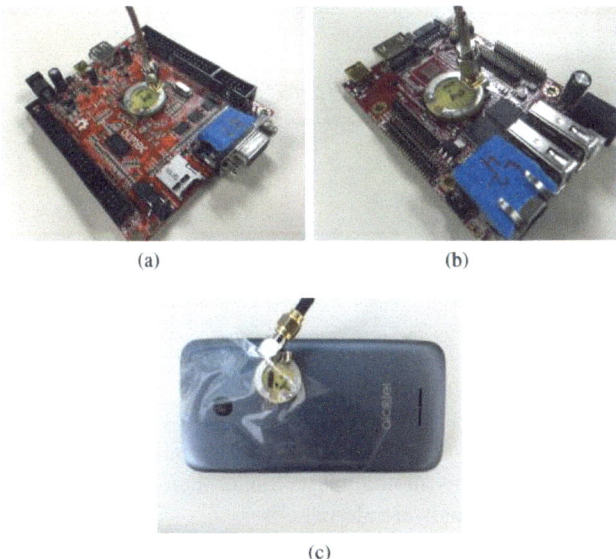

Fig. 8.36 Experimental setups for the measurements [50]. (**a**) A13-OlinuXino-Micro. (**b**) A20-OlinuXino-LIME2. (**c**) Alcatel Ideal

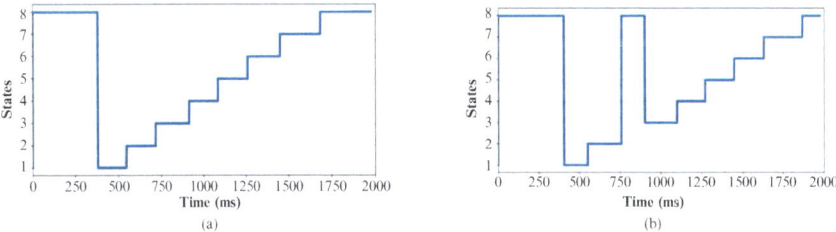

Fig. 8.37 State transition diagrams while only a single core is active [50]. (**a**) The states while profiling the system when Bit_count is running. (**b**) The states while profiling the system when Bit_count with a malware is running

Monitoring When Only One of the Cores Is Active

We first collect training signals and train our neural network by utilizing Algorithm 8.1. This is followed by generating state transition diagrams (see Fig. 8.34). Figure 8.37 shows the monitoring results while Bit_count is executed on a single core of the Alcatel phone. In this figure, we consider two scenarios: benign execution and malware-affected execution. The current state for a benign execution is charted in Fig. 8.37a, where seven of the states correspond to hot regions of the application, while state 8 is an *extra* state that is entered when none of the states allowed by the Markov model have a CNN output above the threshold. Since the execution follows a sequence, staying in each hot region until its work is completed and then transitioning to the next one, correct program execution creates a stairs-like

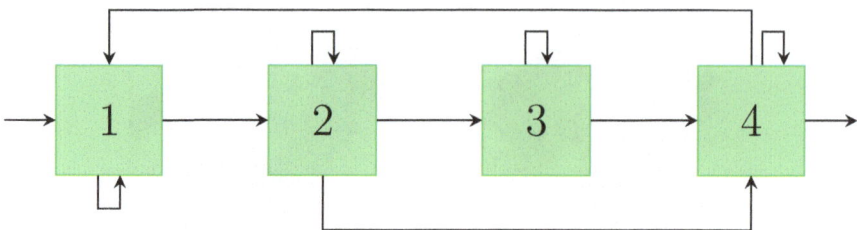

Fig. 8.38 *Proof-of-concept implementation:* State transition diagram of a four-region A/B alternation program [50]

pattern. Figure 8.37b, which corresponds to an execution where, after the second hot region is completed, malware is briefly executed before execution continues normally (into the third hot region). For this execution, *State-8* is entered between *State-2* and *State-3*. Since this anomaly lasts longer than t_M (which we set to 2 ms), this results in a raising the malware alert.

For clarity, we have only shown examples with simple-sequence state transitions (Fig. 8.31). However, that our method can use any Markov Model. To illustrate this, we designed a program that uses our A/B alternation code (which was discussed in detail in Chapter 4) with different alternation frequencies to implement four different hot regions, where the possible transitions between these regions are according to the state diagram (Markov Model) in Fig. 8.38. This program can generate an infinite number of possible execution sequences (because after state 4 it can go back to state 1). This illustrates the main advantage of using a Markov Model to represent expected program behaviors—we do not need to have observed all possible sequences during training, which would be impossible for this program.

The training signals are collected only from a few executions of the code, and accurate monitoring can be achieved if the training provides samples from each state, regardless of the path taken in each training execution. For this particular program, this was achieved after only a few training runs. An example of monitoring, for a run in which the actual sequence of hot regions was 1-2-4-1-2-3-4, is shown in Fig. 8.39. We observe that the state transitions identified during monitoring correctly follow the actual hot-region transitions.

We have performed experiments on all three devices with these programs. We obtain a 0% false negative rate (i.e., malware was detected in all runs that had malware) for all of the devices. However, the rate of false positives (malware alert in a malware-free run) is 0.1, 0.2, and 0.3% for OLinuXino-A13, OLinuXino-A20 and Alcatel, respectively. These ratios are calculated by dividing inaccurate "idle" samples to total number of samples ($> 10^3$). Please note that the tradeoff between false positives and false negatives can be adjusted by changing the time parameters of the monitoring algorithm. Specifically, in this experiment the thresholds were chosen to maximize detection of actual malware, but it is also possible to avoid false positives (false alerts) if a small false-negative (undetected malware) rate is tolerable.

8.6 MarCNNet: A Markovian Convolutional Neural Network for Malware...

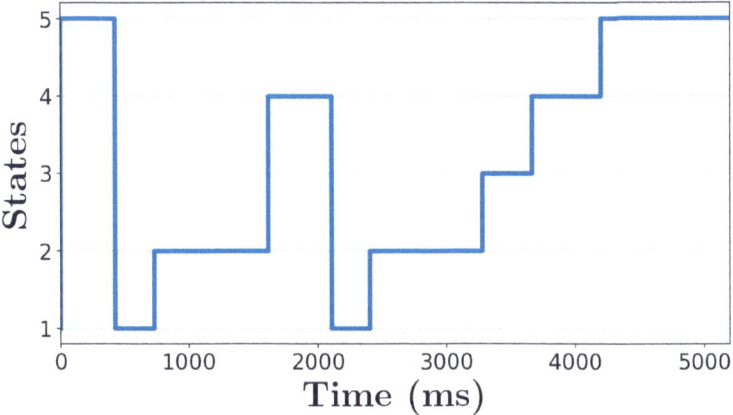

Fig. 8.39 State transitions while profiling a the four-region A/B alternation program [50]

Monitoring When Multiple Cores Are Active

In this section, we consider the scenario when multiple cores are active. For illustrative purposes, we use a scenario where one core on the Alcatel phone runs Bit_count, another core runs Basicmath, and the remaining two cores are idle. Monitoring starts before any program begins and ends after both programs have completed. In these "idle" regions, the proposed model does not report any malware because these "idle" states are implicitly added before the first state and after the last state of any application's Markov Model. First, Figs. 8.40 and 8.41 show the output values of the neural network trained on each of the two applications and the output of the state-transition identification for each of the two applications that are running concurrently. From the neural-network output values (left side of each figure) we observe two main risks to the accuracy of the overall monitoring, namely that multiple neurons can fire at the same time, and that a neuron other than the expected one can fire.

These problems would indeed cause inaccuracy (and false reporting of malware) in this experiment if we ignored the Markov Model for each application. However, with the Markov Model we output the correct transitions (right side of each figure).

The main observations regarding the experiment are:

- Although the training is performed by utilizing single-core-measurements, the proposed methodology can track and profile each program when other programs are active on the other hand, and
- The neural network is able to identify regions in the application for which it was trained, even in the presence of emanations caused by other concurrently-running programs.

As the second step, we investigate the behavior of the model when malware is injected into one of the programs. For this, we have two experiments: one that

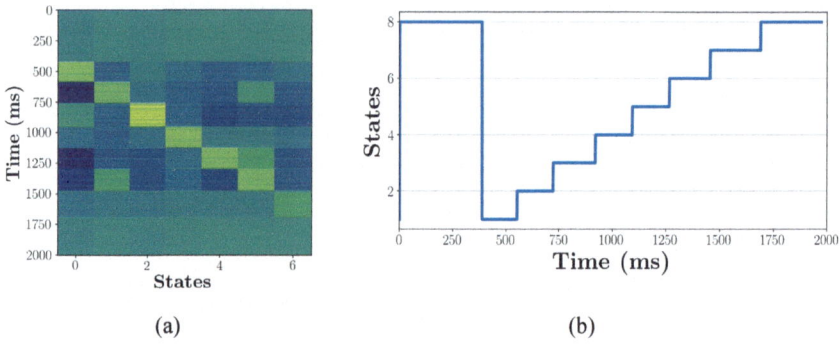

Fig. 8.40 Profiling based on the CNN and Markov Model for Bit_count [50]

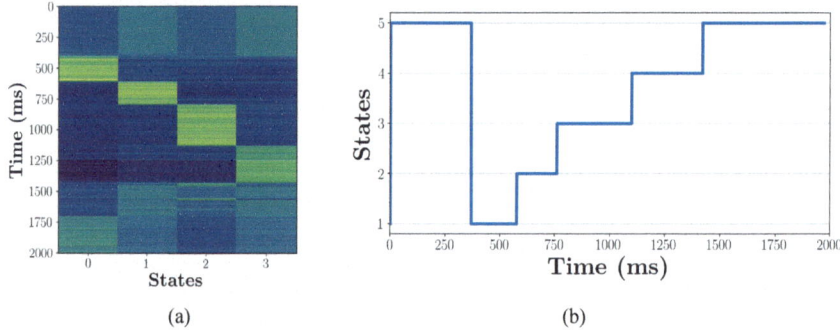

Fig. 8.41 Profiling based on the CNN and Markov model for Basicmath [50]

has malware-free Bit_count and malware-afflicted Basicmath, and another experiment with malware-afflicted Bit_count and malware-free Basicmath. Figure 8.42 shows the spectrograms from these two experiments, with a dotted rectangle indicating the interval during which malware is active in each experiment.

Figure 8.43 shows our framework's state identification output for each application in the first experiment (Basicmath has malware), and Fig. 8.44 shows these outputs for the second experiment (Bit_count has malware). In both experiments, malware (and the afflicted application) is correctly reported.

As a final example, we investigate the behavior of the model when two instances of the same application are executed concurrently. This is the worst case scenario for our framework because the same frequency components are produced by the two executing instances of the same application. We observe that our framework can still detect the malware Fig. 8.45 because the distortion of the malware on the spectrum causes information loss for the CNN to extract features. This time, however, the framework is unable to identify which of the two instances has malware.

8.6 MarCNNet: A Markovian Convolutional Neural Network for Malware... 205

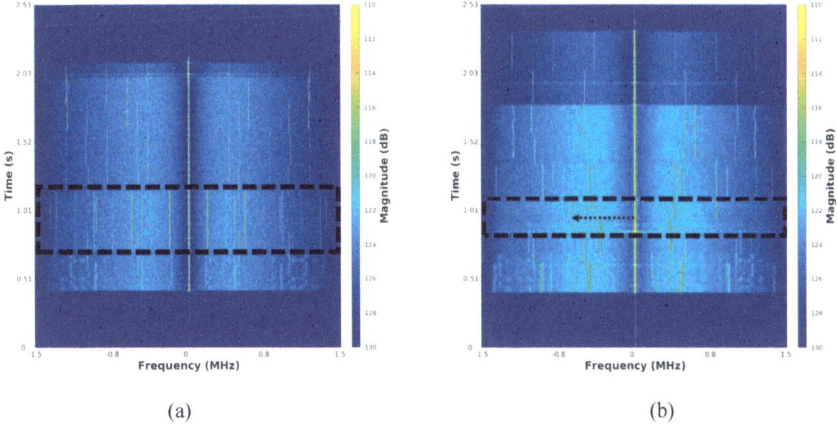

Fig. 8.42 Hot regions when one of the programs has a malware [50]

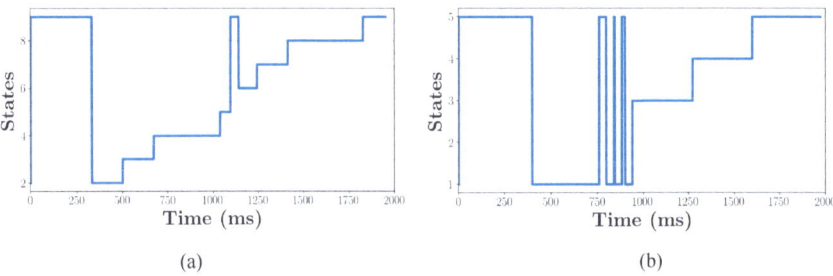

Fig. 8.43 Profiling based on the CNN and Markov model when Basicmath has malware [50]

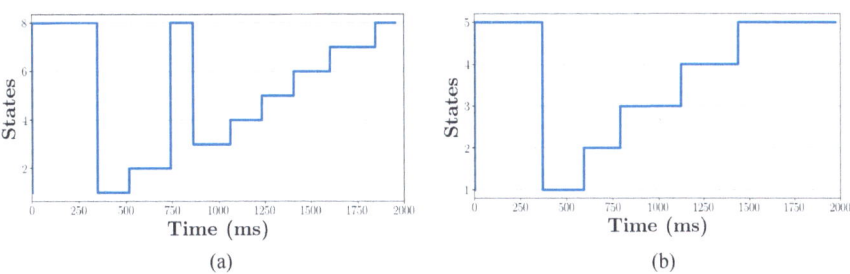

Fig. 8.44 Profiling based on the CNN and Markov model when Bit_count has malware [50]. (**a**) First parallel unit. (**b**) Second parallel unit

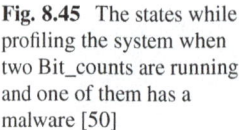

Fig. 8.45 The states while profiling the system when two Bit_counts are running and one of them has a malware [50]

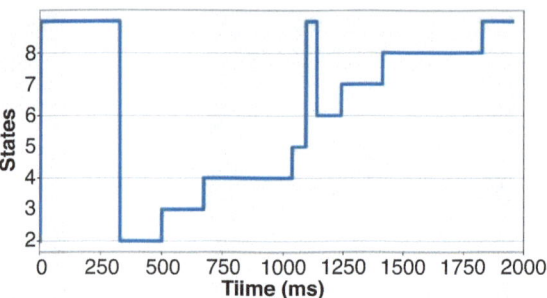

Table 8.5 The performance of the framework for different devices when multiple cores are active (%) [50]

	# of active cores							
	1		2		3		4	
	FP	FN	FP	FN	FP	FN	FP	FN
A13-OLinuXino-Micro	0.1	0	–	–	–	–	–	–
A20-OLinuXino-LIME2	0.2	0	0.6	0	–	–	–	–
Alcatel ideal	0.3	0	0.5	0	1.2	0	2	0

The main observations about these monitoring experiments are as follows:

- Our framework reports that malware is executing with high accuracy.
- Our framework correctly reports which of the programs has the malware, except when the malware affect the entire spectrum or when both a malware-afflicted and a malware-free instance of the same application are executing concurrently and are in similar stages of their execution when the malware executes.
- The system can continue monitoring the application even after the part related to malware has been executed, i.e., it can be used to detect a later occurrence of the same or different malware.
- Combining Markov and CNN Models enables accurate detection of malware and identification of the region in the program that executes just before and just after malware is executed .

Finally, Table 8.5 shows the overall false positive (FP) and false negative (FN) rates for experiments in which different numbers of active cores are used. We observe a false negative rate of 0% in all experiments, with a false positive rate of less than 2%. As already explained, the tradeoff between the false positive and false negative rates can be adjusted, e.g., to reduce false positives at the cost of creating some false negatives.

References

1. Lewansoul learm 6dof full metal robotic arm. Retrieved on April 2019 from https://www.amazon.com/LewanSoul-Controller-Wireless-Software-Tutorials/dp/B074T6DPKX.
2. Pid controller soldering iron code and project. https://github.com/sfrwmaker/soldering_controller. Last Accessed: 2018-08-01.
3. Horn antenna datasheet. https://www.com-power.com/ah118hornantenna.html, 2015 (accessed Nov. 5, 2017).
4. Arduino servo library. https://www.arduino.cc/en/Reference/Servo, accessed April 2019).
5. Software defined radio sdr. https://www.ettus.com/product/details/USRP-B200mini, accessed May 3, 2017.
6. Ossama Abdel-Hamid, Abdel-rahman Mohamed, Hui Jiang, Li Deng, Gerald Penn, and Dong Yu. Convolutional neural networks for speech recognition. *IEEE/ACM Transactions on audio, speech, and language processing*, 22(10):1533–1545, 2014.
7. Tigist Abera, N. Asokan, Lucas Davi, Jan-Erik Ekberg, Thomas Nyman, Andrew Paverd, Ahmad-Reza Sadeghi, and Gene Tsudik. C-flat: Control-flow attestation for embedded systems software. In *Proceedings of the 2016 ACM SIGSAC Conference on Computer and Communications Security*, CCS '16, pages 743–754, New York, NY, USA, 2016. ACM.
8. Alcatel. Alcatel ideal / streak specifications. https://www.devicespecifications.com/en/model/e2ef3d31, accessed March 26, 2018.
9. N.S. Altman. An introduction to kernel and nearest-neighbor nonparametric regression. *The American Statistician*, 46(3):175–185, 1992.
10. Nicoló Andronio, Stefano Zanero, and Federico Maggi. Heldroid: Dissecting and detecting mobile ransomware. In *International Workshop on Recent Advances in Intrusion Detection*, pages 382–404. Springer, 2015.
11. Tyler Bletsch, Xuxian Jiang, Vince W. Freeh, and Zhenkai Liang. Jump-oriented programming: A new class of code-reuse attack. In *Proceedings of the 6th ACM Symposium on Information, Computer and Communications Security*, ASIACCS '11, pages 30–40, New York, NY, USA, 2011. ACM.
12. Robert Callan, Farnaz Behrang, Alenka Zajic, Milos Prvulovic, and Alessandro Orso. Zero-overhead profiling via EM emanations. In *Proceedings of the 25th International Symposium on Software Testing and Analysis, ISSTA 2016, Saarbrücken, Germany, July 18-20, 2016*, pages 401–412, 2016.
13. R.J.G.B. Campello, Moulavi D., and J. Sander. Density-based clustering based on hierarchical denisty estimates. In *Advances in Knowledge Discovery and Data Mining. PAKDD 2013. Lecture Notes in Computer Science*, 2013.
14. Scott A. Carr and Mathias Payer. Datashield: Configurable data confidentiality and integrity. In *Proceedings of the 2017 ACM on Asia Conference on Computer and Communications Security*, ASIA CCS '17, pages 193–204, New York, NY, USA, 2017. ACM.
15. Michael Chui, Markus Löffler, and Roger Roberts. The internet of things. *McKinsey Quarterly*, March 2010.
16. John Demme, Matthew Maycock, Jared Schmitz, Adrian Tang, Adam Waksman, Simha Sethumadhavan, and Salvatore Stolfo. On the feasibility of online malware detection with performance counters. In *Proceedings of the 40th Annual International Symposium on Computer Architecture*, ISCA '13, pages 559–570, New York, NY, USA, 2013. ACM.
17. Ken Dunham. Evaluating anti-virus software: Which is best? *Information Systems Security*, 12(3):17–28, 2003.
18. Matthew R. Guthaus, Jeffrey S. Pingenberg, Dap Emst, Todd M. Austin, Trevor Mudge, and Richard B. Brown. Mibench: A free, commercially representative embedded benchmark suite. In *Proceedings of the IEEE International Workshop on Workload Characterization*, 2001.
19. Yi Han, Sriharsha Etigowni, Hua Liu, Saman Zonouz, and Athina Petropulu. Watch me, but don't touch me! contactless control flow monitoring via electromagnetic emanations.

In *Proceedings of the 2017 ACM SIGSAC Conference on Computer and Communications Security*, CCS '17, pages 1095–1108, New York, NY, USA, 2017. ACM.
20. Intel. A guide to the internet of things infographic. https://www.intel.com/content/www/us/en/internet-of-things/infographics/guide-to-iot.html, 2017 (accessed Feb. 2, 2018).
21. Brent ByungHoon Kang and Anurag Srivastava. Dynamic malware analysis. In *Encyclopedia of Cryptography and Security, 2nd Ed.*, pages 367–368. 2011.
22. Haider Adnan Khan, Nader Sehatbakhsh, Luong Nguyen, Milos Prvulovic, and A. Zajic. Malware detection in embedded systems using neural network model for electromagnetic side-channel signals. *J Hardw Syst Security*, 08 2019.
23. Haider Adnan Khan, Nader Sehatbakhsh, Luong N. Nguyen, Robert L. Callan, Arie Yeredor, Milos Prvulovic, and Alenka Zajic. Idea: Intrusion detection through electromagnetic-signal analysis for critical embedded and cyber-physical systems. *IEEE Transactions on Dependable and Secure Computing*, 18(3):1150–1163, 2021.
24. Steve Lawrence, C Lee Giles, Ah Chung Tsoi, and Andrew D Back. Face recognition: A convolutional neural-network approach. *IEEE transactions on neural networks*, 8(1):98–113, 1997.
25. Yann LeCun, Yoshua Bengio, et al. Convolutional networks for images, speech, and time series. *The handbook of brain theory and neural networks*, 3361(10):1995, 1995.
26. Yann LeCun, Bernhard Boser, John S Denker, Donnie Henderson, Richard E Howard, Wayne Hubbard, and Lawrence D Jackel. Backpropagation applied to handwritten zip code recognition. *Neural computation*, 1(4):541–551, 1989.
27. Yannan Liu, Lingxiao Wei, Zhe Zhou, Kehuan Zhang, Wenyuan Xu, and Qiang Xu. On code execution tracking via power side-channel. In *Proceedings of the 2016 ACM SIGSAC Conference on Computer and Communications Security*, CCS '16, pages 1019–1031, New York, NY, USA, 2016. ACM.
28. Alireza Nazari, Nader Sehatbakhsh, Monjur Alam, Alenka Zajic, and Milos Prvulovic. Eddie: Em-based detection of deviations in program execution. In *Proceedings of the 44th Annual International Symposium on Computer Architecture*, ISCA '17, pages 333–346, New York, NY, USA, 2017. ACM.
29. L.N. Nguyen, CL. Cheng, F.T. Werner, Prvulovic M., and Zajic A. A comparison of backscattering, em, and power side-channels and their performance in detecting software and hardware intrusions. *Journal Hardware System Security*, pages 150–165, 2020.
30. Olimex. A13-olinuxino-micro user manual. https://www.olimex.com/Products/OLinuXino/A13/A13-OLinuXino-MICRO/open-source-hardware, 2016 (accessed Feb. 1, 2018).
31. OlinuXino A13. https://www.olimex.com/Products/OLinuXino/A13/A13-OLinuXino/open-source-hardware.
32. OlinuXino A20. https://www.olimex.com/Products/OLinuXino/A20/A20-OLinuXino/open-source-hardware.
33. Meltem Ozsoy, Caleb Donovick, Iakov Gorelik, Nael Abu-Ghazaleh, and Dmitry Ponomarev. Malware-aware processors: A framework for efficient online malware detection. In *2015 IEEE 21st International Symposium on High Performance Computer Architecture (HPCA)*, Feb 2015.
34. Meltem Ozsoy, Khaled N Khasawneh, Caleb Donovick, Iakov Gorelik, Nael Abu-Ghazaleh, and Dmitry Ponomarev. Hardware-based malware detection using low-level architectural features. *IEEE Transactions on Computers*, 65(11):3332–3344, Nov 2016.
35. D. Quarta, M. Pogliani, M. Polino, F. Maggi, A. M. Zanchettin, and S. Zanero. An experimental security analysis of an industrial robot controller. In *2017 IEEE Symposium on Security and Privacy (SP)*, pages 268–286, May 2017.
36. Gregg Rothermel, Sebastian Elbaum, Alex Kinneer, and Hyunsook Do. Software-artifact infrastructure repository. http://sir.unl.edu/portal, 2006.
37. Jonathan Salwan and Allan Wirth. Ropgadget: Gadgets finder for multiple architectures. https://github.com/JonathanSalwan/ROPgadget, 2011 (accessed Feb. 1, 2018).

38. N. Sehatbakhsh, M. Alam, A. Nazari, A. Zajic, and M. Prvulovic. Syndrome: Spectral analysis for anomaly detection on medical iot and embedded devices. In *2018 IEEE International Symposium on Hardware Oriented Security and Trust (HOST)*, pages 1–8, April 2018.
39. N. Sehatbakhsh, A. Nazari, A. Zajic, and M. Prvulovic. Spectral profiling: Observer-effect-free profiling by monitoring em emanations. In *2016 49th Annual IEEE/ACM International Symposium on Microarchitecture (MICRO)*, pages 1–11, Oct 2016.
40. Nader Sehatbakhsh, Alireza Nazari, Monjur Alam, Frank Werner, Yuanda Zhu, Alenka Zajic, and Milos Prvulovic. Remote: Robust external malware detection framework by using electromagnetic signals. *IEEE Transactions on Computers*, 69(3):312–326, 2020.
41. Hovav Shacham. The geometry of innocent flesh on the bone: Return-into-libc without function calls (on the x86). In *Proceedings of the 14th ACM Conference on Computer and Communications Security*, CCS '07, pages 552–561, New York, NY, USA, 2007. ACM.
42. Nitish Srivastava, Geoffrey Hinton, Alex Krizhevsky, Ilya Sutskever, and Ruslan Salakhutdinov. Dropout: a simple way to prevent neural networks from overfitting. *The journal of machine learning research*, 15(1):1929–1958, 2014.
43. TrendMicro. Persirai: New internet of things (iot) botnet targets ip cameras. http://blog.trendmicro.com/trendlabs-security-intelligence/persirai-new-internet-things-iot-botnet-targets-ip-cameras/, 2017 (accessed Feb. 1, 2018).
44. Ubuntu. Lightdm. https://wiki.ubuntu.com/LightDM, 2017 (accessed Feb. 1, 2018).
45. Arun Viswanathan, Kymie Tan, and Clifford Neuman. Deconstructing the assessment of anomaly-based intrusion detectors. In *Proceedings of the 16th International Symposium on Research in Attacks, Intrusions, and Defenses - Volume 8145*, RAID 2013, pages 286–306, New York, NY, USA, 2013. Springer-Verlag New York, Inc.
46. Bas Wijnen, Emily J. Hunt, Gerald C. Anzalone, and Joshua M. Pearce. Open-source syringe pump library. *PLOS ONE*, 9(9):1–8, 09 2014.
47. Carsten Willems, Thorsten Holz, and Felix C. Freiling. Toward automated dynamic malware analysis using cwsandbox. *IEEE Security & Privacy*, 5(2):32–39, 2007.
48. Baki Berkay Yilmaz, Nader Sehatbakhsh, Milos Prvulovic, and Alenka Zajic. Communication model and capacity limits of covert channels created by software activities. *IEEE Transactions on Information Forensics and Security*, 15:776–789, 2019.
49. Baki Berkay Yilmaz, Elvan Mert Ugurlu, Alenka Zajic, and Milos Prvulovic. Instruction level program tracking using electromagnetic emanations. In *Cyber Sensing 2019*, volume 11011, page 110110H. International Society for Optics and Photonics, 2019.
50. Baki Berkay Yilmaz, Frank Werner, Sunjae Y. Park, Elvan Mert Ugurlu, Erik Jorgensen, Milos Prvulovic, and Alenka Zajic. Marcnnet: A markovian convolutional neural network for malware detection and monitoring multi-core systems. *IEEE Transactions on Computers*, pages 1–14, 2022.
51. Igal Zeifman, Dima Bekerman, and Ben Herzberg. Source code for iot botnet mirai released. https://krebsonsecurity.com/2016/10/source-code-for-iot-botnet-mirai-released, September 2016 (accessed Oct. 10, 2017).
52. Zhi-Kai Zhang, Michael Cheng Yi Cho, Chia-Wei Wang, Chia-Wei Hsu, Chong-Kuan Chen, and Shiuhpyng Shieh. Iot security: Ongoing challenges and research opportunities. In *Proceedings of the 2014 IEEE 7th International Conference on Service-Oriented Computing and Applications*, SOCA '14, pages 230–234, Washington, DC, USA, 2014. IEEE Computer Society.

Chapter 9
Using Analog Side Channels for Program Profiling

9.1 Introduction

Program profiling involves measuring some aspect(s) of software behavior as that software executes, i.e., it is a type of dynamic (runtime) analysis of software. One of the most common kinds of program profiling is to count how many times each part of the program code has been executed, and use that information to identify often-executed (*hot*) parts of the code. Knowledge of hot regions of code is very important for many other tasks, e.g., programmers often focus their program optimization efforts on hot regions because performance improvements in those parts of the code tend to also produce overall performance improvement for the application.

Profiling is typically implemented by adding software probes (instrumentation) to a program's source or binary (executable) code. These probes either write information about events of interest to a log, or they update statistics about these events.

For some systems, especially in real-time and cyber-physical domains, the program has deadlines to meet, and may choose to use different program code depending on how much time it has remaining, e.g., it may suspend some non-essential activity, or it may use code that is less efficient on average but has better executing time guarantees, if it is at risk of not meeting deadlines. In such systems, program execution can change significantly when instrumentation changes the performance characteristics of the program—this creates something akin to the *observer effect* in quantum physics, where the act of measuring a property changes the very property that is being measured. Profiling is an important activity for software developers, so many processors (especially higher-end ones) include hardware support that reduces (but does not eliminate) the execution time overheads (and the "Observer's Effect") of profiling, although this increases the cost of the processor even for systems that are not used for software profiling.

Profiling is also difficult in systems that have very limited resources. The processor performance and/or memory capacity may be barely sufficient for the

© The Author(s), under exclusive license to Springer Nature Switzerland AG 2023
A. Zajić, M. Prvulovic, *Understanding Analog Side Channels Using Cryptography Algorithms*, https://doi.org/10.1007/978-3-031-38579-7_9

system's primary function and cannot accommodate profiling overheads. The system's power source (e.g., energy harvesting) may not be able to support the added energy consumption to collect, store, process, or transmit the profiling data, and the processor often has no advanced hardware support for profiling because it would significantly add to its (extremely low) cost.

Unfortunately, many Internet-of-Things (IoT) devices have both types of limitations, i.e., they are real-time/cyber-physical systems, and they operate under severe cost and energy limitations. To alleviate these problems, a new approach has been proposed recently [10, 28, 30, 36, 41]—profiling of program execution without instrumenting or otherwise affecting the profiled system. This new approach collects profiling information in a highly accurate and completely non-intrusive way by leveraging electromagnetic (EM) side-channel emanations. Because this approach generates profiling information without interacting with or modifying the profiled system, it offers the potential to profile a variety of software systems for which profiling was previously not possible. In addition, the ability to collect profiles by simply placing a profiling device next to the system to be profiled can also provide be advantageous (compared to traditional instrumentation-based approaches) in many traditional contexts.

In this chapter we discuss various signal processing approaches and techniques for program profiling using analog side-channels. We first describe Spectral Profiling [30], which monitors electromagnetic (EM) emanations that are unintentionally produced by the profiled system, looking for spectral "lines" produced by periodic program activity (e.g., loops). This approach allows us to determine which parts of the program have executed at what time and, by analyzing the frequency and shape of the spectral "line", obtain additional information such as the per-iteration execution time of a loop. If finer-grain information about program execution is needed, e.g., which basic blocks or individual instructions have been executed, time-domain analysis of side-channel signals is more advantageous [10, 28, 36]. Several time-domain techniques will be discussed later in this chapter.

9.2 Spectral Profiling

Spectral Profiling has two phases, training and profiling. In the training phase, we run the application with known training inputs to identify which spectra correspond to which part of the program, and also to identify the valid orderings between the parts of the program. In the profiling phase, we run the application with unknown inputs, record how the spectrum changes over time, and combine that with the information from training to detect which part of the program is executing at each point in time.

Spectral Profiling's recognition of activity is based on recognizing the corresponding spectrum. Any spectrum fundamentally corresponds to the signal observed over some interval of time (window), and the duration of this window represents a trade-off between temporal resolution and frequency resolution. Temporal resolution corresponds to being able to tell where exactly some program activity begins and ends. Fundamentally, a spectrum that corresponds to some time window "blurs together" activity for the entire window, so spectra collected with a short window allow more precise identification of the time when program activity has changed. This means that, to improve temporal resolution, we should use spectra collected over very short intervals of time. However, the number of frequency bins (i.e., the frequency resolution) in the spectrum is proportional to the duration of the time interval, so a spectrum collected over a very brief interval "lumps together" similar frequencies into one frequency bin. This means that two program activities that have spectral "spikes" with similar frequencies cannot be told apart when using short-window spectra because the spectrum only has one bin for the entire frequency range that contains both spikes.

Thus the time window should be short enough to capture relevant events in the profiled execution, e.g., it should be shorter than the duration of most loops—intuitively, attribution of execution time to specific code in the application will be performed at the granularity that is similar to the size of the window. In our *Basicmath* benchmark example, we use a window of 1 ms with 75% overlap between consecutive windows, which provides attribution with 0.25 ms granularity and precision between 0.25 ms and 1 ms.

9.2.1 Training Phase

The goal of the training phase is to collect spectral signatures for all regions of the program, and also to identify the possible/probable sequence in which regions of code were executed in a particular program's execution.

When we collect spectra during training, we face a dilemma. We can run the program just like we do for profiling (no modifications to the code, no change to the system, no collection of any information on the profiled system). The spectra obtained that way will then be the same as the spectra obtained during profiling, for the same regions of code, except when spectra produced by a region of code (e.g., a loop) are input-dependent. However, when training spectra are collected this way, we do not know which spectrum corresponds to which part of the code.

Alternatively, we can instrument the program (or use interrupt-based sampling) to record which part of the code executes at what time, but the spectra collected in such execution are distorted by such changes and will poorly match the corresponding spectra during profiling.

Our current approach to resolving this dilemma is to first use instrumentation to measure the average per-iteration execution times for each loop, then re-run the program with the same training inputs but without the instrumentation to get the undistorted spectra. We then use the per-iteration execution times and the frequencies of spikes in the spectrum to create spectrum-to-loop mappings.

Finding Per-Iteration Execution Time

We place instrumentation at the beginning and end of the loop body, which allows us to get the current time-stamps when the iteration begins and ends, subtract the two, and thus get the execution time of the loop body (for that one iteration). As the application executes, these per-iteration times are stored in memory, along with information about which loop they correspond to. When the application ends, we use this information to compute the average per-iteration execution time for each loop instance. Note that we are not interested in the total execution time of the loop, nor do we directly use this information in our profiling. The per-iteration time is only used to identify the frequency at which the corresponding spectrum should have a spike. For example, if the average per-iteration time for a specific loop is T, we will expect the corresponding spectrum to have a spike at a frequency that is relatively close to $f = 1/T$.

In our *Basicmath* example, there are four different loops in the source code. Each of the four loops is instrumented to collect per-iteration execution time T, and Fig. 9.1 shows, for each loop, a histogram of frequencies $f = 1/T$ that correspond to these per-iteration execution times (i.e., Fig. 9.1a corresponds to loop 1, Fig. 9.1b corresponds to loop 2 and so on). This provides us with the approximate frequency at which to expect a spectral "spike" for each loop, along with information about the width and shape of each spike.

Finding Spectral Signatures for Each Loop

After calculating the frequency of each loop, we re-run the application with same inputs but without any instrumentation or profiling-related activity on the profiled device, and record the spectra for each time window. In each spectrum, we identify the spikes, then compare their frequencies and shapes to the histogram obtained from the previous (instrumented run). The matches are imperfect because instrumentation perturbs the execution time of a loop's iteration, and thus changes the frequency and shape in the histogram. However, our matching is highly accurate because frequencies that correspond to different loops tend to differ more than the instrumentation-induced errors, and also because the error introduced by instrumentation is usually in the same direction (increases the per-iteration execution time), and also because our matching approach utilizes the fact that the two runs used the same inputs and thus have the same sequence of loops. For example, Loop 3 and Loop 4 have relatively similar frequencies, but because we

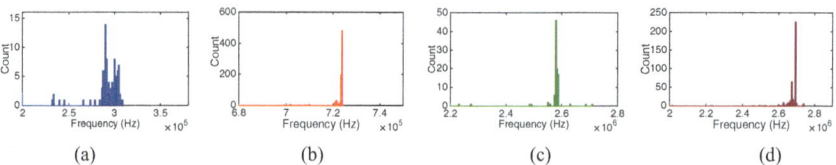

Fig. 9.1 Histogram of frequencies that correspond to per-iteration execution time ($f = 1/T$) for four different loops in Basicmath benchmark for small (training run) inputs [30]

9.2 Spectral Profiling

Table 9.1 Measured and calculated frequency for loops in "Basicmath" application [30]

Loop number	Frequency (measured)	Frequency (calculated)
Loop 1	289.12 KHz	289.1 KHz
Loop 2	720.3 KHz	721 KHz
Loop 3	2.628 MHz	2.577 MHz
Loop 4	2.733 MHz	2.69 MHz

know that Loop 3 is likely to have a lower frequency than, and be executed before, Loop 4, the spectra corresponding to these loops can still be correctly "assigned". Additionally, after successfully assigning spectra to the loops, we will also have the sequences of the "assigned" loops.

Table 9.1 shows the list of frequencies for four loops in *Basicmath*. The "measured" column in the table shows the actual frequency of the loop (i.e., in the instrumentation-free run), and the "calculated" column shows the average frequency calculated from the instrumentation-enabled histogram. The relative error between the calculated and measured frequency for these loops is up to 2%, but we can still easily match them. Also note that the frequency error introduced by instrumentation increases as the frequency increases. This is because instrumentation has more effect on tight loops (short time per iteration, i.e., high frequency).

After matching spectra to loops, we pre-process the spectrum that corresponds to each loop to identify the "spectral signature" for the loop. In our implementation, the signature is a list of frequencies for the strongest spikes in the spectrum, after removing spikes that appear in all spectra (e.g., for EM signals, for example, this eliminates spikes caused by radio stations, etc.). Note that the signature is not just one number that corresponds to the fundamental frequency of the loop. Some loops have a group of spikes instead of one spike, because their per-iteration execution time takes several discrete values (with some variation around each of them). In most cases, the spectrum also contains not only the spikes that correspond to the per-iteration execution time (fundamental frequency), but also spikes at multiples (harmonics) of that frequency. These additional spikes help differentiate spectra that correspond to different loops, so the signature we use includes all spikes whose magnitude is sufficiently above the noise floor.

9.2.2 Profiling Phase

In training, we identified the spectral signature for each loop, and we have also identified the possible/probable sequences of loops—essentially, which loops can execute immediately after which other loops. Profiling consists of running the application with unknown inputs and obtaining profiling information about those runs.

Matching of Loop Spectra

Because the profiling inputs are different from training inputs, it is natural to wonder if the spectrum of a loop will change. We have found that many loops, primarily innermost loops, have spectra that are nearly identical to those found in training. Intuitively, the spectrum changes when the per-iteration execution time changes, and in many loops only the number of iterations changes significantly from input to input, but the work of each iteration (and the statistics of branches and architectural events) remain similar. We call these Loops with Input-Independent Spectra (LIIS), and for these loops the spectrum can be matched to the corresponding spectrum from training.

During profiling, we use the same time window we used during training. For each time window during profiling, we obtain the spectrum for that window, identify the spikes in the spectrum (the spectral signature) and compare that signature to the signatures obtained during training. The comparison is performed by attempting to match the peaks in the profile-time and training-time signature. For each peak in the profile-time signature, we find the closest peak (according to frequency) in the training-time signature. If that closest frequency differs too much, the peak remains unmatched. If the closest frequency is very similar, the peak is counted as matched. After attempting to match each peak, the number of successfully matched peaks is used as the similarity metric between the signatures.

If the similarity is high between a profile-time spectral signature and the best-matching training-time signature of a loop, we attribute the execution during that profile-time window to that loop. For the vast majority of time windows that belong to LIIS loops, this similarity is very high and the execution is correctly attributed to the correct LIIS loop.

However, it is possible that none of the profile-time signatures matches the observed signature well enough. This happens primarily because the spectrum of some loops does change with frequency. For example, a command-line flag may cause every iteration of the loop to take one path in one execution and a significantly different path in another execution. For these loops, the spectrum still indicates that a loop is executing (spikes in the spectrum) and when the loop begins and ends (spikes appear at one time and disappear later) but the spectrum during profiling no longer matches any of the spectra from training.

9.2.3 Sequence-Based Matching

To attribute execution time to these Non-LIIS loops (and report their per-iteration execution time during the profiling run), we rely on the model of possible loop-level sequences constructed during profiling. Sequence-based matching begins after LIIS matching is completed for LIIS loops. The spectra from time windows that remain unmatched after LIIS matching are first clustered according to the same similarity

9.2 Spectral Profiling 217

metric we used to match LIIS loops to spectra from training, i.e., spectra that have many spikes at similar frequencies will be clustered together. At this point we have clusters where each cluster corresponds to a Non-LIIS loop, but we do not yet know which loop in the code this cluster corresponds to.

However, for each "mystery spectrum" we know that it should be matched to a region of code that is not a LIIS loop, and the LIIS loop spectra observed before and after the "mystery spectrum" tell us which loops have been executed before and after each instance of the "mystery" loop whose cluster we are considering. Fortunately, the model of the application's loop-to-loop transitions restricts the possibilities for matching so that usually only a single Non-LIIS loop remains as a possible match. When there are multiple possible matches, i.e., the "mystery spectra" in a cluster could possibly belong to more than one Non-LIIS loop, we match the cluster to the Non-LIIS loop whose training signature has the highest average similarity to the spectra in the cluster.

Figure 9.2 shows the profiling-time spectrogram of the *Basicmath* application. The bold line at 1.008 MHz is the clock signal. The periodic program behavior amplitude-modulates this signal, and the straight lines to the right of this line represent the upper sideband of the modulated signal, i.e., they have the spectrum that corresponds to program behavior but that spectrum is shifted upward in frequency by the clock frequency [12]. In this execution the four loops are executed one after the other (shown by arrows). In this case all four loops were matched through LIIS matching, but if any one of the was not matched through LIS matching, it would still be successfully matched through sequence-based matching.

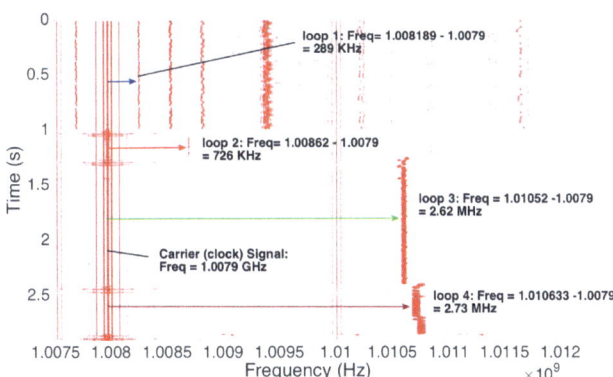

Fig. 9.2 Spectrogram for *Basicmath* benchmark with large (profiling run) inputs [30]

9.2.4 Illustration of Program Profiling Results

This section presents our experimental results, first for profiling of execution in real systems through EM emanations, and then for profiling through power signals generated through cycle-accurate simulation. We tested 13 different applications from the MiBench [14] benchmark suite on both of these platforms. The real system we used is A13-OLinuXino-MICRO [20], a single-board Linux computer which has an ARM Cortex A8 processor [4]. The cycle-accurate simulator we used is SESC [25].

To demonstrate the feasibility and effectiveness of Spectral Profiling, we used it to profile applications running on a single-board computer (A13-OLinuXino), which has a 2-issue in-order ARM Cortex A8 processor with 32 kB L1 and 256 KB L2 caches, and uses Debian Linux as its operating system (OS). Our Spectral Profiling for this system uses electromagnetic (EM) emanations that are received by a commercial small electric antenna (PBS- E1) [3] that is placed next to the profiled system's processor, where the clock signal has the strongest Signal-to-Noise ratio (SNR).

A spectrum analyzer (Agilent MXA N9020A) is then used to record the spectra of the signals collected by the antenna. A spectrum analyzer can be relatively costly (several tens of thousands of dollars), but we elected to use a spectrum analyzer primarily because it provides calibrated measurements, and already has support for automating measurements and for saving and analyzing measured results. In additional experiments, we observed similar spectra with less expensive (<$1,000) commercial software-defined radio receivers.

We apply Spectral Profiling to all 13 applications from the automotive, communications, network, and security categories in the MiBench suite. We used a 1 ms window size with 75% sof overlap between windows in all applications, except for GSM, where we used a 0.5 ms window to improve temporal resolution for attribution of execution time for short-lived loops.

For training, markers were manually inserted before and after each loop in these applications, and each marker reads and records the current clock cycle count from the *ARM Performance Counter Unit* (ARM-PMU) [5], which provides information similar to the x86 "rdtsc" instruction [22]. The training runs are repeated with several different command line flags, in order to identify the sequence of loops that can occur in each application. We note that insertion of markers and identification of possible sequences can both be accomplished automatically by a compiler (identification of loop nests and their connectedness in the control flow graph), but such automated compiler-based marker insertion and generation of the loop-level sequence graph would require additional engineering effort that was not needed for this proof-of-concept implementation.

After training, for which we used the small input set [14], we perform actual profiling with the original unmodified code (no markers) and with the large input set [14]. The accuracy we measure is defined as the fraction of execution time for which our method correctly identifies the loop that is currently executing. This accuracy is not 100% because of (1) ***miss-attribution***, during which our algorithm

9.2 Spectral Profiling

matches the spectrum to a different loop (i.e., loop A is actually executing, but the algorithm matches the spectrum to loop B instead) at loop B, and (2) ***non-attribution***, during which our algorithm finds that the spectrum is too different from loop spectra observed in training, so it leaves such intervals un-attributed. Non-attribution is typically a result of computation whose spectrum varies widely depending on inputs, or activity that has no recognizable spectral signature (e.g., loops whose per-iteration time varies a lot from iteration to iteration).

Figure 9.3 (left) shows the breakdown of profiled execution time into time that was accurately attributed, time that was miss-attributed, and non-attributed time. In all benchmarks except GSM, our method correctly attributes during at least 90% of the execution time, with the arithmetic mean at 93%. Miss-attribution occurs during less than 4% of the execution time, except in QSORT, where miss-attribution occurs during 8% of the time. The larger miss-attribution for QSORT occurs primarily because the std::qsort library function does not have a stable signature, so it is often miss-attributed. It is quite possible that the variation in std::qsort spectra is a result of having multiple loops in that function. Unfortunately, our manual marker insertion did not include library code so our scheme treats the entire std::qsort function as a single entity and expects it to have the same spectrum throughout its execution. We expect that this problem can be overcome with compiler-supported loop instrumentation that includes library code. Overall, the arithmetic mean for miss-recognition is 2.26%.

To confirm that the ability to do Spectral Profiling is a result of a fundamental connection between repetitive program behavior and periodic physical signals produced as a side-effect of that computation, we apply Spectral Profiling to the spectra of power signals produced through cycle-accurate architectural simulation in SESC [25], a cycle accurate simulator which integrated CACTI [24] and WATTCH [8] power models. Table 9.2 provides more details for the simulated

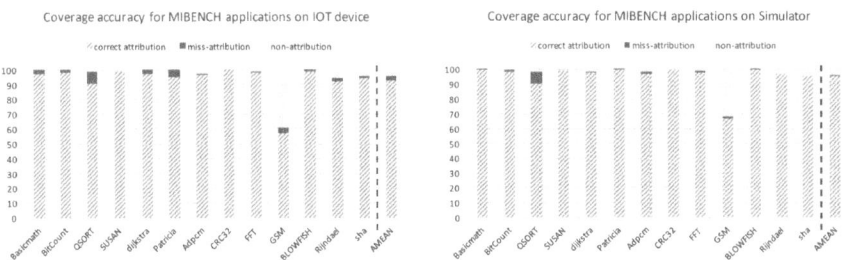

Fig. 9.3 Correct attribution (striped portion) as a percentage of the overall profiled execution time [30]

Table 9.2 Configurations used in simulation [30]

Configuration	Clock rate	Pipeline	Pipeline width	L1 cache	L2 cache
A (simple processor)	50 MHz	In-order	1	4 KB	–
B (modern processor)	1.8 GHz	OoO	4	32 KB	64 MB

configuration. We used this configuration to show that even in completely different configuration, i.e., different clock rate (1.008 GHz vs. 1.8 GHz), different pipeline (In-order vs Out-of-order), etc., Spectral Profiling is still effective and is not machine or architecture dependent.

Figure 9.3 (right) shows the accuracy results for the same applications and with the same breakdown we used for the real-system results. The accurate attribution percentage has an arithmetic mean of 98% for our simulated results, which is slightly higher than for our real-system results. The main reasons for this higher accuracy are that simulation-produced power signals are free of noise, and that they have a single-cycle resolution. In contrast to that, the EM signals received from real systems are subjected to radio-frequency noise, measurement error, and frequency-dependent distortion (for some EM frequencies the real system acts as a better "transmitter" than for others). The arithmetic mean for miss-recognition in this case is 1.19% which is slightly better than for the real system.

9.2.5 Loops with Input-Independent Spectra

As discussed in previous section, some loops produce spectral "spikes" whose frequency does not change significantly with changing inputs, while others have input-dependent spectra. Recall that the frequencies of the "spikes" depend on the loop's average per-iteration time and not on the number or iteration executed in the loop, so loops with input-independent spectra (which we abbreviate as LIIS) tend to be innermost loops. Conversely, loops with input-dependent spectra tend to be outer loops whose inner loops have input-dependent iteration counts, or loops that have a set of control flows which can change the per-iteration execution time of the loop. The reason we are interested in LIIS loops is that, during profiling, their execution can be directly recognized from the spectrum. The remaining loops may still be correctly attributed, but that attribution consists of (1) identifying stable spectral patterns (the spectrum has "spikes" that remain the same for a while, which indicates that the system is likely executing a single loop) and (2) using the program's possible loop-level sequences (learned from training), i.e., the knowledge of which loops may possibly execute immediately after other loops, to attribute that activity to a specific non-LIIS loop. In addition to accuracy in terms of attributing execution time, we have also evaluated accuracy in terms of identifying when each instance of each loop starts and ends. These results, presented in Fig. 9.4, will be discussed in detail in Section 9.2.6.

Figure 9.5 shows how much of the profiled execution time is attributed through each of these mechanisms (LIIS and Sequence). On average, 82% of the execution time is attributed through LIIS, and some applications spend nearly all of their execution time in LIIS loops. However, in several applications (especially Susan and SHA) most of the execution time is attributed through the sequence mechanism,

9.2 Spectral Profiling

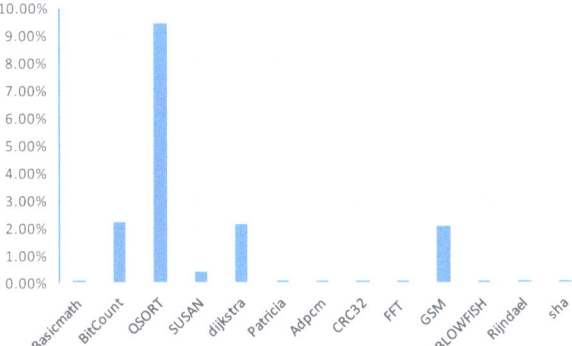

Fig. 9.4 Standard error for loop start/end times, normalized to loop duration [30]

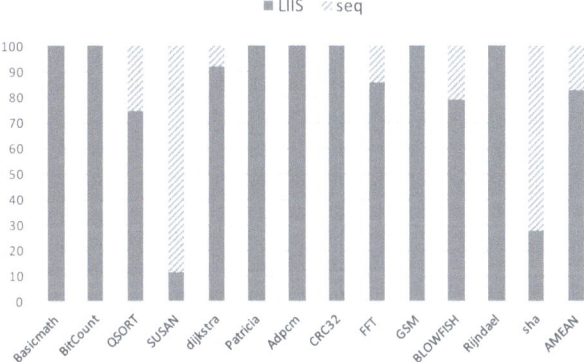

Fig. 9.5 Profiled time attributed through LIIS and Sequence mechanisms [30]

and almost all of this attribution is correct (see Fig. 9.3). In general, we observed that Sequence-based attribution of profiled time is highly accurate, as long as the execution contains enough LIIS recognitions to constrain the set candidate loops to which the (non-LIIS) spectrum may possibly be attributed.

9.2.6 Accuracy for Loop Exit/Entry Time Profiling

In addition to overall accuracy with which execution time is attributed to specific loops, we measured how accurate our method is at determining the exact time when a loop is entered and exited. Specifically, if in the profiled run some loop "Loop1" starts at time t_1, and our Spectral Profiling implementation identifies t_1' as the start time, then the difference between t_1' and t_1 is the deviation (error) for this "sample" in this experiment. The samples in this experiment are loop start and end times for all dynamic instances of all loops executed in the application, and for each application we report the standard deviation across these samples, normalized to the average duration of the loop. Because we need the actual start/end times for each loop to compute the error, we only perform this measurement in simulation (where we can get the actual loop entry/exit times without changing the timing of the execution itself). The per-application results of this experiment are shown in Fig. 9.4, and across all benchmarks, the average of these normalized standard deviations is 1.42%.

9.2.7 Runtime Behavior of Loops

In addition to providing useful information about which regions of the code are hot and how much time is spent in each region, Spectral Profiling can also exploit the shape of the spikes in the spectrum to tell us the runtime behavior of the each loop. For example, sharp spikes indicate that almost all iterations of the loop take same amount of time. Conversely, having a wide spike, or a group of spikes, indicates that different iterations of the loop have different execution times. Such variation in per-iteration execution time could occur in outer loops when the number of iterations in their inner loops varies, and even in inner loops due to architectural events (e.g., cache misses, branch miss-predictions, etc) or differences in control flow among loop iterations. Identifying loops with unusually large per-iteration performance variation may help programmers identify performance problems, e.g., problems caused by unexpectedly large number of architectural events, unexpectedly frequent use of a long and seemingly unlikely program path within the loop body, etc.

To illustrate how Spectral Profiling can help understand performance of a loop, Fig. 9.6 shows the spectrogram (how the spectrum changes over time, where the spectrum is displayed horizontally and elapsed time is shown from bottom to top) for one loop in the "Blowfish" benchmark, for the real system EM signal without any markers. We also show a histogram of actual per-iteration execution times in this loop, obtained by the markers during the execution with same inputs with markers. As seen in this figure, the duration and intensity of spikes in the frequency spectrum

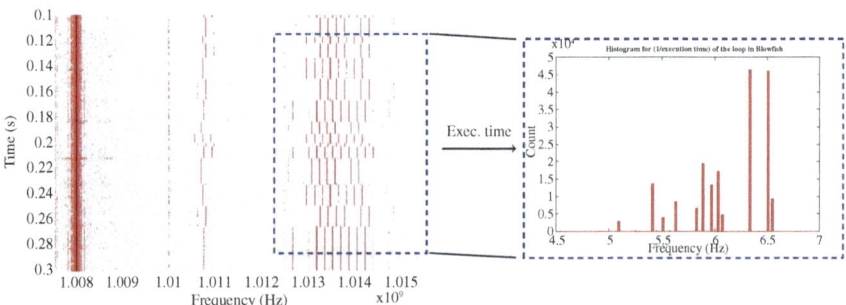

Fig. 9.6 Spectrogram and per-iteration execution time for a loop in *Blowfish* [30]

9.2 Spectral Profiling

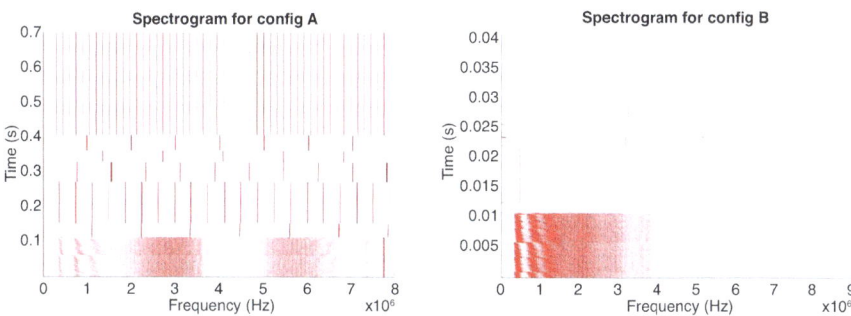

Fig. 9.7 Spectrograms from two different system configurations for *Bitcnt* benchmark for large size input [30]

indicate how often (and when during the loop) different per-iteration execution times occur. In this loop, the variation in per-iteration execution time is caused by cache misses on one memory access instruction and branch mispredictions on two difficult-to-predict branch instruction in the loop body.

9.2.8 Effects of Changing Architecture

To show that the ability to benefit from Spectral Profiling is architecture-independent, Fig. 9.7 shows the spectrogram for a same application, using the same inputs, on two different simulated systems (shown in Table 9.2). In this run, the application executes seven loops. The first loop is a nested loop, so its signature is poorly defined at lower frequencies (which correspond to the outer loop), with a sharp spike that corresponds to the inner loop at around 7.8 MHz in Config A and 81MHz in Config B. The remaining six loops each have a well-defined frequency, and can be clearly seen in spectrograms for both Config A and Config B. The vertical lines in the Config B spectrogram appear weaker because the spectral power of each spectral spike is distributed across a narrow frequency band as the out-of-order execution engine introduces slight variation among per-iteration execution times in each loop. However, spikes are still easily identifiable in both spectrograms, and allow us to attribute execution time to each of these seven loops, and also to determine their per-iteration execution time and its variation.

9.2.9 Effects of Window Size

Ideally, the profiled periodic activity lasts much longer than the window we use to compute the spectrum, so the "blurring" at the beginning and end of the activity

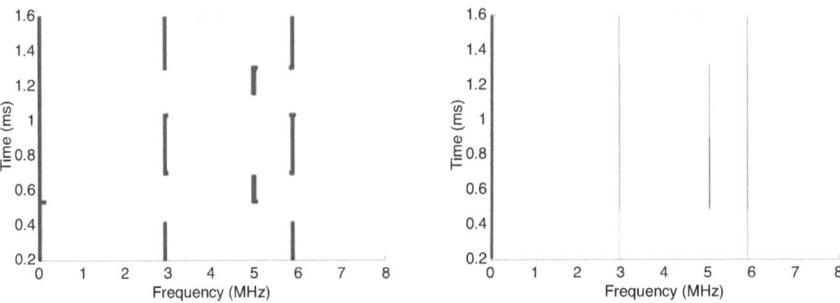

Fig. 9.8 Spectrograms from two different size of window for *Blowfish* benchmark for large size input. Left figure is for 50 μs window, and right is for 500 μs window [30]

introduces an error that is negligible relative to the duration of the activity. However, windows that are too short provide low spectral resolution, i.e., they make it more difficult to tell different spectra apart. Consequently, the window size is a compromise between these two considerations. To illustrate this, Fig. 9.8 shows the spectrogram when the window is 500 μs and when it is 50 μs. The application spends most of its time in two loops, one with a frequency close to 3 MHz (the spectrogram also shows its harmonic that is close to 6 MHz), and the other loop has a frequency close to 5 MHz. Both spectrograms are derived from simulation-based power signals for the same simulation run. As can be seen in the figure, the shorter window allows us to clearly identify transitions between these loops. The longer window, however, sometimes (e.g., for 0.6–1.2 ms on the spectrogram) only indicates that both loops are active during the interval. However, note that the vertical lines in the short-window spectrogram are thicker—this indicates that the frequency bins are wider, i.e., there is less spectral resolution. In this particular example, the frequencies of the loops are far apart, so even the 50 μs window provides enough spectral resolution to distinguish them. This indicates that the accuracy of Spectral Profiling can likely be improved by dynamically choosing the window size.[1]

9.3 Time-Domain Analysis of Analog Side Channel Signals for Software Profiling

As described thus far in this chapter, frequency-domain analysis of analog side channel signals can be used for program profiling. However, while frequency-

---

[1] Recall that our accuracy results are based on experiments where the window size is constant.

domain methods can be effective for efficient coarse-grain program profiling, that advantage is not present for profiling at the level of basic blocks or individual instructions, where time-domain analysis can be applied [10, 13, 28, 36]. The advantage of using the time-domain signal is that, in addition to timing of the execution, fast-changing amplitude changes in the signal can be efficiently leveraged to distinguish sequence of instructions executed. On the other hand, this approach is time consuming—a large number of samples are needed for meaningful analysis. However, it should be noted that program analysis typically does not require real-time monitoring, i.e., the signal can be recorded and analyzed at the pace that is required for the analysis, so the choice of analysis is largely based on the tradeoff between how fine-grained the analysis should be and how much computation is deployed for the analysis. This is unlike, for example, malware detection from the previous chapter, where the ability to prevent data exfiltration, damage to cyber-physical system, etc. depends on the delay between when the monitored system emanates the signal and the time when the analysis of that part of the signal is completed. Therefore, in the rest of this chapter, we present several possible approaches for time-domain program profiling, and the choice of which of these analysis to use depends largely on the level of details that is needed from the analysis.

9.4 ZoP: Zero-Overhead Profiling via EM Emanations

The first attempt to profile code using side channels, ZOP (Zero-Overhead Profiling) [10], computes fine-grained profiling information in a highly accurate and completely non-intrusive way by leveraging electromagnetic (EM) emanations generated by a system as the system executes code. Figure 9.9 illustrates how ZOP works on a simple example. It shows several waveforms recorded during a short fragment of program execution. All of these waveforms start at the same static location in the program, and each follows one of two paths depending on whether the $true$ or the $false$ path of a conditional statement is followed. In particular, the dashed waveforms correspond to execution along the $true$ (conditional branch instruction is "taken") path, whereas the two dotted waveforms correspond to execution along the $false$ path (branch instruction is "not taken"). It is clear from Fig. 9.9 that, while there are some differences between these "training" waveforms that correspond to the same path, these differences are smaller than those between the $true$ and $false$ paths. To determine which path was taken in the "unknown" (solid) waveform using only its EM signal, we calculate the correlation coefficient between that unknown waveform and each of the candidate recorded waveforms. From these correlation coefficients, we are able to determine with high confidence that the branch was taken in the unknown execution, as the branch-taken examples correlate much better with it than the branch-not-taken examples.

ZOP first measures the EM emanations produced by the system to be profiled as the system processes inputs whose execution path is known (*training* phase).

Fig. 9.9 Examples of waveforms collected by measuring EM emanations produced by several executions [10]

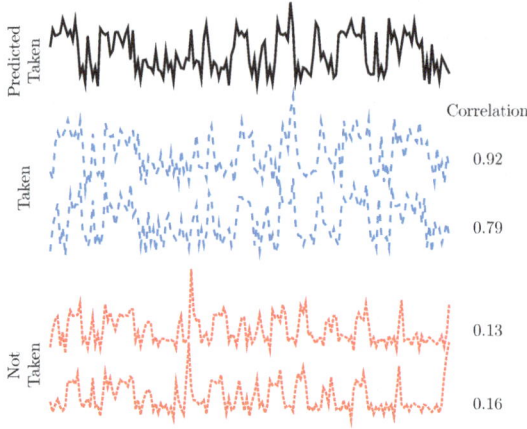

Fig. 9.10 High-level view of our approach [10]

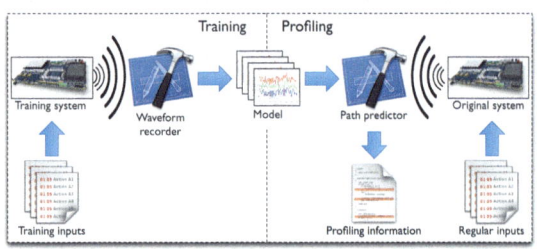

This allows ZOP to build a model of the waveforms produced by different code fragments. ZOP then collects emanations from a new, unknown execution and infers which parts of the code are being executed at which times (*profiling* phase). This inference is accomplished by matching the observed unknown emanations to emanations from the training phase that are known to be generated by particular code fragments.

Figure 9.10 shows a high-level overview of our general approach. As the figure shows, ZOP has two main phases. In the *training phase*, ZOP runs instrumented and uninstrumented versions of the program against a set of training inputs, records EM emanations for these executions, and builds a model that associates the recorded waveforms with the code subpaths that generated them. In the *profiling phase*, ZOP records the EM waveform generated by an execution of a vanilla (i.e., completely uninstrumented) version of the program, finds the closest match between this waveform and the waveforms in the training model, and uses the matching subpaths to predict the overall path taken by the execution being profiled. ZOP implements these two high-level phases in the steps and substeps shown in the workflow portrayed in Fig. 9.11. In the next sections, we explain the different steps and substeps in this workflow in detail.

9.4 ZoP: Zero-Overhead Profiling via EM Emanations

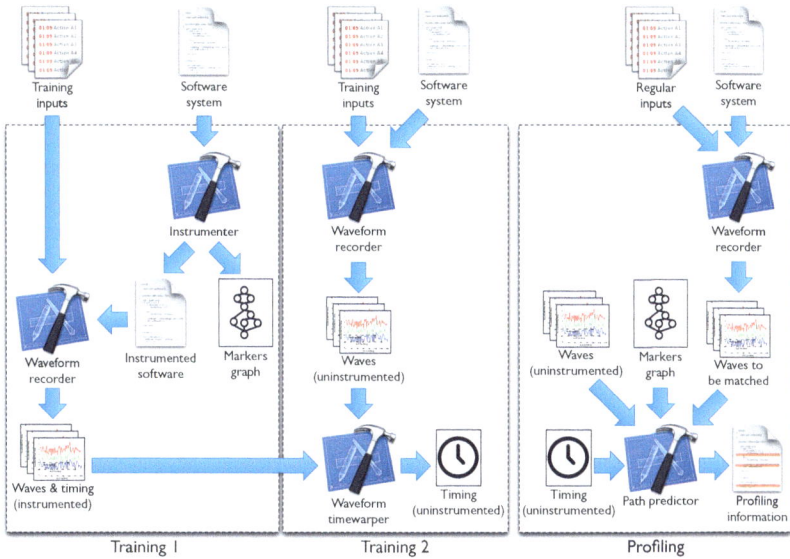

Fig. 9.11 Workflow of ZOP (Note that we repeat some elements to reduce clutter, improve clarity, and better separate the different steps of the approach; that is, multiple elements with the same name represent the same entity [10])

9.4.1 Training 1

The left part of Fig. 9.11 shows the Training 1 phase in the ZOP. During Training 1, ZOP runs an instrumented version of the system with a set of training inputs. This step is needed to build a graph model of the program's states, to determine the timing of each subpath, and to establish the correspondence between subpaths and the EM waveforms they generate. We refer to the instrumentation points as "markers" since they are used to "mark" the time of each executed instrumentation point in the EM waveform. In order to ensure optimal placement of these markers for generating accurate profiling information, the level of granularity of the inserted instrumentation points (markers) is critical.

In general, matching the EM emanations waveform from an unknown execution path to example waveforms for known execution paths is not a simple task. Matching complete program executions is clearly not an option, as it would require observing all possible executions to build a model. An ideal model would, in fact, be one that learns the waveform for each processor instruction independently, as this would make path recognition easiest. While some research techniques match waveforms on an instruction-by-instruction basis [17, 33] for non-profiling applications, such techniques have only been applied to the simplest of processors and has not yet been successfully applied to path profiling.

Based on our experience and preliminary investigation, we contend that longer subpaths must be considered for this matching to be successful in more complex processors, where at any point in time the processor has numerous instructions in various stages of execution, which makes the instruction-by-instruction recognition impractical. Therefore, in our approach, we consider acyclic paths, as defined by Ball and Larus [6], as the basic profiling unit. (Intuitively, acyclic paths are subpaths within a procedure such that every complete path in the procedure can be expressed as a sequence of acyclic paths.) In other words, ZOP learns the waveforms generated by the execution of acyclic paths exercised by the training inputs and then tries to recognize these paths based on their waveforms during profiling. The acyclic paths provide a level of the granularity that simultaneously (1) keeps the marker-to-marker paths short enough that a reasonable number of training examples can represent all the possible marker to marker waveform behaviors and (2) keeps the training instrumentation overhead low enough that the instrumentation itself does not drastically affect the execution waveforms.

The *Instrumenter* module starts by computing the acyclic paths in the code [6]. For every identified path in the source code, it adds markers in the source code to identify such paths. (Typically, the markers are placed at the beginning and end of each path.) The instrumentation locations are similar in spirit to those of lightweight program tracing approaches, such as [19].

The example code shown in Fig. 9.12 consists of a C function called putsub, which is a slightly simplified version of a function present in one of the programs we used in our evaluation. Marker positions for this example function are shown in Fig. 9.13. Each time a marker() is encountered, the marker ID (e.g., A,B,C, etc.) and the time elapsed since the start of the program are recorded in an array. To illustrate this with an example, consider an execution of putsub() that takes the path ABDEF. The recorded values would show the time when A was encountered, followed by the time when B was encountered, and so on. For each training input, ZOP runs the instrumented code and records the EM waveform. It then "marks" the EM waveform with the current program location each time a marker is encountered.

```
1  void putsub(char* lin, int s1, int s2, char* sub) {
2    int i = 0;
3    while (sub[i] != ENDSTR) {
4      if (sub[i] == DITTO) {
5        int j = s1;
6        while (j < s2)
7          fputc(lin[j++], stdout);
8      } else
9        fputc(sub[i], stdout);
10     i++;
11   }
12 }
```

Fig. 9.12 Uninstrumented putsub() function [10]

9.4 ZoP: Zero-Overhead Profiling via EM Emanations

```
1  void putsub(char* lin, int s1, int s2, char* sub) {
2    int i = 0;
3    marker(A);
4    while (sub[i] != ENDSTR) {
5      marker(B);
6      if (sub[i] == DITTO) {
7        int j = s1;
8        while (j < s2) {
9          marker(C);
10         fputc(lin[j++], stdout);
11       }
12     } else        {
13       marker(D);
14       fputc(sub[i], stdout);
15     }
16     i++;
17     marker(E);
18   }
19   marker(F);
20 }
```

Fig. 9.13 Instrumented putsub() function [10]

With this information ZOP could find, for instance, all the start and end times for the instances of the AB subpath in the training executions and extract the portions of the EM waveforms for these times. It is important to stress that instrumentation is only used during the Training 1 phase, and the program profiled during the Profiling phase is unmodified and uninstrumented. It is also worth noting that, while the location of the instrumentation points for putsub() results in a unique basic block subpath between each pair of instrumentation points, this is not a requirement for our approach.

The *Markers Graph* models the possible paths between marker code locations. As an example, Fig. 9.14 shows a graph derived from the putsub() function in Fig. 9.13. The graph's nodes are the markers for putsub(), and a directed edge occurs from marker X to marker Y if the program can reach Y from X without reaching another marker in between. While this graph shows a single edge between X and Y, there may be thousands of training examples for each such two marker subpath. Therefore, to predict the whole execution path, we need to not only predict the next marker but also the time the execution took to get from X to Y.

The *Waveforms and Timing* block of Training 1 contains the recorded waveform examples for subpaths in the program for which the correspondence between an execution's waveform and the code path taken is known. These waveforms, however, are affected by the computations done by the instrumentation, so they are

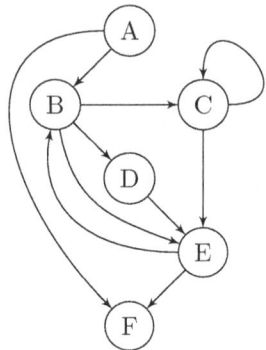

Fig. 9.14 Marker graph for the putsub() example [10]

not suitable for matching uninstrumented code during profiling. Therefore, ZOP's next step is to collect waveforms for the same training inputs, this time without instrumentation, and identify the times in these instrumentation-free waveforms that correspond to marker positions in the code (even though the uninstrumented code has no markers at these positions).

9.4.2 Training 2

The middle part of Fig. 9.11 shows the Training 2 phase of the ZOP approach. In this phase we run an *uninstrumented* version of the code with the same set of inputs used in Training 1, collect the waveforms for these executions, and perform matching to determine the points in these new waveforms that correspond to marker positions in the corresponding waveforms from Training 1. This results in waveforms generated by uninstrumented execution, but in which we do not know which part of the waveform corresponds to which marker-to-marker part of the program code. These waveforms must be compared to those observed during profiling to infer which part of the code is executing at each point in the profiling run. To do this, we must infer the timing of the uninstrumented code, i.e., we must determine which part of the instrumentation-free (Training 2) signal corresponds to which part of the instrumentation-marked (Training 1) signal and thus, transitively, to determine which portions of the waveforms collected during profiling correspond to which subpaths in the instrumentation-free program code.

This two-phase training approach has the key property that, while the device/environment used for Training 2 must be similar to that used for Profiling, the device/environment used for Training 1 can differ from that used for Training 2 and Profiling. For example, ZOP could perform Training 1 on a development board with more resources and flexibility, to facilitate the required instrumentation, and then perform Training 2 and Profiling on a production system that does not have

9.4 ZoP: Zero-Overhead Profiling via EM Emanations

the resources or flexibility to handle instrumentation (since neither of these phases requires instrumentation). Training 2 could then be done on a production system by software developers, whereas Profiling could be done directly on a deployed system, while real users interact with it. Furthermore, future work may allow ZOP to skip Training 1 altogether by deriving the timing information (marker times) via signal processing and machine learning.

9.4.3 Inferring Timing for the Uninstrumented Code Using Time Warping

The key to identifying which uninstrumented (Training 2) waveform corresponds to which part of the code is that, for each training input, we have executed the code twice, once with the instrumented program and once with the uninstrumented program. This means that the path through the code is the same for both executions, and that the EM signals for the two executions will tend to be similar at points that correspond to execution between markers, but one of the signals (the one from Training 1) has additional (marker instrumentation) activity inserted, along with some distortion of the signal at the transitions between instrumentation and "real" program activity. An example matching between instrumented and uninstrumented execution waveforms for the same training inputs is shown in Fig. 9.15. The longer red waveform (at the top of the figure) corresponds to an execution of the instrumented code, and the vertical solid black lines show the (known) timing of the markers as recorded by instrumentation. The shorter waveform (at the bottom of the figure) corresponds to an uninstrumented execution, where timing of the markers is not known because the code is not instrumented. Note that the instrumented and uninstrumented waveforms share many of the same features, but there are also significant differences (see, for instance, the DE and BC paths). These differences are often larger than the differences between two unique dynamic instances of the same subpath, so profiling accuracy would be poor if ZOP simply used (instrumented) waveforms from Training 1 to match to signals collected during (uninstrumented) profiling.

To systematically determine which part of the Training 1 signal corresponds to which part of the Training 2 signal for the same input, a technique such as dynamic time warping [31] can be used. In general, time warping between two signals can cut out parts of the top signal (shifting later samples of this signal to fill the gap made by the cut-out) in such a way that the remaining samples of the top signal are as similar as possible to the bottom signal. After time warping, ZOP knows which points in the instrumentation-free waveform correspond to the marker points in the instrumented-run waveform, as shown in Fig. 9.15.

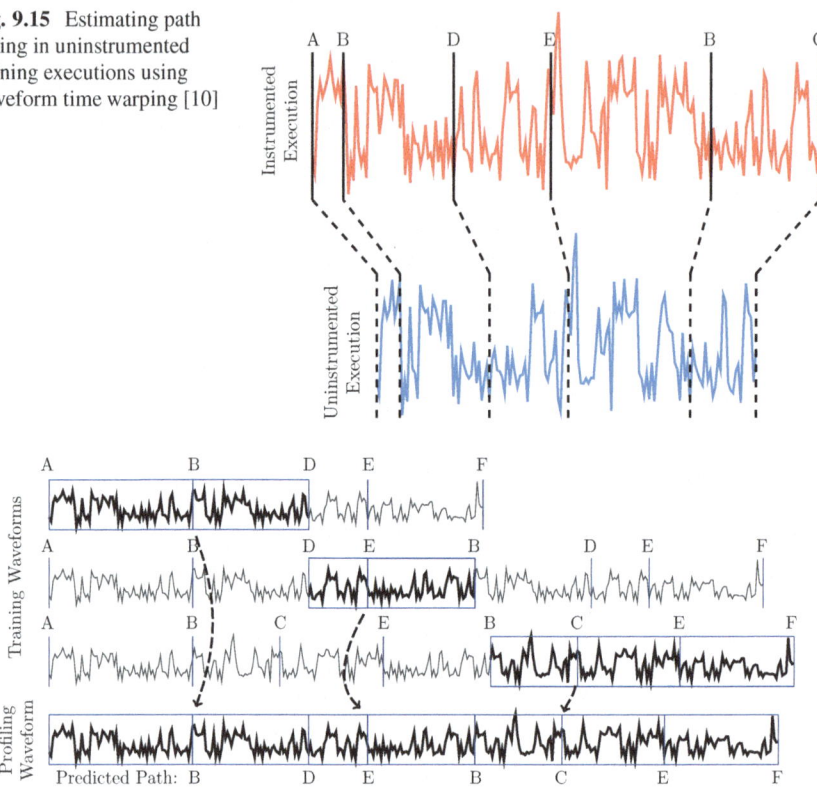

Fig. 9.15 Estimating path timing in uninstrumented training executions using waveform time warping [10]

Fig. 9.16 Predicting an execution path through putsub() by matching training waveform segments to an execution waveform [10]

9.4.4 Profiling

The right column of Fig. 9.11 shows the Profiling phase of ZOP. In the Profiling phase, we run the uninstrumented program with to-be-profiled inputs, record the EM waveforms produced, and compare these waveforms to the waveforms collected (and annotated with marker information) in Training 2.

The Training 1 and 2 phases of ZOP yield waveforms and marker timing information for the set of training inputs used in the uninstrumented program, as well as the markers graph. When a particular short subpath occurs during profiling, the resulting waveform will be similar to a training waveform of that same short subpath. To predict, for example, the execution path taken by the putsub() function, we run the uninstrumented version of putsub() with a to-be-profiled input and record the waveform shown at the bottom of Fig. 9.16.

To illustrate how our *Path Predictor* works, here is one example. For the profiling waveform shown in Fig. 9.16, we start with no information about the path taken.

9.4 ZoP: Zero-Overhead Profiling via EM Emanations

According to the markers graph, the profiling execution must start with marker A at the beginning of the waveform. The next marker encountered can be either B or F according to the marker graph. We use the Pearson correlation coefficient [39] to compare the profiling waveform with the three training waveforms in Fig. 9.16. All three training waveforms start with an AB subpath which very closely matches the start of the profiling waveform. Although it is not shown, assume that we have another training example with the AF path and this AF waveform does not match the profiling waveform. Then we can infer that the profiling execution starts with the AB path and that B occurs at the same time in the execution as it does in the training executions. There are two possible next subpaths from B, either BD or BC. Examining all the training waveform sections for BD and BC, it is clear that the profiling waveform matches the BD section in the top training waveform more closely than it does the BC section in the third training waveform. Therefore we can infer that the profiling execution takes the BD path. From D, the only possible next marker is E, so we find the most closely matching DE waveform and update our predicted path to ABDE. From E the code encounters either F or B next. Comparing the EF and EB waveforms, it is clear the profiling execution has taken the EB path next. We repeat this waveform matching and path updating process until we reach the exit marker F. This process predicts the ABDEBCEF path.

Figure 9.16 and its description captures the essence of the training and path prediction algorithm but some refinements are needed to achieve adequate performance. Consider what happens when an incorrect prediction is made. For example, assume we incorrectly selected ABCE at the start of the profiling waveform instead of the correct path ABDE. In such a case, not only is the subpath through C wrongly predicted, but also, even though we have predicted D correctly as the next marker, the time of the D marker is too early. When we match the training subpaths starting at D assuming this incorrect time for D, the training waveforms may no longer match the profiling waveform well. Blindly selecting the most closely matching next subpath is not guaranteed to result in the most closely matching waveforms for the entire execution. Such errors tend to compound and the predicted execution path may diverge from the actual execution path indefinitely. This issue may be even worse when an incorrect marker is predicted and the predicted path and the actual path diverge for a long time following the incorrect decision. To address these issues, we need to model the search for the optimal execution path more precisely.

When we reach a marker X at a particular time t in the profiling waveform we compare all the training subpath waveforms starting at X against the profiling waveform starting at t and assign a score to each training example. We use the correlation coefficient as the similarity metric between the section of the profiling waveform starting at t and the training subpath waveform. Therefore for each training example we get a correlation value, a next marker, and the time of the next marker (i.e., the start time t plus the duration of the training subpath).

We can think of the search for the optimal execution path through the program as a tree search. The root node is the entry marker (marker A in Fig. 9.13) and each child node has an edge for each training subpath example starting at that node's marker. Each node in this tree has a marker (e.g., A, B, C, etc.) and a starting time t in

Fig. 9.17 Example of path prediction through tree search [10]

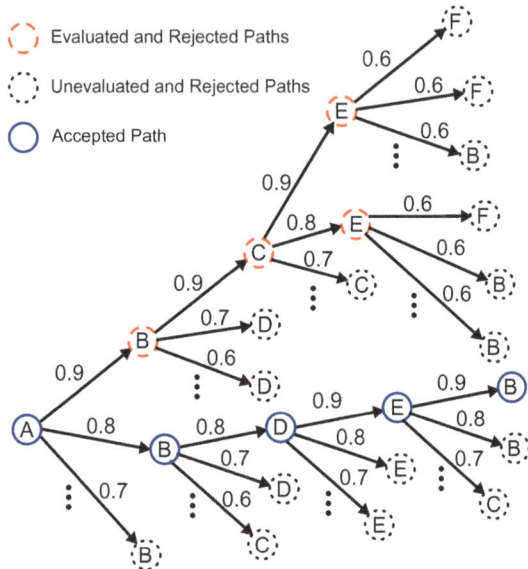

the profiling waveform. Each edge corresponds to a single subpath waveform, from training, and has three properties: a duration (the duration of the training example), a correlation between the training waveform and the profiling waveform starting at time t, and the marker at the end of the subpath in this training example. According to these definitions, a search tree for an example execution waveform of putsub() can be made as shown in Fig. 9.17. Each edge in the figure denotes a training example subpath and its waveform. The edge weights shown are the correlation values for each edge's training example (only the highest correlated subpath edges are shown).

The branching factor for these trees is large because each node may have thousands of training examples. To simplify the search, we employ the following heuristic. The goal of the heuristic is to find a root-to-leaf path whose edges all have correlation greater than a chosen value C_{th}. To evaluate a node, we calculate the correlation of each next edge and sort the nodes in order of decreasing correlation. If the edge with the maximum correlation is greater than C_{th} we continue searching along this edge. Otherwise, we indicate this node as rejected and backtrack along our path so far (i.e., toward the root node). As we backtrack we stop at the first node that has an edge to an unevaluated node with correlation greater than C_{th} and search forward along this edge. In Fig. 9.17, $C_{th} = 0.75$, and the search algorithm follows the red dashed nodes from A to B to C to E along the top-most edges. No edge from this E node exceeds C_{th}, so the algorithm backtracks to C and then continues forward to the E node with $C_{th} = 0.8$. Again, no edge from here exceeds C_{th}, so the algorithm backtracks along C to B to A and moves forward along the accepted blue path.

9.4 ZoP: Zero-Overhead Profiling via EM Emanations

Because ZOP identifies paths by matching short subpaths between marker pairs, and because it backtracks when unsuccessful, it can recognize paths that it never observed during training.

This heuristic clearly results in a path where each edge is greater than C_{th}, if such a path exists, since ZOP only follows edges with correlation greater than C_{th}. However, it is not guaranteed that all paths that meet our selection criteria correspond to the correct whole program execution path. It is also not guaranteed that a path meeting the selection criteria exists.

It is worth noting that many sophisticated heuristic algorithms exist for searching through trees with similar properties [7, 9, 15, 27], so we believe that future research in this area can greatly improve the overall path prediction accuracy.

Two minor refinements are required to make this algorithm practical. First, when correlating the training examples against the profiling waveform it is necessary to correlate the waveforms several times with slight misalignments between the training waveforms and the profiling waveforms and use the best result of all the alignments. This is because the current position in the program is always an estimate, so by trying several different alignments and selecting the best alignment, ZOP can keep track of the current program location with better accuracy. The second refinement is that the training waveforms for each edge are extended beyond the time position of the next marker so that all the training waveforms starting at a given marker have the same length. This is done by finding the training example for each marker with the longest duration and extending the other training example waveforms for this marker to the same length. This is required to allow fair comparisons between training examples, which would otherwise have different lengths (shorter signals are more likely to be more highly correlated due to random chance than longer signals). This approach has the added benefit that (with some preprocessing) all the training waveforms for a given starting marker can be correlated (with several different alignments) against a profiling waveform using a single matrix multiplication which greatly reduces runtime.

Removing nodes before they are evaluated can greatly decrease runtime because the evaluation of each node in this tree is expensive and the tree branches out quickly. Some nodes can be rejected quickly without sacrificing much accuracy. For example, suppose two edges W and Z start at a node B and represent BD training waveforms with nearly identical durations. This repetition is common because executions of the same subpath often have roughly the same runtime. Suppose W has higher correlation to the profiling waveform than Z. We can immediately reject Z without evaluating it because the D node following W and the D node following Z occur at the same time in the profiling waveform (since W and Z start at the same time and have the same duration). If W is evaluated and rejected, evaluating Z would just re-evaluate an identical D node with nearly the same start time.

We can eliminate more edges by observing that many marker sequences do not correspond to valid execution paths. To see this, recall that the path prediction execution paths are interprocedural and that we allow an edge from any marker X to a marker Y if an X to Y transition is possible in the profiled program. Then consider a function which contains a single marker F. This function is called from two points

A and B in the program, returning at points C and D respectively. Then, the only valid marker sequences for this function call would be AFC and BFD. The algorithm described so far would however also evaluate the impossible paths AFD and BFC. Ideally, a fully constrained grammar of all possible paths would be generated to limit the search to possible marker sequences. This grammar could enumerate the set of valid next markers from any node in the search tree. This grammar would be difficult to generate, so instead we keep a function call stack for the currently evaluated execution path and any next marker which would be inconsistent with the call stack is rejected. Note this is a very weak constraint and only eliminates the impossible AFD and BFC sequences when A and B are in different functions.

In the final step of ZOP, we construct the paths for the profiling inputs from a set of predicted markers provided by the previous steps. Every consecutive pair of predicted markers represents a set of basic blocks that are executed between two markers by a training input. First, for every training input and every two consecutive pair of markers we extract the basic blocks that are executed between them. Once ZOP collects the basic blocks between each pair of markers, it uses this information to generate the predicted whole-program basic block path from the sequence of predicted markers. The profiled acyclic paths can be easily identified and counted from this whole-program path.

9.4.5 *Empirical Evaluation*

To assess the usefulness and effectiveness of our approach, we developed a prototype tool that implements ZOP and performed an empirical evaluation on several software benchmarks. (For simplicity, in this section we use the name ZOP to refer to both the approach and its implementation, unless otherwise stated.)

For our evaluation, we used a NIOS II processor on an Altera Cyclone II FPGA. This processor has many of the features of modern complex computer systems (e.g., a 32 bit RISC MIPS-like architecture, a large external DRAM, separate instruction and data caches, dynamic branch prediction) while also providing features that were extremely useful for developing our understanding of how program execution affects the system's EM emanations (e.g., programmable digital I/O pins, access to programmable logic, and cycle-accurate program tracing). For the evaluation, we did not use any FPGA-specific features.

We leveraged LLVM [29] to detect the acyclic paths in the code, identify instrumentation points, and insert instrumentation. We then used LLVM's C backend to generate instrumented C source code. GCC then compiled this source code to a NIOS binary. Both the original and instrumented source code are standard C code and could be compiled and run on any modern architecture.

To observe EM emanations, we used a magnetic field probe (a small inductor in series with a tuning capacitor) that was placed directly over the processor's decoupling capacitors. We used an Agilent MXA N9020A spectrum analyzer to demodulate and record the signal.

9.4 ZoP: Zero-Overhead Profiling via EM Emanations

Table 9.3 Benchmark statistics [10]

Benchmark	LOC	Markers	Training set size	Profiling set size
print_tokens	571	48	240	400
Schedule	415	36	284	400
Replace	563	54	299	400
Total	1549	138	823	1200

For evaluation, we selected three programs in the SIR repository [32]: replace, print_tokens, and schedule. Table 9.3 shows, for each benchmark, its name, its size, the number of markers added during training, and the number of inputs we used during the training and profiling phases.

The decision to use only a few relatively small benchmarks was largely due to limitations of the system we used and of our measurement setup. The system we used in our evaluation, for instance, does not have an operating system. To automate measurements, we thus had to modify the stand-alone programs we profiled so that their `main()` function was called repeatedly from a wrapper executable. Because standalone programs tend to depend on data memory being initialized to zero when `main()` is called and typically do not clean up memory before exiting, this introduced issues that required manual effort for each program. Furthermore, we had to use LLVM's C backend to generate instrumented C code that was recompiled on the target system (in a real application ZOP would directly instrument binaries), which also created problems and required extensive manual checking. In addition, our general purpose measurement setup resulted in long measurement times, which favored the use of shorter executions of smaller programs. In general, avoiding larger programs allowed us to perform fairly extensive manual checks of our results, which helped us gain confidence in their correctness and, most importantly, let us get a deeper understanding of the issues involved in our approach and how to address them.

We selected the inputs for profiling and training as follows. For profiling, we selected inputs that achieved high path coverage, so as to demonstrate that ZOP can accurately profile a large number of different paths. As for the training set, ideally we would want an input set that exercises all the possible behaviors (in terms of EM emissions) of marker-to-marker subpaths; ZOP would then be able to identify complete paths by concatenating these short subpaths. As a more realistic proxy for this set, we selected training inputs that achieved branch coverage and then added a random set of extra inputs (see next paragraph). It is worth noting that production-quality software often already provides test suites with high branch coverage. Most importantly, for the purpose of training, much-easier-to-create unit tests for individual procedures could also be used.

Specifically, we performed our input selection by starting with the existing set of inputs in the SIR repository [32]. For each benchmark, we randomly split the inputs for that benchmark into two equally-sized disjoint sets: *training superset* and *profiling supersets*. This guarantees that the inputs used for training are completely independent of those used for profiling. From the training superset, we randomly selected a minimal subset of inputs that achieved the same branch coverage as the complete set. We then added 150 extra inputs randomly selected from the training superset to increase the chances of having different paths covered by different numbers of inputs, so as to be able to study how the characteristics of the training inputs affect ZOP's accuracy and answer RQ2. We selected 150 as the number of extra training inputs based on earlier experiments, as that number is not excessively large and yet can provide a higher variety in coverage. We call the resulting set the *training set*. To determine the set of inputs for profiling (i.e., the *profiling set*), we randomly selected a subset of the profiling superset that achieved the same acyclic-path coverage as the complete set and then added random inputs to get to 400 inputs, which was the largest number of inputs we could measure in the amount of time we had available.

9.4.6 Illustration of Program Profiling Results Using ZOP

We evaluate how well the overall program path identified by ZOP matches the actual overall path executed by the program. To obtain the ground-truth paths for this comparison, we have also executed each program with the same inputs used in our ZOP experiments, using the (instrumentation-based) approach by Ball and Larus [6]. For each benchmark and each profiled input, we then compare the number of times each acyclic path was executed in the ground-truth results and in the output of ZOP. We then calculated the average accuracy of ZOP for each benchmark using the following formula:

$$\text{accuracy} = \frac{\sum_{i=1}^{n} g_i a_i}{\sum_{i=1}^{n} g_i},$$

where

$n = $ number of acyclic paths per benchmark.,

$g_i = $ actual number executions of acyclic path i (ground truth),

$z_i = $ ZOP (predicted) number of executions of acyclic path i,

$a_i = \min\left(\frac{g_i}{z_i}, \frac{z_i}{g_i}\right) = $ accuracy for acyclic path i.

Therefore, when ZOP underestimates the number of executions of a path, the accuracy is computed as $a_i = \frac{z_i}{g_i}$, whereas when ZOP overestimates the number of executions of a path, the accuracy is computed as $a_i = g_i/z_i$. $a_i = 0$ when $z_i = 0$. To give equal weight to each path execution, each a_i is weighted by g_i.

9.4 ZoP: Zero-Overhead Profiling via EM Emanations

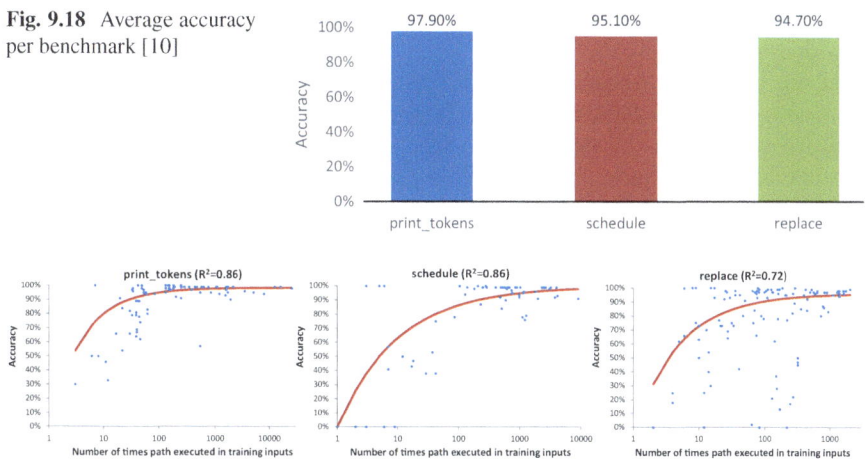

Fig. 9.18 Average accuracy per benchmark [10]

Fig. 9.19 Number of training examples vs accuracy for print_tokens, schedule, and replace [10]

Figure 9.18 shows the path profiling accuracy results. As the table shows, ZOP's estimates are fairly accurate. On average, ZOP correctly predicts 94.7% of the paths for replace, 97.9% for print_tokens, and 95.1% for schedule. In other words, the profiling information computed by ZOP *without any instrumentation* is always over 94% accurate.

We also computed how the accuracy of ZOP's path count estimates is affected by the number of times each path is exercised by the training set. We show these results in Fig. 9.19. Each data point in this figure represents the accuracy of ZOP's estimate for a single static acyclic path in the indicated benchmark (i.e., a single a_i value). For each benchmark, the figure also shows a fit for a saturating power curve[2] for each benchmark and the curve's goodness of fit (i.e., R^2). We chose this type of curve because, among all simple curves we tried, including linear, quadratic, exponential, etc. it produces (by far) the best goodness-of-fit. A logarithmic scale is used for the x-axis to more directly show the effect of increasing the number of training path instances by an order of magnitude.

For the print_tokens and schedule benchmarks in Fig. 9.19, the accuracy is poor when the acyclic path is executed less than 100 times, but greatly improves beyond this point. In fact, the vast majority of paths with more than 100 occurrences during training have nearly perfect accuracy. This is promising, as it implies that paths can be identified accurately by a relatively small number of inputs that cover them. Moreover, it also implies that accuracy can be improved by adding more training inputs.

---

[2] The curve is $y = a - bx^c$ where x is the number of dynamic instances, y is accuracy, and a, b, and c are constants chosen (for each benchmark separately) to produce the best fit.

The replace benchmark manifests a slightly different behavior. While the accuracy does increase with the number of times the paths are covered during training, there are several paths with more than 100 training examples that have accuracy below 80%, and even a few below 50%. In general the accuracy for replace does not improve as quickly as for the other two benchmarks as a function of the number of executions of a path during training.

The path prediction algorithm traverses a program's marker graph as shown in the example in Fig. 9.17. This traversal results in the evaluation (and possibly selection) of many impossible paths. The technique that we use to navigate the graph is context sensitive but does not distinguish between different call sites that invoke the same callee from within a procedure. Therefore, the algorithm could reach a callee from a given call site within a procedure and return to a different call site within the same procedure. This is particularly problematic in programs in which this situation occurs frequently and may lead to imprecision and poor predictions.

Finally, for 3 of the 400 inputs in replace's profiling set, the path prediction algorithm got "lost" while exploring the marker graph. As mentioned while describing our approach, if the waveforms collected during training do not closely match the waveform collected during profiling for more than a short time, the predicted and actual control flows can diverge beyond recovery. Once this happens, the remainder of the prediction for that input is completely incorrect. This condition only happened for the replace benchmark and only for three inputs, but it indicates that, in addition to providing more training to ZOP in this application, an analysis method which can recover from predicting an incorrect path may, either alone or in combination with ZOP, further improve results. We next describe one such method.

9.5 P-TESLA: Basic Block Program Tracking Using Electromagnetic Side Channels

One of the challenges ZOP has in tracking program execution is that it is using a tree structure and needs to backtrack every time infeasible path is encountered. Furthermore, ZOP requires longer code sequences to verify program execution which limits fine-grain code execution verification. To overcome this limitation, we have also developed a program tracking approach, which we call P-TESLA, that exploits electromagnetic (EM) emanations to reconstruct a detailed (basic-block-level) execution path with high accuracy. To accomplish this, we need to overcome the following challenges: (1) train a signal emanation model that associates signal patterns (or signatures) with fine-grain code segments or subpaths, and (2) represent the test signal using these signal patterns to reconstruct the program execution path. Specifically, we use a an improved two-step training process that exploits instrumented training to annotate the uninstrumented training signals, and identify which signal snippets correspond to which code segments. We also then use a novel signal matching technique that efficiently establishes correspondence between

9.5 P-TESLA: Basic Block Program Tracking Using Electromagnetic Side...

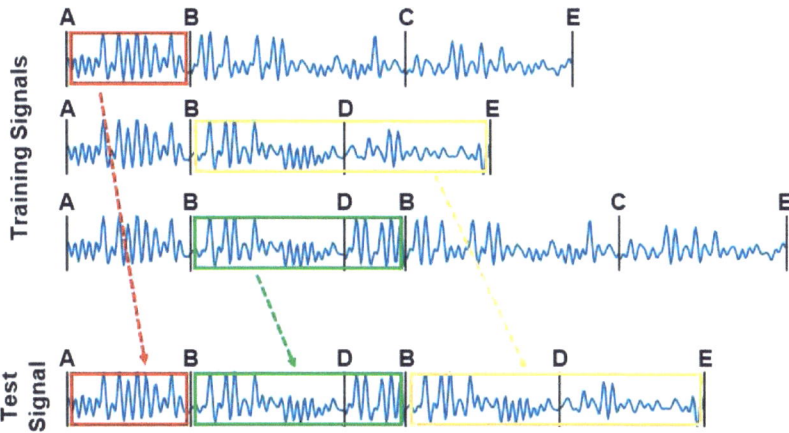

Fig. 9.20 The test signal is matched with the training signals. The signal matching is shown using color-matched boxes and dashed arrows. The underlying program sub-paths of the matched signal snippets (e.g., red, green, and yellow boxes) are concatenated to reconstruct the program execution path

the test signal and the training signals, and exploit this signal correspondence to reconstruct the execution path.

A high-level overview of this system is shown in Fig. 9.20. In the training phase, the system first executes an instrumented version of the program, and records the corresponding EM emanations. The instrumented program also outputs a marker sequence and their execution time-stamps that indicate the program execution path. Next, the system executes an uninstrumented (unaltered) program with the same program input, and compares the uninstrumented EM signal with the instrumented EM signal to map (uninstrumented) signal fragments to the underlying program subpaths. We call this mapping *Virtual Marker Annotation*, as it achieves the effect of having markers, but without any code instrumentation. This process uses the fact that, given the same program input, the instrumented and the uninstrumented programs follow the same program path, and thus the same marker sequence. This process also uses instrumented markers and time-stamps for efficient and precise signal mapping. Next, in the testing phase, the system monitors the EM side-channel signal caused by the execution of the vanilla (i.e., uninstrumented and unaltered) version of the program. It then matches the test signal with the training signals. For instance, in Fig. 9.20, different parts of the test signal are matched with different signal snippets from the training signals. Finally, the system concatenates the underlying program sub-paths of the matched signal snippets to reconstruct the program execution path. In Fig. 9.20, red, green, and yellow boxes correspond to program sub-paths A–B, B–D–B, and B–D–E, respectively. Thus, the reconstructed program path is A–B–D–B–D–E.

9.5.1 Signal Preprocessing: Amplitude Demodulation

To monitor program execution, the system first receives the emanated EM signal through an antenna, performs amplitude demodulation of the received signal, and then digitizes the demodulated analog signal using an analog-to-digital converter. The digitized signal is next scale-normalized before any further signal analysis. These preprocessing steps are applied to both training and testing phases.

First, we demodulate the received signal $r(t)$ at CPU clock frequency f_c as

$$x_a(t) = |r(t) \times e^{j2\pi f_c t}|, \tag{9.1}$$

Here, $x_a(t)$ is the amplitude demodulated analog signal, and t denotes the time. The demodulated signal $x_a(t)$ is then passed through an anti-aliasing filter with bandwidth B, and sampled at a sampling period T_s, i.e.,

$$x_d(n) = x_a(nT_s). \tag{9.2}$$

Here, $x_d(n)$ denotes the sampled signal at sample index n. The anti-aliasing filter cancels unwanted signals with frequencies beyond $f_c \pm B$. Note that, the sampling period T_s is determined by the well known Nyquist criterion $\frac{1}{T_s} > 2B$. Next, we scale normalize $x_d(n)$

$$x(n) = \frac{x_d(n)}{max(x_d(n))}. \tag{9.3}$$

The normalization ensures that the system is robust against changes in amplitude of the EM signals (e.g., due to change in the antenna's distance, position, etc.). The scale normalized signal $x(n)$ is then used for further signal analysis by the system in the training and testing phases.

Instrumented Training

As in the training for ZOP, in P-TESLA the training signals for each part of the code need to be marked with what code was executed when each part of the signal was collected. Thus, for basic-block profiling, P-TESLA is also first trained with instrumented program executions and then with uninstrumented program runs, and then the marked points in the instrumented-execution signals are projected onto the uninstrumented-execution training signals. However, to support profiling that is finer-grained than in ZOP, in P-TESLA the positions of markers in the signal need to have less timing uncertainty, and this has led to using different markers in the instrumented training and a different approach to marker projection onto the uninstrumented-training signals. Specifically, each marker now has a unique identification number (ID) that identifies its position in the program's control-flow-graph (CFG). The marker function execution records the **marker ID** along with the **execution timestamp** (i.e., the clock count when the function was executed). Thus, the markers act as program execution checkpoints that partition the CFG into smaller code-segments which we refer to as marker-to-marker code-segments or subpaths.

We insert these markers in strategic program locations. The marker insertion is dictated by the following criteria: (1) any program execution control-flow path must be uniquely and unambiguously represented by a sequence of marker-to-marker subpaths, and (2) all marker-to-marker subpaths must be acyclic and intra-procedural. Based on these criteria, we inserted markers in the following code locations: entry and exit nodes of functions, loop heads, and target nodes of go-to statements.

CFG partitioning helps to provide training coverage for program execution. Specifically, any practical program has a large number of feasible program execution paths. In fact, due to cyclic paths in CFG, programs can have infinite number of unique execution paths. Hence, it is neither practical nor possible to provide training for all unique execution paths for any practical program. However, the markers enable us to represent any execution path as a concatenation of marker-to-marker subpaths. Thus, instead of providing training for all unique execution paths, we provide training that covers all marker-to-marker subpaths. Note that the number of marker-to-marker subpaths is limited, and can be exercised using a relatively fewer number of strategic executions.

The markers also enable us to annotate the monitored EM signal. The marker execution time-stamps help to establish a correspondence between executed code segments (i.e., marker-to-marker subpaths) and the signal fragments they generate. It is important to emphasize that, while the markers provide an abstract partitioning, the program execution (for both training and testing) follows a single contiguous trace (from program's start to end), and generates a continuous EM signal. Consequently, it is not obvious which marker-to-marker subpath generated which part of the emanated signal. So, we use the marker execution time-stamps to annotate the start and the end of each marker-to-marker subpath in the monitored EM signal.

At the beginning of the program execution, we reset the processor's Time Stamp Counter (TSC) to zero. The markers record the TSC values as time-stamps, which then indicate the time-interval (in clock-cycles) from the program's start. We convert these time-stamps to sample-index using the following equation

$$n = round(\frac{t \times f_s}{f_c}). \qquad (9.4)$$

Here, n indicates the sample-index corresponding to the timestamp t, f_s is the sampling rate of the monitored signal, and f_c is the clock frequency of the monitored CPU. The rounding operation ensures that the resultant n is an integer value.

We also identify the program's start (i.e., sample-index $n = 0$) in the signal. To facilitate the automatic detection of program's start (demonstrated in Fig. 9.21), we insert a for-loop just before the beginning of the program execution. The for-loop executes a periodic activity (e.g., increments a loop counter variable) and generates a periodic and identifiable signal pattern. The end of this periodic pattern indicates

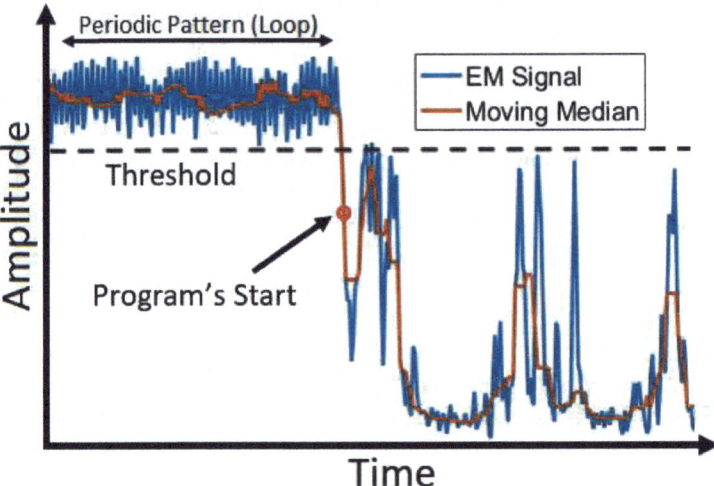

Fig. 9.21 Automatic detection of program's start: the end of the periodic pattern (for-loop) indicates the program's start. We identify this when the moving average of the EM signal drops below the threshold

the end of the for-loop (i.e., the start of the program). Furthermore, at the beginning of the program execution, the program is loaded into the system memory. This leads to memory accesses, which in turn stalls the processor and causes a dip in the signal amplitude [13]. We identify this transition (from the end of for-loop to the beginning of program execution) when the moving median of the signal drops below a predefined threshold (as shown in Fig. 9.21). This acts as the reference point (i.e., sample-index $n = 0$). All markers are then annotated according to their sample-indices.

Figure 9.22 demonstrates an annotated instrumented signal with each marker represented by a vertical red line. Markers m_0, m_1, m_2, and m_3 are placed at sample-index n_0, n_1, n_2, and n_3 respectively. The marker ID sequence (e.g., m_0, m_1, m_2, ...) indicates the program execution path, while execution time-stamps or sample-index sequence (e.g., n_0, n_1, n_2, ...) identify the start/end of the marker-to-marker code-segments. Thus, marker annotation establishes a correspondence between code-segments and emanated EM signal snippets. For instance, in Fig. 9.22, the signal snippet between sample-index n_0 and n_1 corresponds to the execution of the code-segment or subpath between marker m_0 and m_1.

While the instrumented training helps us to partition the CFG and to annotate the signal, the instrumentation alters the original signal emanation patterns. Thus, the instrumented training signals and corresponding signal emanation models cannot be used in the testing phase, in which the device executes a vanilla version (i.e., unaltered and uninstrumented) program. Specifically, the instrumentation adds overheads to the original program (i.e., function calls that record marker ID and execution time-stamps). The execution of these overhead codes (i.e., marker

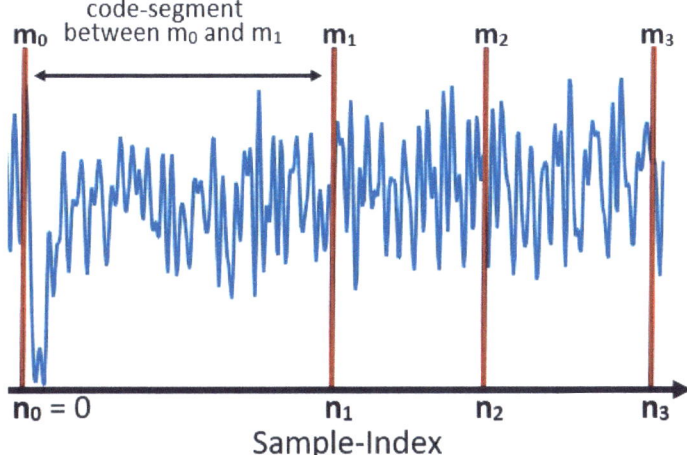

Fig. 9.22 Markers (vertical red lines) are placed on the signal according to their execution timestamps. The signal snippet between two consecutive markers corresponds to the EM emanation from the marker-to-marker code-segment

functions) requires additional computation (and computational time), and in turn, causes extraneous EM emanations that are irrelevant to the original program.

To evaluate the impact of instrumentation, we investigate the EM signature of the marker functions. We identify the marker functions in the instrumented signal using their time-stamps. We then crop out and compare these signal snippets. Figure 9.23 overlays 100 signal snippets corresponding to the marker function execution. We observe that the marker functions emanate very similar signal patterns. This is expected as the marker functions execute the same code segment. However, the beginning and the end of the marker signals demonstrate a marked variability. This is due to the microprocessor's instruction pipeline architecture that overlaps multiple instructions during execution. Thus, the processor's EM emanation at any instance depends on all instructions that are moving through different stages of the pipeline, rather than just one single instruction. Consequently, the execution of the same marker function may demonstrate signal variability towards the beginning and the end of the function call depending on the variations in the preceding and the following code segments (i.e., other instructions in the pipeline).

Likewise, the marker functions themselves also affect the EM emanations of the adjacent code segments. In the instrumented execution, all marker-to-marker code-segments are separated by marker functions. As such, the EM emanation patterns from the code-segments are altered at the boundaries due to the "cross-over" effect from the marker functions. For larger code segments (e.g., consisting of a few hundred instructions), the duration of the emanated signal is much larger compared to the altered boundaries. Thus, the impact of the instrumentation is trivial. However, for smaller code segments such as basic blocks consisting of only a few instructions, instrumentation can alter the overall signal emanation pattern significantly. Thus,

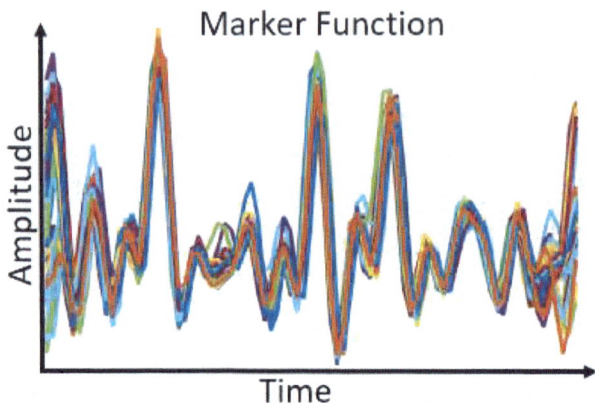

Fig. 9.23 EM signals corresponding to marker function execution

cropping out the marker signal-snippets from the instrumented signals would not replicate the uninstrumented signals. Instead, we exploit uninstrumented executions to create a better signal emanation model.

Uninstrumented Training

The system is next trained with uninstrumented program executions. However, before we can use the uninstrumented training for program execution monitoring, we must first annotate that signal. Unlike the instrumented executions, the uninstrumented executions do not have markers or time-stamps. Thus, we cannot directly annotate or identify which signal snippet corresponds to which code segment. Instead, we compare uninstrumented execution with the instrumented execution to identify and demarcate the marker-to-marker code segments in the signal. We call these demarcations "virtual markers" as they play the same role as the marker functions, albeit, without adding any overhead code or altering the original program or its signal emanation patterns.

Virtual Marker Annotation

In the instrumented execution, code segments are separated by marker functions. Each marker function execution records a pair of information m and t, where m represents the marker ID that indicates the execution point in the CFG, and t is the execution time-stamp. We then convert the time-stamp t to its equivalent sample-index n (using Eq. (9.4)). If the program executes k marker-to-marker code-segments, the instrumented execution records a sequence of $k + 1$ markers (including the starting and the ending markers). Thus, instrumented execution outputs a maker ID sequence $M = \{m_0, m_1, m_2, ..., m_k\}$ that M uniquely identify the program execution path and the corresponding sample-index sequence $N = \{n_0, n_1, n_2, ..., n_k\}$ that indicates which signal snippet corresponds to which code segment. Thus, the task of virtual marker annotation is to generate marker ID

9.5 P-TESLA: Basic Block Program Tracking Using Electromagnetic Side...

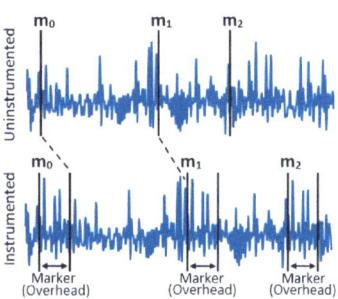

Fig. 9.24 EM signals corresponding to the uninstrumented (top) and the instrumented (bottom) program executions. The dotted lines indicate the correspondence between the uninstrumented and the instrumented signal

sequence M' and sample-index sequence N' for the uninstrumented training signal, without actually using instrumentation or marker functions.

To annotate the virtual markers, we execute the instrumented and the uninstrumented programs with the same input. Thus, the executions follow identical paths through CFG (i.e., execute the same marker-to-marker code segments in exact same order). This ensures the marker ID sequence is identical for instrumented and uninstrumented executions (i.e., $M' = M$). However, due to the overhead computations (i.e., marker functions), the timestamps or sample-indices for the virtual markers are significantly different (i.e., $N' \neq N$).

To estimate the virtual marker sample-index sequence N', we compare the uninstrumented and instrumented EM signals (Fig. 9.24). We notice that the execution of the same code segment (e.g., m_0-m_1-m_2) requires more computational time in the instrumented version due to the marker function overheads. Thus, we estimate the sample-indices for the virtual markers by adjusting for the overhead computational time using the following equation.

$$n'_i = n'_{i-1} + (n_i - n_{i-1} - n_{oh}) \quad \text{for } i \in \{1, \ldots, k\}. \tag{9.5}$$

Here, n'_{i-1} and n'_i indicate the sample-indices for the $(i-1)$-th and i-th virtual marker in the uninstrumented signal, n_{i-1} and n_i indicate the sample-indices for the $(i-1)$-th and i-th marker in the instrumented signal, and n_{oh} is the overhead computational time (in samples) for the marker function execution. Thus, $(n_i - n_{i-1} - n_{oh})$ is the overhead-subtracted execution time for the i-th code segment. Note that, sample-index n'_0 is 0 (indicating the starting point of the program), and we iteratively estimate $n'_1, n'_2, \ldots, n'_k$.

While Eq. (9.5) gives a good initial estimation for the virtual marker annotation, it does not account for the execution-to-execution hardware variabilities such as cache hits or misses that may lead to variabilities in computational time. To mitigate this issue, we fine-tune the initial sample-index estimations by matching uninstrumented signal with its instrumented counterpart.

First, we identify the signal snippet corresponding to a given code segment in the instrumented signal using its time-stamps. Let $x(n)$ be the instrumented signal with n indicating its sample-index. Thus, the signal snippet between sample index

$n = n_{i-1}$ and $n = n_i$ corresponds to the i-th code segment (i.e., the subpath between markers m_{i-1} and m_i). We then exclude or crop-out the first n_{oh} samples from this signal snippet as they correspond to the marker function **not** the original code segment. In Fig. 9.24, the dotted lines indicate the correspondence between the uninstrumented and instrumented signals. Thus, this overhead-subtracted signal snippet $s_i(n)$ acts as the EM signature or template for the code segment. We search for this signal template by sliding it across the uninstrumented signal $y(n')$. We limit our search within $\pm d$ samples of the initial estimations (i.e., between sample-index $n' = n'_{i-1} - d$ to $n' = n'_i + d$). This makes the search computationally efficient, and also helps to avoid false signal matches. At each search position, we compute the Euclidean distance between the template and the uninstrumented signal. We then choose least Euclidean distance match for updating the initial estimations

$$\hat{l} = \arg\min_{l} e(l) \quad \text{for } l \in [-d, d]. \tag{9.6}$$

Here, $e(l)$ is the Euclidean distance between the template $s_i(n)$ and the uninstrumented signal $y(n')$, and l indicates the shift from the initial estimated n'_i. Thus, $\hat{l}$ is the shift corresponding to the best match. Finally, we update initial estimated n'_i using the following equation

$$n'_i := n'_i + \hat{l}. \tag{9.7}$$

This iterative process is depicted in Algorithm 9.1. We note that, although this improved virtual marker annotation was designer for P-TESLA, it could be used to create finer-grain virtual markers for ZOP (or any other scheme that needs virtual markers).

Input: $x(n)$, $y(n')$, $N = \{n_1, n_2, \ldots, n_k\}$
Output: $N' = \{n'_1, n'_2, \ldots, n'_k\}$
initialization: set $n'_0 = 0$;
for $i := 1$ *to* k **do**
  | Estimate n'_i (Eq. 9.5);
  | Find best match (Eq. 9.6);
  | Update n'_i (Eq. 9.7);
end

Algorithm 9.1: Virtual marker annotation

9.5.2 Program Execution Monitoring

To reconstruct the program execution path, P-TESLA compares the device's EM emanation with the (uninstrumented) training signals and predicts the control-flow execution path. The path prediction involves two steps. In step 1, we match the

9.5 P-TESLA: Basic Block Program Tracking Using Electromagnetic Side...

monitored signal with the training signals to establish a signal correspondence. In step 2, we exploit this signal correspondence to predict the program execution path by using the training signal annotations (i.e., the virtual markers). We discuss these steps with further details in the following paragraphs.

Signal Matching

To establish signal correspondence, we match fixed-length windows from the monitored signal to the training signals, and then adjust the window-size according to the signal similarities. The signal matching process is demonstrated in Fig. 9.25. First, we extract a fixed-length initial window W of size L from the monitored signal. We then slide W across all training signals to find the best (i.e., the least Euclidean distance) match. This establishes a window-to-window signal correspondence (shown with a dashed arrow in Fig. 9.25). Next, we compare the samples that follow these windows. In Fig. 9.25, the initial window and its subsequent samples are overlaid on the matched window and its subsequent samples using red dots. We then iteratively extend the signal correspondence as long as the overlaid monitored signal is similar to the underlying training signal. Specifically, in each iteration, we compare the D subsequent samples and compute the sample-to-sample squared difference. If the mean squared difference is below a predefined threshold θ, we update the matched window size: $L := L + D$, and keep comparing the next D samples. Otherwise, we terminate the window extension process. We then again extract the next unmatched window from the monitored signal, match it across all training signals, and adjust the window-size. This process goes on until we establish signal correspondence for the entire monitored signal.

This approach for adaptive window extension is computationally more efficient than that of fixed-length window matching, in which smaller fixed-length windows from the test signal are matched against the training signals. In our experiments, P-TESLA was 13.37 times faster than fixed-length-window matching. This is because

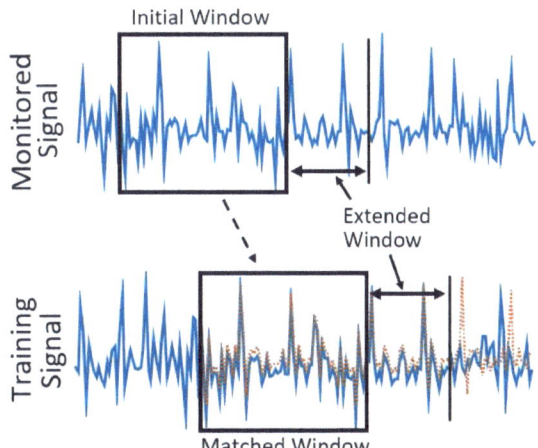

Fig. 9.25 Signal matching process: dashed arrow indicates the correspondence between fixed-length windows in monitored and training signals. Window size is extended based on signal similarities, up to the point where the training signal (blue line) starts to deviate significantly from the monitored signal (overlaid red dots)

window matching is computationally costly as it matches each window against all training signals. In contrast, P-TESLA initiates the search using a small fixed-length window, and then gradually extends the window size. The window extension is computationally much cheaper than window matching, as it matches the signal to one training signal, **not** to all training signals.

Furthermore, the time complexity for the window search is directly proportional to the window size L. Thus, smaller window leads to faster search. However, if the window is too small, the match becomes unreliable. In addition, smaller values for D enable finer adjustment of the window size. However, too small a value for D may lead to early termination of the window extension due to a few noisy samples. In our experiments, we used $L = 64$ and $D = 8$. The parameters L and D were chosen using a validation set.

Path Reconstruction

We next exploit the correspondence between the monitored and the training signals to reconstruct the execution path. Figures 9.26 and 9.27 illustrate the path reconstruction process through a simplified example. Figure 9.26 shows the program CFG where the nodes represent the markers and the edges represent the marker-to-marker subpaths. The training signals and the monitored signal are shown in Fig. 9.27. The (virtual) markers are annotated on the training signals with vertical black lines, and indicate that training signal 1 corresponds to program path $Start - A - C - A - C - End$, while training signal 2 corresponds to $Start - A - B - End$. Note that, for this simple CFG, these two training executions are sufficient to provide coverage for all marker-to-marker subpaths (i.e., edges on the graph). However, most applications require a large number of executions (e.g., hundreds or even thousands) for high code coverage.

In Fig. 9.27, we indicate the correspondence between monitored and training signals using color-matched windows and dashed arrows. For instance, red windows

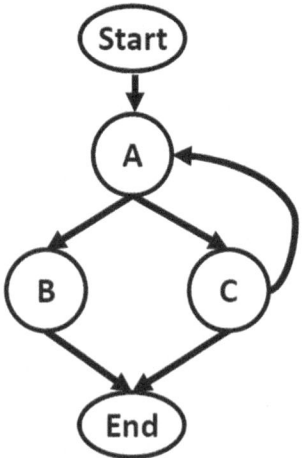

Fig. 9.26 Program control-flow graph

9.5 P-TESLA: Basic Block Program Tracking Using Electromagnetic Side...

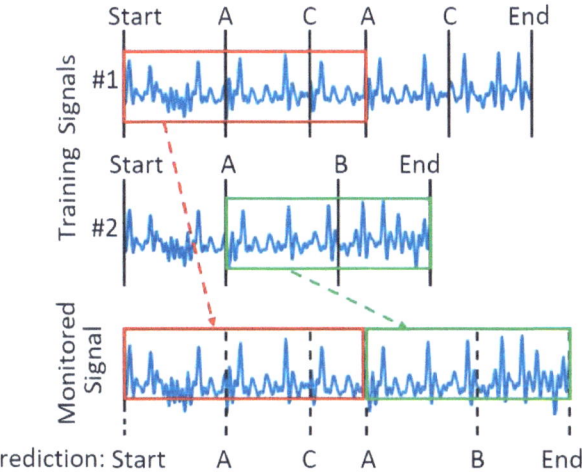

Fig. 9.27 Program execution path prediction using signal correspondence

in the training and the monitored signals demonstrate similar signal patterns, and so do the green windows. This signal correspondence enables us to reconstruct the monitored signal by concatenating matched-windows (e.g., red and green windows) from different training signals. More importantly, the signal similarity or correspondence implies that the matched windows correspond to the same program subpath. Thus, we reconstruct the program execution path for the monitored signal by concatenating the program subpaths corresponding to the matched training windows. For instance, the red window (in training signal 1) corresponds to the program subpath $Start - A - C - A$, and the green window (in training signal 2) corresponds to the program subpath $A - B - End$. Therefore, we concatenate these subpaths to reconstruct the execution path. In Figs. 9.26 and 9.27, the reconstructed execution path ($Start - A - C - A - B - End$) is indicated with dashed vertical lines.

Experimental Evaluations

We evaluate the system by monitoring two different devices as they execute three different benchmark applications. The evaluation matrix, the benchmark applications, and the experimental results are discussed in the following sections.

To evaluate system, we compute the edit distance between the actual execution path and the reconstructed execution path. Specifically, we use Levenshtein distance [18] that computes the minimum number edits (insertions, deletions or substitutions) required to change the reconstructed marker sequence into the actual marker sequence. We then compute the path reconstruction accuracy as:

$$\text{Accuracy} = 1 - \frac{\text{Edit Distance}}{\text{Length of Actual Marker Sequence}}, \quad (9.8)$$

We further compare the actual and the reconstructed timestamps by computing and reporting the absolute timing difference between the actual and the reconstructed markers. Note that edits are excluded from this comparison, as there are no timestamps for the edited (e.g., inserted or deleted) markers.

Benchmark Applications

We selected 3 benchmark applications (*Print Tokens*, *Replace*, and *Schedule*) from the SIR repository [26], which are commonly used to evaluate techniques that analyze program execution. Table 9.4 provides some more details about these applications.

Each of these applications has many inputs, each taking a different execution path through the CFG of the application. We used disjoint sets of inputs for training and testing. For each application, we randomly selected 500 inputs for training, and 100 for testing.

We evaluate system by executing these applications on two different devices: (1) FPGA device and (2) IoT device.

First, we monitored an Altera DE-1 prototype (Cyclone II FPGA) board. This device has a 50 MHz NIOS II soft-processor. We placed a magnetic probe near the device to collect the EM side-channel signal. We then used an Agilent MXA N9020A spectrum analyzer to observe and demodulate the EM emanations. The demodulated signal is next passed through an anti-aliasing filter with 5 MHz bandwidth. Finally, we sampled the filtered signal at 12.8 MHz sampling rate, and analyzed the digitized signal using the system.

Table 9.5 summarizes the mean accuracy. We observe that P-TESLA achieves excellent accuracy for monitoring all three benchmark applications, with roughly 99% accuracy for *Print Tokens* and *Replace*, and near-perfect accuracy for *Schedule*.

We also report the mean timing difference of the predicted timestamps in Table 9.6. For *Print Tokens* and *Schedule* the mean timing difference is less than

Table 9.4 Benchmark applications statistics

Benchmark	LOC	Basic blocks
Print tokens	464	178
Replace	495	245
Schedule	579	175

Table 9.5 Mean accuracy for FPGA

Benchmark	Path prediction accuracy
Print tokens	98.7%
Replace	99.1%
Schedule	99.8%

Table 9.6 Mean timing difference for near-field FPGA experiments

Benchmark	Mean timing difference
Print tokens	0.98 samples
Replace	4.13 samples
Schedule	0.88 samples

9.5 P-TESLA: Basic Block Program Tracking Using Electromagnetic Side...

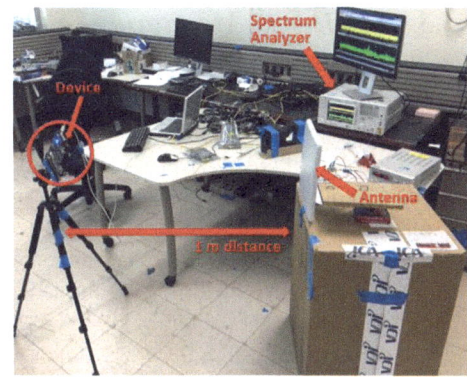

Fig. 9.28 Experimental setup: monitoring from 1 m distance

Table 9.7 Mean accuracy at 1 m for FPGA

Benchmark	Path prediction accuracy
Print tokens	95.44%
Replace	95.14%
Schedule	99.40%

Table 9.8 Mean timing difference at 1 m for FPGA

Benchmark	Mean timing difference
Print tokens	3.78 samples
Replace	5.56 samples
Schedule	1.32 samples

1 sample. However, for *Replace*, the mean timing difference is roughly 4 samples. Note that, at the experimental sampling rate (12.8 MHz), 1 sample is equivalent to 78.125 ns. Thus, all timing estimations are very precise.

Monitoring from a Distance We further evaluate P-TESLA by monitoring the FPGA device from a distance of 1 m using a panel antenna. Figure 9.28 shows the experimental setup. We summarize the mean accuracy in Table 9.7. P-TESLA achieves better than 95% accuracy on all three benchmarks, and for *Schedule* it achieves 99.4% accuracy.

Table 9.8 shows the mean timing difference for the predicted marker timestamps. While the timing differences are slightly higher than that of with probe, the predicted timestamps are still very precise (within a few samples).

While P-TESLA demonstrates excellent performance from 1 m distance, we notice slight degradation in accuracy compared to that of with probe (i.e., at 1 cm distance). This degradation is due to the lower SNR at distance, and can be improved by using high-gain antennas and/or low-noise amplifiers.

9.5.3 IoT Device Monitoring

We demonstrate the robustness of P-TESLA by monitoring an A13-OLinuXino IoT development board. This device has a 1 GHz Cortex A8 ARM processor [20]. Unlike the FPGA device, A13-OLinuXino runs on a Debian Linux operating system. We collected the EM side-channel signal by placing a magnetic probe near the microprocessor. The signal was recorded and demodulated using a spectrum analyzer (Agilent MXA N9020A). We then pass the signal through an anti-aliasing filter with 20 MHz bandwidth, and sample it at 51.2 Msps.

Table 9.9 shows the accuracy of P-TESLA for monitoring the IoT device. P-TESLA demonstrates high accuracy on all three benchmark applications; 94.15% on *Print Tokens*, 96.85% on *Replace*, and 95.91% on *Schedule*. This is somewhat lower accuracy then for the FPGA, which is expected because the A13-OLinuXino has a much faster processor (1 GHz compared to FPGA's 50 MHz), and that makes fine-grained execution monitoring more challenging. Furthermore, the operating system on A13-OLinuXino leads to more variations between training and testing executions. This, in turn, can cause performance degradation.

The timing differences reported in Table 9.10 demonstrate that the predicted time-stamps are also quite precise. The mean timing difference for all three benchmark applications is within 10 samples. Note that, at 51.2 Msps), each sample is equivalent to 19.5 ns.

Monitoring from a Distance We also evaluate the system by monitoring the IoT device from a distance. For this, we placed a slot antenna 1 m away from the device.

Table 9.11 shows that the system achieves roughly 90% mean accuracy on all three benchmarks, and Table 9.12 shows the mean timing difference. We again notice some performance degradation at distance, mainly due to lower SNR.

Table 9.9 Mean accuracy for near-field IoT device experiments

Benchmark	Path prediction accuracy
Print tokens	94.15%
Replace	96.85%
Schedule	95.91%

Table 9.10 Mean timing difference for near-field IoT device experiments

Benchmark	Mean timing difference
Print tokens	7.68 samples
Replace	10.33 samples
Schedule	6.92 samples

Table 9.11 Mean accuracy at 1 m distance for IoT device

Benchmark	Path prediction accuracy
Print tokens	89.87%
Replace	90.79%
Schedule	90.40%

Table 9.12 Mean timing difference at 1 m distance for IoT device

Benchmark	Mean timing difference
Print tokens	11.40 samples
Replace	33.66 samples
Schedule	7.53 samples

Table 9.13 Mean accuracy at different SNR

	Signal-to-noise ratio		
Device	30 dB	20 dB	10 dB
FPGA	97.97%	96.14%	75.38%
IoT	93.20%	91.73%	71.77%

Table 9.14 Mean timing difference at different SNR

	Signal-to-noise ratio		
Device	30 dB	20 dB	10 dB
FPGA	0.99 samples	5.76 samples	8.91 samples
IoT	8.03 samples	9.17 samples	12.73 samples

Monitoring at Different Signal-to-Noise Ratios

To further evaluate the practicality of the system, we measure its performance at different signal-to-noise ratios by applying additive white Gaussian noise (AWGN) to the monitored EM signal. For this experiment, we execute and monitor the benchmark application *Print Tokens* on both FPGA and IoT devices. Table 9.13 shows accuracy results from these experiments. For the FPGA device, P-TESLA achieves 97.97% mean accuracy at 30 dB SNR. We also notice a slight degradation in system performance (96.14% mean accuracy) at 20 dB SNR. However, at 10 dB SNR, we observe significant degradation (75.38% mean accuracy). Likewise, we observe a similar trend for the IoT device, for which system achieves 93.20%, 91.73%, and 71.77% mean accuracy at 30 dB, 20 dB, and 10 dB SNR, respectively.

For reference, we received signals with better than 30 dB SNR for both FPGA and IoT devices by placing magnetic probes next to the processors. For remote monitoring, we received signals with roughly 20 dB SNR, using antennas and low-noise amplifiers (LNA) at 1 m distance. Also, note that, we did not use any EM shielding, and that experiments were performed while other electronic devices (e.g., laptops, cellphones, measurement instruments, etc.) were active inside the room.

We also report the mean timing difference at different SNR in Table 9.14. We notice that, for both FPGA and IoT devices, the timing differences increase as SNR gets lower.

9.6 PITEM: Permutations-Based Instruction Tracking via Electromagnetic Side-Channel Signal Analysis

Thus far in this chapter, we have shown that program profiling using analog side-channels is possible, and we have shown that such profiling can be performed at the levels of loops, acyclic paths, and basic blocks [10, 28, 30]. However, even when profiling at the basic-block granularity, when deciding which path through the program code the specific execution has taken, these profiling approaches rely on the overall difference between basic blocks that are eligible for execution at that point. Now we will investigate time-domain analysis for profiling program execution at the granularity of one or a few instructions[36, 42], i.e., we consider how to correctly profile execution even when the difference among the eligible basic blocks is very small. This happens, for example, when profiling a program that executes an if-then-else where the "then" and the "else" part of the code differ in only one of a few instructions.

Here we present a system that allows us to track instruction-level execution in the presence of noise. This framework consists of two major steps:

- Identifying groups of instructions that are referred to as *instruction types* that have similar EM signatures,
- Tracking all possible orderings, i.e., permutations, of these *instruction types* and therefore, monitoring program execution at *instruction-type* granularity.

To make the definition of these terms more clear, we present a simple example. Assume that we have a processor with only four instructions: ADD, SUB, MUL and STR. Let ADD and SUB operations have similar EM signatures, while MUL and STR have signatures that are different from both each other and from ADD/SUB. Based on the EM signatures, the first part of our framework is to identify 3 *instruction types* that we label with letters: *type* A (ADD and SUB), *type* B (MUL), and *type* C (STR). The second part of our framework detects which ordering of these instruction types has been executed. For example, a code block that has the instructions ADD-MUL-STR in order can be recognized as permutation "ABC", whereas a code block that executes the instructions MUL-STR-ADD can be recognized as "BCA". Generating EM signatures for sequences rather than single instructions addresses the pipeline effect to a great extent, as these sequences represent the overall EM emanations for longer periods of time. This allows this framework to be applicable for tracking program activity even for devices with complex processor architectures and higher clock frequencies.

9.6.1 Determining Instruction Types by Using EM Side Channel

In this section, we describe the procedure to identify groups of instructions with similar EM side channel signatures, which we call *instruction types*. Since different architectures are implemented differently on micro-architecture level, these *instruction types* differ for different architectures. Figure 9.29 presents an outline of the procedure step by step. These steps are explained in detail in the following sections.

9.6 PITEM: Permutations-Based Instruction Tracking via Electromagnetic... 257

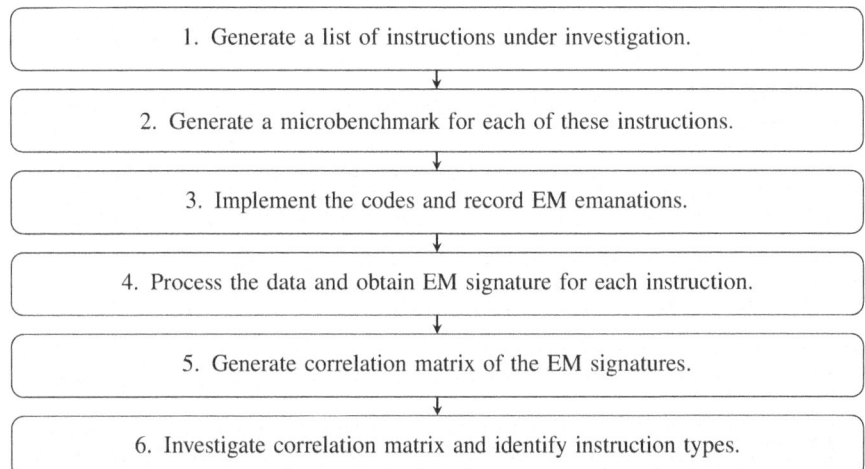

Fig. 9.29 Flowchart of determining instruction types [36]

9.6.2 Generating a List of Instructions Under Investigation

This step includes examining the instruction set for the given processor and selecting the desired and applicable instructions to investigate. The applicability of the instructions is based on the micro-architecture of the processor.

9.6.3 Generating Microbenchmarks

After instruction selection, we generate a microbenchmark for each instruction whose pseudo-code is given in Fig. 9.30. One should note that this work uses an *instrumented* measurement setup where an input/output (I/O) pin is set to "high" (logical 1) just before the code under observation is executed, and reset to "low" (logical 0) after the code under observation finishes execution. Therefore, the input/output pin signal is used to find the starting and ending points of the region of interest.

For most processors, a single instruction does not generate a distinct variation in EM signals [11]. To overcome this, the instruction under interest is repeated N times to magnify the EM signature. Note that this repetition is realistic because the pseudo-code structure is only used in determining *instruction types* step and it is not utilized in any testing scheme. The starting and ending markers are preceded and succeeded by two empty loops, respectively. This for-loop structure allows for a fair comparison since they make sure that the same instructions are pipelined before and after all instructions under interest.

```
while true
    for
        %empty for loop
    end

    % Set I/O Pin to High

    for N times
        %Instruction Under Interest
    end

    % Reset I/O Pin to Low

    for
        %empty for loop
    end
end
```

Fig. 9.30 Pseudo-code for instruction type detection setup [36]

9.6.4 Recording EM Emanations for Microbenchmarks

After microbenchmark implementation, EM emanation measurements are performed. The measurement includes two synchronized channels: the EM emanation signal, and the I/O pin signal. For better localization, a near-field antenna with proper antenna gain and size should be utilized. The choice of the antenna size is based on their ability to capture relevant EM emanations from the processor and reject the interference from other parts of the device.

9.6.5 Data Processing to Obtain EM Signatures

In this step, the recorded signals for EM emanations and the I/O pin signals are processed to obtain EM signatures of each instruction. Figure 9.31 presents an overview of this data-processing. First, the input I/O signal is filtered with a moving median filter to overcome overshooting. Then, the signal is normalized to account for a possible DC offset. Next, the amplitude values are quantized to their binary representations by using a 3 dB threshold. The binary stream of data is smoothed by removing outliers and the starting and ending points are determined.

The input EM signal contains unintentional EM emanations that are amplitude modulated (AM) to the periodic signals present on the board [12]. Among these modulations, the one around the first harmonic of the processor clock frequency is both the strongest and the most informative. Therefore, the input EM signal is first down-converted using the clock frequency, and then low-pass filtered to reduce measurement noise at higher frequencies. However, interference due to non-program activity related periodic activities such as voltage regulators, memory refresh, etc. might still occur within the filtered bandwidth. Some of these interferences are so strong that they dominate the EM signature. To eliminate them, we identify and store

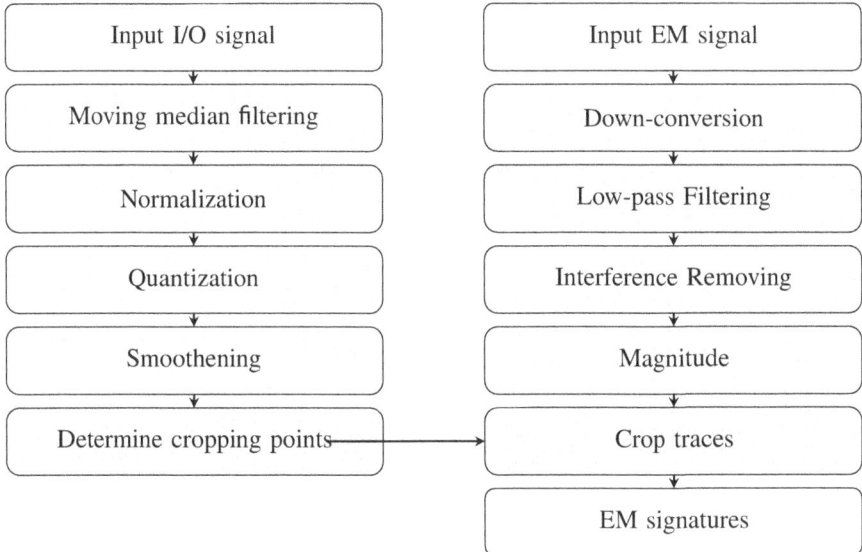

Fig. 9.31 Flowchart of data processing [36]

the frequencies at which these peaks occur from a recording that is collected when the board is idle. Next, we remove these peaks in the "interference removing" step by using band-stop filters centered around these interference frequencies. Since the modulation around the clock frequency is not necessarily conjugate-symmetric in the frequency domain, the resulting signal can be complex-valued. The amplitude of this complex-valued signal contains the shape information whereas the phase carries time-shift information. Since our objective is to determine the shape, we only keep the magnitude. Finally, the processed EM signal is cropped into chunks. The cropping points are obtained from the input I/O signal. These cropped chunks are EM signatures that represent the EM waveforms generated from the repetition of instructions N times.

9.6.6 Generating Correlation Matrix

Next, we generate a correlation matrix that represents similarity between the generated EM signatures. The degree of dissimilarity between two time domain signals can be measured in several ways: L1 norm, L2 norm, cross-correlation etc. The power consumed by the devices fluctuates during run-time and this creates a DC offset in the measured EM emanations. Since L1 norm and L2 norm measure the sum of point-wise distances, it is an error-prone measure in the presence of DC offsets. Therefore, these distance measures are not suitable without normalization. On the other hand, cross-correlation measures similarity of the waveforms, which is a suitable similarity metric for our purposes. The execution of some instructions takes longer than the others, and their EM signatures are longer in length. To account for different lengths, while correlating two signals, the correlation is performed by sliding the shorter waveform over the longer one, and the highest correlation that is found during this sliding is taken to be the cross-correlation of these two waveforms. This cross-correlation is performed for all EM signature pairs to generate the correlation matrix.

9.6.7 Identifying Instruction Types

The objective of this step is to find instruction types, i.e., the groups of instructions that have similar EM signatures. The correlation matrix of the EM signatures shows how similar the signatures are to each other. A subjective method to find the *instruction types* from the correlation matrix is visual inspection. However, this approach is prone to misclassifications due to its subjective nature. As an objective method, we propose to utilize *hierarchical (agglomerative) clustering*. This clustering technique is a bottom-up algorithm that starts with treating each sample as a separate cluster and merges these clusters pair-wise until all samples are merged into a single cluster [21]. This generates the *cluster tree (dendrogram)*, a sequence of clusterings that partition the dataset [40]. Unlike other clustering algorithms, such as *K-means*, this method does not require a random centroid initialization or an a priori number of clusters. This is especially useful here, as we do not have a priori knowledge about the number of *instruction types*. We can decide on the number of clusters by observing the *dendrogram*. The main disadvantage of this algorithm is its large time complexity ($O(N^2 \log(N))$), and space complexity ($O(N^2)$) [16, 34]. Since the number of instructions under investigation is typically not a large number, the dataset size is relatively small, therefore, the time and space costs are affordable.

The output of the correlation matrix presents the similarity measure between the signatures. However, *hierarchical clustering* is based on the dissimilarities between samples. Therefore, by using [37], we convert the cross-correlation values, ρ, to distance values, d, as follows,

$$d = \sqrt{2(1-\rho)}. \tag{9.9}$$

9.6.8 Detecting Permutations of Instruction Types

The next step is to detect which sequence of *instruction types* is encountered in a testing scenario. The most straightforward way to do so is to find the EM signature of the *instruction types* when they are executed once (instead of N times), and use

9.6 PITEM: Permutations-Based Instruction Tracking via Electromagnetic...

these signatures as a dictionary while testing. However, this approach has two major drawbacks:

1. Most processors have a pipelined architecture, where the execution of an instruction is divided into different stages, and execution of consecutive instructions each in a different stage is overlapped. Since all stages have transistor switching activity during their operation, all stages behave as an EM emanation source. Therefore, the measured EM emanation at the antenna is a combination of the EM waves radiated from different stages; it is not possible to isolate the EM signature of an individual instruction because the difference generated by a single-instruction is significantly smaller than the difference generated by instructions surrounding it [11].
2. Tracking a single instruction, even if it is executed in isolation, would require several samples per clock cycle and perfect cropping of the start and end points. The required sampling rate becomes very costly for devices with high clock frequencies. Moreover, misalignment of the cropping points would lead to significant loss in tracking performance.

Instead, we generate EM signatures for all short-duration permutations of the *instruction types* and track these permutations. Note that this approach addresses the aforementioned problems for the following reasons.

1. By generating EM signatures for the permutations, we observe the overall effect of the permutation block. Certain interactions caused by different orderings of these instructions and the impact of the pipeline are embedded into the EM signature. Although it is not possible to isolate the impact generated by each of these instructions, we obtain an EM signature that covers their aggregate impact. Furthermore, by using permutations instead of single instructions, we have a longer waveform and the significance of the surrounding instructions becomes less dominant.
2. Having a sequence of instructions results in a longer EM waveform, therefore, we can afford to have fewer samples per clock cycle compared to the single instruction case. Furthermore, the impact of miss-croppings becomes less evident since the duration of miss-cropping length is small relative to the total EM signal length.

One should keep in mind that inclusion of permutations is very helpful to address the pipeline effect but there are a few issues that cannot be addressed with this scheme:

- The EM signature still experiences the pipeline impact in the beginning and at the end due to the instructions that come before and after the permutation, respectively.
- For some processors, the pipeline length might be longer than the length of the permutation, and this limits the capability of the permutation to represent the emanation coming from all pipeline stages.

Also, note that, in addition to pipeline structure, there are other factors such as cache accesses, interrupts, etc. that impact EM emanations. Cache accesses and interrupts have much larger signatures [13] than single instructions and they need to be detected and removed as shown in [13] before single-instruction tracking is applied. However, since our benchmark codes are relatively short, we do not experience these impacts and do not need to remove them.

Finally, note that the number of the permutations is $K!$, where K is the number of clusters or *instruction types*. For large K, the number of permutations becomes a large number which leads to a costly measurement and large memory usage. Therefore, the choice of K from the *cluster tree* should be made carefully.

The steps of detecting permutations of *instruction types* are explained step-by-step as follows.

9.6.9 Picking an Instruction to Represent Each Instruction Type and Generating Microbenchmarks for Permutations

Since the instructions from the same *instruction type* have similar EM signatures, the most common instruction from each *type* is chosen to represent its *type*. *Instruction types* are labeled with the first K capital letters of the alphabet.

As mentioned earlier, with K clusters, we need to generate $K!$ benchmarks that include all permutations. For example, if there are 4 identified *instruction types* (A, B, C and D), the permutations should include: $ABCD, ABDC, \ldots, DCBA$.

Execution of most programs and embedded systems go through loops during the operation. These loops include execution of the same instruction sequences several times. Furthermore, a program spends most of the execution time in functions that are called many times successively. Considering this repetition-based nature of the most programs, we propose to investigate the impact of repetition of instruction blocks on tracking performance. In particular, we create EM signatures for different repetitions of the same instruction block. For ease of reference, we use the notation $(ABCD)_N$, where $(ABCD)$ is the investigated permutation block, and N is the number of permutation block repetition. Note that the ultimate goal is to track permutation blocks with $N = 1$. However, due to the repetitive structure of code implementations, N could be different than 1. For example, a certain permutation might appear within a loop that is repeated several times and this a priori knowledge can be used to improve tracking performance.

The pseudo-code for the new microbenchmark structure is shown in Fig. 9.32.

After implementing the pseudo-codes, EM emanations from the device are measured. Two measurements are taken for each microbenchmark at different times. The waveforms obtained in the first measurement are used for *training*, whereas the second measurement waveforms are used for *testing*.

```
while true
  for
    %empty for loop
  end

  % Set I/O Pin to High

  for N times
    -First instruction in permutation order,
    -Second instruction in permutation order,
    .
    .
    .
    -Kth instruction in permutation order.
  end

  % Reset I/O Pin to Low

  for
    %empty for loop
  end
end
```

Fig. 9.32 Pseudo-code for permutation detection setup [36]

9.6.10 Training: Generating Templates for Each Permutation

In this part, we generate templates for each permutation and these templates are used in the testing phase for prediction. A general overview of training can be found in the left hand part of Fig. 9.33. By using the same procedure that is described in Sect. 9.6.5, several EM signature traces are obtained for each permutation from *Measurement 1* recordings. These traces from the same permutation are aligned using cross-correlation. Then, they are cropped so that all of them have the same length. Finally, the EM template for the corresponding permutation is generated by using the point-wise average of the aligned and cropped EM signature traces.

9.6.11 Testing: Predicting Testing Measurements Using Templates

A general overview of testing can be found in the right hand part of Fig. 9.33. Testing traces are obtained by applying the procedure described in Sect. 9.6.5 to *Measurement 2* recordings. These testing traces are labeled with their permutation order. The prediction step implements a matched-filter-like structure where the filters are the normalized versions of EM templates generated in Sect. 8.6.3 [35]. We use such a structure because a matched filter is the optimum receiver in terms of maximizing the signal-to-noise ratio (SNR) when the received signal is corrupted by additive random noise [23]. Another advantage of such a structure is that the matched filter finds the best offset between the received signal and the templates on its own without requiring synchronization. After correlating the testing trace with all

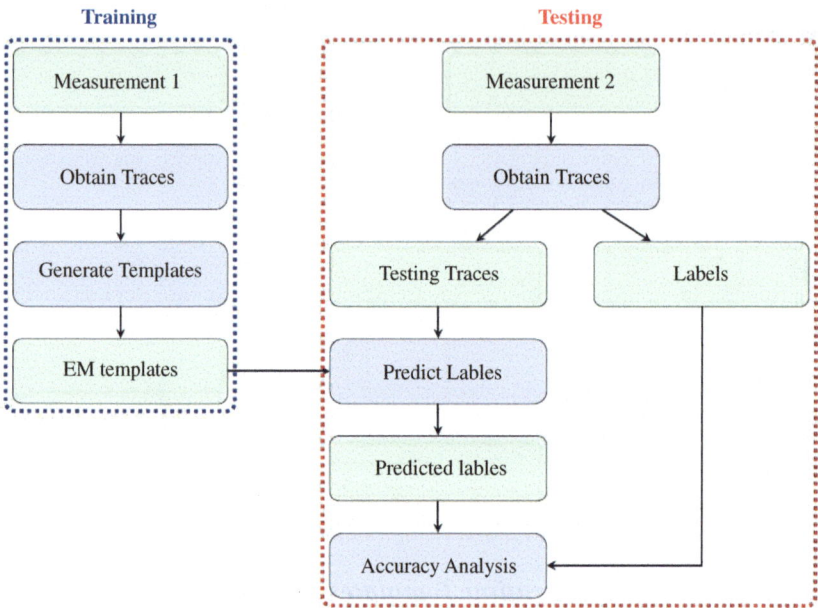

Fig. 9.33 Overview of training and testing [36]

filters, the permutation of the trace is predicted as the template whose corresponding filter gives the highest correlation.

9.6.12 Results

In this section, we explain our experimental setup and provide the results for *instruction type* determination and permutation tracking of *instruction types*.

Experimental Setup
To demonstrate the feasibility of the proposed methodology, we experiment on two devices. The first device is Intel's DE1 Altera FPGA Board that has Altera NIOS-II (soft) processor [2]. This processor is a general purpose RISC (reduced instruction set computer) processor that implements Nios-II architecture with 6 pipeline stages. The operating clock frequency is 50 MHz and this board does not have a present operating system. The second device is A13-OLinuXino, which is a low-cost embedded Linux mini-computer that has ARM Cortex A8 processor that operates at 1 GHz clock frequency. The processor implements ARMv7-A architecture and is an in-order, dual-issue, superscalar microprocessor with a 13-stage main integer pipeline [20]. For the rest of this discussion, we refer to the first and second devices as the DE1 device, and the A13 device, respectively.

9.6 PITEM: Permutations-Based Instruction Tracking via Electromagnetic... 265

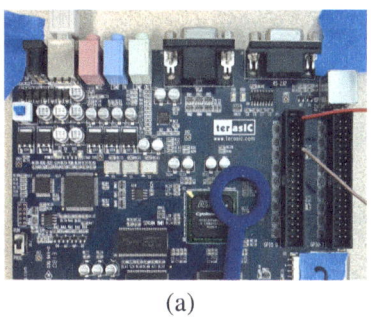
(a)

(b)

Fig. 9.34 Experimental setups used for EM emanation recordings [36]. (**a**) DE1 device. (**b**) A13 device

To record the EM emanations, we use Aaronia's H3 near-field magnetic probe for the DE1 device and H2 near-field magnetic probe for the A13 device [3]. These probes are chosen so that the resonance frequency of the probes are aligned with the clock frequencies of the devices. We locate the probes on the pin edges of the processors as shown in Fig. 9.34 using the results presented in [38]. Also note that two GPIO pins of these devices are utilized to record the marker signal. Measurements are obtained with a Keysight DSOS804A high-definition oscilloscope at a 10 Gbps [1].

Instruction Type Determination Results

For this proof-of-concept evaluation, after inspecting the instruction set for both devices, we select 17 instructions per each device for investigation that are listed in Fig. 9.35 with their corresponding abbreviations. These are frequently used instructions including mathematical operations such as addition and multiplication; logical operations such as AND, OR; and memory access instructions such as load and store, etc.

After selecting the instructions, we generate microbenchmarks for each of them that include $N = 10$ and $N = 100$ times repetitions for the DE1 and A13 devices, respectively. The reason for the higher N value of the A13 device microbenchmarks is the higher operating clock frequency of the device. Next, we record EM emanations and obtain EM signatures for different instructions as described in Sect. 9.6.5. Figures 9.36 and 9.37 present examples of the obtained EM signatures of different instructions for the DE1 and A13 devices, respectively. Note that each subfigure plots several EM emanation traces obtained for different executions of the given instruction. One can easily observe that these traces are highly aligned indicating that the EM signatures of the instruction sequences are relatively similar for successive executions.

In Fig. 9.36, we observe that some instructions, such as LDW and MULI, have very different EM signatures that differ both in length and shape, whereas some instructions, such as ADD and SUB, have EM signatures that are similar in both length and shape. Similar conclusions can be drawn from Fig. 9.37, as

DE1 Device			A13 Device		
	Assembly Code	Description		Assembly Code	Description
ADDI	addi r20, r20, 171	Add immediate value to register	ADDI	add r3, r3, #1	Add immediate value to register
ADD	add r20, r20, r16	Add two register values	ADD	add r3, r3, r3	Add two register values
SUBI	subi r20, r20, 173	Subtract immediate value from register value	SUBI	sub r3, r3, #1	Subtract immediate value from register value
SUB	sub r20, r20, r16	Subtract two register values	SUB	sub r3, r3, r3	Subtract two register values
LDW	ldw r20, 0(r21)	Load 32-bit word from memory to register	LDR	ldr r3, [r1]	Load a word from memory to register
STW	stw r20, 0(r21)	Store word from register to memory	STR	str r3, [r1]	Store a word from register to memory
MUL	mul r20, r20, r16	Multiply two register values	MUL	mul r3, r3, r3	Multiply two register values
MULI	muli r20, r20, 173	Multiply immediate value with register value	SMULL	smull r3, r2, r3, r3	Multiply two 32-bit signed values
DIV	div r20, r20, r20	Divide two register values	UMULL	umull r3, r2, r3, r3	Multiply two 32-bit unsigned values
OR	or r20, r20, r16	Bitwise logical OR operation of register values	ORR	orr r3, r3, r2	Bitwise logical OR operation of register values
ORI	ori r20, r20, 173	Bitwise logical OR operation of register and immediate values	ORRI	orr r3, r3, #1	Bitwise logical OR operation of register and immediate values
MOVI	movi r20, 173	Move immediate value to register	MOVI	mov r3, #1	Move immediate value to register
MOV	movi r20, r16	Move register value to register	ANDI	and r3, r3, #1	Bitwise logical AND operation of register and immediate values
AND	and r20, r20, r20	Bitwise logical AND operation of register values	AND	and r3, r3, r2	Bitwise logical AND operation of register values
CMPEQ	cmpeq r20, r20, r16	Compare if register values are equal	CMP	cmp r3, r0	Subtract register values to compare
XOR	xor r20, r20, r16	Bitwise logical XOR operation of register values	CMPI	cmpi r3, #1	Subtract immediate value from register value to compare
XORI	xori r20, r20, 173	Bitwise logical XOR operation of register and immediate values	NOP	nop	No operation

Fig. 9.35 Investigated instructions [36]

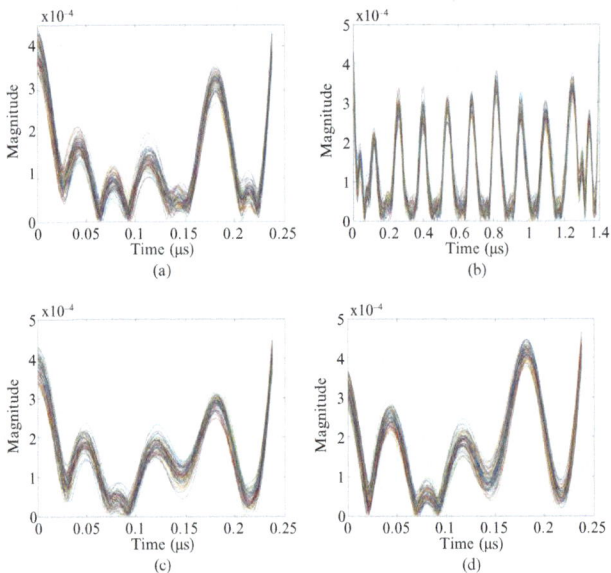

Fig. 9.36 Obtained EM signatures of several instructions for the DE1 device [36]. (**a**) LDW. (**b**) MULI. (**c**) ADD. (**d**) SUB

well. One should note that these conclusions validate our initial assumption that some instructions have similar EM signatures while some have very different EM signatures. We also observe that the instructions with similar EM signatures are those that have similar physical implementations, such as addition and subtraction.

As can be seen in Fig. 9.37, the EM signature of LDW is significantly different than MUL, ADDI and SUBI. This difference reflects the significance of the

9.6 PITEM: Permutations-Based Instruction Tracking via Electromagnetic...

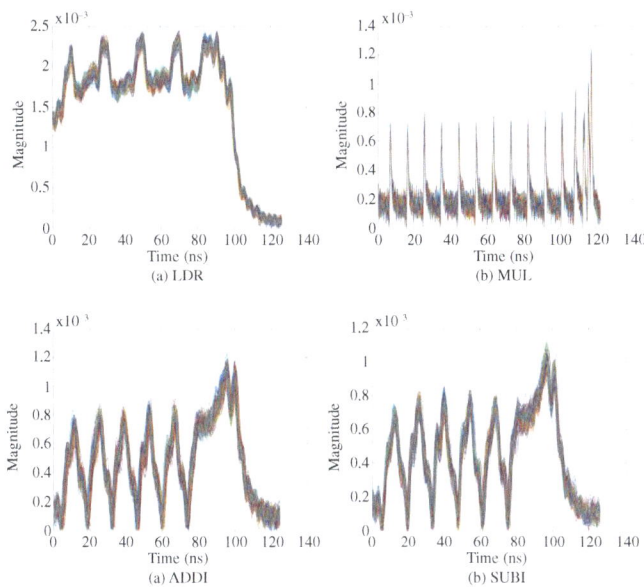

Fig. 9.37 Obtained EM signatures of several instructions for the A13 device [36]. (**a**) LDR. (**b**) MUL. (**c**) ADDI. (**d**) SUBI

additional memory-access pipeline present in the A13 device. Although the EM signatures are obtained by repeating the same instruction N times, we are not observing the repetition of the same pattern N times in the EM signature. We note that the beginning and ending parts of the signatures are significantly different than the middle parts, where we can observe the same kind of pattern repetition. This observation emphasizes the significance of the pipeline effect caused by the instructions that come before and after the instruction sequence. From these observations, we conclude that the EM signatures are indicative of the pipeline structure. These observations also confirm that each stage of a pipeline emits EM signals while executing instructions.

After obtaining the EM signatures, we generate the correlation matrices for the DE1 and A13 devices as shown in Fig. 9.38. In Fig. 9.38, brighter colors indicate higher correlation (and similarity). Then, we convert this correlation matrix to a distance matrix and cluster these instructions using *average-link hierarchical clustering* described in Sect. 9.6.7. The resulting *dendrograms* are presented in Fig. 9.39. The bottom part of the *dendrograms* starts with each instruction in its own cluster, and as it goes to the top, these clusters are merged pair-wise so that all instructions are in one cluster at the top. By investigating the correlation matrices and *dendrograms*, we set the number of clusters, K, to be 4 in both cases as for both cases when K is set to 4, resulting clusters include instructions that are similar in operation such as MULI and MUL. The clusters are indicated with different branch

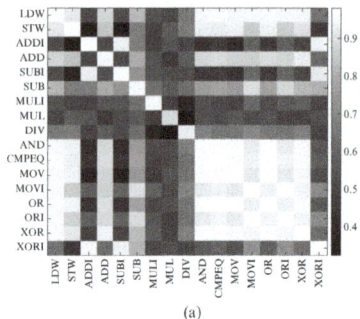
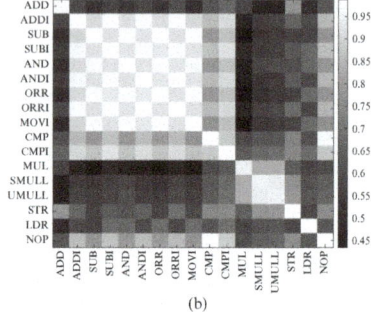

Fig. 9.38 Correlation matrices for DE1 and A13 devices [36]. (**a**) DE1 device. (**b**) A13 device

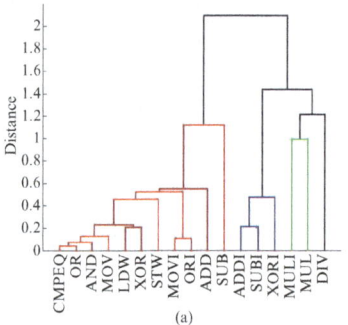
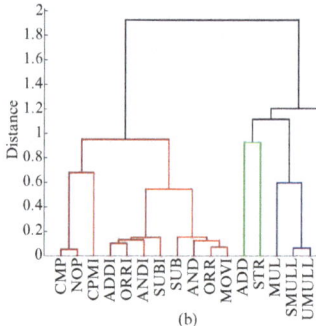

Fig. 9.39 Dendrogram of the instructions obtained with hierarchical clustering for DE1 and A13 devices. Different colors represent the clusters [36]. (**a**) DE1 device. (**b**) A13 device

colors. Red, blue, green and black colors represent A, B, C and D type clusters, respectively.

For DE1 device, A-type instructions are memory-access operations (LDW, STW) as well as the 1 clock-cycle arithmetic and logic operations that use register sources (ADD, SUB, AND, CMPEQ, MOV, OR, XOR), and also two instructions (MOVI, ORI) that use immediate values. B-type instructions are arithmetic and logic operations that use immediate values (ADDI, SUBI, and XORI). C-type instructions are the multiplication operations (MULI and MUL), and the D-type instruction type has only the DIV instruction.

For the A13 device, A-type instructions include all 1 clock-cycle arithmetic and logic operations that are either using register values or immediate values. One exception is the ADD instruction that uses register values, which is clustered as a C-type instruction along with the store instruction (STR). B-type instructions are the multiplication operations (MUL, SMULL, UMULL), and D-type instructions are the load operation (LDR).

As it has been discussed earlier, the clusters reflect the structure of the pipeline. For example, load and store operations include the memory address location calcu-

9.6 PITEM: Permutations-Based Instruction Tracking via Electromagnetic...

lations, which are addition and subtraction operations. Therefore, these instructions have the same type as addition and subtraction for the DE1 device. However, A13 implements a separate memory-access pipeline, which results in separate clusters for load and store operations. Similarly, we note that different variants of multiplication operations are clustered in the same group for both devices, likely because they all use the same physical multiplier unit within the processor.

9.6.12.1 Permutation Tracking Results

After determining the *instruction types*, as representative of each type, we pick the most common instruction from each type, but choosing them randomly from each type leads to the same results as well. These instructions are LDW, ADDI, MULI, and DIV for the DE1 device, and ADDI, MUL, STR, and LDR for the A13 device representing A, B, C, and D, respectively. Using these instructions, we generate microbenchmarks, record EM emanations and generate EM signatures for different N values.

Figures 9.40 and 9.41 present example EM signatures for different permutations and N values obtained from the DE1 and A13 devices. For $N = 1$, the plotted EM

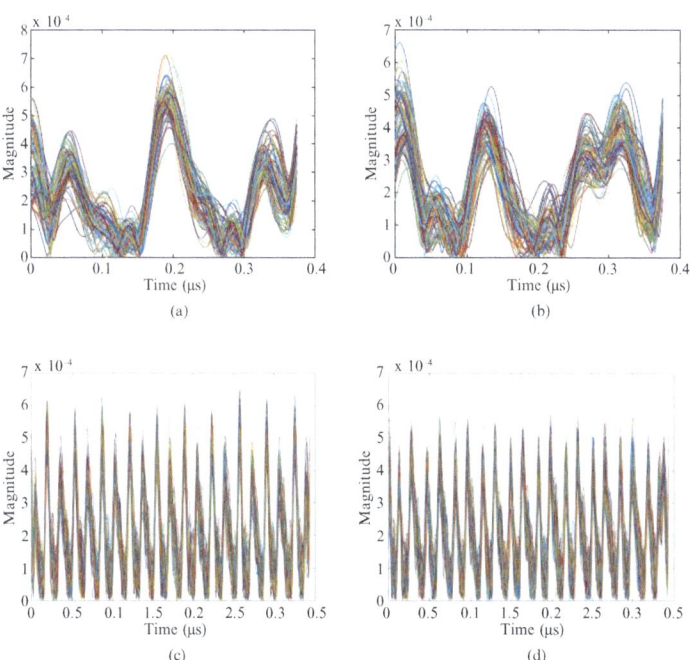

Fig. 9.40 Sample EM signatures of permutations with different N values for the DE1 device [36]. (**a**) $(ABCD)_1$. (**b**) $(DCBA)_1$. (**c**) $(ABCD)_{10}$. (**d**) $(DCBA)_{10}$

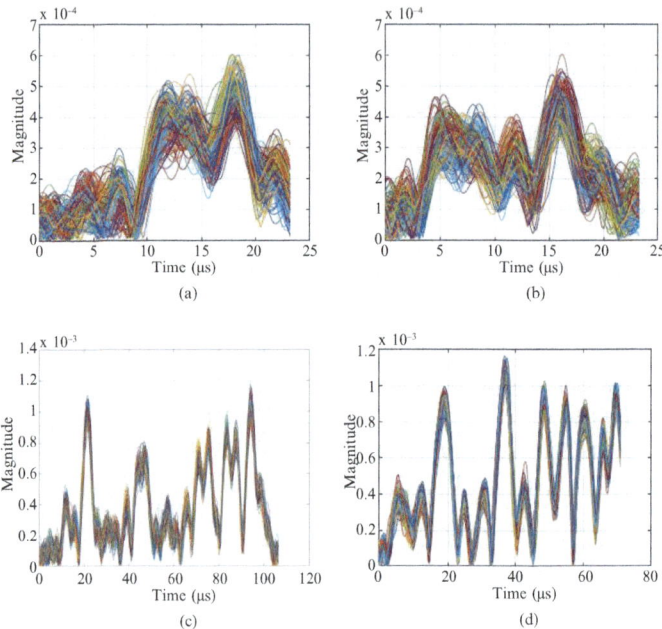

Fig. 9.41 Sample EM signatures of permutations with different N values for the A13 device [36]. (**a**) $(ABCD)_1$. (**b**) $(DCAB)_1$. (**c**) $(ABCD)_{10}$. (**d**) $(DCAB)_{10}$

signatures of the permutations $(ABCD)_1$ and $(DCBA)_1$ are visually different from each other for both devices. However, we note that, when $N = 1$, the EM signatures for the A13 device has a large variance among different executions, whereas this variance is much smaller for $N = 10$.

In Fig. 9.40, when $N = 10$, we can observe a pattern that is repeated 10 times, but this repetition is not present in Fig. 9.41 when $N = 10$. This difference can be explained by the difference of the pipeline lengths: DE1's pipeline length (6 stages) is shorter than A13's pipeline length (13 stages). Therefore, a permutation block of length 4 instructions is better at representing the contributions from the pipeline stages of the DE1 device and get the EM signature into a steady state, whereas the EM signature for A13 does not reach the steady state with 10 repetitions.

Note that, for both devices, we cannot visually identify the individual A, B, C, and D *instruction types* from the EM signatures. As discussed earlier, this is because a single execution of an instruction does not create significant variation in the signature, and, due to the pipeline, the variation generated by execution of a single instruction is distributed among different stages of the pipeline. To test this, while executing the permutation only once ($N = 1$), we repeat each instruction within the block 10 and 100 times for the DE1 and A13 devices, respectively. The EM signatures obtained with this repetition are shown in Figs. 9.42 and 9.43. In Fig. 9.42, we see that the instruction blocks A and B that appear in the beginning

9.6 PITEM: Permutations-Based Instruction Tracking via Electromagnetic...

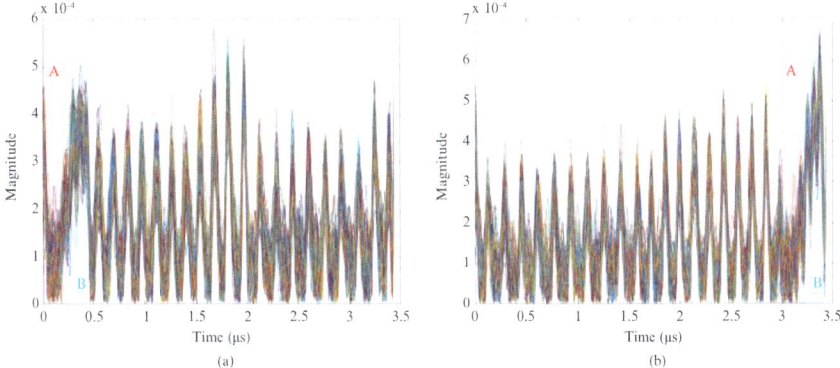

Fig. 9.42 Sample EM signatures when instruction types are repeated 10 times within the permutation block for the DE1 device [36]. (**a**) $(A_10B_{10}C_{10}D_10)$. (**b**) $(D_10C_{10}A_{10}B_10)$

of ABCD permutation can be identified at the end of DCAB permutation. Note that identifying C and D precisely is still not possible. In Fig. 9.43, we can identify all instruction type blocks clearly as indicated in the figure. These results show that, although single execution of an instruction does not generate an identifiable waveform pattern, several consecutive executions can result in distinct waveforms. Note that this is just an observation, and that it cannot be directly used for testing purposes because enforcing the repetition of the same instruction within the blocks is not very realistic for many programs and, therefore, is not practically applicable.

We generate templates for each permutation using *Measurement 1*, and use these templates to predict the permutation order of the snippets obtained from *Measurement 2*. To evaluate the deviation of our classification for different test sets, we perform 10 independent tests for each configuration. For each test, we use 100 snippets per permutation, resulting in a total of 2400 testing snippets per test, and 24,000 testing snippets for all tests.

"Bandwidth" is a *hyperparameter* in our application, which needs to be selected during training. One common approach to select *hyperparameter* values is *cross-validation*, where the training data set is partitioned into training and validation sets, and the best *hyperparameter* values are selected based on the performance on the validation set. Note that the testing set is kept completely independent during this process. Although *cross-validation* is a valid approach for our scheme to set the "bandwidth" value, we aim to analyze the impact of the utilized low-pass filter bandwidth on the accuracy. Therefore, instead of setting "bandwidth" value using *cross-validation*, we report the accuracy for different bandwidth values as shown in Fig. 9.44. Accuracy is calculated as the ratio of number of the correctly identified permutations to the total number of testing traces.

The value for every point in Fig. 9.44 is obtained by averaging 10 independent test trials and the error bar around the marker indicates the standard deviation across

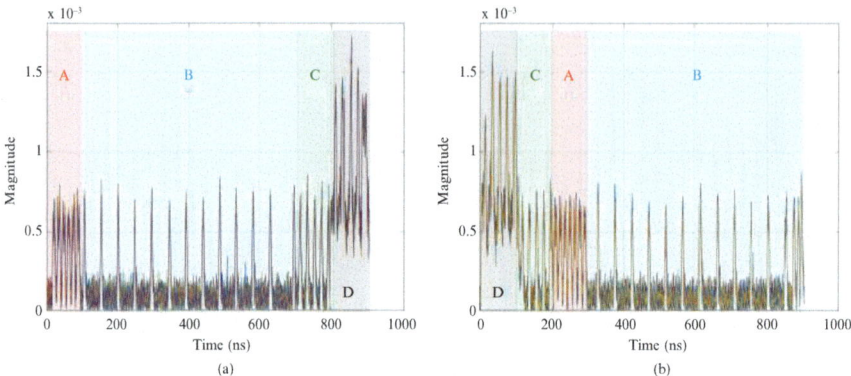

Fig. 9.43 Sample EM signatures when instruction types are repeated 100 times within the permutation block for the A13 device [36]. (**a**) $(A_100B_{100}C_{100}D_10)_1$. (**b**) $(D_100C_{100}A_{100}B_100)_1$

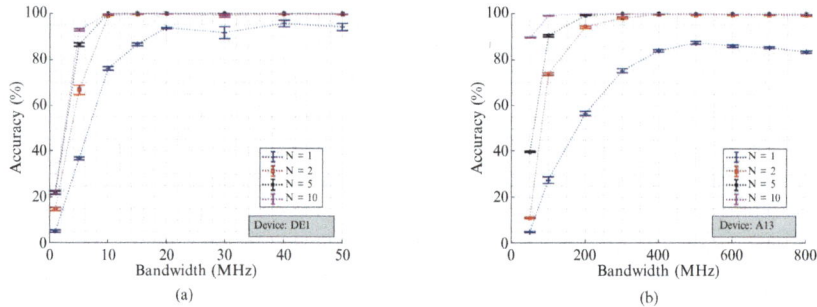

Fig. 9.44 Impact of N (number of permutation-block repetitions) on accuracy [36]

these trials. Since the standard deviation for different trials is low, most of these points are concentrated around the average value.

In Fig. 9.44, we observe that the highest accuracies for $N = 1$ are 95.67 and 87.35% for DE1 and A13, respectively. For higher values of N, the accuracy significantly improves. We also note that for $N > 1$, accuracy increases with increased bandwidth and converges to 100% beyond 10 MHz for DE1, and beyond and 400 MHz for A13 devices.

To investigate the $N = 1$ case, we provide the confusion matrices for a single test in Fig. 9.45. Note that both matrices are mostly diagonally dominant. When we investigate the correlation matrix for the DE1 device, we realize that the permutations "DC<u>AB</u>" and "DC<u>BA</u>", which only differ by the order of A and B instructions, are the two classes that are mostly confused with each other. Similarly, for A13 device, "<u>AB</u>CD" is confused with "<u>BA</u>CD", "<u>AC</u>DB" is confused with "<u>CA</u>DB", and "D<u>AC</u>B" is confused with "D<u>CA</u>B". All these confusions are between

9.6 PITEM: Permutations-Based Instruction Tracking via Electromagnetic... 273

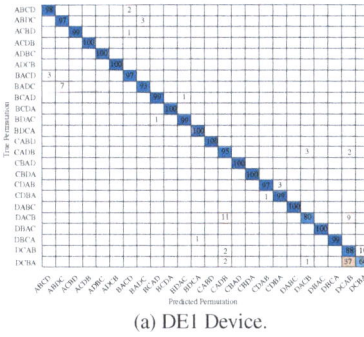
(a) DE1 Device.

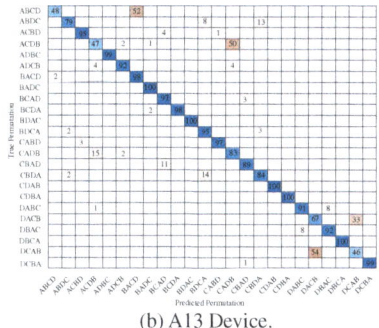
(b) A13 Device.

Fig. 9.45 Confusion charts for $N = 1$ [36]

two classes that differ in the order of only two instructions, while the order and the location of the other two instructions are the same.

Increasing N from 1 to 2 significantly improves accuracy, to 100%. This shows the trade-off between the number of repetitions and the accuracy. Including repeated versions of the permutation blocks is significantly increasing the detection accuracy at the expense of limiting the applicability of the proposed method. Therefore, if the program under investigation is repetitive in nature, e.g., because of loops, this method can be modified for better performance.

9.6.13 Further Evaluation of Robustness

In this section, we test robustness and extensions of our method. First, we evaluate its performance for different SNR levels, and then we extend its tracking ability to *instructions* rather than *instruction types*.

9.6.14 Performance for Different SNR Levels

To evaluate performance for different SNR levels, we assume that the measurement channel is corrupted by additive white Gaussian noise (AWGN). To estimate the SNR of the measured signal, we need to estimate the noise floor. One simple approach to measure the noise floor is to record the EM emanations during the idle state. However, this is not a valid approach because the power consumption during the idle state is lower than program execution due to the power optimizations. Lower power consumption leads to weaker EM emanations, which disqualifies this approach for SNR estimation. Therefore, we obtain measurements at locations with strongest EM emanations and use the power of the signals obtained from these measurements as the referenced signal power, which includes both the signal and the measurement noise. After obtaining these traces, we introduce additional additive white Gaussian noise with different noise powers to the testing traces. Since we are introducing additional noise to the signal that is already corrupted by measurement noise, we refer to the ratio between the measured signal power and this additional noise power as relative SNR.

Let $\mathbf{x}$ be a vector representing a testing trace with a length of L samples, and x_i be the sample values of this trace such that $\mathbf{x} = \begin{bmatrix} x_1 & x_2 & \dots & x_L \end{bmatrix}$. To add the noise, we first calculate P_S, the signal power of the vector, as

$$P_s = \frac{1}{L} \sum_{i=1}^{L} |x_i|^2. \tag{9.10}$$

Note that P_s is the power of the measured signal and, therefore, it features contributions from both the signal itself and the measurement noise. Then, we add AWGN to each sample of $\mathbf{x}$ and obtain the new signal $\mathbf{y}$ that is corrupted by additional AWGN. The elements of $\mathbf{y}$, y_i, are obtained as

$$y_i = x_i + z_i \quad \text{for } 1 \leq i \leq L, \tag{9.11}$$

where $z_i \sim \mathcal{N}(0, P_s/SNR_{lin}^{rel})$ are independent and identically distributed realizations of a random variable that has a zero-mean Gaussian (normal) distribution with P_s/SNR_{lin}^{rel} variance, and SNR_{lin}^{rel} represents the relative SNR in linear scale.

One should note that we add AWGN to all testing traces, but not to the EM templates from tracking. Then, we perform the same testing procedure for different relative SNR values with 40 MHz (DE1) and 500 MHz (A13) low-pass filter bandwidths. The results are shown in Fig. 9.46, where we observe that the accuracy decreases for low relative SNR levels, as expected. Accuracy performance converges for relative SNR values higher than 15 dB. We observe that the permutations with larger N are generally more resistant to AWGN. Finally, we note that DE1 device is more resistant to additional AWGN. We believe the reason for this is the relative simplicity of the DE1 device compared to A13.

Permutations of Instructions from the Same *Instruction Type*

Here we extend the applicability of the permutation tracking from tracking different instruction types to tracking instructions of the same instruction type. To do so, we pick 4 instructions from instruction type A for both devices. For the DE1, we pick ADD, AND, MOV, and CMPEQ; for the A13, we pick ADD, AND, MOV, and CMP. Note that these instructions use the same destination registers, register sources and immediate values, therefore we minimize the variation that might be caused by data values or register differences. Using these 4 instructions, we repeat the experiments and report accuracy in Fig. 9.47. Note that, $N = 1$ has low accuracy because these instructions are from the same type, so their permutations have very similar shapes. This low accuracy confirms that instructions were clustered into types correctly.

If we can afford to repeat the permutation sequence for several times, i.e., increase N, accuracy increases. In fact, for $N = 100$, accuracy for the DE1

Fig. 9.46 Impact of relative SNR on accuracy for permutations of different instructions for different N values [36]

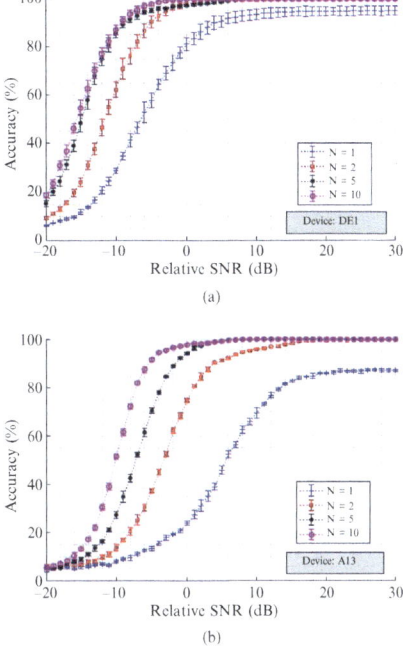

Fig. 9.47 Impact of N value on accuracy for permutations of instructions from the same instruction type [36]. (**a**) DE1 device. (**b**) A13 device

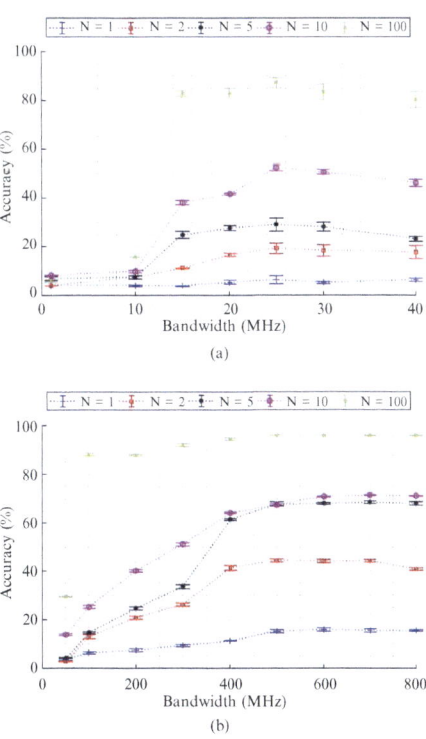

and A13 devices get as high as 87.5 and 95.78%, which are relatively high accuracies considering the similarity of the instruction signatures. To be able to determine the differences between the instructions from the same type, we need more observations. This again points to the trade-off between the accuracy and the number of observations.

References

1. Dsos804a high-definition oscilloscope: 8 ghz, 4 analog channels.
2. May 2019.
3. AARONIA. Datasheet: Rf near field probe set dc to 9ghz. http://www.aaronia.com/Datasheets/Antennas/RF-Near-Field-Probe-Set.pdf, 2016 (accessed Apr. 6, 2017).
4. ARM. Arm cortex a8 processor manual, April 2016. Retrieved April 3, 2016 from "https://www.arm.com/products/processors/cortex-a/cortex-a8.php".
5. ARM. Arm performance monitor unit. http://infocenter.arm.com/help/index.jsp?topic=/com.arm.doc.ddi0388f/Bcgddibf.html, accessed April 2, 2016.
6. Thomas Ball and James R. Larus. Efficient path profiling. In *Proceedings of the 29th Annual ACM/IEEE International Symposium on Microarchitecture*, MICRO 29, pages 46–57, Washington, DC, USA, 1996. IEEE Computer Society.
7. Ezio Biglieri, Dariush Divsalar, Marvin K Simon, Peter J McLane, and John Griffin. *Introduction to trellis-coded modulation with applications*. Prentice-Hall, Inc., 1991.
8. David Brooks, Vivek Tiwari, and Margaret Martonosi. In *ACM/IEEE International Symposium on Computer Architecture*, ISCA-27, pages 83–94, 2000.
9. Cameron B Browne, Edward Powley, Daniel Whitehouse, Simon M Lucas, Peter I Cowling, Philipp Rohlfshagen, Stephen Tavener, Diego Perez, Spyridon Samothrakis, and Simon Colton. A survey of monte carlo tree search methods. *Computational Intelligence and AI in Games, IEEE Transactions on*, 4(1):1–43, 2012.
10. Robert Callan, Farnaz Behrang, Alenka Zajic, Milos Prvulovic, and Alessandro Orso. Zero-overhead profiling via EM emanations. In *Proceedings of the 25th International Symposium on Software Testing and Analysis, ISSTA 2016, Saarbrücken, Germany, July 18–20, 2016*, pages 401–412, 2016.
11. Robert Callan, Alenka Zajic, and Milos Prvulovic. A practical methodology for measuring the side-channel signal available to the attacker for instruction-level events. In *47th Annual IEEE/ACM International Symposium on Microarchitecture, MICRO 2014, Cambridge, United Kingdom, December 13–17, 2014*, pages 242–254, 2014.
12. Robert Callan, Alenka Zajic, and Milos Prvulovic. FASE: finding amplitude-modulated side-channel emanations. In *Proceedings of the 42nd Annual International Symposium on Computer Architecture, Portland, OR, USA, June 13–17, 2015*, 2015.
13. Moumita Dey, Alireza Nazari, Alenka Zajic, and Milos Prvulovic. Emprof: Memory profiling via em-emanation in iot and hand-held devices. In *2018 51st Annual IEEE/ACM International Symposium on Microarchitecture (MICRO)*, pages 881–893, 2018.
14. Matthew R. Guthaus, Jeffrey S. Pingenberg, Dap Emst, Todd M. Austin, Trevor Mudge, and Richard B. Brown. Mibench: A free, commercially representative embedded benchmark suite. In *Proceedings of the IEEE International Workshop on Workload Characterization*, 2001.
15. Richard E Korf. Depth-first iterative-deepening: An optimal admissible tree search. *Artificial intelligence*, 27(1):97–109, 1985.
16. Takio Kurita. An efficient agglomerative clustering algorithm using a heap. *Pattern Recognition*, 24(3):205 – 209, 1991.

17. Mehari Msgna, Konstantinos Markantonakis, and Keith Mayes. Precise instruction-level side channel profiling of embedded processors. In *Information Security Practice and Experience*, pages 129–143. Springer, 2014.
18. Gonzalo Navarro. A guided tour to approximate string matching. *ACM computing surveys (CSUR)*, 33(1):31–88, 2001.
19. Peter Ohmann and Ben Liblit. Lightweight control-flow instrumentation and postmortem analysis in support of debugging. In *Automated Software Engineering (ASE), 2013 IEEE/ACM 28th International Conference on*, pages 378–388. IEEE, 2013.
20. Olimex. A13-olinuxino-micro user manual. https://www.olimex.com/Products/OLinuXino/A13/A13-OLinuXino-MICRO/open-source-hardware, 2016 (accessed Feb. 1, 2018).
21. Mahamed Omran, Andries Engelbrecht, and Ayed Salman. An overview of clustering methods. *Intell. Data Anal.*, 11:583–605, 11 2007.
22. Gabriele Paoloni. White paper: How to benchmark code execution times on intel ia-32 and ia-64 instruction set architectures. Technical report, Intel Corporation, September 2010.
23. John G Proakis and Masoud Salehi. *Fundamentals of communication systems*. Pearson Education India, 2007.
24. G. Reinman and N. Jouppi. Cacti 2.0: An integrated cache timing and power model. *Technical Report*, 2000.
25. Jose Renau, Basilio Fraguela, James Tuck, Wei Liu, Milos Prvulovic, Luis Ceze, Smruti Sarangi, Paul Sack, Karin Strauss, and Pablo Montesinos. SESC simulator, January 2005. http://sesc.sourceforge.net.
26. Gregg Rothermel, Sebastian Elbaum, Alex Kinneer, and Hyunsook Do. Software-artifact infrastructure repository. *URL* http://sir.unl.edu/portal, 2006.
27. Wheeler Ruml. *Adaptive tree search*. PhD thesis, Citeseer, 2002.
28. Richard Rutledge, Sunjae Park, Haider Khan, Alessandro Orso, Milos Prvulovic, and Alenka Zajic. Zero-overhead path prediction with progressive symbolic execution. In *2019 IEEE/ACM 41st International Conference on Software Engineering (ICSE)*, pages 234–245, 2019.
29. Colin Schmidt. Low Level Virtual Machine (LLVM), Feb 2014. Retrieved on April 1 from https://github.com/llvm-mirror/llvm.
30. Nader Sehatbakhsh, Alireza Nazari, Alenka Zajic, and Milos Prvulovic. Spectral profiling: Observer-effect-free profiling by monitoring em emanations. In *2016 49th Annual IEEE/ACM International Symposium on Microarchitecture (MICRO)*, pages 1–11, 2016.
31. Pavel Senin. Dynamic time warping algorithm review. 2008.
32. Software-artifact infrastructure repository. http://sir.unl.edu/.
33. Daehyun Strobel, Florian Bache, David Oswald, Falk Schellenberg, and Christof Paar. SCANDALee: a side-channel-based disassembler using local electromagnetic emanations. In *Proceedings of the 2015 Design, Automation & Test in Europe Conference & Exhibition*, pages 139–144. EDA Consortium, 2015.
34. Rose Helena Turi. *Clustering-based colour image segmentation*. Monash University PhD thesis, 2001.
35. G. Turin. An introduction to matched filters. *IRE Transactions on Information Theory*, 6(3):311–329, 1960.
36. Elvan Mert Ugurlu, Baki Berkay Yilmaz, Alenka Zajic, and Milos Prvulovic. Pitem: Permutations-based instruction tracking via electromagnetic side-channel signal analysis. *IEEE Transactions on Computers*, pages 1–1, 2021.
37. Stijn van Dongen and Anton J. Enright. Metric distances derived from cosine similarity and pearson and spearman correlations, 2012.
38. Frank Werner, Derrick Albert Chu, Antonije R Djordjević, Dragan I Olćan, Milos Prvulovic, and Alenka Zajic. A method for efficient localization of magnetic field sources excited by execution of instructions in a processor. *IEEE Transactions on Electromagnetic Compatibility*, 60(3):613–622, 2017.

39. R.J. Wherry. *Contributions to correlational analysis*. Academic Press, 1984.
40. Yee Leung, Jiang-She Zhang, and Zong-Ben Xu. Clustering by scale-space filtering. *IEEE Transactions on Pattern Analysis and Machine Intelligence*, 22(12):1396–1410, 2000.
41. Baki Berkay Yilmaz, Elvan Mert Ugurlu, Frank Werner, Milos Prvulovic, and Alenka Zajic. Program profiling based on markov models and em emanations. In *Cyber Sensing 2020*, volume 11417, page 114170D. International Society for Optics and Photonics, 2020.
42. Baki Berkay Yilmaz, Elvan Mert Ugurlu, Alenka Zajic, and Milos Prvulovic. Instruction level program tracking using electromagnetic emanations. In *Cyber Sensing 2019*, volume 11011, page 110110H. International Society for Optics and Photonics, 2019.

Chapter 10
Using Analog Side Channels for Hardware Event Profiling

10.1 Introduction

Memory profiling approaches are designed to help programmers and system developers characterize the memory behavior of a program, and this information can then be used by the compiler and/or programmer to improve performance [30, 46]. In general, simulation, hardware support, and program instrumentation are the three main methods of profiling cache behavior. Cache simulation can help identify general locality problems that would cause performance degradation on cache-based systems in general [45], but is typically much slower than native execution and fails to model the complex architectural details of modern on-chip chases and their interaction with memory. Hardware support typically takes the form of hardware performance counters, which are counting actual microarchitectural events as they occur without significant performance overheads due to counting itself [47]. However, an interrupt is needed to attribute a counted event to a specific part of the code, which would cause major performance degradation if every event is attributed. Instead, a sampling method is used, typically by interrupting each time the count reaches some pre-programmed threshold T, to provide statistical attribution of the events [3, 9, 23, 47]. This creates a trade-off between (1) the granularity at which events are attributed, and (2) the overhead introduced by profiling and the distortion of results by profiling activity itself [25, 26, 44]. Program instrumentation methods can also be used for memory profiling [1, 24], but they have a similar trade-off between precision/granularity and overhead/disruption. Problems associated with hardware counters and/or program instrumentation are exacerbated in real-time and cyber-physical systems, where programs have to meet real-time deadlines, and thus often change behavior depending on their own performance. In such scenarios, overhead introduced by monitoring may result in substantially changing the very behaviors the profiling is trying to characterize.

Debugging and profiling of interrupts has also posed a major challenge to researchers and developers. Interrupts are asynchronous to program control flow, i.e., they can occur at any time during program execution. This makes it next to impossible for software or simulation to reproduce realistic scenarios that involve interaction between interrupts and program execution, so testing and debugging of interrupts, interrupt handlers, and interrupt interaction with programs and the operating system often adds significantly to overall development costs for embedded systems [8]. Existing profilers perform analysis of performance metrics mostly by instrumenting application and system code, which captures program-generated effects, but fails to capture *unscheduled* events [33] such as interrupts.

To address these issues, we have proposed EMPROF, a new approach to profile memory [11], and PRIMER [12], a new method of profiling interrupts, both leveraging electromagnetic (EM) emanations from devices. By continuously analyzing these EM emanations, EMPROF identifies where in the signal's timeline each period of stalling begins and ends, allowing it to both identify the memory events that affect performance the most (LLC misses) and to measure the actual performance impact of each such event (or overlapping group of events). Because EMPROF is completely external to the profiled system, it does not change the behavior of the profiled system in any way, and requires no hardware support, no memory or other resources, and no instrumentation on the profiled system. Similarly, PRIMER analyzes the device's external EM side-channel signal in real-time, without any interference with the device's program execution, providing a detailed analysis of not only the overall overhead created by interrupts, but also their distribution over time (i.e., exact occurrence of interrupts in program execution time-line). As the CPU follows a generic procedure to handle such asynchronous system events, our approach can be generalized and applied to all non-deterministic interrupts across multiple platforms. More details on both methods are presented in following sections.

10.2 Memory Profiling Using Analog Side Channels

The main idea of EMPROF is that the processor's circuitry exhibits much less switching activity when a processor has been stalled for a while, which manifests in low signal levels in analog side channels. Thus we can profile the performance impact of memory activity by observing when these low-activity intervals occur and how long they last.

In general, a processor has two types of stalls. Front-end stalls occur when the front-end of the processor cannot fetch instructions, e.g., because of a miss in the instruction cache (I$), and the processor is fully stalled when it completes instructions it already fetched but still cannot fetch new ones. This typically occurs for hundreds of cycles when the I$ miss is also a miss in the Last-Level Cache (LLC, which is typically unified for instructions and data) and thus experiences the main memory latency. Back-end stalls occur when the processor cannot retire an instruction for a while, e.g., because of a data cache (D$) miss, and the processor is fully stalled when it runs out of one or more of its resources (such as ROB entries, load-store queue entries, etc.) and can no longer fetch new instructions. This

10.2 Memory Profiling Using Analog Side Channels

typically occurs for hundreds of cycles when the D$ miss is also an LLC miss and thus experiences the main memory latency.

In a sophisticated out-of-order processor, the fully-stalled condition can be averted for tens of cycles because the processor usually already has many tens of instructions in various stages of completion when an I$ miss occurs, and because the processor stall has plentiful resources when the D$ miss occurs. The exact number of cycles during which the processor can keep busy depends on the program and the processor's microarchitectural state at the point of the miss, but the latency of a last-level cache (LLC) miss can be in the hundreds of cycles, so after a while the fully-stalled condition still occurs and then lasts for many (tens or hundreds of) cycles. Most resource-constrained devices, such as IoT and hand-held devices, use simple in-order cores that need less power and produce less heat [14, 29]. Compared to more sophisticated cores, these simpler cores can avert a full stall for significantly fewer cycles. However, even in-order cores are typically designed to attempt to execute more than one instruction per cycle (i.e., they are superscalar cores) and to have more than one memory request in progress (i.e., they exploit memory-level parallelism that allows them to benefit from multiple read channels to multi-banked LLCs), so their stalls stall vary in duration.

It should be noted that non-stalled cycles during a cache miss have far less impact on performance than stalled cycles, and that multiple cache misses that are in progress concurrently result in fewer overall number of stalled cycles than if these misses were handled with no overlap. The overall goal of profiling the LLC misses is typically to assess their performance impact on different parts of the program (e.g., to guide performance optimization efforts), so profiling that accounts for the processor's stalled time during these misses, rather than simply counting these misses, is typically more relevant to these overall goals. For example, the impact of variability in memory latency on performance prediction is studied in [20, 28], and hardware counters capable of counting stall cycles due to LLC misses (as opposed to counting the number of these misses) are used to estimate this latency [13]. Therefore, our focus will be primarily on accounting for stalled cycles and attributing them to parts of the application, but for clarity of presentation we will often use the term *miss* to refer to a sequence of stalled cycles that are all caused by one LLC miss or by several highly-overlapped LLC misses. Also, we will often use the term *miss latency* to refer to the number of stalled cycles that are all caused by one LLC miss or by several highly-overlapped LLC misses.

10.2.1 Side-Channel Signal During a Processor Stall

Variations in the EM signal are mostly caused by changes in circuit activity, and each cycle in a long stall is similar to other long-stall cycles (very little is changing in the state of the processor), so during a long stall the signal will be relatively stable from cycle to cycle, and the signals from one long stall will be similar to signals from other long stall, while being very different from signals that come from

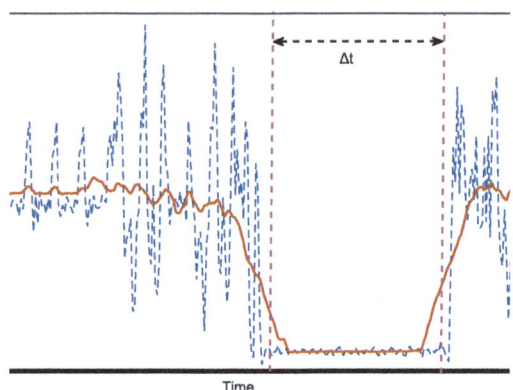

Fig. 10.1 Change in EM emanation level caused by processor stall, with signal magnitude shown in dashed blue and its moving average shown as solid red [11]

a busy processor. Our actual measurements confirm this, with the dashed blue line in Fig. 10.1 showing the magnitude of a time-domain signal acquired with a 40 MHz bandwidth around the clock frequency (1.008 GHz) of the ARM Cortex-A8 processor on the A13-OLinuXino-MICRO IoT single-board computer [32]. Also shown (solid red line) in Fig. 10.1 is the moving average of the signal. The part of the signal that corresponds to the processor stall is between the two dotted vertical lines in Fig. 10.1, and the duration of this part of the signal is Δt. This duration can be determined from the signal by multiplying the number of samples in that interval with the duration of a sample period, and the number of cycles this stall corresponds to can be computed by multiplying Δt with the processor's clock frequency.

To understand the relationship between side-channel signals and the various types of stalls, we begin with signals obtained by modeling power consumption in a cycle-accurate architectural simulator SESC [34], because the simulator can provide us with the ground-truth information about the exact cycle in which the miss occurs, the cycle in which the resulting full stall begins, and the cycle in which the stall ends. Our results are based on simulations of a 4-wide in-order processor, with two levels of caches with random replacement policies, which mimics the behavior of the processors encountered in many IoT and hand-held devices. We collect the average power consumption for each 20-cycle interval, which corresponds to a 50 MHz sampling rate for a 1 GHz processor.

To observe how the side-channel signal is affected by different types of cache misses, a small application was created that performs loads from different cache lines in an array. The size of the array can be changed in order to produce cache misses in different levels of the cache hierarchy. Two signals that correspond to the same part of the code are shown in Fig. 10.2. The signal in Fig. 10.2a corresponds to an L1 D$ miss that is an LLC hit, so it was a very brief stall, during which the processor core consumes very little power because none of its major units are active. In contrast, Fig. 10.2b corresponds to an LLC miss, with an order-of-magnitude longer low-power-consumption period that corresponds to a much longer stall.

10.2 Memory Profiling Using Analog Side Channels

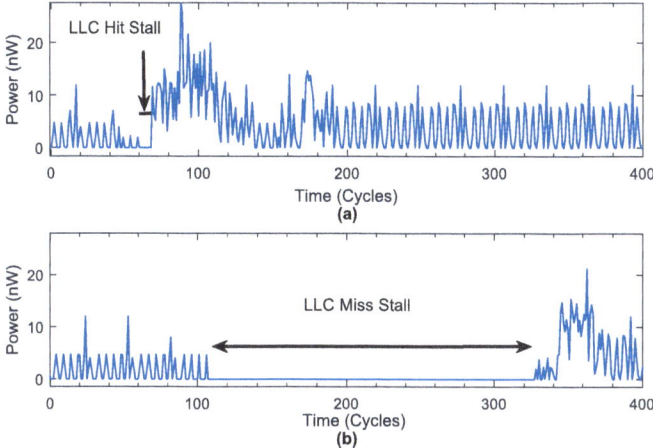

Fig. 10.2 (a) LLC hit stalls and (b) LLC miss stalls in SESC simulator [11]

When the sampling rate is reduced, each data point (sample) in the signal corresponds to an increased number of cycles, and thus shorter stalls become increasingly difficult to find in the signal. The long stalls that correspond to LLC misses, however, still contain multiple signal samples that allow the stall to be identified. The reduction in the signal's sampling rate does, however, reduce the measurement resolution of the duration of those stalls—e.g., with one signal sample per 20 processor cycles, the duration of the stall can be "read" from the signal only in 20-cycle increments.

We have confirmed that the power signal produced by SESC drops to its full-stall level not when the miss actually occurs, but a number of cycles afterwards, when the processor runs out of *useful* work. As explained before, this occurs either when the processor suffers an I$ miss and eventually completes the instructions it did fetch prior to that, or when the processor has a D$ miss, finishes all the work that could be done without freeing any pipeline resources, and eventually cannot fetch any more instructions until the miss "drains" out of the pipeline.

10.2.2 Memory Access in Simulated Side-Channel Signal

Figure 10.3a shows a signal that corresponds to several consecutive LLC misses. When the first miss occurs, the processor still has a lot of resources and its activity continues largely unaffected by the miss. During this activity, several other LLC misses occur and eventually the first miss ends while the processor is still busy and has not stalled yet. This allows the processor to fetch more instructions, staying busy until the second LLC miss ends, etc. until eventually the processor encounters a miss during which it does run out of resources and is forced to stall. Since the first

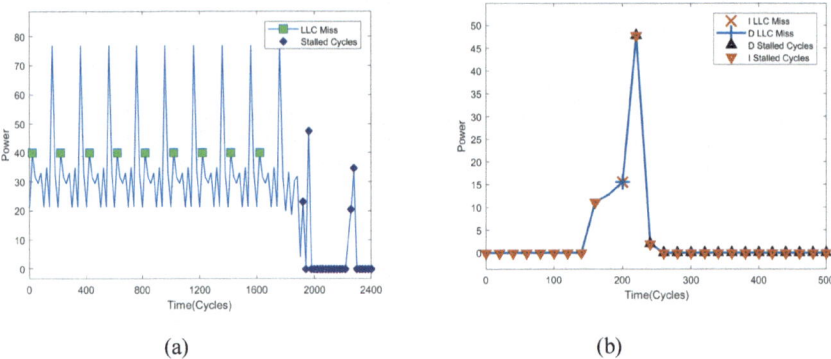

Fig. 10.3 Some LLC miss events produce no individually attributable stalls: (**a**) Stalls can be avoided for some LLC misses. (**b**) Stalls for coinciding misses overlap [11]

several LLC misses in this example do not cause any stall activity, a detector of LLC misses based on identifying the resulting stalls in the signal will fail to detect those misses, which will cause under-counting of LLC miss events. However, since those misses do not introduce stalls, their impact on performance is much lower compared to stall-inducing misses, so the signal-based reporting of miss-induced stalls in this example would still be close to the actual performance impact of these misses.

Another situation in which the number of LLC misses would be under-reported involves overlapping LLC misses. One example of such a situation is shown in Fig. 10.3b, where an I$ access and a D$ access are both LLC misses that overlap and the processor is stalled during that overlapped time. When analyzing the side-channel signal to identify stalls caused by LLC misses, this usually results in reporting only one stall, which would be counted as one LLC miss. If the goal of the profiling is to actually count LLC misses, rather than account for their impact on performance, this will cause under-counting of LLC misses. However, this situation is a typical example where the performance penalties of multiple LLC misses overlap with each other, and the resulting overall performance impact is not much worse than if there was only one LLC miss, so the signal-based reporting of how many long stalls occur and the performance impact of each such stall still very accurately tracks the actual performance impact of the LLC misses.

10.2.3 Memory Access in Measured EM Side-Channel Signals

We now run the same small application on the Olimex A13 OLinuXino MICRO [32], an IoT prototyping board. We use a near-field magnetic probe to measure the board's EM emanations centered around the processor's clock frequency. Even though we are using EM emanations from a real system, rather than power consumption from a simulated system, stalls produced by cache misses

10.2 Memory Profiling Using Analog Side Channels

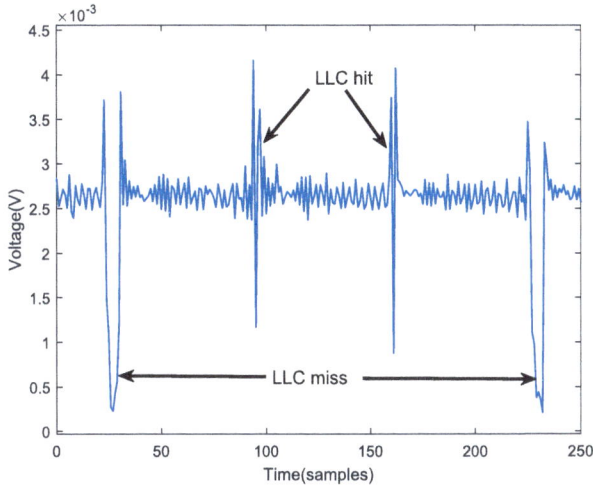

Fig. 10.4 LLC hit and miss from physical side-channel signal of Olimex board [11]

in this system still produce signal patterns (Fig. 10.4) similar to that observed for the simulator—stalls cause a significant decline in signal magnitude. The stalls produced by most LLC misses last around 300 ns, with small variations that are likely due to variations in how much work the processor can do until it stalls after encountering an LLC miss.

However, the stalls caused by LLC misses exhibit several behaviors that were not encountered in the simulator because it used a simplified model of the main memory. One such behavior, shown in Fig. 10.5, occurs when an LLC miss occurs while the memory is performing its periodic refresh activity.

We observed that a stall for an LLC miss that coincides with a memory refresh lasts approximately 2–3 μs, and this situation occurs approximately at least every 70 μs for the on-board H5TQ2G63BFR SDRAM chip [19]. Since these stalls do affect program performance and (especially) the tail latency of memory accesses, we count them (and account for their performance impact) separately when reporting our experimental results.

10.2.4 Prototype Implementation of EMProf

Because cache misses occur at a time when the processor is busy, the signal shape at the point of the miss changes significantly depending on the surrounding instructions and the processor's schedule for executing them. Therefore, rather than attempting to recognize the signal that corresponds to the misses themselves, which would require extensive training for each point in the code where the misses can occur, EMPROF detects the signal activity that corresponds to a stalled processor.

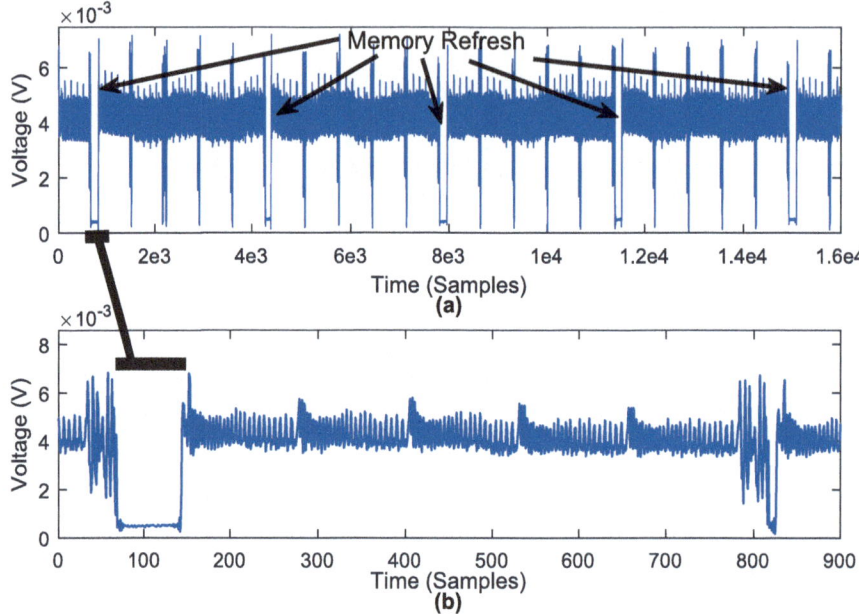

Fig. 10.5 Memory refresh in Olimex IoT device, with (**a**) memory refresh replacing LLC miss, (**b**) zoomed-in memory refresh [11]

The first step in this detection is to normalize the signal to compensate for the variation among different devices, measurement setups, etc. For example, we have found that even small changes in probe/antenna position can dramatically change the overall magnitude of the received signal. However, this change largely consists of a constant multiplicative factor that is applied to the entire signal. Similarly, the voltage provided by the profiled system's power supply can vary over time. The impact of power supply variations on the EM emanations is largely that signal strength changes in magnitude over time. EMPROF compensates for these effects by tracking a moving minimum and maximum of the signal's magnitude, using them to normalize the signal's magnitude to a range between 0 (which corresponds to the moving minimum) and 1 (which corresponds to the moving maximum). EMPROF then identifies each significant *dip* in the signal whose duration exceeds a threshold. The threshold is selected to be significantly shorter than the LLC latency but significantly longer than typical on-chip latencies.

One of the key advantages of EMPROF is that it can efficiently identify LLC-miss-induced processor stalls, without any a-priori knowledge about, or training on, the specific program code that is being profiled. This enables EMPROF to profile execution equally well regardless of the program analysis tools and infrastructure available for the target system and software. One specific example of this versatility is that EMPROF can be used for memory profiling of the system's boot from its very beginning, even before the processor's performance monitoring features are initialized, and before the software infrastructure for saving the profiling information is loaded.

10.2.5 Validation of EMProf

EMPROF's goal is to identify (in the side-channel signal) each LLC miss that stalls the processor, and to measure the duration of that stall. Since existing profiling methods do not operate at such level of detail, validation of EMPROF is a non-trivial problem. One problem is that hardware performance counters, such as `LLC-load-misses`, count all LLC miss events, including those that cause no stalls and those that overlap with other misses. Another problem is that their accuracy is reduced by either high interference with the profiled execution (when sampling the counter value often) or significant sampling error (when sampling the counter value rarely), so they are not very effective for relatively brief periods of execution. One illustration of this is that occurs when using `perf`[23] on Olimex A13-OLinuXino-MICRO to count LLC misses for a small application that was designed to generate only 1024 cache misses; The number of misses reported by `perf` had an average of 32,768 and a standard deviation of 14,543!

Thus, we validate EMPROF's results in two ways. First, we engineer a microbenchmark that generates a desired number of LLC misses, and then we compare EMPROF's reported LLC miss count to an a-priori-known number of misses. Second, we apply EMPROF to the power side-channel signal generated through cycle-accurate simulation, and then compare EMPROF's results to the simulator-reported ground truth information about where the misses occur in the signal's timeline and where the resulting stall begins and ends.

Experimental Setup

To demonstrate EMPROF's ability to profile LLC misses, we ran the engineered microbenchmark on Android-based cell phones—Samsung Galaxy Centura SCH-S738C [36] and Alcatel Ideal [5], and on the Olimex A13-OLinuXino-MICRO [32] IoT prototype board. More detail on these three devices is provided in Table 10.1.

In our experimental setup, the signals are received using a small magnetic probe and then recorded using a Keysight N9020A MXA spectrum analyzer [21].

Table 10.1 Specifications of experimental devices [11]

	Alcatel	Samsung	Olimex
Processor	QS[a] MSM8909	QS[a] MSM7625A	AW[b] A13 SoC
Frequency	1.1 GHz	800 MHz	1.008 GHz
#Cores	4	1	1
ARM core	Cortex-A7	Cortex-A5	Cortex-A8

[a]Qualcomm Snapdragon
[b]Allwinner

```
1   // perform page touch
2   for(#pages_to_be_used)
3       load(page(cache_line_0))
4
5   exec_blank_loop()  // empty for loop
6
7   // perform memory loads
8   // TM: Total Misses requested
9   while(num_accesses != TM)
10      page = rand()
11      cache_line = rand()
12      addr = page*PAGE_SIZE + cache_line*CACHE_LINE_SIZE
13      load(addr)
14      // CM: Consecutive Misses occurring in groups
15      if(num_accesses % CM == 0)
16          micro_function_call()
17      num_accesses++
18
19  exec_blank_loop()  // empty for loop
```

Fig. 10.6 Pseudocode of the microbenchmark [11]

The spectrum analyzer was used for initial studies mainly because it has built-in support to visualize signals in real-time, because it offers a range of measurement bandwidths, and because it can perform basic time-domain analysis.

Validation by Microbenchmarking

We implement a microbenchmark (Fig. 10.6) that generates a known pattern of memory references leading to LLC misses. The access pattern of the microbenchmark can be adjusted to produce LLC misses in groups, where the number of LLC misses in a group and the amount of non-miss activity between groups can be controlled. We refer to the total number of expected LLC misses as *TM* and the number of LLC misses consecutively in groups as *CM*. For example, if *TM*=100 and *CM*=10, the microbenchmark creates 10 groups of 10 consecutive LLC misses, each group separated by a micro-function call.

First, every page is accessed once to avoid encountering page faults later. Then the microbenchmark executes a tight `for` loop with no memory accesses. The signal that corresponds to this loop has a very stable signal pattern that can be easily recognized, which allows us to identify the point in the signal where this loop ends and the part of the application with LLC miss activity begins.

Next, the microbenchmark executes a section of code that contains the LLC misses. The access pattern accesses cache-block-aligned array elements (so that each access is to a different cache block), with randomization designed to defeat any stride-based pre-fetching that may be present in the processor. Finally, another tight `for` loop allows us to easily identify the point in the signal at which the carefully engineered LLC miss activity has ended.

Figure 10.7a shows the overall EM signal from one execution of this microbenchmark on the Olimex A13-OLinuXino-MICRO board. The "memory accesses" portion of the signal includes many LLC misses, which occur in groups of 10, i.e., *CM*=10. A zoom-in that shows the signal for one such group of 10 LLC misses is shown in Fig. 10.7b.

10.2 Memory Profiling Using Analog Side Channels

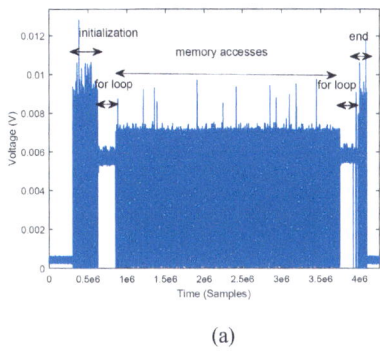
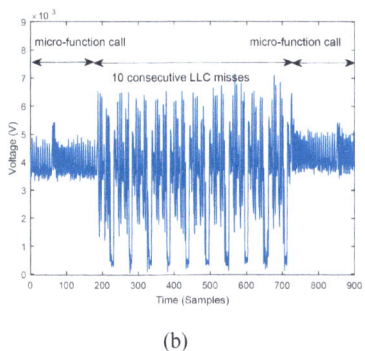

(a)   (b)

Fig. 10.7 EM signal from single microbenchmark run on Olimex device: (**a**) Entire microbenchmark run, (**b**) Zoomed-in LLC miss section with $CM=10$ [11]

Table 10.2 Accuracy of EMProf for microbenchmarks on Alcatel cell phone, Samsung cell phone and Olimex IoT device [11]

Benchmark		Devices		
#TM	#CM	Alcatel	Samsung	Olimex
256	1	99.61%	99.22%	99.61%
256	5	100%	99.61%	99.22%
1024	10	99.41%	99.61%	99.51%
4096	50	99.83%	99.71%	98.98%

Because the number of misses in the "memory accesses" section of the microbenchmark is known, and because this section can easily be isolated in the signal, we can apply EMPROF to this section and compare the LLC miss count it reports to the actual LLC miss count the section is known to produce. The results of this comparison were used to compute EMPROF's accuracy shown in Table 10.2. On all microbenchmarks the accuracy of EMPROF's LLC miss counting is above 99%, with an average of 99.52%.

10.2.6 Validation by Cycle-Accurate Architectural Simulation

To evaluate EMPROF's ability to measure the duration of each LLC-miss-induced stall, and to assess the impact of overlapping LLC misses, we use a simulator configuration that mimics the processor and cache architecture of the Olimex A13 OLinuXino-MICRO board. The simulator is enhanced to produce a power consumption trace that will be used as a side-channel signal in EMPROF, and also to produce a trace of when (in which cycle) each LLC miss is detected and when the resulting stall (if there is a stall) begins and ends.

We first run the same microbenchmarks we ran on the Olimex board and compare the two signals (Fig. 10.8). We observe that, although one signal is produced by the simulator's (unit-level) accounting of energy consumption and the other is

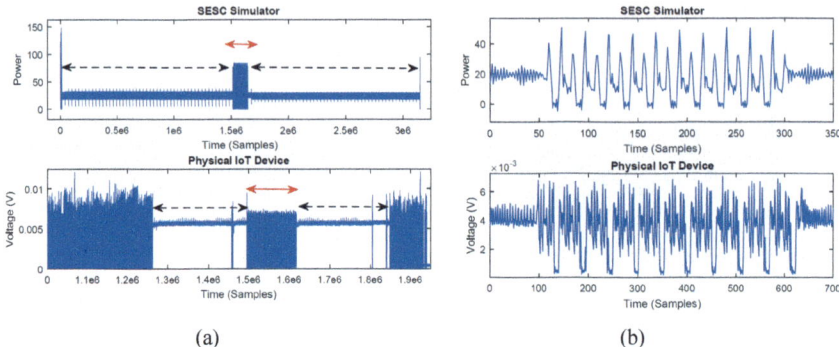

Fig. 10.8 Comparison of signals from the SESC simulator and actual Olimex A13-OLinuXino-MICRO IoT device. (**a**) An entire microbenchmark run. Solid red arrows indicate the LLC-miss-intensive part, while dashed black arrows indicate the identifier loops. (**b**) Zoom-in to the LLC-miss-intensive part with *CM*=10 [11]

produced by actually receiving EM emanations from a real processor (Fig. 10.8b), the relevant aspects of the two signals are similar in nature—the loops used to mark the beginning and ending of the "memory accesses" section in the microbenchmark are clearly visible and can be easily identified in the signal, and the LLC misses themselves exhibit similar behavior in the signal (Table 10.3). The most prominent difference between the two overall signals is that the start-up and tear-down involves much more activity on the real system than on the simulator (Fig. 10.8a), primarily because the simulation begins at the entry point and ends at the exit point of the microbenchmark's executable, whereas in the real system the start-up and tear-down also involves system activity, e.g., to load and prepare the executable for execution before any of the executable's instructions are executed.

Having determined that the simulator's power signal is a reasonable proxy of the real system's EM signal for the purposes of EMPROF validation, we proceed to run the microbenchmarks and several SPEC CPU2000 benchmarks in the simulator, analyze the resulting power signals in EMPROF, and compare EMPROF's results to the ground-truth results recorded by the simulator. The results of this comparison are shown in Table 10.4 in terms of accuracy of the reported LLC miss count and the reported LLC-miss-induced stall cycles.

10.2.7 Validation by Monitoring Main Memory's Signals

Finally, we wanted to systematically verify that the stalls reported by EMPROF coincide with actual memory activity. Intuitively, when a memory request is getting fulfilled by the main memory, the activity level in main memory should increase. Similar to how we receive the EM emanations from the processor, we can also receive the EM emanations from main memory. By receiving the two signals

10.2 Memory Profiling Using Analog Side Channels

Table 10.3 Comparison of total dips from SESC simulator vs detection algorithm [11]

Benchmark	Simulator	EMProf	% detected
ubench 256 1	734	717	97.7
ubench 256 5	734	718	97.8
ubench 256 10	734	719	98.0
ubench 256 50	734	719	98.0
ubench 1024 1	1527	1513	99.1
ubench 1024 5	1527	1512	99.0
ubench 1024 10	1527	1518	99.4
ubench 1024 50	1530	1518	99.27
ubench 4096 1	4707	4692	99.7
ubench 4096 5	4707	4692	99.7
ubench 4096 10	4710	4698	99.8
ubench 4096 50	4710	4697	99.8
ammp	3286	x	x
bzip2	6581740	6895724	104.8
crafty	579	x	x
equake	4633305	4950480	106.8
gzip	336757	336896	100.04
mcf	1669598	1671107	100.09
parser	1511009	1512353	100.09
twolf	4274	4310	100.84
vortex	1619197	1634684	100.96
vpr	2891	2891	100

Table 10.4 Accuracy of EMProf on simulator data for benchmarks [11]

Benchmark		Miss accuracy (%)	Stall accuracy (%)
#TM	#CM	Microbenchmark	
256	1	97.7	99.3
256	5	97.8	99.3
1024	10	99.4	99.9
4096	50	99.8	99.8
SPEC CPU2000			
ammp		99.67	98.4
bzip2		95.2	99.96
crafty		99.31	100
equake		93.2	98.5
gzip		99.96	99.8
mcf		99.9	99.5
parser		99.9	99.7
twolf		99.16	100
vortex		99.04	99.9
vpr		100	99.24

Fig. 10.9 Measurement setup for simultaneous monitoring of the processor's emanations and memory signals [11]

simultaneously and checking whether the dips in the processor's signal coincide with memory activity, we can gain additional confidence that the dips detected by EMPROF are indeed a consequence of LLC misses.

Unfortunately, not every device is amenable to such dual-probing—the processor and memory tend to be physically close to each other, so the physical placement of both probes is a challenge. However, we found that on the Olimex board, the processor and memory are relatively spaced apart, allowing simultaneous probing with no interference. We additionally use a passive probe to measure the CAS pin activity off a resistor, and this experimental setup is shown in Fig. 10.9.

Figure 10.10a shows the magnitude of processor's and memory's side-channel signal for CM=10 to the main memory that are separated by non-memory activity. We observe that an LLC miss causes the processor's signal magnitude to drop significantly (Fig. 10.10b), while the memory request caused by that LLC miss results in a sudden burst of memory activity.

Finally, one may intuitively expect that the memory's EM signal might be a better indicator of LLC misses than the processor's signal. Unfortunately, this is not true— while it is true that memory activity occurs on each LLC miss, memory activity also occurs due to DMA transfers, DRAM refreshes, and many other reasons that are completely unrelated to LLC misses. Furthermore, by measuring stalls in the processor's EM signal, we obtain information that is more relevant for performance optimizations: how often the processor *stalls* due to LLC misses and how long these stalls last—recall that the processor stays busy for some time during an LLC miss.

10.2 Memory Profiling Using Analog Side Channels

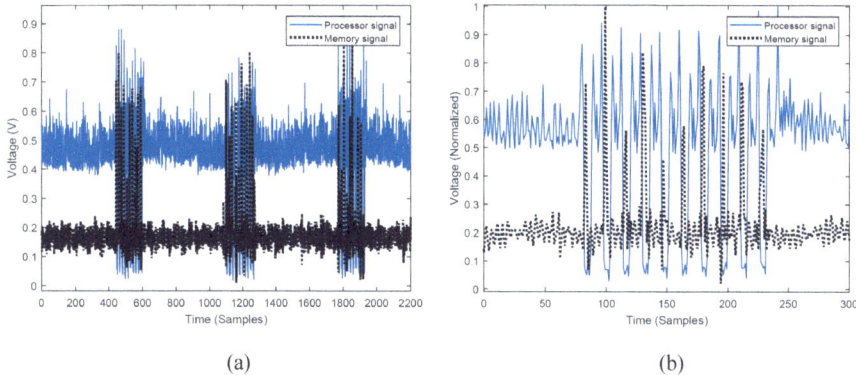

Fig. 10.10 EM side-channel emanations from the processor (solid blue) and memory (dotted black) for a microbenchmark with *CM*=10. (**a**) Three groups of LLC misses with periods of non-LLC-miss activity in between. (**b**) Zoom-in showing a single group of LLC misses*CM*=10 [11]

10.2.8 Experimental Results

To evaluate the effectiveness of EMPROF on real-world applications, it was additionally applied to ten SPEC CPU2000 benchmarks [17] on the two Android cellphones and one IoT device. The integer SPEC CPU2000 benchmarks were chosen as they display a range of realistic memory behaviors [46].

The final results of SPEC CPU2000 benchmarks were recorded using a near-field magnetic probe, by first down-converting the signal using a ThinkRF Real Time Spectrum Analyzer WSA5000 [43] and then digitizing the signal using a computer fitted with two PX14400 high speed digitizer boards from Signatec [40]. This was needed because the N9020A MXA has a limit on how long it can continuously record a signal, and most SPEC benchmarks execution times significantly exceed this limit.

Profiling Results

Table 10.5 shows the results for the total number of LLC cache misses and the number of stall cycles (as a percentage of the benchmark's execution time), as reported by EMPROF for each benchmark on each target device.

The number of LLC misses is much lower for the Alcatel cellphone than for the other two devices, mainly because the LLC in that phone is 1 MB, while Olimex and Samsung device only have 256 KB LLC. Samsung device's processor has a hardware prefetcher, so it is able to avoid some of the LLC misses that occur in the Olimex device, and therefore the number of misses Olimex can be expected to be higher than in the Samsung phone even though their LLCs have the same size. Additionally, the processor in the Olimex board has a higher clock frequency than the processor in the Samsung phone, while their main memory latencies (in

Table 10.5 Statistics of total LLC misses and total percentage latency in execution time obtained from EMProf for Alcatel cell phone, Samsung cell phone and Olimex IoT device [11]

Benchmark		Total LLC misses			Miss latency (% total time)		
		Devices			Devices		
		Alcatel	Samsung	Olimex	Alcatel	Samsung	Olimex
#TM	#CM	Microbenchmark					
256	1	257	254	255	0.92	3.57	9.44
256	5	256	255	258	1.15	4.06	10.10
1024	10	1030	1020	1029	1.00	4.19	9.88
4096	50	4103	4084	4138	2.05	4.55	10.25
SPEC CPU2000							
ammp		20511	280141	174142	2.42	7.59	9.06
bzip2		451103	3353123	5932148	2.15	1.04	6.59
crafty		314385	901153	675781	2.44	0.38	1.52
equake		627734	2138132	5925404	2.68	0.61	7.49
gzip		152264	1001392	513256	1.11	0.39	1.21
mcf		303365	895295	546714	5.22	7.18	3.28
parser		365412	1934334	2318384	2.19	3.39	8.63
twolf		35086	1014888	228465	1.03	4.84	3.00
vortex		235667	897317	533784	3.63	2.03	2.9
vpr		4970	100750	203130	0.09	0.23	0.6
Average		251049.7	1251652.5	1705120.8	2.3	2.77	4.43

nanoseconds) are very similar, creating more stall time per miss in the Olimex board and also allowing fewer LLC misses to be completely hidden by overlapping them with useful work in the processor.

While the total stall cycles provide an indication of how performance is affected by LLC misses overall, a key benefit of EMPROF is that it also provides information about the stall time of each LLC miss. Figure 10.11 shows the histogram of LLC miss latencies observed on the three devices for the SPEC CPU2000 benchmark *mcf*. Most stalls are brief in duration, mostly due to the processor's ability to keep busy as the miss is being serviced. However, a significant number of stalls last hundreds of cycles, and we observe that, compared to the IoT board, the two phones have a thicker "tail" in the stall time histogram.

10.2.9 Effect of Varying Measurement Bandwidth

During our initial study, we tested the effect of a varying measurement bandwidth from 20, 40, 60, 80, and 160 MHz. Low measurement bandwidth may not produce enough samples to exemplify the cache miss feature, and with increasing measurement bandwidth, an increase in sampling rate correlates to a better and more accurate detection of LLC miss and its associated latencies. Figure 10.12

10.2 Memory Profiling Using Analog Side Channels

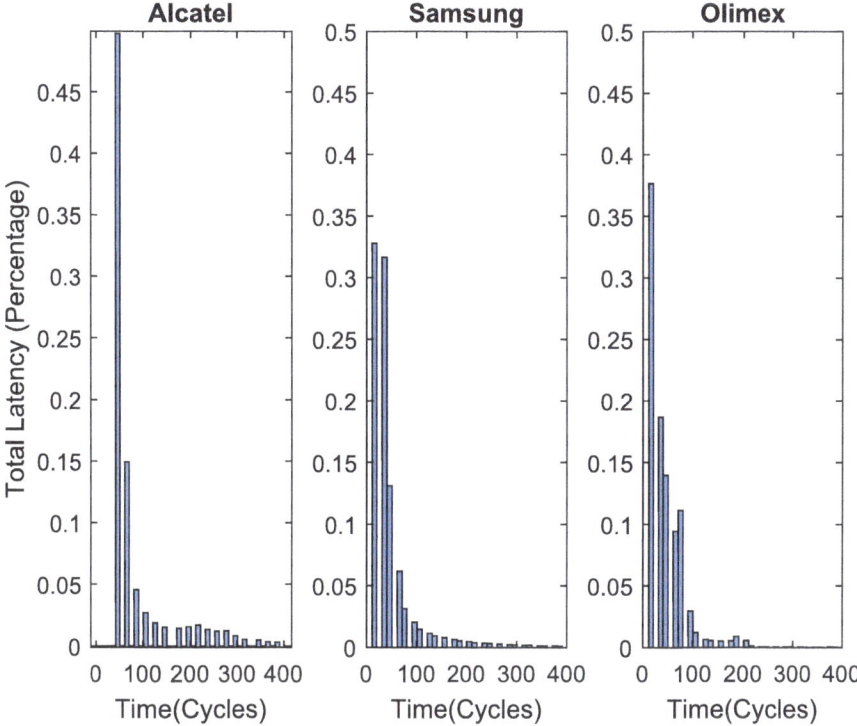

Fig. 10.11 Histogram of stall latencies obtained for SPEC CPU2000 *mcf* benchmark for Olimex IoT device, Alcatel cell phone and Samsung cell phone [11]

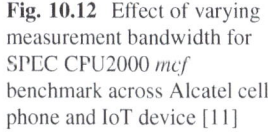

Fig. 10.12 Effect of varying measurement bandwidth for SPEC CPU2000 *mcf* benchmark across Alcatel cell phone and IoT device [11]

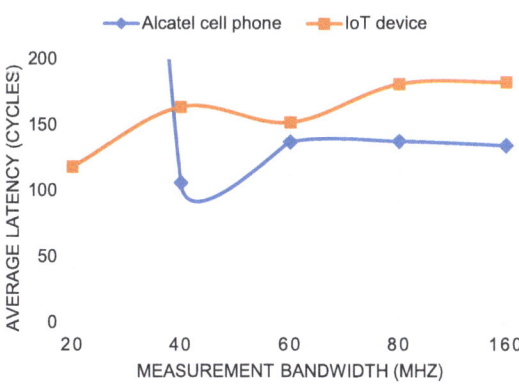

illustrates this effect for Alcatel cell phone and A13-OLinuXino-MICRO running SPEC CPU2000 *mcf* benchmark. In the IoT device, a lower sampling rate mostly results in less accurate stall latency determination. However, for the Alcatel cell phone the lowest sampling rates also prevent detection of many LLC-induced stalls, such that at 20 MHz EMPROF detects only the very few stalls that have extremely

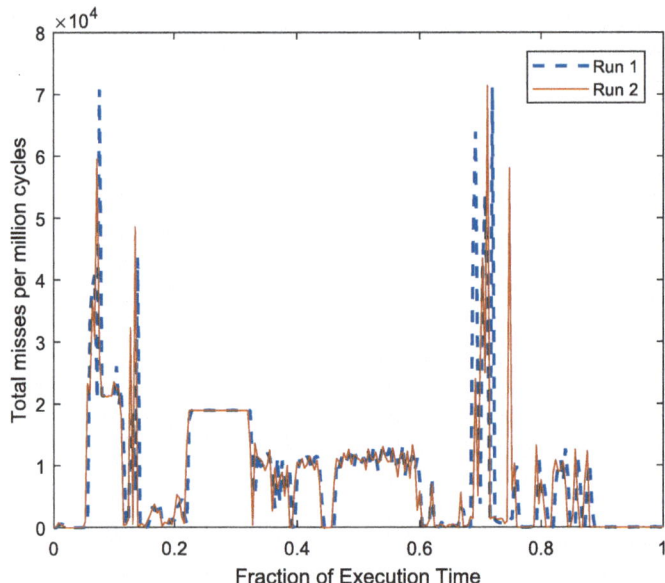

Fig. 10.13 Boot sequence EMPROF profiling for two distinct runs on IoT device [11]

long durations (their average duration is 1100 clock cycles). For both devices, the average stall time stabilizes at 60 MHz or more of measurement bandwidth, indicating that bandwidth equivalent to only 6% of the processor's clock frequency is sufficient to allow identification of LLC-miss-induced stalls in the signal.

Profiling the Boot Sequence
One of the most promising aspects of EMPROF is its ability to profile hard-to-profile runs, such as the boot sequence of the device. Figure 10.13 shows the rate of total LLC misses as time progresses for two boot-ups of the IoT device. These results can be used to, for example, decide whether memory locality optimization should be considered as a way to speed up the boot of the device.

10.2.10 Code Attribution

While accurate profiling of miss-induced stalls provides very valuable insight into performance, it would be even more helpful for developers to be able to identify in which parts of the code these LLC misses and stalls are happening. Ideally, such attribution of misses to code would also be based on the EM signal, to retain the zero-interference advantages of EMPROF. Several methods that attribute parts of the signal to application code have been reported in the literature, including Spectral Profiling [39] and EDDIE [31], which can attribute parts of the signal to

Table 10.6 Observable loops in SPEC CPU2000 *parser* benchmark [11]

Region	Function	Total miss	LLC miss rate (per million cycles)	Mem stall cycles (%)	Avg. miss latency (cycles)
A	read_dictionary	292,296	2667.71	1.63	218.71
B	init_randtable	34,805	317.66	0.18	211.85
C	batch_process	1,840,246	16,795.47	10.17	217.72

the code at the granularity of loops, and ZOP [10], which can achieve fine-grain attribution of signal time to code but requires much more computation. By applying one such scheme and EMPROF to the same signal, the LLC miss stall discovered in the signal by EMPROF's could be attributed to the application code in which they occur. To illustrate this, Fig. 10.13 shows how the spectrum (horizontal axis) for the SPEC CPU2000 benchmark *parser* changes over time (vertical axis). There are three distinct regions that can be observed in this spectrogram, which correspond to three functions in *parser*. To illustrate how signal-based attribution would be used in conjunction with EMPROF, we (manually) mark the transitions between these function in the signal as shown by horizontal dashed lines in Fig. 10.13, and attribute misses in each part of the signal to the corresponding function. Table 10.6 shows EMPROF's results with this attribution. As we see, the batch_process function should be the main target for optimizations that target LLC misses—it occupies the largest fraction of execution time, it suffers the highest LLC miss rate, and it has the highest fraction of its execution time spent on stalls caused by these LLC misses. We note that this only serves to illustrate how this attribution would work—finer-grain analysis (such as Spectral Profiling) of the signal would identify multiple loops in some of these functions and thus likely provide more precise (and informative) results than those shown in Table 10.6.

10.3 Profiling Interrupts Using Analog Emanations

System events are unpredictable changes to the flow of execution in a processor, due to sudden/unexpected changes in the system or within the executing code itself [7, 15]. They are categorized into exceptions (program/software-generated errors) and interrupts (external sources e.g., I/O and clocks). Even though interrupts are heavily used to handle input/output and real-time events in embedded devices, their impact on performance can be particularly hard to predict—it depends on the exact time when the interrupt service routine (ISR) executes and on which specific part of the application (or system) code was executing at the time when execution was diverted to the ISR.

Traditional interrupt profilers have depended upon heavily modifying the system to monitor interrupts. Such approaches either add a lot of program instrumentation or insert kernel-based modifications for trace debugging and replays. Figure 10.14

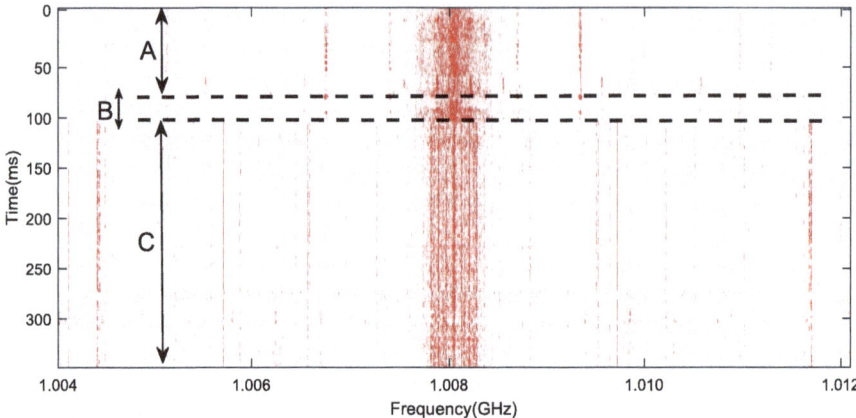

Fig. 10.14 Delay introduced by instrumentation in a pacemaker application [12]

demonstrates the overhead of instrumentation on a pacemaker application running on a low-power IoT device, where the red pulses indicate the generated paced pulse in the case of an irregular heartbeat. Despite adding minimal profiling instrumentation code to log the interrupt instances, introduced a delay of ∼88 ms in the generation of the paced pulse, which could have serious consequences as time-critical circuitry depends upon this output. Additionally, the logging activity itself introduces a RAM overhead of ∼16%. These methodologies are thus known to interfere with the native application events beyond recognition.

As discussed in Chap. 8, EM emanations can be used for monitoring program execution, e.g., to identify malware and other anomalous activities in devices [16, 31, 38]. These techniques compare the side-channel signal to previously safe and correct executions of the same applications to report any deviations in their code execution's spectral signatures to perform a coarse-grain analysis. In Chap. 9, we have discussed how EM side-channels can be used for identifying and modeling function-granularity code segments [10, 39]. Here we show how EM side channels can be used for profiling of interrupts.

10.3.1 Overview of PRIMER

In this section, we delve into details about interrupts and their EM signatures, and how PRIMER is designed to detect these signatures and distinguish them from normal program execution and from each other. To simplify the explanation of the PRIMER detection algorithm, we will first focus on a popular IoT STM32 board, which is 32-bit microcontroller based on ARM Cortex-M4 processor [41], and then we will apply our methods to additional devices and systems.

10.3 Profiling Interrupts Using Analog Emanations

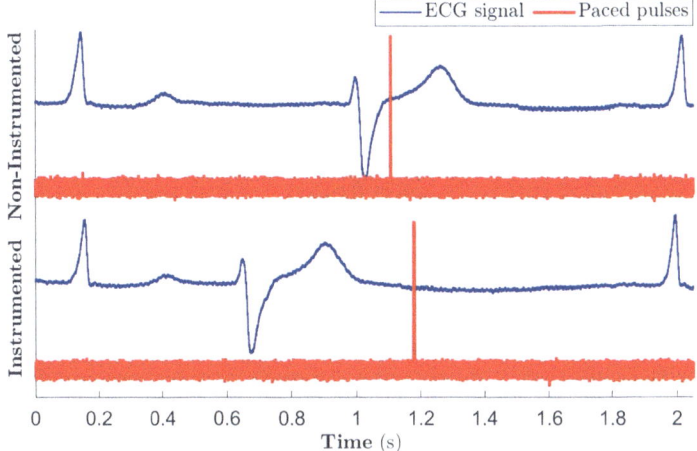

Fig. 10.15 EM signatures of ISRs of various interrupts in time-domain [12]

To observe EM signals from the STM32F446RE IoT board, it was configured to operate at a clock frequency of 16 MHz and the EM signals were acquired using a passive near-field probe with a 20 MHz bandwidth. Microbenchmarks were designed for five types of commonly used interrupts—timer, UART transmit and receive, ADC and GPIO. Known compute-only repetitive code was added before and after the ISR that helped isolate and extract the EM signature for a specific type of interrupt.

To generate timer interrupts, a 32-bit timer was used with a prescalar set to scale down the internal clock from 16 MHz to 800 Hz. The timer interrupts were then observed for varying counter periods in the down-count mode. The baud rate for UART serial communication was set at 9600 and the ISRs were recorded for varying lengths of data transmitted. The on-board 12-bit resolution ADC was configured to trigger every time a counter overflowed a predetermined value, and the corresponding ISR was extracted accordingly. For the purpose of observing GPIO interrupts, the on-board push button was triggered asynchronously and the EM signature of the GPIO ISR was extracted.

Figure 10.15 shows the time-domain EM side-channel signatures of all these interrupts when measured at the processor clock frequency. Each of these signatures comprises of interrupt controller activity, the IRQ handler routine for that interrupt, and any processor hardware events associated with that interrupt. Any user-level ISR callback is kept empty to ensure that the signature is unperturbed due to the application. The EM signatures of these interrupts were found to be very distinct from one another, both in the length of the ISR and the overall shape, but very similar across multiple instances of the same interrupt.

Let us attempt to build a detection algorithm to profile UART transmit interrupts for a microbenchmark that is continuously transmitting the contents of a data buffer containing 5 bytes of data over the serial interface. In this case, the interrupt occurs at the beginning of transmission of the data buffer. As the baud rate used is 9600 with a standard 8N1 mode, each byte transmission will take $1/9600 * 10 = 1.04167$ ms. This is because a single byte transmission comprises of 1 start bit, 8 data bits and 1 stop bit, i.e., a total of 10 bits. Using this information, given the time interval between two UART TX interrupts T_{tx} (assuming continuous transmissions), the size of transmitted data N_{tx} can be determined by

$$N_{tx} = \frac{T_{tx}(ms)}{1/9600 * 10} = \frac{T_{tx}(ms)}{1.04167(ms)}, \quad (10.1)$$

As these interrupts do not have a distinct spectra, existing spectral-based profilers will either treat them as a deviation and disregard, or may cause a false positive in the anomaly detection mechanism. A more naive approach for detecting these interrupts, inspired by EMProf, would be to apply primitive thresholding to identify the location of the UART interrupts. However, we found that such an algorithm led to non-interrupt activity to be also marked as interrupts, thereby introducing false positives. Upon further inspection, we identified that this algorithm counted every byte transmission as an interrupt too, which made it incorrectly report an interrupt occurring every 1.04 ms. Thus, no matter the size of data buffer, this algorithm always incorrectly reports $N_{tx} = 1$. Thus, a more sophisticated approach is needed to detect these complex EM signatures.

10.3.2 Training PRIMER

As we found that the interrupts have prominent EM signatures, we first capture some of these patterns from microbenchmarks to create a set of templates. As the training phase was performed on the same signal from which the templates were extracted, the exact occurrences and locations of the interrupt are known for validation.

However, utilizing all the interrupts of the training signal as templates increases the complexity of the detection algorithm. To improve the detection coverage and decrease the complexity, new training templates are added by performing a parametric analysis in an iterative manner. For that, we first calculate the *recall rate* of the algorithm, which is defined as the ratio of the number of instances correctly identified to the total number of expected instances, with the current set of training templates.

The training set is initialized with a single template chosen randomly. The recall analysis is then performed by varying threshold values of Pearson correlation coefficient and Gaussian distance as shown in Fig. 10.16. A new template (that goes undetected in the previous step) is added to the training set until the recall is one for both metrics. After obtaining a training set which achieves zero false positive and a recall rate of one, the cutoff thresholds are set to the minimum coefficient values that satisfy these conditions.

10.3 Profiling Interrupts Using Analog Emanations

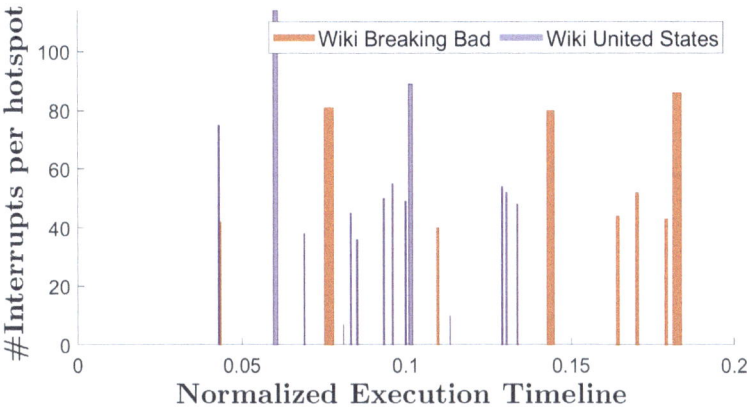

Fig. 10.16 Parametric sweep of Gaussian and correlation coefficients for creating training template dictionary [12]

Algorithm 10.1 Testing Phase of Detection Algorithm

1: **for** each template **do**
2:     find degree of correlation with testing signal;
3:     find Gaussian distance with testing signal;
4: **end for**
5: **for** each sample in testing signal **do**
6:     compute best matches with Pearson correlation;
7:     compute best matches with Gaussian distance;
8: **end for**
9: **if** (correlation > correlation threshold) or (Gaussian distance > Gaussian threshold) **then**
10:     select candidate sample;
11: **end if**

12: **for** each candidate **do**
13:     DTW(candidate, best matching template);
14:     **if** warping is within allowed deviations **then**
15:         select estimated event match;
16:         select associated latency;
17:     **end if**
18: **end for**
19: **Result** Estimated event matches, latencies

Algorithm 10.1 provides an overview of the designed detection algorithm. The testing scheme consists of two parts. In the first phase, correlation and Gaussian distances are found for the signal-under-test with the training templates. For each sample in the test signal, the corresponding template having the best match is logged. As every sample in a test signal is actually a candidate for the starting point of a pattern, therefore, the sample space increases as the size of the test signal increases. To reduce the sample space and complexity, we perform thresholding on the obtained matches to create a population of probable candidates. The thresholds applied are 1% lower than the cutoff thresholds obtained during the training parametric analysis to allow patterns with small variations to be considered as

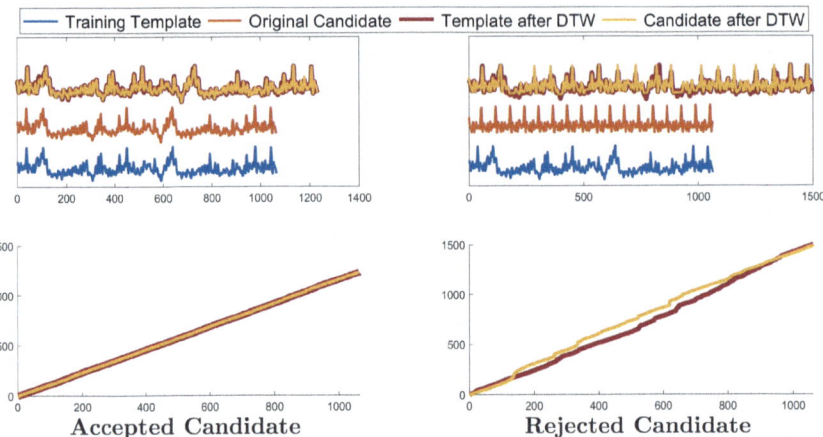

Fig. 10.17 Result of mDTW—effect of DTW on the signals (top) and the resulting degree-of-warp (bottom), for an actual interrupt (accepted candidate, left) and a non-interrupt part of execution (rejected candidate, right) [12]

candidates. This operation helps to decrease the number of negative samples in the population and to report more robust accuracy results.

The second phase of testing comprises of a *modified dynamic time warping* algorithm, which we will refer to as *mDTW*. Here, dynamic time warping (DTW) is performed for each candidate with its corresponding best-matched template found in the first phase. In DTW, each of the samples in both the signals are stretched (repeated) to obtain the best possible match. However, due to *unbounded stretching*, some non-similar candidate may be treated as a positive match to the template. To avoid this, we add *bounds* to the warping of the candidate and the template as follows.

For illustration, we will focus on two candidates obtained after Phase 1 as shown in Fig. 10.17 as the "original candidate". The result of DTW results in warped template and warped candidate, shown as the overlapped signals in the top figures. Now, we create vectors of length of original template where each element corresponds to the number of points inserted for each corresponding sample of the original signal for both warped template and warped candidate. For example, if after warping, the warped signal had sample#2 repeated thrice, the corresponding vector value for sample#2 will be 3. This vector provides us with information about per sample stretching, and hence we term it as the *Degree of Warp (DoW)*. The DoW for the warped candidates and templates are shown by yellow and red lines in bottom sub-figures of Fig. 10.17 respectively.

We then calculate the standard deviation of the absolute difference between the DoWs of candidate and template, to allow small constant change but at the same time avoid large deviations across multiple samples. The threshold for the standard deviation is set to 1% of the length of the original template. This ensures that those candidates that had less stretching due to small variations in their EM signature are

10.3 Profiling Interrupts Using Analog Emanations

still accepted, and also the fact that extreme training coverage is not required to record every template to account for minor variations, thus reducing the template dictionary size. However, a candidate that had very large deviations and that had more intermittent stretching as compared to the warping in the training template will result in the candidate to be rejected from being considered as an interrupt, as this means that the candidate just had some small parts of the signal that were similar to the template, thus resulting in a higher correlation coefficient.

This is illustrated in Fig. 10.17, where the candidate that was accepted had very small deviations in its DoW as compared to the DoW of its best-matched training template. Even though some samples were inserted in both the candidate and the template to match the signals, their DoW variations were nearly negligible and also the final result of the warped candidate and template majorly overlap. Hence, when the standard deviation of the difference between their DoWs is computed, it is within the 1% deviations allowed, thereby being accepted as an interrupt. This was indeed correct as the candidate was actually an interrupt.

In comparison, the candidate that is rejected in Fig. 10.17 just has some features similar to the template. In order to have the best match with the training template with DTW, the candidate underwent far more stretching in the middle section, as can be seen in the considerable difference between the DoWs of the two signals. This also resulted in some non-matching features in the warped candidate and template. Thus, when the standard deviation of the difference is computed, it is beyond the allowed deviations, thereby making mDTW reject this candidate. This was again correct as the candidate belongs to another part of signal that just had some patterns similar to the template, but actually was not an interrupt.

As example of why mDTW is helping, we note that after Phase 1 of the testing algorithm on our UART transmit microbenchmark, we had more interrupts than were actually happening with $N_{tx} = 4$. After applying mDTW, we achieved 100% accuracy in identifying the interrupt locations and successfully rejecting the false positive candidates from Phase 1, resulting in correctly determining $N_{tx} = 5$.

10.3.3 Validation

Now that we have found the typical patterns of the different interrupts and extracted training templates for all the different types of interrupts, let us validate the results obtained after running the detection algorithm on different test signals. It is fairly easy to corroborate the total number of interrupts being detected, but in order to thoroughly validate PRIMER, we need a mechanism that can confirm the exact locations of the interrupts as well. Out of the five interrupts chosen to be analyzed on the STM32F4 device, the timer interrupt and the UART transmit interrupt provided us with means to validate the locations as well without any additional instrumentation.

Fig. 10.18 Detecting locations of timer interrupts using PRIMER, and zoomed-in instance of one of the locations on STM32F4 IoT board [12]

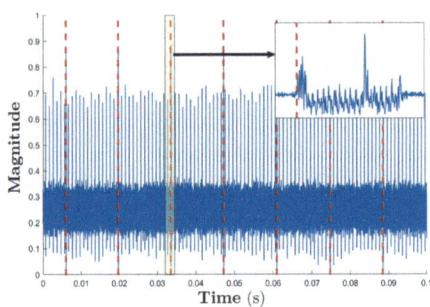

First we can determine the buffer size of the data being transmitted N_{tx} based on the time interval obtained between two consecutive UART transmit interrupts using Eq. (10.1). Then, the frequency of timer interrupts can be determined as the prescalar and counter register values are known while programming. Thus, the duration between two consecutive timer interrupts can be calculated as

$$T_{timer} = \frac{counter + 1}{\frac{F_{clock}}{prescalar}} = \frac{counter + 1}{\frac{16\,\text{MHz}}{20000}} = \frac{counter + 1}{800}(s), \quad (10.2)$$

By comparing the expected programmed values of T_{timer} and N_{tx} with the locations obtained from PRIMER, accuracy of our detection algorithm was calculated.

Figure 10.18 shows an example of how PRIMER's detection algorithm identifies interrupts. It shows a section of the microbenchmark execution generating timer interrupts that are programmed to be 0.0137 s apart. The vertical red dashed lines show the instances of the detected interrupts, as can be seen in the zoomed-in instance of a single such detected timer interrupt, and we can see that PRIMER is able to accurately identify the locations of the timer interrupts.

As we have the validation mechanism established for both the timer and UART transmit interrupts, we evaluate the accuracy of PRIMER for varying values of down-counter values for timer interrupts (i.e., varying values of T_{timer}), and varying sizes of transmit data buffer (N_{tx}), where training has been performed on just one instance of each of these interrupts. To further evaluate the effectiveness of PRIMER, we also test the same configuration of microbenchmarks used on another STM32F4 board and test the robustness of our designed detection algorithm, as shown in Table 10.7. The results reported are computed after averaging the interrupt time intervals in 1 s runs for each interrupt configuration.

Based on the results from Table 10.7, we can infer that PRIMER is able to accurately determine the locations of the interrupts as we achieve 100% accuracy in determining the both the timer interrupt frequency and the transmit data buffer size. This indicates that PRIMER is now ready to be applied and evaluated under various scenarios.

10.3 Profiling Interrupts Using Analog Emanations

Table 10.7 Validating accuracy of detected locations of timer and UART transmit interrupts across two STM32F4 IoT devices [12]

Interrupt type	Specification T_{timer} or N_{tx}	Accuracy Board 1	Board 2
Timer	0.0137 s	100%[a]	100%
	0.0637 s	100%	100%
	0.1261 s	100%	100%
	0.2511 s	100%	100%
UART transmit	5 bytes	100%[a]	100%
	10 bytes	100%	100%
	15 bytes	100%	100%
	20 bytes	100%	100%

[a] Trained configuration

Fig. 10.19 STM32F4 board with blue near-field probe experimental setup [12]

10.3.4 Results

Profiling Interrupts on STM32F4

PRIMER's effectiveness was evaluated across various types of interrupts on the STM32F4 IoT board, which is based on STM32F446RE chip with a 16 MHz system clock, ARM Cortex M4 core with an extensive range of I/O and peripherals and communication interfaces [41]. A small magnetic probe, as shown in Fig. 10.19, is used to acquire the clock amplitude-modulated EM signal with a 20 MHz bandwidth, which is then recorded and saved using a Keysight N9020A MXA spectrum analyzer [21]. A spectrum analyzer is used primarily because it allows for signals to be observed, analyzed, and visualized in real-time. However, the signal acquisition requirements (bandwidth, signal-to-noise ratio, etc.) needed for our prototype implementation are well within the capabilities of smaller, less sophisticated (and much less expensive) software-defined radios (SDRs), so that is what we envision would be used for practical applications of our approach.

To compute the accuracy of PRIMER across different kinds of interrupts, the number of times the application executed an interrupt handler was recorded and stored in a device register and the value was logged at the end of the execution. For every run, PRIMER's accuracy was calculated against this register value. Table 10.8 shows the accuracy of PRIMER in counting interrupts across five different kinds of interrupts, namely, timer, UART transmit, UART receive, ADC and push button GPIO. For the timer interrupt different T_{timer} values were used, for the UART transmit interrupt N_{tx} was varied with fixed and a random value, the UART receive was triggered by sending data from a local computer to the IoT device, the ADC interrupt was recorded every time an ADC conversion was triggered for a 12-bit resolution for a predetermined value, and the on-board push button was pressed asynchronously multiple times to trigger GPIO interrupt.

Each of these interrupts were trained for the first configuration of each interrupt on Board 1 as indicated in Table 10.8. Hence, these results always have 100% accuracy. The obtained interrupt templates and the thresholds were then applied not only across other configurations of that interrupt on the same device (Board 1), but also on all the configurations on another device of the same type (Board 2). We can see from Table 10.8 that we are able to detect nearly all, if not all, of the interrupts using PRIMER with very high accuracy.

Insights from PRIMER on IoT Applications

Now that we have PRIMER set up for five different interrupts on the STM32F4 IoT device, it would be interesting to see how it performs on a real IoT applications and what insights we can obtain. To demonstrate this, we first designed a typical

Table 10.8 Accuracy of PRIMER for detecting total interrupts for 5 types of interrupts across two STM32 devices [12]

Interrupt type	Specification	Accuracy	
		Board 1	Board 2
Timer	0.0137 s	100%[a]	100%
	0.0637 s	100%	100%
	0.1261 s	100%	100%
	0.2511 s	100%	100%
UART transmit	5 bytes	100%[a]	98.4%
	10 bytes	98.9%	97.9%
	15 bytes	100%	98.44%
	20 bytes	100%	97.9%
	random 20 bytes	97.9%	97.9%
UART receive	5 bytes	100%[a]	100%
	10 bytes	100%	100%
ADC	12 bit	99.27%[a]	99.35%
GPIO	Push button	100%[a]	100%

[a] Trained configuration

10.3 Profiling Interrupts Using Analog Emanations

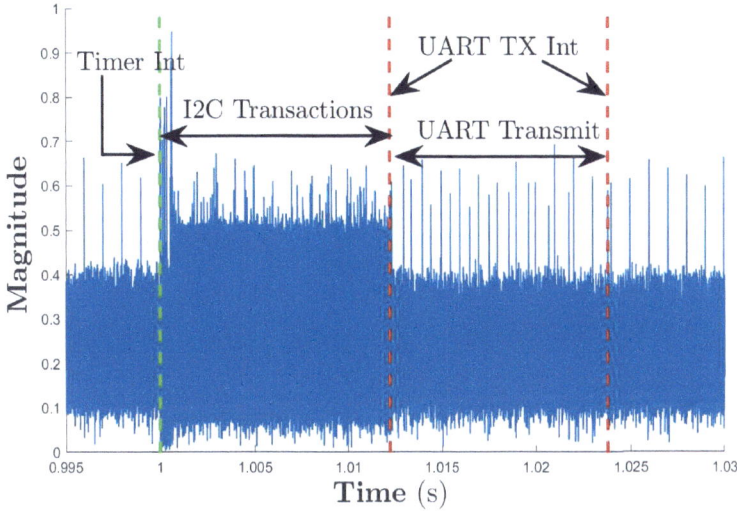

Fig. 10.20 Insights from PRIMER's results on a data logging IoT application [12]

IoT application, where the device reads the temperature from Microchip MCP9808 digital temperature sensor [27] and communicates this information to a host.

The application was designed as follows. At 1 s intervals, the IoT device communicates with the temperature sensor using I²C communication interface. Once the temperature data is ready and available, the IoT device sends the temperature data to the host computer using UART serial interface a couple of times to ensure proper logging.

Each of these repetitive 1 s segments consists of one timer interrupt followed by a series of I²C transactions to read the temperature data, upon whose completion two UART transmissions are made in interrupt mode. This can be seen in Fig. 10.20 where the vertical green dashed line is the location of timer interrupt and the two vertical red dashed lines represent the UART transmit (TX) interrupt locations detected by PRIMER. Upon analyzing these locations, we found that the timer interrupt locations (T_{timer}) are indeed 1 s apart, and N_{tx} was determined as 11 bytes using Eq. (10.1), which is again correct as the typical data transmitted was of the format "24.2509 C\r\n". Please note that even though the ISR now had user callback code present, this is executed only after the entire driver-level interrupt library finishes, thus the EM signature of the interrupt obtained during the training phase still accurately identifies the location of the interrupt.

PRIMER can be particularly useful in applications that are time-critical, and in situations where physical debugging of the embedded device may not be feasible, such as in the medical field. PRIMER's effectiveness in such applications can be seen in the designed prototype of a pacemaker on this low-power board as follows.

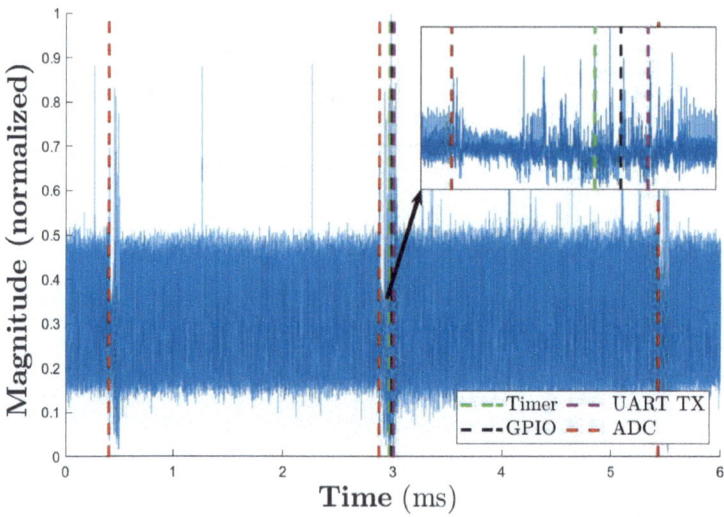

Fig. 10.21 Insights from PRIMER's results on a pacemaker application [12]

The device acquires analog ECG signal from external circuitry and uses its internal ADC to convert it to a digital signal that can be used by an internal comparator and quatizer to detect valid heartbeats. A synchronously operating timer simultaneously checks for valid heartbeats at regular intervals. In the occasion of an irregular heartbeat, the quantizer fails to detect the heartbeat, causing the timer to overflow, generate a paced pulse (which connects to an external analog circuitry), and also communicate this critical information (we envision a wireless form of communication that typically uses AT commands for interacting with the wireless module, hence UART is used for the prototype).

Figure 10.21 clearly shows the sequence of these interrupts unfolding in the event of an irregular heartbeat as identified by PRIMER. In a case of failure, PRIMER will also be able to infer the anomalous stage, which can be then quickly fixed. In this way, such insights can be provided by using PRIMER for profiling and monitoring interrupts completely remotely without the measurement mechanism perturbing the application execution.

Profiling Interrupts on MSP430

As we have established that PRIMER works very well with a variety of interrupts on the STM32F4 IoT device, to test its efficacy on different architectures, we use another popular embedded device, the MSP430G2 Launchpad, which is based on the MSP430G2553 16 MHz 16-bit MCU and a variety of I/O and peripherals [42].

As MSP430 operates in the low power mode, an interrupt wakes up the device to switch it from low-power to normal execution mode, performs the required operations and then enters low-power mode again. After isolating the ISR from the wake-up and sleep, PRIMER is trained with the timer interrupt pattern. We also log

10.3 Profiling Interrupts Using Analog Emanations

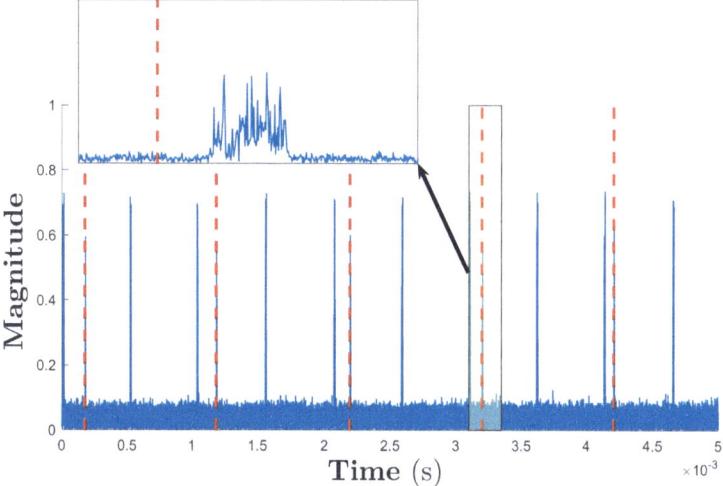

Fig. 10.22 Detecting timer interrupts on MSP430 device using PRIMER [12]

the value of the number of times the device executes the timer interrupt handlers for validating the total number of interrupts observed.

PRIMER obtained 100% accuracy in finding the total number of timer interrupts based on the logged register value. The average T_{timer} was also correctly determined as 1 ms. Figure 10.22 shows an example, with locations of detected timer interrupts (as detected by PRIMER) shown as vertical lines. The zoomed-in section shows the EM signature of one of the detected timer interrupts.

These results show that PRIMER's non-intrusive nature coupled with its ability to profile and distinguish between different interrupts on different hardware architectures makes it a valuable tool for interrupt profiling.

10.3.5 Applications

In this section, we examine how PRIMER can be used to profile interrupts (and gain some interesting insights) for a more complex IoT device. Specifically, we used the Olimex A13-OLinuXino-MICRO IoT prototype board [32], which is based on the Allwinner A13 SoC chip [6], with a 1 GHz ARM Cortex A8 core (ARMv7-A architecture) 32 kB DL1 and 32 kB IL1 caches, and a 256 kB L2 cache. The board's operating system (OS) is Debian Linux 7 (Wheezy). The EM emanations were collected using a near-field probe and recorded using a spectrum analyzer [21].

For brevity, we focus on one exception (software-generated) and one interrupt (hardware-based), namely page fault and network interrupt respectively. Page fault exception was selected as it is a very commonly occurring exception and will occur at least once at the beginning of the program execution. Besides, there are existing profilers that will help us instrument the program to evaluate against some "ground-truth". If profiled successfully, it would prove the effectiveness of PRIMER in identifying even program-generated interrupts. Of all the classes of device interrupts, I/O interrupts have the maximum latency, among which network interrupts have been shown to be the most expensive [37]. Besides, CPS and IoT devices depend heavily on network-related activities and insights on network interrupts for performance improvements or device monitoring will be beneficial.

10.3.6 Profiling Exceptions: Page Faults

If a page is not present in the device cache or main memory when it is referenced, a page fault occurs and the entire requested page is moved from the disk to main memory.

As page faults are a type of OS event, we can modify a profiler that profiles software events to find them. For our purpose, we use the `perf` Linux profiling tool, which is a Linux-based profiler that can instrument CPU performance counters to sample hardware events, and also has a kernel interface that can measure software events using kernel counters [23]. `Perf record` command was used to collect profiles pertaining to a single process, and the data collected was analyzed using `perf report` command. `Perf` operates based on event-based sampling, wherein a sample is recorded upon sampling counter overflow. The report generated is then analyzed based on the type of event sampled. PERF_RECORD_SAMPLE events are the ones relevant to the execution of the application as they contain the sample of the actual hardware counter or in our case, a software event. Samples are timestamped with an unsigned 64-bit word, which records elapsed nanoseconds since system uptime, which can be used to find the exact locations of the faults in the execution timeline of the application.

We devise a microbenchmark that generates a known number of page faults, one at a time. This is achieved by memory-mapping a very large file, and then performing single-byte loads for each page, thus creating a page fault for each of those loads. In order to separate these page fault in time for ease of identification, a compute-only function is called between each pair of loads and the distinct EM signature of this function allows us to easily isolate the signal that corresponds to each of the page faults automatically. During training, these signals are used as examples to create templates for EM signature of page faults.

We next perform experiments where the microbenchmark is modified to create 1000 page faults individually (one at a time), then in pairs (bursts of 2), and finally in bursts of 5, and our technique is used to identify page faults in each of these executions. Table 10.9 shows the results of this experiment. First, the code is instrumented to record the timestamp for each of these page faults, which is used as the ground-truth to evaluate the recall and precision of EM-based detection of page

10.3 Profiling Interrupts Using Analog Emanations

Table 10.9 PRIMER's accuracy for identifying page faults in microbenchmarks [12]

#page faults	#consecutive	Instrumented runs		Non-instrumented recall
		Recall	Precision	
1000	1	95.5%	100%	99.5%
1000	2	97%	100%	98.3%
1000	5	97.3%	100%	98.4%

faults. Note that the measured precision is 100% in this experiment, i.e., no false positives.

As this run included the additional overhead of sampling using `perf` simultaneously, it is important to understand the effects of instrumentation. As we now know where in the EM signal timeline the page faults are scheduled to appear, we can run the microbenchmark but without `perf`. We observed that the overall shape of the page fault patterns obtained with and without instrumentation is preserved. However, the typical length of page fault with instrumentation is seen to be ~3.7 μs, whereas the length with no sampling overhead is ~1.5 μs, amounting to a ~60% overhead. This is a significant amount of overhead overwhelming the native execution altogether. Using `perf` to report timestamped samples for a non-instrumented run will not be an accurate representation of the page fault locations in execution timeline.

As PRIMER was effective on the instrumented run, we now apply it on non-instrumented data and report how well the results correlate in execution timeline. Since no ground-truth is available, we assume that precision in this experiment is also 100% and report only the recall percentage in Table 10.9. We also confirmed, by inspecting the results manually, that no page faults were reported in the parts of the signal that correspond to the compute-only-function.

Now that we have found the typical patterns of page fault and extracted templates, we ran PRIMER on a real application. However, due to a large amount of contiguous memory being allocated (which is a more realistic OS behavior), this led to some of the page fault patterns to be not as well-defined and lose some features, thereby leading to reporting fewer number of page faults compared to `perf`. We observed that by lowering the measure of correlation we do find more page fault patterns, but at the cost of very high computing times with mDTW. Thus, we add another pass to PRIMER where it lowers the correlation coefficient of the area around which it finds multiple page fault patterns by 10%, which will allow us to find more of the page fault patterns when they happen together. We refer to such a region as *hotspot*. The hotspots are determined by performing a Gaussian analysis on the distance between two consecutive estimated page faults and treating the non-outliers as a part of the same hotspot. We believe that hotspots provide us with more meaningful information of identifying an important activity affecting and stalling the processor, than obtaining individual page fault locations as will be demonstrated in the subsequent sections.

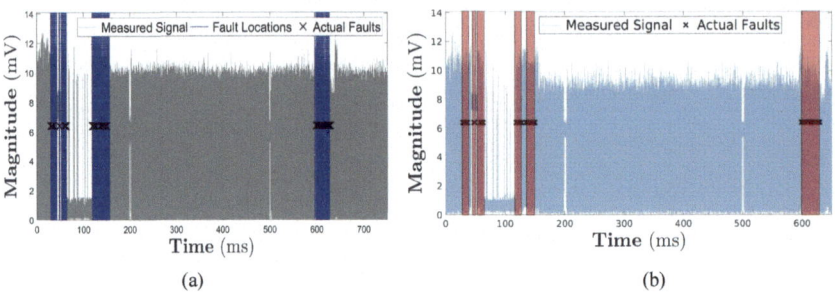

Fig. 10.23 Detected page faults in instrumented *mcf* SPEC2000 benchmark execution (lines) and actual fault locations from timestamps (crosses): (**a**) Identified page fault locations, (**b**) Corresponding hotspot regions [12]

We illustrate this using an instrumented *mcf* SPEC2000 benchmark application run. Please note that to align the timestamps from perf, the actual application run is preceded by a small custom application with known locations of page faults followed by a sleep subroutine. This was used just to align the timestamps and did not contribute to the results reported. Figure 10.23a compares the locations of page faults detected against the expected timestamps. We report the recall, where the total expected instances are obtained from perf sampling data. The recall for this run was found to be ∼80% with no false positives. Next, based on the page faults reported by PRIMER, we identify the hotspots during *mcf* application run. We classify extremely closely-spaced page faults part of a single hotspot. Figure 10.23b illustrates the hotspots found along with their respective widths. We find 3 hotspots arising from *mcf* (the first 4 arise from alignment; the last 2 in figure are 2/3 hotspots).

To quantify our results and accuracy, we run a parametric analysis on the effect of recall on accuracy of finding hotspots. We randomly remove reported page faults to reduce the recall and calculate the accuracy of finding clusters. The results of this analysis is shown in Fig. 10.24. The "accuracy" here is defined as the measure of finding *each* hotspot. We observe that by having a *minimum recall of 30% for each of the hotspots*, the corresponding hotspot will be found.

Hotspots provide us with more information than just reporting the number of page faults and their locations. Figure 10.25 shows the total number of page faults in the hotspot and the "width" of the hotspot, which defines the hotspot region. The values are normalized for ease of comparison. We can see that for the three hotspots in *mcf*, the number of interrupts within a hotspot does not necessarily correlate with the width of the hotspot region, like we see for hotspots 1 and 3. Upon further analysis, we observed that there are more consecutive page faults in close proximity in hotspot 1 as compared to hotspot 3, where the page faults are more scattered with some non-interrupt activity between consecutive page faults. This results in higher interrupt density for hotspot 1 in comparison to hotspot 3. Obtaining such

10.3 Profiling Interrupts Using Analog Emanations

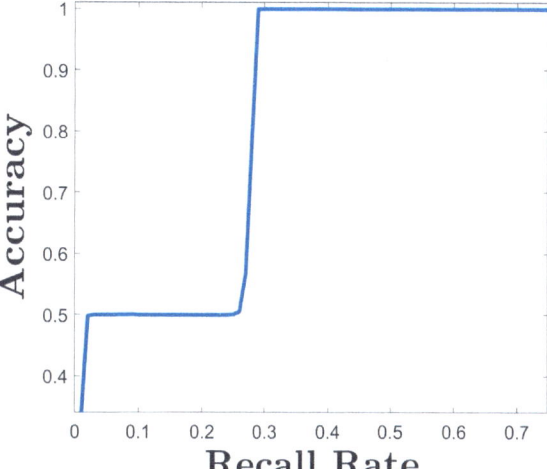

Fig. 10.24 Accuracy of finding a hotspot for varying recall rate [12]

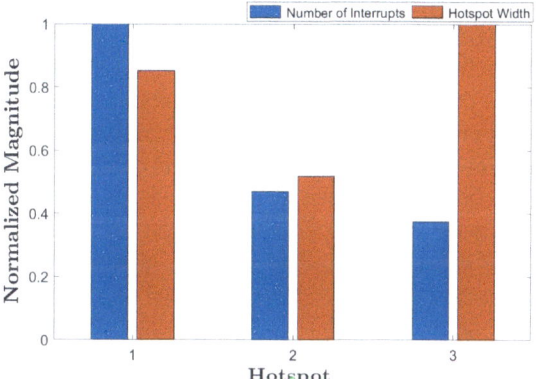

Fig. 10.25 Example of insights from hotspot data [12]

insights from the results reported by PRIMER without adding any instrumentation to the device is one of the unique advantages of our profiler.

Now that we have achieved adequate accuracy to report hotspots, we run PRIMER on *mcf* non-instrumented code. Even though the signatures of page faults in the two cases change due to the overhead of event sampling for each event by perf, as the coarse-grain number of page faults detected during the timeline of program execution for both the runs correlate, this means that we are able to find page faults even in the non-instrumented execution with the same recall rate.

We then perform an experiment where we ran a subset of SPEC CPU2000 benchmarks, which was selected as a representative of a variety of memory access patterns [35]. As ground-truth, we use the perf profiling tool to collect timestamps of page faults and these results are reported as "Instrumented Runs" in Table 10.10. In addition to the recall percentage, we also report the number of hotspots (periods of repeated page fault activities) identified in each run (worst-case out of 5 runs

Table 10.10 PRIMER results for identifying page faults for SPEC CPU 2000 workloads on Olimex IoT board [12]

SPEC	Instrumented runs			Non-instrumented runs
	Recall	Hotspots found		Reported
		Expected	Reported	
bzip2	47.47%	5	6[a]	6[a]
gzip	90.64%	4	4	4
mcf	79.74%	3	3	3
parser	21.24%	9	10[a]	9
perlbmk	23.23%	3	3	3
vortex	66.15%	11	10[b]	9[b]

[a] Overcount
[b] Undercount

shown in table to explain variations; remaining runs had 100% accuracy), along with the number of hotspots identified from `perf` data. We observe that, out of the five workloads, *gzip* and *mcf* had relatively high recall for individual page faults, and we accurately identified each hotspot. In *perlbmk*, although the recall was low, each of the three hotspots was still correctly identified. In *bzip2* and *parser*, the number of reported hotspots was higher than expected, not because of false positives (there were none), but because page faults in the middle of a long-lasting hotspot were not identified, causing the same actual hotspot to be divided into two separate hotspots. Finally in *vortex*, PRIMER fails to report one of the hotspots—that particular hotspot had < 10 actual page faults, only 3 of which were detected, which has prevented that part of the execution to be labeled as a hotspot as it did not meet the minimum requirement to qualify as a hotspot (≥ 5 faults in a region needed to define hotspot).

Table 10.10 also shows hotspot-reporting results for runs using the same inputs but without `perf` ("Non-Instrumented Runs"). Although we have no ground-truth data for these runs, in these highly-deterministic applications we expect that the number of page faults and their grouping are similar to their instrumented counterparts. We indeed find that PRIMER reports similar numbers of hotspots in these `perf`-free runs and their occurrences correlate in timeline.

10.3.7 Profiling Interrupts: Network Interrupts

As Olimex IoT device has an SoC with integrated Ethernet controller, a USB network adapter was connected to enable WiFi capabilities. Thus, the CPU remains busy for the entire duration of the network transaction, as it is responsible for both transmission and reception of network packets. This in turn implies that the CPU will conduct data transfer operations to send out and bring in data to the chip.

To observe this transaction in EM side-channel signal, we acquire EM emanations from the execution of a simple networking application—`ping` networking

10.3 Profiling Interrupts Using Analog Emanations

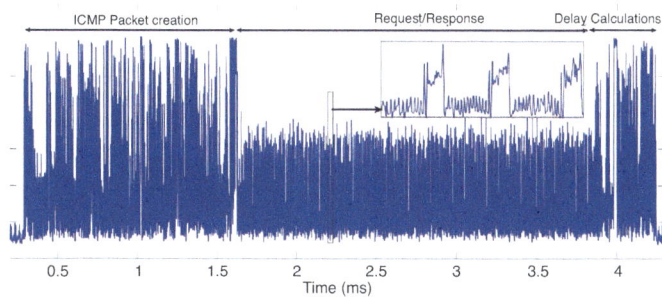

Fig. 10.26 EM emanations obtained from an IoT device requesting Google homepage's destination IP address using ping. A zoomed-in repetitive section of the request/response phase used for training is also shown [12]

utility. ping uses the Internet Control Message Protocol (ICMP) to send echo request packets to the target host and waits for the echo response from the host.[1]

Figure 10.26 shows a sample run of ping attempting to check if the homepage of Google is reachable. The various stages of ping execution are annotated in the EM signal timeline. Our region of interest is the segment that has the actual network frame being sent out and received, which corresponds to the Request/Response section in Fig. 10.26. A zoomed small part of the Request/Response stage in the figure shows three of the data transfer patterns repeating back to back. Based on our preliminary analysis, this data transfer is what consumes majority of the network interrupt event time, irrespective of the protocols or encryption mechanisms being used. As the data transfer section remains common across all network interrupt triggers no matter the application used, we will be focusing on this region of transfers to identify network activity. PRIMER was trained to find these patterns corresponding to network activity.

Since ping simply determines if the requested destination server was reached or not, we wanted to examine an application that is involved with actual website data transfers during networking activity. We chose a popular command line tool curl used for transferring data with URLs.[2] It supports various network protocols to transfer data and is used commonly in cars, television sets, routers, printers, media players, etc., and can build on many platforms.

Figure 10.27a shows a simple activity of curl trying to access the homepage of Facebook. In this process, we report a total of 3076 network interrupts. However, this number on its own does not give us enough information about the transactions that occurred in this process. As each part of data transfer has the same EM signature and occur back-to-back, hence we cluster the network interrupts together to form a

---

[1] ping (8)—Linux man page [https://linux.die.net/man/8/ping].

[2] Everything curl: command line tool and library for transferring data with URLs (since 1998) [https://curl.haxx.se/].

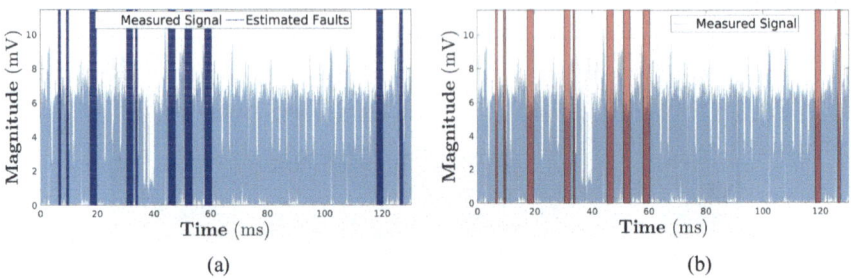

Fig. 10.27 Detected network interrupts for curling Facebook homepage: (**a**) Identified network interrupt locations, (**b**) Hotspot determination from detected interrupts [12]

Table 10.11 PRIMER detection algorithm results for identifying network interrupts from curl requesting websites [12]

Website homepage	Training #Hotspots	Testing 1 FP	Testing 1 TP	Testing 2 FP	Testing 2 TP	Testing 3 FP	Testing 3 TP
Google	4	0%	96.3%	0%	100%	0%	100%
Facebook	25	0%	100%	0%	100%	0%	100%
Github	14	0%	100%	0%	100%	0%	100%
Wikipedia	28	0%	100%	0%	100%	0%	100%
Youtube	36	0%	100%	0%	100%	0%	100%

hotspot. Each hotspot detected signifies one data transfer activity, which is more informative than reporting the total number of patterns found. Hence, we use a similar strategy here as well to categorize the patterns into a hotspot as shown in Fig. 10.27b.

To quantify the results obtained, we further analyze the network transaction packets (which translates to a hotspot here) when different website servers are requested using curl. As these websites handle heavy traffic, they have their own mechanism of redirects to handle network traffic. Table 10.11 shows how the number of hotspots vary with different websites requests on the trained device. It can be clearly seen that each website has a varying number of hotspots. To demonstrate robustness of PRIMER, we use the same trained data to identify network interrupt hotspots for the trained device and two more devices of the same type but with the EM measurements performed several days later. We found no false positives for any of the tested devices, and mostly 100% true positives. The only reason PRIMER missed one hotspot was because the number of interrupts in that region was very small and had a lower recall leading to <10 interrupts being identified in that region, and this did not qualify as a hotspot.

Figure 10.28a shows the detected hotspots in the execution timelines of three devices, as they request the Wikipedia main page using curl network utility. The overall occurrences of the hotspots largely correlate across the devices proving that such applications have distinct hotspot patterns. The minor variations in distances between consecutive hotspots due to other background activities can still

10.3 Profiling Interrupts Using Analog Emanations

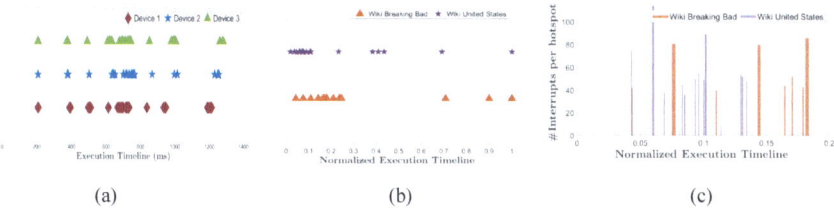

Fig. 10.28 Hotspot-based insights from PRIMER's results for requesting Wikipedia pages using curl utility: (**a**) Hotspots in execution timeline for three IoT devices (of the same type) requesting Wikipedia homepage, (**b**) hotspots in execution timeline for 2 different Wikipedia articles on the trained IoT device, and (**c**) hotspot density in (partial) runtime for 2 different Wikipedia subpages on the trained device [12]

be tolerated. Additional analysis can be done on the report generated by PRIMER that can learn the occurrence pattern of the hotspots and attribute it to accessing a specific website if a match is found. Figure 10.28b shows a comparison of the hotspots obtained in the normalized execution timeline of the application. The figure shows two different Wikipedia articles accessed: *Breaking Bad* and *United States*. We can see that accesses to different subpages within a website cause very different network interrupt patterns, as the network data transfers depend upon the content of the page, which varies too. This can be seen very clearly in Fig. 10.28c, which shows not only the occurrences of the hotspots in execution timeline of these webpage accesses, but also information about the lengths of the hotspots (i.e., effectively illustrating the "hotspot density") for the beginning section of Fig. 10.28b. This kind of information can only be provided by actively monitoring router traffic or by use of sophisticated software, which most embedded devices are not built to run. By monitoring the data analyzed by PRIMER, for the first time, such detailed study can be done without any additional instrumentation in the device.

Another insight that we can obtain is regarding the network protocol being used to transfer data. All of the above results were obtained by using curl use HTTP1.0 protocol. HTTP1.0 does not officially require a host header, however for every request/response pair, it needs to create a new connection for the transaction and close the connection when it ends [22]. This requires extra network packets to be sent to open and close the connection for each transaction. As HTTP1.0 suffers from such efficiency problems (known as TCP slow start), HTTP1.1 is generally used as it allows persistent connections.

For a sample run of curl accessing the Facebook homepage, we observed that our algorithm reported a total of 25 hotspots to access with HTTP1.0 whereas 22 hotspots with HTTP1.1. As connections did not have to be closed and opened, this led to fewer transactions in HTTP1.1 as compared to HTTP1.0, which explains the lower number of hotspots observed. Again, the number of hotspots reported along with the hotspot profile can also provide information about the networking protocol being used as well, in addition to insights about the website being accessed. Such insights, obtained in a contact-less manner, are what makes our profiler extremely versatile for profiling and monitoring embedded devices, especially CPS and IoT devices.

10.3.8 Potential Other Uses of PRIMER

In addition to helping debug interactions between interrupt handlers and other code and helping guide optimizations, the information extracted by PRIMER—specifically timing and duration each interrupt handler execution and which code is has interrupted—have other potential uses, such as:

Security and Monitoring With CPS and IoT devices requiring continuous internet transactions, the insights provided by PRIMER for network interrupt profiling provides in-depth information about how the hotspots in network transactions vary based on the protocol being used and the websites being accessed. By developing a security monitor based on PRIMER, spurious and malicious internet accesses by these devices can be detected and appropriate measures can be taken accordingly to prevent the device from being compromised. This can also be easily extended to incorporate more system events to build an anomaly/malware detector by identifying unexpected execution paths.

Enhance Side Channel Research Traditional EM emanations-based research work, especially in the area of cryptography, have mentioned presence of interrupts perturbing their signals causing loss of important information [2, 4]. Owing to this, they had to collect thousands of training sets to finally obtain barely hundreds of "clean" signals. As PRIMER provides with the exact location and duration of the interrupt, the interrupt section can be easily cropped out to provide them with a clean signal to work with, thereby reducing the manual overheads of acquiring thousands of measurements to do analysis. For instance, system events occurring within loop sections of code may disrupt the EM signature of a functional loop in frequency spectra. Using PRIMER as a pre-processor for the signals to be fed to other profilers such as Spectral Profiling and EDDIE [31, 39] will increase the accuracy of these profilers.

Application Fingerprinting Combined with additional mechanism to characterize known websites and protocols, similar to what [18] achieves by using cache side-channel and instrumenting browser source code, our approach can be used for fingerprinting websites. The information provided by PRIMER in this case, acts like a side-channel based on which, leakage estimations can be done. PRIMER can be used as a defense mechanism in this case, which allows developers to fix their software to conceal their side-channel information.

References

1. ACM. *Memspy: Analyzing memory system bottlenecks in programs*, volume 20, 1992.
2. Dakshi Agrawal, Bruce Archambeault, Josyula R. Rao, and Pankaj Rohatgi. The em side-channel(s). In *Revised Papers from the 4th International Workshop on Cryptographic Hardware and Embedded Systems*, CHES '02, pages 29–45, London, UK, UK, 2003. Springer-Verlag.
3. Prathima Agrawal, Aaron Jay Goldberg, and John Andrew Trotter. Interrupt-based hardware support for profiling memory system performance, June 16 1998. US Patent 5,768,500.
4. Monjur Alam, Haider Adnan Khan, Moumita Dey, Nishith Sinha, Robert Callan, Alenka Zajic, and Milos Prvulovic. One&Done: A Single-Decryption EM-Based Attack on OpenSSL's Constant-Time Blinded {RSA}. In *27th {USENIX} Security Symposium ({USENIX} Security 18)*, pages 585–602, 2018.
5. Alcatel. Alcatel ideal / streak specifications. https://www.devicespecifications.com/en/model/e2ef3d31, accessed March 26, 2018.
6. Allwinner. A13 Datasheet. https://linux-sunxi.org/images/e/eb/A13_Datasheet.pdf, March 2012, accessed February 2021.
7. ARM. ARM Cortex-A Series: Programmer's Guide for ARMv8-A, 24 March, 2015. ARM DEN0024A (ID050815).
8. Bruno Bouyssounouse and Joseph Sifakis. *Embedded Systems Design: The ARTIST Roadmap for Research and Development*, volume 3436. Springer, 2005.
9. Shirley Browne, Jack Dongarra, Nathan Garner, Kevin London, and Philip Mucci. A scalable cross-platform infrastructure for application performance tuning using hardware counters. In *Supercomputing, ACM/IEEE 2000 Conference*, pages 42–42. IEEE, 2000.
10. Robert Callan, Farnaz Behrang, Alenka Zajic, Milos Prvulovic, and Alessandro Orso. Zero-overhead profiling via EM emanations. In *Proceedings of the 25th International Symposium on Software Testing and Analysis, ISSTA 2016, Saarbrücken, Germany, July 18–20, 2016*, pages 401–412, 2016.
11. Moumita Dey, Alireza Nazari, Alenka Zajic, and Milos Prvulovic. Emprof: Memory profiling via em-emanation in iot and hand-held devices. In *2018 51st Annual IEEE/ACM International Symposium on Microarchitecture (MICRO)*, pages 881–893, 2018.
12. Moumita Dey, Baki Berkay Yilmaz, Milos Prvulovic, and Alenka Zajic. Primer: Profiling interrupts using electromagnetic side-channel for embedded devices. *IEEE Transactions on Computers*, pages 1–1, 2021.
13. Stijn Eyerman and Lieven Eeckhout. A counter architecture for online DVFS profitability estimation. *IEEE Transactions on Computers*, 59(11):1576–1583, 2010.
14. Michael K Gowan, Larry L Biro, and Daniel B Jackson. Power considerations in the design of the alpha 21264 microprocessor. In *Proceedings of the 35th annual Design Automation Conference*, pages 726–731. ACM, 1998.
15. Part Guide. Intel® 64 and IA-32 architectures software developer's manual. *Volume 3B: System Programming Guide, Part*, 2, 2011.
16. Yi Han, Sriharsha Etigowni, Hua Liu, Saman Zonouz, and Athina Petropulu. Watch me, but don't touch me! contactless control flow monitoring via electromagnetic emanations. In *Proceedings of the 2017 ACM SIGSAC Conference on Computer and Communications Security*, pages 1095–1108. ACM, 2017.
17. John L Henning. SPEC CPU2000: Measuring CPU performance in the new millennium. *IEEE Transactions on Computer*, 33(7):28–35, 2000.
18. Taylor Hornby. Side-channel Attacks on Everyday Applications: Distinguishing Inputs with FLUSH+ RELOAD. *BackHat 2016*, 2016.
19. Hynix. 2Gb DDR3 SDRAM Lead-Free and Halogen-Free (RoHS Compliant) H5TQ2G63BFR. http://www.farnell.com/datasheets/1382727.pdf?_ga=2.38941163.99566001.1522095243-708330394.1522095243, accessed March 20, 2018. "Rev 0.5".

20. Tejas S. Karkhanis and James E. Smith. A First-Order Superscalar Processor Model. In *Proceedings of the 31st Annual International Symposium on Computer Architecture*, ISCA'04, pages 338–, Washington, DC, USA, 2004. IEEE Computer Society.
21. Keysight. N9020a mxa spectrum analyzer. https://www.keysight.com/en/pdxx202266-pn-N9020A/mxa-signal-analyzer-10-hz-to-265-ghz?cc=US&lc=eng, accessed March 26, 2018.
22. Balachander Krishnamurthy, Jeffrey C Mogul, and David M Kristol. Key differences between HTTP/1.0 and HTTP/1.1. *Computer Networks*, 31(11-16):1737–1751, 1999.
23. Linux. Linux kernel profiling with perf. https://perf.wiki.kernel.org/index.php?title=Tutorial&oldid=3520, accessed March 2018.
24. Chi-Keung Luk, Robert Cohn, Robert Muth, Harish Patil, Artur Klauser, Geoff Lowney, Steven Wallace, Vijay Janapa Reddi, and Kim Hazelwood. Pin: Building customized program analysis tools with dynamic instrumentation. In *Proceedings of the 2005 ACM SIGPLAN Conference on Programming Language Design and Implementation*, PLDI '05, pages 190–200, New York, NY, USA, 2005. ACM.
25. Allen D. Malony, Daniel A. Reed, and Harry A. G. Wijshoff. Performance measurement intrusion and perturbation analysis. *IEEE Transactions on parallel and distributed systems*, 3(4):433–450, 1992.
26. Allen D Malony and Sameer S Shende. Overhead compensation in performance profiling. In *European Conference on Parallel Processing*, pages 119–132. Springer, 2004.
27. Microchip Technology Inc. MCP9808 Datasheet, 2011, accessed February 4, 2021. DS25095A, ISBN: 978-1-61341-739-3.
28. Rustam Miftakhutdinov, Eiman Ebrahimi, and Yale N Patt. Predicting performance impact of DVFS for realistic memory systems. In *Proceedings of the 2012 45th Annual IEEE/ACM International Symposium on Microarchitecture*, pages 155–165. IEEE Computer Society, 2012.
29. T. Mudge. Power: a first-class architectural design constraint. *IEEE Transactions on Computer*, 34(4):52–58, April 2001.
30. Priya Nagpurkar, Hussam Mousa, Chandra Krintz, and Timothy Sherwood. Efficient remote profiling for resource-constrained devices. *ACM Transactions on Architecture and Code Optimization (TACO)*, 3(1):35–66, 2006.
31. Alireza Nazari, Nader Sehatbakhsh, Monjur Alam, Alenka Zajic, and Milos Prvulovic. Eddie: Em-based detection of deviations in program execution. In *Proceedings of the 44th Annual International Symposium on Computer Architecture*, ISCA'17, pages 333–346, New York, NY, USA, 2017. ACM.
32. OlinuXino A13. https://www.olimex.com/Products/OLinuXino/A13/A13-OLinuXino/open-source-hardware.
33. Hewlett Packard. CXperf User's Guide, First Edition, June 1998. Customer Order Number B6323-90001.
34. Jose Renau, Basilio Fraguela, James Tuck, Wei Liu, Milos Prvulovic, Luis Ceze, Smruti Sarangi, Paul Sack, Karin Strauss, and Pablo Montesinos. SESC simulator, January 2005. http://sesc.sourceforge.net.
35. Suleyman Sair and Mark Charney. Memory behavior of the SPEC2000 benchmark suite. Technical report, Technical report, IBM TJ Watson Research Center, 2000.
36. Samsung. Samsung galaxy centura sch-s738c user manual with specs. https://www.cnet.com/products/samsung-galaxy-centura-sch-s738c-3g-4-gb-cdma-smartphone/specs/, accessed March 26, 2018.
37. Lambert Schaelicke, Al Davis, and Sally A McKee. Profiling I/O interrupts in modern architectures. In *Proceedings 8th International Symposium on Modeling, Analysis and Simulation of Computer and Telecommunication Systems (Cat. No. PR00728)*, pages 115–123. IEEE, 2000.
38. Nader Sehatbakhsh, Alireza Nazari, Monjur Alam, Frank Werner, Yuanda Zhu, Alenka Zajic, and Milos Prvulovic. Remote: Robust external malware detection framework by using electromagnetic signals. *IEEE Transactions on Computers*, 69(3):312–326, 2020.

References

39. Nader Sehatbakhsh, Alireza Nazari, Alenka Zajic, and Milos Prvulovic. Spectral profiling: Observer-effect-free profiling by monitoring em emanations. In *2016 49th Annual IEEE/ACM International Symposium on Microarchitecture (MICRO)*, pages 1–11, 2016.
40. Signatec. Px14400a – 400 ms/s, 14 bit, ac coupled, 2 channel, xilinx virtex-5 fpga, pcie x8, high speed digitizer board. http://www.signatec.com/products/daq/high-speed-fpga-pcie-digitizer-board-px14400.html, accessed March 26, 2018.
41. STMicroelectronics. STM32 Nucleo-64 Boards Data Brief. https://www.st.com/resource/en/data_brief/nucleo-f446re.pdf, October 2020, accessed February 4, 2021. DB2196 - Rev 14.
42. Texas Instruments Inc. *MSP-EXP430G2 LaunchPad Development Kit User's Guide*, July 2010–Revised March 2016, accessed February 4, 2021. SLAU318G.
43. ThinkRF. Rtsa v3 real-time spectrum analyzer. https://s3.amazonaws.com/ThinkRF/Documents/ThinkRF+RTSAv3+TDS+74-0034.pdf, accessed March 26, 2018.
44. Vincent M Weaver. Self-monitoring overhead of the Linux perf_ event performance counter interface. In *2015 IEEE International Symposium on Performance Analysis of Systems and Software (ISPASS)*, pages 102–111. IEEE, March 2015.
45. Josef Weidendorfer, Markus Kowarschik, and Carsten Trinitis. A Tool Suite for Simulation Based Analysis of Memory Access Behavior. In Marian Bubak, Geert Dick van Albada, Peter M. A. Sloot, and Jack Dongarra, editors, *Computational Science - ICCS 2004*, pages 440–447, Berlin, Heidelberg, 2004. Springer Berlin Heidelberg.
46. Qiang Wu, Artem Pyatakov, Alexey Spiridonov, Easwaran Raman, Douglas W. Clark, and David I. August. Exposing Memory Access Regularities Using Object-Relative Memory Profiling. In *Proceedings of the International Symposium on Code Generation and Optimization: Feedback-directed and Runtime Optimization*, CGO'04, pages 315–, Washington, DC, USA, 2004. IEEE Computer Society.
47. M. Zagha, B. Larson, S. Turner, and M. Itzkowitz. Performance Analysis Using the MIPS R10000 Performance Counters. In *Supercomputing'96: Proceedings of the 1996 ACM/IEEE Conference on Supercomputing*, pages 16–16, Jan 1996.

Chapter 11
Using Analog Side Channels for Hardware/Software Attestation

11.1 Introduction

Intuitively, an embedded device (or any other hardware/software system) is trusted when there is belief that it will do what it is expected to, i.e., that the device's hardware is what we think it is and that the software on that device is what we think it is. Establishing this trust for a device that is initially untrusted is a very important problem, and practical solutions for this rely on *attestation*. Attestation is a procedure that involves the initially untrusted device (prover) and an already trusted device (verifier), and the goal of the attestation procedure is for the verifier to attain confidence in the integrity of program code, execution environment, data values, hardware properties, etc. in the prover system. Attestation typically relies on a *challenge-response* paradigm [22], where the verifier asks the prover to perform specific calculations, which typically involve measurement (e.g., checksum) of the prover's code, data, etc., which the verifier checks against expected values for a "clean" (trustworthy) system. The verifier considers the prover's integrity to not be compromised if *(i)* the checksum provided by the prover matched with the *expected* value computed by the verifier, AND *(ii)* the verifier believes that the computation that produced the response itself has not been tampered with, e.g., by producing expected values without computing them from the prover's actual code/data.

In high-end modern processors, the assurance that the response computation itself was not tampered with is typically provided by using a hardware-supported Trusted Execution Environment (TEE) [2, 10, 11, 15], which uses dedicated hardware (e.g., SGX, TPM, etc.) within the prover. In low-end processors and/or embedded systems, however, there are form factor, battery life, and other constraints that prevent the use of hardware-supported TEEs or other hardware supports. Instead, a popular approach, *Software Attestation* [9, 12, 17, 22–24], asks the prover to compute the checksum using ordinary software execution, while the verifier attains some confidence about the absence of tampering through measurement of the request-to-response time (i.e., using the execution-time side channel).

In such an attestation procedure, the verifier sends a request (challenge) to the prover to compute a checksum of the prover's program memory. When the verifier receives the prover's response, it compares the received checksum to the expected value, which is computed over the verifier's "clean" copy of what should be in the prover's program memory. To prevent a compromised system from simply sending the correct response value back without actually computing it, the verifier's request (challenge) includes parameters with which the checksum computation must be computed, so the prover's checksum computation can only begin after the challenge has been received. Additionally, the verifier measures the time between when it has sent the request and when it has received the response. This measure time is compared to the expected request-to-response time, in an effort to detect various scenarios in which the prover's checksum computation is tampered with. The two main scenarios that allow a compromised prover (i.e., one whose program memory contains malicious changes) to send back a correct response involve (1) keeping (in data memory) a "clean" copy of maliciously modified program-memory regions, and (2) forwarding the challenge to another (more powerful) system that is controlled by the attacker, computing the correct response on that system and sending it back to the prover, which can then forward it to the verifier. The response-time verification relies on the assumptions that:

1. Checksum computation over the actual program memory would be faster than computation that must check which memory to use for each part of the checksum computation. A key consideration here is that the authentic code that implements the checksum is optimal—if the attacker can create a faster version of that code, the time savings can be used to implement the region-checking and thus generate the response in time.
2. The added latencies of forwarding the challenge to another system and of forwarding the response back to the prover increase the overall challenge-response time more than any time savings from the increased computation speed on that other system. This assumption is often difficult to maintain in practice, because the allowed challenge-response time must provide for worst-case network latency from the verifier to the prover and then back to the verifier, and the difference between a typical latency and the worst-case latency may be sufficient to accommodate forwarding of the challenge to and response from another system.

Unfortunately, the overall request-to-response time provides only one coarse-grained measurement, and this method is not able to monitor the prover *during* the attestation without imposing a significant performance and cost overheads to the system. This, in turn, makes the software attestation schemes vulnerable to attacks which have very low-latency compared to the overall response time, i.e., where the additional time t_{add} to forge the checksum on a compromised system is short in comparison to the expected response time ($t_{add} \ll t_{good}$). Moreover, due to the network limitations and/or micro-architectural events, this request-to-response time may be noisy, as it includes the round-trip network latency and/or variations caused by the micro-architectural events (e.g., cache miss). To tolerate the variance and reduce the false-positive rate, the verifier must increase the time threshold (t_{th}) that the observed time (t_{resp}) is compared to, and this potentially makes these schemes even more vulnerable to low-latency attacks.

In the rest of this chapter, we describe EMMA [21], a new approach for attestation of embedded devices. EMMA leverages the side-effects of the prover's hardware behavior, but without requiring any specific support from the prover's hardware. This scheme is based on the key insight that *the existing approach of execution-time measurement for attestation is only one example of using externally measurable side-channel information*, and that other side-channels, some of which can provide much finer-grain information about the verifier's computation of the response, can be used. In particular, instead of measuring the overall challenge-to-response time as seen by the verifier, EMMA uses *electromagnetic (EM)* side-channel signals that are unintentionally emanated by the system during attestation. We first study the possible attack scenarios on attestation methods, and then design EMMA such that fully addresses these vulnerabilities. Also, to increase accuracy (and reduce false-positives), we first investigate the different sources of variation (e.g., micro-architectural events), and then carefully design EMMA such that it effectively minimizes these variations while tolerating uncontrollable sources of variations (e.g., environmental). Using an extensive set of measurements, we show EMMA's ability to achieve high accuracy under different attack scenarios while being robust against different sources of environmental variations.

11.2 Software Attestation

The main goal of attestation is to establish a *dynamic root of trust* (DRT) [5, 7, 17] on an untrusted platform. After successful attestation, the code and data within this DRT is assumed to be unmodified, and this can be leveraged to measure the integrity of other parts of the untrusted system, e.g., asking the code that is within the dynamic root of trust to compute a checksum of some other piece of code and/or initiate execution of other code in the system without concerns about tampering with their initial execution environment.

In software attestation, the DRT is instantiated through a *verification function*, a self-checking function that computes a checksum over its own instructions and sends it to the verifier. To establish trust, this checksum has to match with the expected value, and other measurable properties of the checksum calculation itself must pass certain tests. The function typically consists of *(i)* an initialization phase, *(ii)* a main computation loop, and *(iii)* an epilogue.

In existing frameworks, the measured property is the request-to-response time, which is assumed to correspond to the execution time of the checksum computation, and the test consists of checking if the response-time was fast enough. In this work, the measured property is the prover's EM emanations during the checksum computation, and the test consists of verifying that signal against a model of emanations for a legitimate checksum computation.

To attack an attestation framework, the attacker has two options. The first option is to forge the checksum value using classic checksum collision attacks [27, 28]. This attack, however, can be easily defeated by using a sufficiently long checksum [24]. The other, more realistic option, is to modify the checksum code to compute the correct checksum without violating the requirements (e.g., $t_{resp} = t_{good} + t_{add} < t_{th}$). To launch such an attack the adversary has two options. One option is to modify the checksum calculation's main loop to calculate the checksum on another region of the memory, where the unmodified copy of the code is kept. This attack is called a *memory copy* [22, 24, 25] attack. The other option is to modify the prologue/epilogue phases by *(a)* forwarding the challenge to another device; this is called a *proxy* attack [17], or *(b)* by removing the malicious code before calculating the checksum and hiding it in other parts of the memory; this is called a *rootkit* attack [8].

The main challenge for the adversary is that, while changing the checksum calculation requires a small change in the program code, any change (even a single instruction) in the checksum computation phase will be significantly magnified due to a large number of iterations of the loop. Thus, the adversary will be faced with a fundamental choice: a brief period of malicious activity (but with a large change to the program code) in the epilogue/prologue, or spreading malicious activity over a longer time with a small change to the program (in the checksum phase). Hence, an ideal detection framework should be able to detect single-instruction modifications in the checksum loop, and somewhat larger changes in the epilogue/prologue phases.

Given these challenges, to build an effective framework for utilizing EM signals for attestation, two requirements should be met: *(i)* to protect the system against short-term attacks, i.e., attacks on prologue/epilogue phases (e.g., *proxy attacks* [17]), the detection framework should be able to analyze the signal with fine time-resolution (i.e., a time-domain analysis), and *(ii)* to detect small changes during the main checksum computation's loop, the detection framework should be able to detect small changes in the per-iteration time of the loop (i.e., a frequency-domain analysis). Unfortunately, existing side-channel analysis frameworks [6, 13, 18] are unable to satisfy both conditions.

In the following sections we present an EM-monitoring algorithm that can *(a)* analyze the signal in time-domain to detect even small changes before/after the main checksum computation loop begins/ends, and *(b)* check whether the attestation process (during the checksum computation) matches with a known-good model (to ensure that this process is not modified by an adversary) by using a frequency-domain analysis.

11.3 EMMA: EM-Monitoring Attestation

EMMA is designed to determine, with high accuracy, the exact *begin* and *end* time of the checksum calculation, and the *per-iteration* execution time of the main

11.3 EMMA: EM-Monitoring Attestation

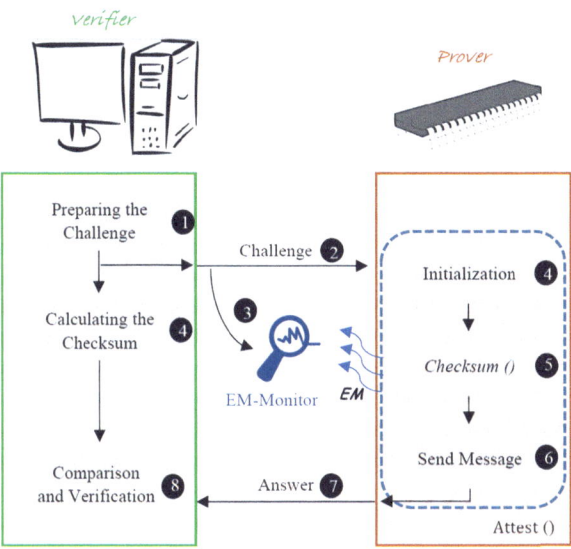

Fig. 11.1 Overview of EMMA framework [21]

checksum loop. An overview of the EMMA framework is shown in Fig. 11.1. This framework consists of a Verifier, $\mathcal{V}$, (e.g., a trusted PC) and a Prover, $\mathcal{P}$, (e.g., an embedded system). The Verifier in EMMA includes, or is connected to, a monitoring system (EM-MON) that can receive and analyze the EM signals unintentionally emanated by $\mathcal{P}$. Attestation begins with $\mathcal{V}$ preparing a challenge locally (❶). The challenge includes a seed value (which will be used later to initialize a Pseudo-Random Number Generator (PRNG) in $\mathcal{P}$), an address range, the total number of iterations for the checksum loop, a random value to initialize the checksum in $\mathcal{P}$, and a random *nonce*. The challenge is then sent to the $\mathcal{P}$ via a communication link (❷) which invokes the *verification function*, attest() by causing an interrupt on $\mathcal{P}$. This function runs at the prover's highest processor privilege level, with interrupts turned off.

Upon sending the challenge to $\mathcal{P}$, $\mathcal{V}$ also starts the "monitoring process" (❸) on EM-MON. Through analysis of EM signals emanated from $\mathcal{P}$, EM-MON determines three critical values about $\mathcal{P}$'s checksum computation, and reports an error if any of them deviates from a known-good model (*reference* model). These tasks/values include *(i)* the delay between reception of the challenge and start of the checksum loop, and the signal signature during this initialization phase, *(ii)* the checksum loop's EM signature (i.e., frequency of spikes), and *(iii)* the total attestation time on $\mathcal{V}$.

These values/tasks are primarily chosen because, fundamentally, there are two critical durations in the execution of the attestation process on the prover: *(a)* the time that is taken from when the command for performing attestation is received until the start of the actual checksum process (because the malware could contact another device during this period, or quickly hide the malicious code before starting the procedure), and *(b)* the time taken for the checksum process itself (because the malware could try to do "extra work" during checksumming to hide the presence of malicious code).

After sending the challenge, the verifier, independently, calculates the "expected" checksum on its own (known-good) copy of the embedded system's program memory (❹). At the same time, the self-checking verification function starts with initializing its local variables based on the received challenge (❹), and then it starts the "checksum calculation", Checksum() (❺). This function is an optimized loop that, in each iteration reads a memory location in a pseudo-random fashion, and updates the checksum based on the content of that memory location. The address range and the total number of iterations of the loop are determined by the challenge.

Once Checksum() is finished, $\mathcal{P}$ forms a response (❻) that includes the calculated checksum and the random nonce which was initially sent by $\mathcal{V}$, and sends this response to $\mathcal{V}$ (❼). The original nonce acts as an identification and helps to increase the overhead for a proxy attack.

Finally, $\mathcal{V}$ compares the information received from $\mathcal{P}$ with the expected value (computed by $\mathcal{V}$) of the checksum, and also compares the results from EM-MON to the expected ones. If they all match, one trial of the attestation protocol is completed successfully (❽). At this point the *dynamic root of trust* has been established, and the verification function can be used to compute the hash value over the prover's other memory contents (entirely or partially). This hash value can then be sent back to the verifier, which in turn, provides the current state of the prover to the verifier. Note that all of these functionalities are considered to be part of the verification function; thus, their code has been used in computing the verification function's checksum, which means that the hash function or the invocation of the executable were also not tampered with.

Verification Function (attest())
This function has three parts: a *prologue* or initialization phase, *checksum computation*, and an *epilogue* which is responsible for sending the response back to the verifier and invoking the executable or hash computation function.

To defeat possible attacks against attestation, this function should be carefully designed to have predictable execution time. Thus, from the computer-architecture perspective, attest() has to be designed such that *(a)* the prologue and epilogue phases are fairly deterministic and sufficiently long such that a simple time-domain analysis can be used to find a potential deviation (due to malicious activities caused by the attacker), and *(b)* the computation loop should have minimum per-iteration variation, and it should not be parallelizable.

In addition to considering these runtime requirements, the checksum computation function should be secured against "static" attack scenarios, namely *pre-computation* and *replay* attacks [24, 25], where the attacker leverages previous challenge-response pairs to calculate the new response. To avoid these attacks, the checksum computation should be a function of the challenge (to prevent replay

11.3 EMMA: EM-Monitoring Attestation

attacks), and the memory address generation in each iteration should be done in a pseudo-random fashion (to prevent pre-computation attacks).

Using these two criteria (i.e., runtime and security requirements), we design our attestation algorithm (mainly based on prior work [17, 22–25]) to satisfy these conditions. Algorithm in Fig. 11.2 shows the checksum's pseudo-code. We will first describe the algorithm and then provide a brief formal security analysis of the code.

Checksum Algorithm

The main checksum loop consists of a series of alternating XOR and ADD instructions. This series has the property of being *strongly-ordered* [25]. A strongly-ordered function is a function whose output is very likely to change if the operations are evaluated in a different order. Using a strongly-ordered function forces the adversary to perform the same operations on the same data in the same sequence as the original function to obtain the correct result. Furthermore, using this sequence prevents parallelization and out-of-order execution since, at any step, the current value is needed to compute the subsequent values.

We use a 160-bit checksum to keep all the registers utilized, and to significantly reduce the checksum collision probability [24]. The checksum is stored as a vector in a set of 8/16-bit general purpose registers (blocks), depending on the architecture of the processor (i.e., AVR, ARM, etc.). To traverse the memory in a pseudo-random fashion, we use a pseudo-random number generator (PRNG). Like previous work, we use a 16-bit T-function [14] to generate these random numbers. Each partial checksum block is also dependent on *(a)* the last two calculated partial sums;

Algorithm 1 The checksum computation algorithm used in EMMA.

1: Initialization:
2: $RNum = seed$
3: Set $MASK$ based on $beginAddress$ and $endAddress$
4: $Offset = beginAddr$
5: $cSum = seed$
6: Checksum: // *checksum main loop* (Checksum())
7: **for** i=1 to *totIter* **do**
8:   **for** j=1 to 10 **do**
9:     $RNum = RNum + (RNum^2 \vee 5) \bmod 2^{16}$
10:    $memAddr = memAddr \oplus RNum$
11:    $memAddr = (memAddr \wedge MASK) + Offset$
12:    $cSum_j = cSum_j + (Mem[memAddr] \oplus cSum_{j-1})$
13:    $cSum_j = cSum_j + (i \oplus PC)$
14:    $cSum_j = cSum_j + (RNum \oplus memAddr)$
15:    $cSum_j = cSum_j + (SR \oplus cSum_{j-2})$

Fig. 11.2 The checksum computation algorithm used in EMMA [21]

this prevents parallelization and pre-computation attack, *(b)* a key (to avoid replay attacks), *(c)* current memory address (data pointer) and PC (if available depending on the architecture); this avoids memory copy attack, *(d)* the content of the program memory; this prevents changing the attestation code, and *(e)* the Status Register (SR), to prevent changes of the interrupt-disable flag.

To avoid variations due to possible branch mispredictions, in the actual implementation of the checksum, the inner 10-iteration loop is fully unrolled, resulting in straight-line code to calculate all ten partial sums. To avoid variations due to cache misses, the $MASK$ is generated such that the data access address range fit, into an L1 cache (if any). Note that $MASK$ is a function of the received challenge (see line 3), so it is the verifier's responsibility to generate a challenge that sets, the $MASK$ properly. Also, to cover the full address range, the verifier can send multiple challenges with different address ranges. Further, the attestation code itself is compactly designed so that it fits into an instruction cache.

Security Analysis

Based on the framework proposed by Armknecht et al. [3], in general, to analyze the security of any software attestation framework, two core components should be analyzed: memory address generator (Gen) and the checksum computation/compression function (ChK).

To avoid attacks that rely on partially completing attestation computation ahead of time, addresses a_i generated by Gen should be "sufficiently random" [3], i.e., a_i should be *computationally indistinguishable* from uniformly random values within a certain time-bound, t_{min}, (assuming that the adversary, $\widetilde{\mathcal{P}}$, does not know the seed in advance). In practice, $\widetilde{\mathcal{P}}$ can use an arbitrary seed value to compute all the possible addresses on its own, making them easily distinguishable from random values. However, it can be shown that to maintain the security, it is only required that $\widetilde{\mathcal{P}}$ cannot derive any meaningful information about a_{i+1} from a_i and the seed without investing a certain minimum amount of time, $t_{compute} \geq t_{Gen}$ [3]. Specifically, we assume that an algorithm with input s that does not execute Gen cannot distinguish $a_{i+1} = Gen(a_i)$ from uniformly random values. This property holds true for T-functions [14], since either the adversary needs to spend the same amount of time as Gen to compute the next address or, alternatively, pre-record all possible (addr, nextAddr) pairs. For a 16-bit T-function, saving all the pairs requires more than 128KB memory, which means to access this data in run-time, the attacker needs to access either the L2 cache or the main memory, thus $t_{compute} = t_{mem}$. In our design (line 9 in Algorithm 10.1) t_{Gen} is only a few cycles (<5) which is clearly much less than $t_{mem} > 20$ in typical low-end processors.

The purpose of the checksum function, ChK, is to map the contents of the prover' memory, to a smaller attestation response, r, which reduces the amount of data to be sent from $\mathcal{P}$ to the verifier $\mathcal{V}$. A mandatory security requirement on Chk is that it should be *hard* for $\widetilde{\mathcal{P}}$ to replace the correct input, S, to Chk with some other value, $S' \neq S$, that yields the same attestation response r (i.e., *second pre-image resistance* of cryptographic hash functions). However, unlike the hash functions where the adversary may know the correct output (and searches for the

11.3 EMMA: EM-Monitoring Attestation

second output), in the software attestation schemes the adversary does not even know the correct (first) response. The reason is that, as soon as $\widetilde{\mathcal{P}}$ knows the correct response, he could send it to $\mathcal{V}$, and would not bother to determine a second one. This leads to a much *weaker* second pre-image resistance (called *blind*) requirement for attestation.

Using this fact, our checksum is designed such that it significantly reduces the chance of collision, while being computationally hard for a second pre-image attack. It can be proven, as shown in [16], that ChK we use provides almost full coverage (i.e., almost all possible numbers in the $[0, 2^{16} - 1]$ range for a 16-bit partial checksum), which, in turn, makes ChK resistant to (blind) pre-image attacks. In fact, using the framework in [3], the probability of a checksum collision in our framework (for 160-bit checksum and $totIter = 100$) is $< 10^{-40}$.

EM-Monitoring

The EM-MON component ensures that the attestation computation in $\mathcal{V}$ is not tampered with. Figure 11.3 shows this monitoring framework. Using an antenna (e.g., a magnetic probe) and a signal acquisition device (e.g., a software-defined-radio), the EM signal is captured and received as a time-series (❶). Depending on the required analysis, either *time-domain* or *frequency-domain* analysis is selected (❷). For analyzing the prologue and epilogue, time-domain is used; for the checksum loop, frequency-domain analysis is used (This decision is made depending on the current state of the FSM, which we will discuss below).

For frequency-domain analysis, the signal is transformed into a sequence of Short *Frequency-Domain* Samples (SFDS) using a Short-Time Fourier Transform (STFT). This transformation divides the EM signal into consecutive, equal-length, and overlapping segments of size t and then compute the STFT for each of these segments to obtain the corresponding SFDS (❷ and ❸-bottom). Segment size, t, has to be chosen such that it provides a balance between the time resolution and EM-MON's computational needs. Also, t should be long enough to capture several iterations of the checksum loop to model the *average* behavior of the loop. Each block in the main checksum loop on Arduino Uno takes about 20 clock cycles, and calculating the entire 160-bit checksum takes ≈ 400 cycles (about 25 µs). Therefore,

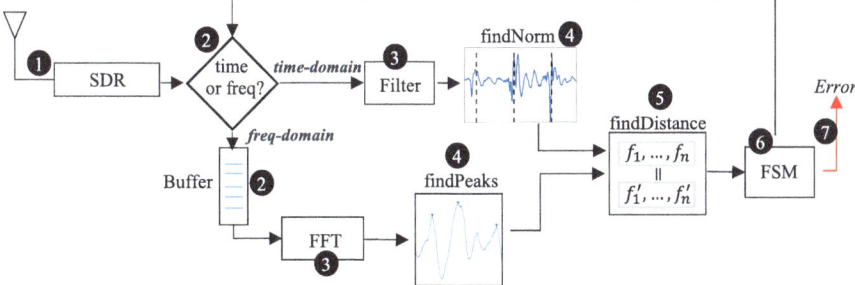

Fig. 11.3 Overview of EM monitoring framework [21]

in this implementation of EM-MON, we use 1 ms segment size with 80% overlap (i.e., each segment corresponds to 40 iterations), and consecutive segments differ only in 8 iterations.

While these numbers provide adequate time-resolution for the checksum computation loop, they would not detect small changes before/after the main loop. Thus, time-domain analysis is used in these regions. The signal is first low-pass filtered and normalized using mean subtraction and scale normalization (❸-top), and then segmented into consecutive, equal-length, and overlapping segments of size t (❹-top). We used $t = 24$ samples in our setup. Consecutive segments differ only in one sample (we used an SDR with 2.4 MHz sampling-rate), to maximize the time-resolution (the time-resolution in this setup is about 6 cycles vs. 3200 cycles for the frequency-domain analysis).

For frequency-domain analysis, each SFDS then goes into the findPeaks() module (❹-bottom) where n spikes are selected and later used as "signatures" for each SFDS. In findPeaks(), the first step in finding a spike is that for each frequency, f, that is of interest in an SFDS, we first compute the corresponding normalized frequency as $f_{norm} = (f - f_{clk})/f_{clk}$, where f_{clk} is the clock frequency for that SFDS. This normalized frequency is expressed as an offset from the clock frequency so that a shift in clock frequency does not change f_{norm} with it and is normalized to the clock, so it accounts for the clock frequency's first-order effect on execution time. We call this technique *"clock-adjustment"*. The criteria for selecting spikes are choosing the n largest amplitude local maxima (peaks), excluding the spike for the clock, that are not part of the *noise*. To find the noise, we record the spectrum once without executing the attestation function and save all the spikes that are 3-dB above the noise floor as *noise*. For our evaluations, we select $n = 7$, to capture the checksum loop's fundamental frequency and its second and third harmonics (in both sidebands). The frequency of the clock is also reported, to prevent attack scenarios where the clock speed is increased to hide malicious activities by speeding up checksum computation.

To check the correctness of the execution, the time-domain segments generated by findNorm() or the frequency peaks (a vector of size n) generated by findPeaks() should be compared to a known "reference model" that is achieved during the secure execution of attest() (❺). Note that the reference model needs to be created only once. For time-domain signals, the reference model is a *dictionary* of segments, each with size $t = 24$ (note that the entire initialization phase takes about 500 μs, i.e., the dictionary size is about 5000 elements). For the frequency-domain, the model is a vector of size n which stores the frequency bins of the computation loop's spikes. We assume that either the manufacturer or the end-user is able to create a correct reference model using a known-good instance of a prover device.

We use Pearson's correlation for the frequency-domain, and cross-correlation for time-domain, as the distance metric. In the time-domain, the entire signal can be reconstructed by concatenating the best-match segments in the dictionary.

11.3 EMMA: EM-Monitoring Attestation

The reconstructed signal is then compared to the original signal (i.e., the signal obtained during the actual execution of the initial phase) using mean-square-error (MSE) method. Finally, to decide whether the signal "matches" with the correct execution, a simple moving average (SMA) filter is used to remove short-term runtime noises (i.e., when only a few samples deviate from the model due to environmental/measurement noise). If the SMA of the error always stays below a a threshold (we used $th_t = 0.1$), the signals are *matched*. In the frequency-domain analysis, the signal is matched if the correlation coefficient is larger than $th = 0.8$. Based on these distance comparisons, findDistance() outputs two boolean values (isMatched_t, isMatched_f) showing whether the signal matches with either the initialization or the computation phase or none of them.

The final stage of EM-MON is a Finite-State Machine (FSM ❻). The default state for the FSM is when EM-MON is waiting for the attestation to start ($state = 0$). Upon receiving a challenge from $\mathcal{V}$, FSM switches to $state = 1$, starts a timer called challengeTimer, and starts time-domain analysis. During this phase, FSM throws an error if findDistance() reports "not-matched" (for time-domain analysis). FSM switches to $state = 2$ once findDistance() reports "match" for the frequency-domain analysis (i.e., this is when the computation loop begins). Also, the FSM throws an error if challengeTimer is larger than a threshold. Checking this value ensures that the system can be protected against *proxy*, *code compression*, and *return-oriented rootkit* attacks, where the attacker needs to spend some (non-negligible) time to set up the attack before actually starting the Checksum().

Note that this is an important and unique feature of EMMA—existing methods are all unable to measure this delay (*even when they are directly connected to the verifier by a cable*), and can only measure the time between sending the challenge and receiving the checksum value.

The output of findDistance() becomes zero (for the frequency-domain analysis) when the checksum loop completes, so the FSM switches to $state = 3$, and checks the challengeTimer once again. This check ensures that the total execution time of attestation does not exceed a threshold which is defined by $initTime + perIteration \times totIter$, where *perIteration* is the checksum loop per-iteration time, *totIter* is the total number of iterations for calculating the checksum, and $initTime$ is a constant.

Finnaly, in $state = 3$, EM-MON starts a timer called checksumTimer and waits for an acknowledge from $\mathcal{V}$ that the checksum is received. At this point, if checksumTimer is too large, FSM again throws an error. Otherwise, it switches back to $state = 0$ and waits for the new attestation challenge. This check ensures that the adversary can not spend any extra time after the checksum calculation is finished and before actually sending the checksum to $\mathcal{V}$. Note that, in all cases, FSM can only transit from state n to $n + 1$ to enforce the correct ordering in attestation.

11.4 Experimental Evaluation

Measurement Setup

To evaluate the effectiveness of our method, we used a popular embedded system, Arduino Uno, with an ATMEGA328p microprocessor clocked at 16MHz. To receive EM signals, the tip of a small magnetic probe [1] was placed about 10 cm above the Arduino's microprocessor (with no amplifier). To record the signal, we used an inexpensive compact software-defined radio (RTLSDRv3 [20], about $30). We recorded the signals at 16 MHz with a 2.4 MHz sampling rate. Note that all of our measurements were collected in the presence of the other sources of EM interference, including an active LCD that was intentionally placed about 15 cm behind the board. A set of TCL scripts were used to control the attestation process. The real-time EM-Monitoring algorithm was implemented in MATLAB2017b.

Implementation

Arduino Uno uses an ATMEGA328p microprocessor, an Atmel 8-bit AVR RISC-based architecture, with a 16KB Program memory and a separate Data memory (unlike most other architectures where a single memory is used both for data and for executable instructions). This micro-controller has 32 8-bit general purpose registers, where the last 6 registers can be combined in groups of two, and form three 16-bit registers (namely X, Y, and Z). The Z register can be used to access/read the program memory using the *LPMZ* assembly instruction. Note that, unlike most of the micro-controller architectures, AVR does not provide direct access to the Program Counter (PC) register, so the value of the PC cannot be used during checksum calculation.

The 160-bit checksum is kept in twenty 8-bit registers ($r0 - r19$). The Z register ($r31 : r30$) is used for reading the program memory ($memAddr$), and the Y register is used to store the random number generated by PRNG. Inputs from the challenge are pushed to the stack prior to invoking `attest()`, and are then read in the *initialization* phase. Registers $r25 : r24$ are used to save the $MASK$ value, $r23 : 21$ are used to save the *nonce*, and $r20$ is used to store the content of the memory. Finally, the X register contains the current index (i). Each partial checksum calculation ($cSUM$) takes 20 cycles on ATMEGA328p. If an extra one-cycle instruction (e.g., ADD) is added to the partial checksum code, the per-iteration time (and the corresponding spike in the frequency domain) changes by about 5%. The initialization phase takes about 500 μs to receive the challenge via a Serial communication port.

Attacks

We evaluate the security of EMMA by implementing different attacks on a software-based attestation framework. These attacks can be divided into *(i)* attacks on the main checksum loop (shown as L-1 and L-2), and *(ii)* attacks on the epilogue/prologue phase (EP-1 and EP-2).

11.4 Experimental Evaluation

L-1: Memory-Copy Attack The most straightforward attack against software-based attestation methods is the *memory-copy attack*, where the adversary has created a copy of the original code *elsewhere* in memory, and the checksum code is modified to use that range of addresses instead of the original ones. Since the challenge sent by $\mathcal{V}$ could request to read any memory line in the program memory address space, potentially including the supposedly "empty" memory space where the "clean" copy of the original code is kept, to avoid detection this modified code must check addresses that are used during checksum computation, and then perform accesses without modification for unmodified memory ranges, redirect them to "clean" copies for modified memory ranges, or use override values for supposedly empty ranges that now actually contain the attacker's data (including "clean" copes of original values from program and data memory).

This checking and redirection of memory requests introduces overheads during checksum computation. Specifically, the adversary needs to change the *memAddr* register (register Z in our implementation) to point to another address in the memory (at least one added instruction). Since *memAddr* value is used in the checksum calculation, that value then has to be restored (another instruction). Note that, in our implementation, since accessing the program memory is only possible through Z, the adversary's only option is changing Z. Even for program/data location that are unchanged by the adversary, the checksum code must suffer overheads of checking (a compare and a branch instruction) that the address falls in a range that still contains original instructions/data. Overall, to implement this attack, the adversary has to add at least two instructions per check-summed location.

One countermeasure to defeat this attack is to fill the unused memory regions with random values that are known to the verifier [29]. Castelluccia et al. [8], however, showed that this defense can be circumvented using a simple *compression* attack, where the unused parts of memory are compressed and stored in non-executable regions. Hence, to provide a stronger security guarantee, in this implementation of EMMA we relax the assumption that "free" space is filled with random values and allow all "free" memory locations to be filled with the same value (e.g., 0xFF). This allows the attacker to store malicious code in an empty region of program memory, and to modify checksum computation so that *LPM Rd, Z* (i.e., load from program memory) for that region of the memory is replaced with *SER Rd* (set *Rd* to 0xFF). In our experiments, *LPM* instruction uses 2 more cycles than *SER* does, thus removing 2 cycles from the 4-cycle performance penalty introduced by the *compare-and-branch* check, that is still needed to determine which region of program memory is being addressed. The reduced additional latency makes the modification of the checksum computation more difficult to detect.

To evaluate our framework, we implemented the Memory-Copy attack, and we trained EMMA on (only one) attack-free instance of attestation. We then applied EMMA to both attack-afflicted and attack-free instances of the attestation. The spectra of the resulting signals (Fig. 11.4) show the spikes that correspond to the original checksum computation loop, and also the spikes that correspond to the modified checksum computation (red), which are shifted closer to the processor clock's fre-

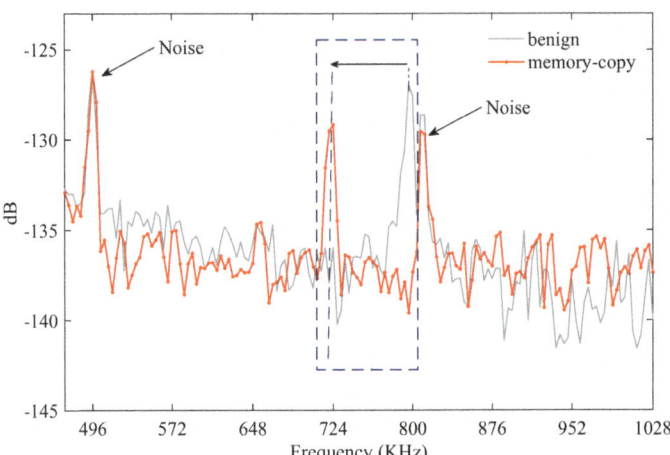

Fig. 11.4 EM spectrum during checksum computation for the original code (gray) and Memory-Copy attack code (red). The x-axis is the frequency offset relative to the processor's clock frequency [21]

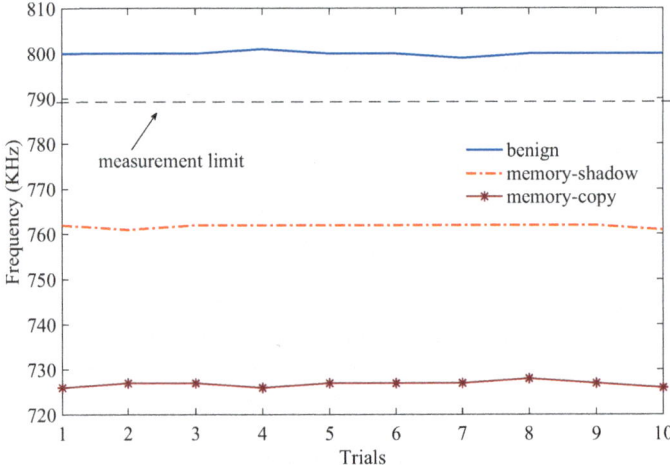

Fig. 11.5 Frequency of the checksum computation loop for (**a**) attack-free, (**b**) memory-shadow attack, and (**c**) memory copy attack code [21]

quency because the per-iteration time of the loop has increased. Figure 11.5 shows the spike's frequency for 20 attestation instances, 10 attack-afflicted, and 10 attack-free, showing a consistent difference among them. We find that EMMA successfully labels all these instances, i.e., all attack-afflicted instances are labeled as attack-afflicted (successful detection), and all attack-free are labeled as attack-free (no false positives). The "measurement-limit" line refers to the threshold that was used in EMMA for the labeling decision.

11.4 Experimental Evaluation

L-2: Memory-Shadow Attack Castelluccia et al. [8] showed that instead of adding a *branch and compare* instructions to compare the address, the adversary can reduce the number of added cycles by one (i.e., 2 instead of 3) by copying the malicious code into a specific "free" area of the program memory (assuming it is not filled with random values). This is called a Memory-Shadow Attack.

This attack leverages the fact that most of the program memory on an embedded system usually remains unused. In this case, if the second half of the program memory (addresses with $MSB = 1$) is empty, all accesses to the last quarter of the memory (addresses $11x \ldots x$) could be redirected to the third quarter (addresses $10xx \ldots x$) by flipping just one bit in the address. The last quarter of the program memory would then be shadowed and could be used to hide the malicious code. To implement this attack, the adversary simply needs to check the highest bit, and if it is one, sets the second-highest bit to zero. This can be implemented in an AVR ISA with two instructions "*sbrs r*31, 7" and "*cbr r*31, 6" which takes two cycles; thus, the overall number of added cycles will be three, which is only one cycle more than the actual checksum loop (recall that the adversary does not need to read from the memory, so that "saves" two cycles).

To further show that adding even one cycle to the main loop would shift the frequency of the loop, Fig. 11.5 shows the frequency shift for adding a simple one-cycle ALU instruction (a *CBR*, clearing a single bit in a register) to the main loop. This confirms that the adversary *cannot add any instructions* to the main loop without such a change being detected by EMMA, a very strong security guarantee! Furthermore, just like we did for the memory-copy attack, we used 10 trials for this attack and showed that its frequency is well below the actual frequency of the main loop (see Fig. 11.5). Just like it did for the memory-copy attack, for the memory-shadow attack EMMA also achieves a 0% false positive rate (no false alarms) and a 100% true-positive rate (all attacks detected).

To further compare EMMA with prior work, Fig. 11.6 shows the detection probability for Memory-Shadow attack, as a function of the number of checksum loop iterations. Intuitively, having more iterations magnifies the overhead of adding extra instructions (cycles) to the loop thus it gets easier to detect. However, to limit runtime variations, the number of iterations is limited by the size of L1 and/or instruction cache. Also, more iterations requires longer computation time which, in turn, increases the overhead (power, device availability, etc.). Thus, ideally the detection framework should be able to detect attacks with a small number of iterations to minimize these overheads and the increase accuracy. As shown in Fig. 11.6, EMMA detects attacks that involve as few as 100 iterations, which is 20x smaller than what can be detected by EDDIE [18], and more than two orders of magnitude smaller than the fastest timing-based approach. The main reason for this dramatic improvement in sensitivity, compared to EDDIE, is that EMMA leverages Pearson correlation as a distance metric instead of using non-parametric tests used in EDDIE [18]. Compared to the timing-based methods, EMMA can provide fine-grain per-iteration monitoring, which enables it to detect small changes. Note that an attacker, who knows that EMMA requires about 100 iterations to detect the

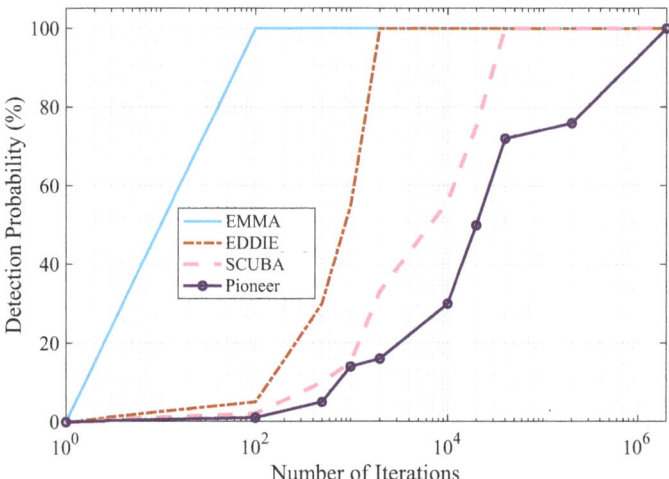

Fig. 11.6 The probability of detecting memory-shadow attack for EMMA and prior work [21]

change, can design a new attack to *selectively* run the memory-shadow attack in some iterations. While the attacker can modify the code in that way, achieving that functionality itself would add a *branch-and-compare* instruction (i.e., to check the iteration number), which, in turn, adds overhead that will be detected by EMMA.

EP-1: Rootkit-Based Attack Another class of attacks leaves the original checksum loop unchanged, but adds work before and/or after it, e.g., by hiding/removing the modified code before the checksum computation and restoring it after the (unmodified) checksum computation. These types of attacks are extremely difficult to detect for existing timing-based schemes because the time overhead introduced by these attacks are typically hundreds of microseconds, which is less than 1% of the overall execution time of the attestation procedure. Moreover, frequency-only detection methods (e.g., EDDIE [18], Zeus [13]), are also unable to efficiently detect these attacks if the change is smaller than the minimum time-resolution of the method (typically around 1 ms). However, for EMMA, these attacks are not difficult to detect, as they add many cycles of work between when the checksum computation is supposed to start and when it actually starts.

As an example of such attacks, we implement the Rootkit-Based Attack [8], which leverages Return-Oriented Programming (ROP) [26]. In this attack, a hook (jump) replaces the first instruction in the attestation. The initiation of attestation results in a jump to the malware's hiding functionality, which deletes the attacker's code (including itself) from program memory, but leaves a series of ROP gadgets so that, after the (unmodified) attestation code sends its response, the malware is re-installed on the device.

The *deleting* procedure is the most time-consuming part of the attack, where the adversary needs to copy the malware-hiding functionality and the modified

11.4 Experimental Evaluation

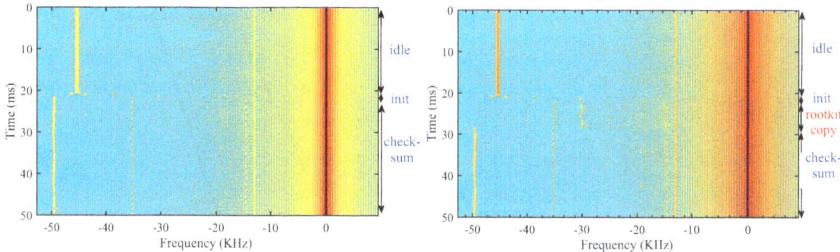

Fig. 11.7 Spectrogram of the attestation code in normal (left) and rootkit-attack (right) runs. Note that the slight differences in colors between the two spectrograms correspond to variations in signal magnitude which are caused by different positioning of the antenna. Such variation is common in practice, but EMMA was designed to be robust to such variation [21]

checksum loop to the data-memory, and replace them with the original code. Figure 11.7 shows the spectrogram of the attestation procedure with and without the rootkit-based attack. As can be seen in the figure, for the normal behavior of the attestation code, initialization takes about 500 μs which includes receiving the challenge and invoking attest(). Note that, based on the initialization time, we set the *threshold* for challengeTimer to 2 ms or 8 samples (i.e., the maximum delay between sending the challenge and starting the checksum main loop is smaller than 2 ms). As illustrated in Fig. 11.7, in the presence of rootkit attack though, there is an extra phase between the initialization and the start of the main checksum loop that takes about 8 ms, which is far larger than the timer's threshold and thus triggers an error caused by checking the challengeTimer. Moreover, we found that, even without using the timer, the time-domain analysis can successfully detect the attack—the time-domain signatures for the rootkit were very different than that of in receiving the challenge.

We evaluate EMMA against this attack, in 10 runs, in which EMMA has successfully detected all instances of the attack.

EP-2: Proxy Attack In a proxy attack, instead of calculating the checksum on its own system, the prover ask another, faster, device (the proxy) to compute the correct checksum within the time limit, which enables malware on the device to go undetected. In this case, similarly to a rootkit-based attack, after receiving the challenge, some this time is used for sending (forwarding) the challenge and any other necessary information (e.g., nonce) required to correctly compute the checksum in the proxy.

The major limitation in the existing methods for detecting proxy attacks is that the adversary can simply hide this attack if $t_{send} \ll t_{threshold}$. EMMA, however, is not limited to only considering the overall attestation time; it can distinguish the initialization phase from checksum calculation very accurately.

Figure 11.8 shows the probability of detecting this attack (as a function of t_{send}) and compares it with other existing methods. As can be seen in this figure, EMMA can

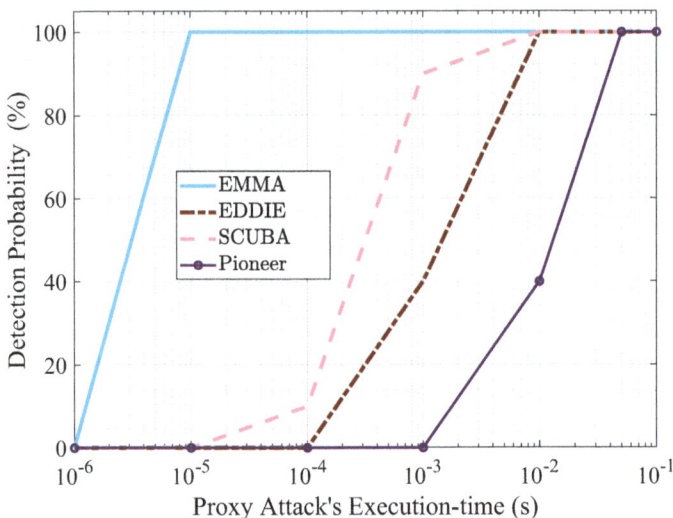

Fig. 11.8 The probability of detecting the proxy attack for EMMA and state-of-the-art [21]

detect attacks that are up to 2 orders of magnitude smaller than attacks in existing work. This is particularly important for IoT devices connected to the network, where the attacker can leverage proxy attacks very easily. Our evaluations show that, for low-end devices such as Arduino, the fastest time the attacker can achieve to forward the challenge to another (colluding) device is about 800 μs (using a WiFi module), which is well within the detectable range of EMMA. To evaluate EMMA against this attack, we used 10 trial runs for the proxy attack, and found that EMMA can successfully detect all instances of the attack with no false positives (for other methods, the accuracy is well below 80%).

11.5 Sensitivity Analysis

Applicability to Other Platforms
To show that EMMA is applicable to other systems, with different processors and/or architectures, and frequency ranges, we tested our checksum main loop, Checksum(), on three other embedded systems: a TI MSP430 Launchpad with a processor clocked at 16 MHz, an STM32 ARM Cortex-M Nucleo Board also clocked at 16 MHz, and an Altera's Nios-II soft-core implemented on a Startix FPGA board clocked at 60 MHz. The criteria for choosing these boards were to pick the embedded systems that are popular and widely used, and have different architectures than Arduino.

11.5 Sensitivity Analysis

Running the same attestation code on these boards, we then confirmed that by using the same setup as in previous section, the EM-MON in EMMA receives EM signals similar to that of for Arduino board (i.e., spikes at the frequency of the loop), and further, we confirmed that our detection algorithm can successfully detect when this loop starts and when it ends by adding the training information (i.e., the position and the number of spikes) for each board to our framework. Finally, to further show that even single added instruction to the loop's code is detectable by EMMA, we added a single-cycle "ADD" instruction to the checksum's assembly code for each of the boards, (i.e., to model a minimal change that malware may require).

We then used EMMA, to label malware-free and malware-afflicted (i.e., runs with the extra instruction) checksum runs (10 each) for all the three boards. Our results showed that, in all cases, EMMA successfully detected the malicious runs.

Applicability to More Complex Systems Next, we tested our framework on a more complex embedded system, A13-OLinuXino Single-Board-Computer [19]. This board has an ARM A8 in-order core clocked at 1GHz, with 2 level of caches, a branch predictor, and a prefetcher. It also runs a Debian Linux OS. We ran our checksum loop on this board and measured the beginning/end time and the loop's per-iteration time.

Our measurements showed that, even though caches, a branch predictor, and a prefetcher introduce some variation in the per-iteration execution time of the loop, in practice this variation is not significant. The checksum code is small, so it fits completely inside the CPU's L1 instruction cache. Thus, once the CPU caches are warmed up, no more cache misses occur. The time taken to warm up the CPU caches is a very small fraction of the total execution time, and so the variance in the execution time caused by caches is negligible.

We also observed that in our code, branch mis-predictions only happen in the last iteration (recall that the inner loop was enrolled), when the checksum computation is finished. This has negligible impact on the overall execution time of the checksum loop. Furthermore, due to random memory access pattern in the checksum loop, results in low prefetcher's accuracy, so the prefetcher has little impact on the execution time of the checksum loop.

To further analyze the effect of having caches, branch predictors, and prefetchers, we used gem5 [4] simulator to simulate the checksum code on an in-order ARM core machine with similar configurations to that of A13-OLinuXino board. Our results showed that, for 1.2 KB checksum code, our code accessed the cache about 7000 times, out of which only 21 accesses were L1 misses (i.e., $>99.5\%$ hit-rate), and only about 800 more cycles (mostly due to L2 misses) were added to the overall execution time (i.e., $<0.01\%$). Note that this extra delay only happens inside the checksum loop, and does not affect the delay for the proxy and/or rootkit attacks since those attacks happen *before* the beginning of the checksum. Furthermore, our results showed a negligible mis-prediction rate for the branch predictor, and $<1\%$ prefetching accuracy for the checksum loop.

To evaluate EMMA, we ran the same experiment (adding an extra "ADD" instruction) discussed in the previous section, and found that our detection algorithm successfully detected all instances of the attack with no false positive.

Simultaneous Monitoring of Multiple Devices
To demonstrate its ability to monitor multiple devices at the same, we used EMMA to monitor eight Arduino devices. For each device, we used a hand-made coil (cost around $3) as the measurement probe and taped it to the board (on the center of the board). We then used SMA cables and 2-way channel splitter/combiner (cost around $4), to connect all the probes to a single SDR (cost $30). The SDR is connected to a computer where EMMA was implemented. We then repeated the measurements in the previous section, by attesting these devices one-by-one. In all the experiments, we saw no significant degradation in the accuracy, and in all cases we were able to detect the attack with perfect accuracy, which validates that the ability of EMMA to monitor multiple devices with the same EM signal receiver. However, when we increased the number of devices to 16, we observed much higher noise and significantly lower SNR, which prevented successfully attestation.

Based on these experiments, with the current setup, EMMA is able to monitor up to 8 devices with no performance loss. The entire setup cost for monitoring is about $80 (i.e., $10 per device), which makes the system a practical approach for monitoring security-critical devices. Furthermore, multiple EMMA setups can be used to monitor a large group of devices where a C&C server can manage the whole process and send and receive the results from individual EMMA setups.

Robustness and Variations Over Time
To test the robustness of our algorithm over time and against environment variations (e.g., temperature, interference, etc.), we repeat the attestation procedure at one-hour intervals, over a period of 24 hours, while keeping the Arduino board and the receiver active throughout the experiment, to observe how the emanated signals vary over time as device temperature (and room temperature) and external radio interference such as WiFi and cellular signals change during the day and due to the day/night transition. At each hour we ran the attestation once (without any malicious behavior). The training data was collected before the first hour of the experiment. The goal was to show how the false positive (FP) rate changes over time. We observed a significant increase in FP rate after hour 2 when we disabled the clock-adjustment feature. However, when this feature is active, EMMA achieved perfect accuracy (i.e., 0% FP). The main reason for why clock-adjustment improves long-term monitoring results so dramatically is that the clock-rate for the Arduino began to drift after about one hour of continuous usage. If it were prevented from accounting for this drift, EMMA would detect the frequency-drift-induced shift in the frequency of the main loop and treat that as an indication that the main loop's code has been tampered with.

References

1. AARONIA. Datasheet: Rf near field probe set dc to 9ghz. http://www.aaronia.com/Datasheets/Antennas/RF-Near-Field-Probe-Set.pdf, 2016 (accessed Apr. 6, 2017).
2. Tigist Abera, N. Asokan, Lucas Davi, Jan-Erik Ekberg, Thomas Nyman, Andrew Paverd, Ahmad-Reza Sadeghi, and Gene Tsudik. C-flat: Control-flow attestation for embedded systems software. In *Proceedings of the 2016 ACM SIGSAC Conference on Computer and Communications Security*, CCS '16, pages 743–754, New York, NY, USA, 2016. ACM.
3. Frederik Armknecht, AhmadReza Sadeghi, Steffen Schulz, and Christian Wachsmann. A security framework for the analysis and design of software attestation. In *Proceedings of the 2013 ACM SIGSAC Conference on Computer & Communications Security*, CCS '13, pages 1–12, New York, NY, USA, 2013. ACM.
4. Nathan Binkert, Bradford Beckmann, Gabriel Black, Steven K. Reinhardt, Ali Saidi, Arkaprava Basu, Joel Hestness, Derek R. Hower, Tushar Krishna, Somayeh Sardashti, Rathijit Sen, Korey Sewell, Muhammad Shoaib, Nilay Vaish, Mark D. Hill, and David A. Wood. The gem5 simulator. *SIGARCH Comput. Archit. News*, 39(2):1–7, August 2011.
5. Ferdinand Brasser, Kasper B. Rasmussen, Ahmad-Reza Sadeghi, and Gene Tsudik. Remote attestation for low-end embedded devices: The prover's perspective. In *Proceedings of the 53rd Annual Design Automation Conference*, DAC '16, pages 91:1–91:6, New York, NY, USA, 2016. ACM.
6. Robert Callan, Farnaz Behrang, Alenka Zajic, Milos Prvulovic, and Alessandro Orso. Zero-overhead profiling via EM emanations. In *Proceedings of the 25th International Symposium on Software Testing and Analysis, ISSTA 2016, Saarbrücken, Germany, July 18-20, 2016*, pages 401–412, 2016.
7. X. Carpent, G. Tsudik, and N. Rattanavipanon. Erasmus: Efficient remote attestation via self-measurement for unattended settings. In *2018 Design, Automation Test in Europe Conference Exhibition (DATE)*, pages 1191–1194, March 2018.
8. Claude Castelluccia, Aurélien Francillon, Daniele Perito, and Claudio Soriente. On the difficulty of software-based attestation of embedded devices. In *Proceedings of the 16th ACM Conference on Computer and Communications Security*, CCS '09, pages 400–409, New York, NY, USA, 2009. ACM.
9. Binbin Chen, Xinshu Dong, Guangdong Bai, Sumeet Jauhar, and Yueqiang Cheng. Secure and efficient software-based attestation for industrial control devices with arm processors. In *Proceedings of the 33rd Annual Computer Security Applications Conference*, ACSAC 2017, pages 425–436, New York, NY, USA, 2017. ACM.
10. R. d. Clercq, R. D. Keulenaer, B. Coppens, B. Yang, P. Maene, K. d. Bosschere, B. Preneel, B. d. Sutter, and I. Verbauwhede. Sofia: Software and control flow integrity architecture. In *2016 Design, Automation Test in Europe Conference Exhibition (DATE)*, pages 1172–1177, March 2016.
11. G. Dessouky, S. Zeitouni, T. Nyman, A. Paverd, L. Davi, P. Koeberl, N. Asokan, and A. Sadeghi. Lo-fat: Low-overhead control flow attestation in hardware. In *2017 54th ACM/EDAC/IEEE Design Automation Conference (DAC)*, pages 1–6, June 2017.
12. R. W. Gardner, S. Garera, and A. D. Rubin. Detecting code alteration by creating a temporary memory bottleneck. *IEEE Transactions on Information Forensics and Security*, 4(4):638–650, Dec 2009.
13. Yi Han, Sriharsha Etigowni, Hua Liu, Saman Zonouz, and Athina Petropulu. Watch me, but don't touch me! contactless control flow monitoring via electromagnetic emanations. In *Proceedings of the 2017 ACM SIGSAC Conference on Computer and Communications Security*, CCS '17, pages 1095–1108, New York, NY, USA, 2017. ACM.
14. Alexander Klimov and Adi Shamir. New cryptographic primitives based on multiword t-functions. In Bimal Roy and Willi Meier, editors, *Fast Software Encryption*, pages 1–15, Berlin, Heidelberg, 2004. Springer Berlin Heidelberg.

15. Patrick Koeberl, Steffen Schulz, Ahmad-Reza Sadeghi, and Vijay Varadharajan. Trustlite: A security architecture for tiny embedded devices. In *Proceedings of the Ninth European Conference on Computer Systems*, EuroSys '14, pages 10:1–10:14, New York, NY, USA, 2014. ACM.
16. Li Li, Hong Hu, Jun Sun, Yang Liu, and Jin Song Dong. Practical analysis framework for software-based attestation scheme. In Stephan Merz and Jun Pang, editors, *Formal Methods and Software Engineering*, pages 284–299, Cham, 2014. Springer International Publishing.
17. Yanlin Li, Jonathan M. McCune, and Adrian Perrig. Viper: Verifying the integrity of peripherals' firmware. In *Proceedings of the 18th ACM Conference on Computer and Communications Security*, CCS '11, pages 3–16, New York, NY, USA, 2011. ACM.
18. Alireza Nazari, Nader Sehatbakhsh, Monjur Alam, Alenka Zajic, and Milos Prvulovic. Eddie: Em-based detection of deviations in program execution. In *Proceedings of the 44th Annual International Symposium on Computer Architecture*, ISCA '17, pages 333–346, New York, NY, USA, 2017. ACM.
19. OlinuXino A13. https://www.olimex.com/Products/OLinuXino/A13/A13-OLinuXino/open-source-hardware.
20. RTL-SDR. v3. https://www.rtl-sdr.com/rtl-sdr-quick-start-guide/, 2016 (accessed April 2019).
21. Nader Sehatbakhsh, Alireza Nazari, Haider Khan, Alenka Zajic, and Milos Prvulovic. Emma: Hardware/software attestation framework for embedded systems using electromagnetic signals. In *Proceedings of the 52nd Annual IEEE/ACM International Symposium on Microarchitecture*, MICRO '52, page 983–995, New York, NY, USA, 2019. Association for Computing Machinery.
22. A. Seshadri, A. Perrig, L. van Doorn, and P. Khosla. Swatt: software-based attestation for embedded devices. In *Proceedings of the IEEE Symposium on Security and Privacy*, pages 272–282, May 2004.
23. Arvind Seshadri, Mark Luk, and Adrian Perrig. Sake: Software attestation for key establishment in sensor networks. *Ad Hoc Netw.*, 9(6):1059–1067, August 2011.
24. Arvind Seshadri, Mark Luk, Adrian Perrig, Leendert van Doorn, and Pradeep Khosla. Scuba: Secure code update by attestation in sensor networks. In *Proceedings of the 5th ACM Workshop on Wireless Security*, WiSe '06, pages 85–94, New York, NY, USA, 2006. ACM.
25. Arvind Seshadri, Mark Luk, Elaine Shi, Adrian Perrig, Leendert van Doorn, and Pradeep Khosla. Pioneer: Verifying code integrity and enforcing untampered code execution on legacy systems. In *Proceedings of the Twentieth ACM Symposium on Operating Systems Principles*, SOSP '05, pages 1–16, New York, NY, USA, 2005. ACM.
26. Hovav Shacham. The geometry of innocent flesh on the bone: Return-into-libc without function calls (on the x86). In *Proceedings of the 14th ACM Conference on Computer and Communications Security*, CCS '07, pages 552–561, New York, NY, USA, 2007. ACM.
27. Xiaoyun Wang, Dengguo Feng, Xuejia Lai, and Hongbo Yu. Collisions for hash functions md4, md5, haval-128 and ripemd, 2004. no lai-xj@cs.sjtu.edu.cn 12647 received 16 Aug 2004, last revised 17 Aug 2004.
28. Xiaoyun Wang, Yiqun Lisa Yin, and Hongbo Yu. Finding collisions in the full sha-1. In Victor Shoup, editor, *Advances in Cryptology – CRYPTO 2005*, pages 17–36, Berlin, Heidelberg, 2005. Springer Berlin Heidelberg.
29. Yi Yang, Xinran Wang, Sencun Zhu, and Guohong Cao. Distributed software-based attestation for node compromise detection in sensor networks. In *Proceedings of the 26th IEEE International Symposium on Reliable Distributed Systems*, SRDS '07, pages 219–230, Washington, DC, USA, 2007. IEEE Computer Society.

Chapter 12
Using Analog Side Channels for Hardware Identification

12.1 Introduction

Computer systems and other electronic devices are typically assembled by a system integrator. These systems contain one or more printed circuit boards (PCBs) that have been procured from various suppliers. These PCBs are themselves produced by board-level integrators, and typically contain a number of integrated circuits (ICs) that are also obtained from various suppliers. The procurement ecosystem for both ICs and PCBs is sophisticated, so the actual manufacturer of an IC very rarely acts as a direct supplier to a PCB integrator, the PCB integrator is very rarely a direct supplier to the system integrator, and the system integrator may not be the direct supplier to the end user of the system. However, for both the system integrator and the end user of the system, it is important to know which actual devices (both IC- and PCB-level) are present in the system. Different devices, even when they are functionally compatible, can differ in other properties, such as the level of trust its manufacturer enjoys, the level of reliability and environmental tolerance the device can be expected to provide, the set of inter-operability issues with specific software and with other devices, and the bugs/vulnerabilities that must be taken into account to ensure that the system functions correctly and securely.

Unfortunately, the complex supply chain for both ICs and PCBs makes it difficult to avoid counterfeits [22], which are thought to represent around 1% of all semiconductor sales [26] and cost legitimate component manufacturers approximately $100 billion in lost sales [28]. Furthermore, even in the absence of malicious intent, one legitimate device can be (and often is) substituted with another legitimate device that the PCB manufacturer may consider to be equivalent. However, some of the properties of these devices may differ, especially when it comes to inter-operability, bugs, vulnerabilities, and reliability. To overcome this problem, it is very important to correctly recognize/authenticate components on a PCB or in a system, so that the appropriate software patches can be applied, and so that tracking and mitigation of reliability and inter-operability issues can be correctly implemented.

Currently, companies and large users (e.g., governments, data center operators, etc.) rely on several different methods for recognition/authentication of electronic parts. The most commonly used methods involve: physical inspection and electrical inspection [20]. Physical inspection methods rely on examining the inside and outside of the components and analyzing their material composition. Electrical inspection methods rely on testing the components' electrical characteristics, performance, and durability through burn-in tests [20]. However, the effectiveness of these methods can vary based on the type of counterfeit, level of intrusiveness during testing, cost, time, and other conditions [22]. Reliable, non-destructive approaches that allow for easy, precise, and cost-effective recognition/authentication of electronic components are needed.

Another approach for recognition/authentication of electronic parts is to rely on electromagnetic (EM) side-channels, i.e., on unintentional emanations that are created as a consequence of transistors switching on a component while it is performing computations [1, 21]. In previous chapters, we have discussed how the EM side channel can be used to detect malware detection, provide software attestation, profile software, identify hardware events during program execution, and in other ways gain insights into the behavior of software as it executes on the hardware. However, analog side channel signals are actually created by hardware activity that results from program execution, and these signals provide information not only about the software that is being executed, but also about the hardware on which the resulting electronic activity is occurring. This is the insight that we will begin to leverage in this chapter.

Specifically, in this chapter we discuss how EM side-channels can be used to recognize/authenticate hardware components integrated onto a PCB. By focusing on components on a PCB, this method provides an opportunity for designers and manufacturers to authenticate PCBs and systems that have already been assembled by third parties [43]. The main practical use of this method is to detect counterfeit ICs based on changes in the EM emanations that comprise the EM side-channel. This specific types of counterfeiting these methods focus on are those that alter the physical design of the component, e.g., when the intended IC has been replaced with a reverse engineered copy, with a lower quality component, or where the IC's design has been tampered with [22].

12.2 Side Channel Signals Carrying Information About Hardware Properties

This section discusses what spectral features are found to be relevant for component recognition and how we can excite electronic components in order to maximize the presence of these side-channel features.

12.2.1 Spectral Features

Our method relies specifically on the unintentionally modulated signals that are emanated by the component. These are some of the strongest signals available in the EM side-channel [6], and they are caused by the device's computational activity unintentionally modulating periodic signals, such as processor and other clocks, that are already present in a device [6]. Computational activity results in the superposition of a time-varying current on the traces inside and connecting the components used for the execution. The magnitude of the emanations depends on the change in power when executing the activity, while modulation frequency is related to the time it takes to execute the repetitive behavior. While multiple types of modulation can occur in a device, this work relies only on amplitude modulated (AM) signals generated by a component [29].

We have observed that any change in the component or program activity affects the properties of the emitted AM signal. For example, the shape and spread of the sidebands in the frequency spectrum are related to the time it takes to execute parts of the program activity. If the execution time varies, the sideband's shape will spread in frequency. By keeping the program activity consistent for all tests, the spectral features of the emanations can be used as a signature for identifying the components. We have experimentally determined that these features include: (1) the overall modulation frequency, (2) the sideband shape and spread, (3) the relative strength of the sideband's fundamental frequency and the higher harmonics, (4) the carrier frequency, and (5) the carrier spread. An example of a signal, with these properties annotated, is shown in Fig. 12.1.

When selecting spectral features to use for component identification, the device needs to be in a known and repeatable state, and the component of interest needs to be active—otherwise, there will be nothing in the spectrum to use for identification. There are several ways to achieve an active and repeatable state. One is to run an excitation program that is carefully crafted to have repeatable and stable behavior, so that any observed changes can be attributed to hardware behavior and not software behavior. This approach was introduced in [44] and detailed in this chapter. Similarly, firmware tends to have stable and repeatable behavior, so it can be leveraged for hardware identification. Finally, we have shown in [25] that even the idle "on" state that typically already exists on each device can be used for hardware identification. These methods perform similarly well, so the choice between them mainly depends on how the method would be used "in the field".

12.2.2 Device Excitation

We use the A/B alternation program from Chap. 4 to create controllable emanations on the device. Generally, the emanations generated by A/B alternation are much stronger than other activities in the spectrum, minimizing the influence of these

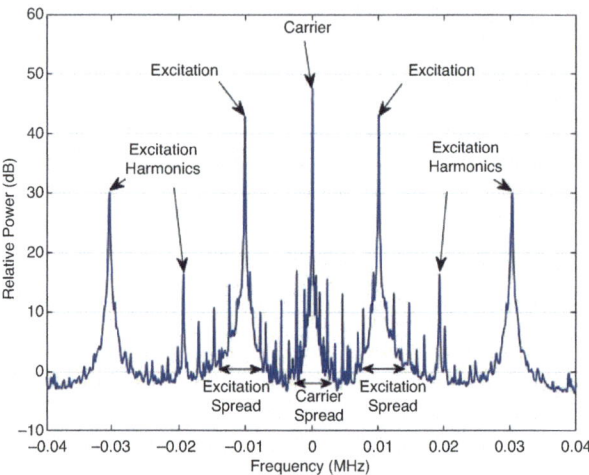

Fig. 12.1 Example of the spectrum produced by LD/ST alternation [43]

other activities. Furthermore, as demonstrated in [42], different choices for A and B activity can emphasize activity in specific components (e.g., processor. memory, etc.) while limiting the influence other components have during measurement. Finally, the A/B alternation can be easily implemented on a variety of devices, making it ideal for authentication of components integrated on multiple types of motherboards. In fact, it has already been implemented on various types of laptops, desktops, cell phones, computer boards, and FPGAs [4, 5, 41, 42], on "bare metal" and under operating systems (OSs) that include several versions of Windows, various flavors and versions of Linux, Android, etc.

An example of the spectrum generated using A/B alternation is shown in Fig. 12.1, where a load (memory read) instruction that hits in the processor's L2 cache is used as activity A, and a store (memory write) that hits in the L2 cache is used as activity B. As an be seen from this spectrum, the alternation frequency is 10 kHz. Since the A/B alternation was designed to produce a duty cycle of 50%, in this spectrum the odd harmonics at 10 and 30 kHz (i.e., at 1x and 3x alternation frequency) are much stronger than even harmonics (at 2x alternation frequency). However, the presence of any signal at ±20 kHz (2x alternation frequency, an even harmonics) indicates the actual duty cycle is not exactly 50%. The shape and spread of the sidebands are caused by variations in the execution time of the excitation program, while the spread of the carrier is caused by the instability of its source. Finally, the power of the sidebands relative to each other and to the carrier are based on three factors: the physical properties of the component, the excitation program, and the location of the measurement probe. *Keeping the excitation program and probe position fixed allows us to observe the physical properties of the components.*

Even if the component is being excited, it can still be difficult to locate useful spectral features. Depending on the program activity, the spectrum can change rapidly in time. As a result, locating useful spectral features, in both time and frequency, can be challenging. Selecting a measurement state where the device's spectrum is relatively constant makes identification easier.

12.3 Signal Compression and Processing

After the component has been excited and its emanations have been recorded, the measured signal is processed to make it easier to locate important features for identification. To eliminate the influence of time variations and noise on the measurements and to improve the efficiency of identification, an updated version of the approach described in [45] is used. This new approach is described below.

During a test, the EM emanations from the excited component are recorded for a period of time, T. The number of samples, M, is equal to the measurement time multiplied by the sample frequency, f_s. To better emphasize the elements of the signal used for identification, the signal is converted to the frequency domain. However, instead of converting the entire signal at once, the measurement is first broken into several segments, N_R. A short-time discrete Fourier transform (STFT) is then applied to each segment. A flattop window is used to improve the accuracy of the relative amplitudes of different frequency components.

Each STFT operation for the kth frequency component is calculated by

$$Y_h[k] = \sum_{n=1}^{N} y[n + (h-1)N_s] w[n-k] \exp(-i2\pi k n/N), \qquad (12.1)$$

where y is the original measurement, h is the STFT operation number, w is the flattop window function, N_s is the number of non-overlapping samples, and N is the window size for the STFT operations. The non-overlap time is the number of samples, N_s, shifted between STFT segments.

The number of elements in each segment, N, affects the resolution of the window. The resolution needs to be high enough for features of the spectrum to be distinguishable. However, if the size of N is too large, it will increase the measurement and processing times. After M, N, and N_s have been defined, the number of STFT operations performed for the data can be calculated by

$$N_R = \text{floor}\left(\frac{M-N}{N_s} - 1\right). \qquad (12.2)$$

To reduce the impact of noise and other time variations in the signal on the spectrum, these N_R STFT operations are averaged together as

$$\overline{Y}[k] = \frac{1}{N_R} \sum_{h=1}^{N_R} |Y_h[k]|, \qquad (12.3)$$

where $\overline{Y}$ is a row vector containing the averaged frequency magnitudes. Here, to eliminate the impact of the starting time on the data, the phase is removed by taking the magnitudes of the STFT results. Afterwards, the strongest component in the spectrum is downconverted to 0 Hz. This component is assumed to be the carrier. Since the exact frequency of the carrier is influenced by factors such as manufacturing variability and temperature, the carrier frequencies of two identical components on separate devices will be slightly different. Shifting the carrier to the center of the spectrum ensures that this difference does not influence the identification. Afterwards, a band-pass filter is applied to the downconverted signal by removing 10% of the bandwidth.

Next the measurements are converted from a linear scale to a decibels scale (dB) using

$$\overline{Y}_{h_{dB}}[k] = 20\log_{10}(\overline{Y}_h[k]). \tag{12.4}$$

This conversion reduces the influence of the strong signal components, while increasing the influence of weaker components in the spectrum. Since the modulation is not intentional, the carrier tends to be significantly stronger than the sideband components, usually tens to hundreds of times stronger. Without this conversion, the carrier has a disproportionate influence on the data. Afterwards, the measurements are standardized by subtracting their mean and dividing by their standard deviation.

Once all the measurements are processed, they are combined in the $H \times N$ matrix

$$\mathbf{Y} = \begin{bmatrix} \overline{Y}_{1_{dB}} \\ \overline{Y}_{2_{dB}} \\ \cdot \\ \cdot \\ \cdot \\ \overline{Y}_{H_{dB}} \end{bmatrix}, \tag{12.5}$$

where H is the number of measurements. In other words, each row represents the averaged frequency components of a measurement. However, the matrix generally contains redundant information because H and N are not equal. To reduce the size of the data and improve efficiency, singular value decomposition (SVD) is applied to the data matrix. As a result, the data matrix is decomposed into the following form:

$$\mathbf{Y} = \mathbf{U \Sigma V}^T, \tag{12.6}$$

where $\mathbf{U}$ is the left singular vectors matrix, $\mathbf{\Sigma}$ is the singular values matrix, and $\mathbf{V}^T$ is the transpose of the right singular vectors matrix. The measurement data is then projected into a new vector space, $\mathbf{Z} \in \Re^{H \times K}$, by

$$\mathbf{Z} = \mathbf{Y V}_K, \tag{12.7}$$

where $\mathbf{V}_K$ has been reduced to $K \times K$ matrix. The size of K is significantly smaller than the length of the original measurements and corresponds to the K largest singular values in $\mathbf{\Sigma}$. These K vectors represent the signal components of the data matrix that contains the directions corresponding to largest K singular values. As a result, the original $H \times N$ dimensional feature space of the measurements has been reduced to $H \times K$ dimensions, without much loss in information about the data.

12.4 Training and Testing Process

The identification process can be broken into training and testing phases. In the training phase, EM signatures from example components are recorded and processed. Here, the identities of the components are already known. If the training components are used on multiple types of devices, only one of the devices needs to be used for training. Each training component is excited using the excitation program, and its emanations are recorded for T seconds. The modulation frequency used for the excitation program is set beforehand and kept constant for all measurements. The carrier frequency of the component is located using either a method such as the ones detailed in [6] and [29], or it can be identified manually (e.g., by obtaining the clock frequency from the device's datasheet or other documentation). The measurements are processed into averaged STFTs and stored in a matrix. If multiple types of components are tested, such as memory and processor, the signatures can be divided into multiple matrices based on the component's function or the carrier frequency. The training matrix or matrices are then decomposed using SVD. The first K columns from the resulting matrix $\mathbf{V}$ are selected. Next, the matrix $\mathbf{V}_K$ is used to project the training data into the new vector space. In other words, assuming the projected training data set is $\mathscr{Y}$, we generate a model ($\mathscr{Y}$, $\mathbf{V}$) that is used in the testing phase.

In the testing phase, the EM emanations generated by one or more unknown components are recorded. As in the training phase, the tested component is excited using the excitation program. The measurements are then processed into averaged STFTs. The resulting signatures are projected into a new vector space using the matrix created from the training data. Afterwards a k-NN algorithm is applied to the projected training and testing data by using the model ($\mathscr{Y}$, $\mathbf{V}$). The k-NN algorithm determines the identity of the tested component based on the standardized Euclidean distance between the projected measurements and the training measurements.

Although k-NN is a fairly simple type of clustering, it is a practical tool for classifying components. As the experiments in the following section demonstrate, it can accurately determine the identities of several types of components.

12.5 Experimental Validation

To demonstrate its effectiveness, the identification method is applied to components from seven types of IoT devices from Olimex and two FPGA development boards. This method is not limited to such devices; we use them here only as examples. All the IoT devices are from the same manufacturer (Olimex) because it presents the most difficult case for identifying components—differences between devices from different manufacturers are even larger, making them easier to identify.

12.5.1 Measurement Setup

Figure 12.2 shows the measurement setup used for the experiments. A small handmade circular coil probe with a 1 mm radius is used to measure the magnetic field emanated by the component being tested. This probe was selected so that its small size would limit the influence of emanations from other nearby components on measurements. The probe is connected to a Keysight M9391A PXIe Vector Signal Analyzer (VSA) for recording the signal. During the measurements, the probe is placed directly on top of the monitored component, where the emanations are the strongest. To ensure consistency between measurements, the probe is positioned using the Riscure EM Probe Station's [31] motorized XYZ table. The table is controlled through a USB port using a laptop. The laptop is also used to control the device under test (DUT) and the VSA through their Ethernet ports. In cases where the DUT does not have an Ethernet port, a USB-to-Ethernet adapter is used. A diagram of the test setup is shown in Fig. 12.3.

During a measurement, all the components are excited using the excitation program described in Sect. 12.2.2. The program generates a 10 kHz excitation signal with a 50% duty cycle by executing an alternating pattern of addition and load instructions. When measuring the external memory components, the array size of the load instruction is set to be much larger (8.4 MB) than the processor's cache to

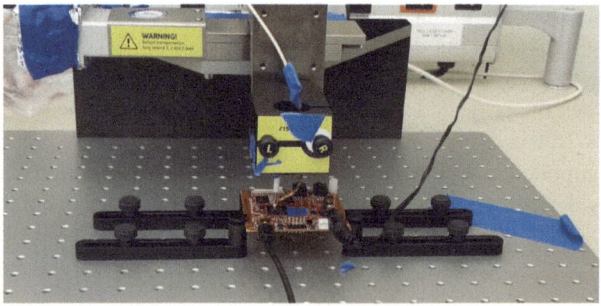

Fig. 12.2 Measurement setup used for the experiments [43]

12.5 Experimental Validation

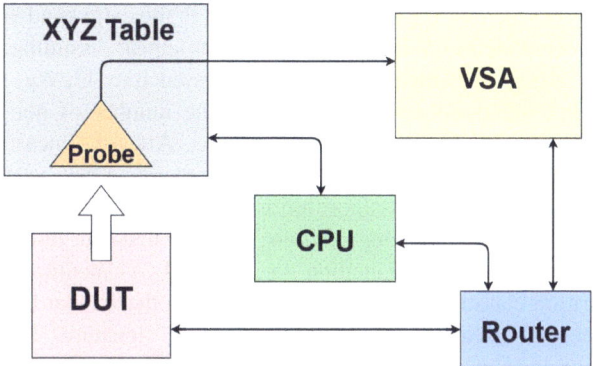

Fig. 12.3 Diagram of the measurement setup [43]

ensure that the external memory is active during the measurements. When exciting a processor, the array size of the load instructions is kept small (8.2 kB) to ensure that the load execution is mostly confined to the processor's cache. This small array size minimizes the influence the external memory has on the signal. When testing multicore processors, the excitation program is executed only on the first core, while the rest are left idle.

While a component is excited, its emanations are recorded for 1 s (T from Sect. 12.3). Each of the seven types of IoT devices had 10 individual units, i.e., they are 10 samples of each type; however, one unit (an A33-MAIN) is removed from the results since it was damaged. Therefore, a total of 69 boards are tested. Furthermore, cross-type testing is conducted because some of the device types have the same components. In these situations, measurements from only one type of devices are used for training. Similarly, in situations where the same component is used multiple times on the same device, measurements from only one instance of the component are used for training.

To account for the limited number of boards, a k-fold cross-validation scheme with five-folds is run for 10,000 iterations. Such schemes are commonly used in cases where the sample size is small since their results have a relatively low bias and variance compared to other cross-validation approaches [3]. The accuracy of correctly identifying each type of component and the overall accuracy of correctly identifying all the components are calculated by averaging the results for each iteration.

Several of the component measurements have interrupts caused by their device's operating systems [19]. To improve consistency and ensure the spectrum is primarily a result of the excitation program, these interrupts are removed from the measurements. The measurement bandwidth is 220 kHz (reduced to 200 kHz after processing). A small bandwidth is used to reduce the amount of interference present in the measurement. More bandwidth may help distinguish different components from each other, but the interference signals present in the extra bandwidth can also make the same components on different device types appear dissimilar.

During processing, each measurement is processed into 43, N_R, STFTs with a length, N, of 20480 before being averaged. The number of non-overlapping samples, N_s, is 4000, which corresponds to 20 ms. After the measurements are processed, the training measurements are used to generate a new vector space for evaluation. Afterwards, the k-NN algorithm is applied to the testing data by using the model ($\mathscr{Y}$, **V**). For the following sections, only the first four dimensions of the new vector space are used for evaluation, i.e., $K = 4$. As mentioned previously, the k-NN algorithm classifies a measurement based on the standardized Euclidean distance between the measurement and each training signature. The differences between the four coordinates of the test and training points are scaled by dividing with 1, 1, 2, and 3, respectively, for the memory and processor and with 1, 1, 3, and 3 for the Ethernet transceivers. The number of dimensions and the scaling factors can be tuned to improve the classification accuracy for a specific set of components.

12.5.2 Test Devices

In the experiments, seven types of IoT devices from Olimex are tested. These devices are the A10-OLinuXino-LIME [10], A13-OLinuXino [11], A13-OLinuXino-MICRO [12], A20-OLinuXino-LIME [13], A20-OLinuXino-LIME2 [14], A20-OLinuXino-MICRO [15], and A33-OLinuXino [16]. For simplicity, these devices will be referred to as A10-LIME, A13-MAIN, A13-MICRO, A20-LIME1, A20-LIME2, A20-MICRO, and A33-MAIN for the rest of this work. All the devices run Linux OS provided by their manufacturer. The only change made to the devices was installing the excitation program.

Pictures of tested IoT devices are shown in Fig. 12.4. All the IoT devices are part of Olimex's OLinuXino open source hardware product line. They are convenient options for these experiments since Olimex has provided detailed information about each device (such as the schematic, parts list, and PCB layout) on their website [24].

Two additional memory components (MEM5 and MEM6) are tested to demonstrate the impact of projecting the measurements into the new feature space. These extra components are integrated into DE0-CV Cyclone V [7] and DE1 Cyclone II development boards [8], shown in Fig. 12.5. These components and their motherboards are significantly different from the Olimex devices, both in functionality and physical properties. The differences result in the projected data from the development boards being easily distinguishable from the IoT measurements. Therefore, these components are not included in later subsections. Correctly classifying similar, yet physically different components is more challenging, especially if the components are integrated onto similar motherboards.

The reason motherboards from the same manufacturer increase the difficulty is that manufacturers commonly use the same components and parts of PCB layouts in multiple designs to save time and money. These similarities can influence the

12.5 Experimental Validation

Fig. 12.4 Pictures of the seven IoT devices (not to scale) [43]

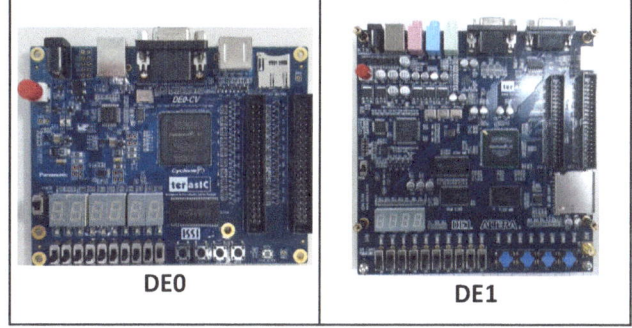

Fig. 12.5 Pictures of the DE0-CV Cyclone V and DE1 Cyclone II development boards (not to scale) [43]

parts of the spectrum not related to the component of interest. The A10-LIME and A20-LIME1 are examples of reused PCB layouts. The PCB for the A10-LIME is an older revision of the PCB for the A20-LIME1. In this situation, the traces on both motherboards will have similar emanation properties since they are almost identical in shape and composition. If one motherboard is made by a different manufacturer, the emanation properties change since the physical configurations of the traces and the material properties of the PCB would change. However, in either case, the signal properties of the emanations from the traces and the components still depend on the component and the program activity.

The following experiments focus on identifying the external memory, processor, and Ethernet transceiver components present on each device. Some of these components are not present on all the devices, while, in other cases, some devices use the same components as others. A complete list of the components, the devices they are present on, and the measurement frequency are provided in Table 12.1. For simplicity, the components will be referred to by their label in the table. More information about the devices and their components can be found on the component manufacturer's websites (referenced in the table).

Table 12.1 List of tested components [43]

Label	IC name	IC type	Devices	Carrier frequency (MHz)	Source
MEM1	K4B4G1646Q-HYK0	4Gb DDR3 SDRAM	A10-LIME	384	[32]
MEM2	H5TQ2G83FFR	2Gb DDR3 SDRAM	A13-MAIN U1 and U2	408	[34]
MEM3	H5TQ2G63BFR	2Gb DDR3 SDRAM	A13-MICRO	408	[33]
MEM4	MT41K256M16HA-125:E	4Gb DDR3 SDRAM	A20-LIME1, A20-LIME2 U2 and U3, A20-MICRO U2 and U3	384	[27]
MEM5	IS42S16320D-7TL	512Mb SDR SDRAM	DE0	100	[23]
MEM6	A2V64S40CTP-G7	64Mb SDR SDRAM	DE1	50	[9]
PROC1A	Allwinner A10 (Cortex-A8)	SoC (Processor)	A10-LIME	1008	[35]
PROC1B	Allwinner A13 (Cortex-A8)	SoC (Processor)	A13-MAIN, A13-MICRO	1008	[36]
PROC2A	Allwinner A20 (Cortex-A7)	SoC (Processor)	A20-LIME1, A20-LIME2, A20-MICRO	960	[37]
PROC2B	Allwinner A33 (Cortex-A7)	SoC (Processor)	A33-MAIN	960	[38]
ETH1	RTL8201CP	Ethernet Transceiver	A10-LIME	25	[30]
ETH2	LAN8710A-EZC-TR	Ethernet Transceiver	A20-LIME1, A20-MICRO	25	[40]
ETH3	KSZ9031RNXCC-TR	Ethernet Transceiver	A20-LIME2	25	[39]

12.5.3 Measurement Projection

Before discussing the method's performance, we examine the impact of projecting the measurement data into the new feature space. As an example, three measurements from all six types of memory components are projected into a new 3D feature space (shown in Fig. 12.6). All measurements are recorded for 0.2 s with

12.5 Experimental Validation

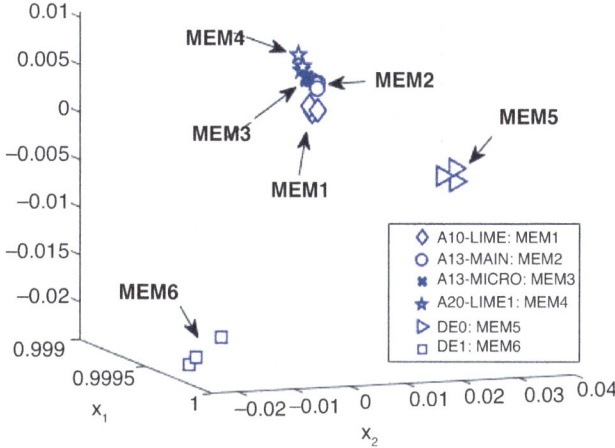

Fig. 12.6 Example of the memory measurements projected into the new feature space [43]

a bandwidth of 1 MHz and an excitation frequency of 100 kHz. For simplicity, the example measurements are used for generating the feature space before being projected into it.

In Fig. 12.6, the points from MEM5 and MEM6 are relatively far away from the other components. Their isolation is the result of their measurements having significantly different spectral features, which is translated (through projection) into different coordinates in the feature space. These differences are the result of the first four memory components being radically different from the last two. While all six are SDRAM, the first four are DDR3 with operating frequencies greater than 300 MHz, while the last two are SDR (single data rate) with operating frequency below 200 MHz.

On the other hand, points from MEM1 through MEM4 are clustered closer together, so they are difficult to *visually* distinguish from one another. The reason for this clustering is that the components are similar in functionality to one another. Furthermore, the fact the motherboards are from the same manufacturer likely helped their similarities. However, the closer the projected points from different types of components are to one another, the greater the risk of them being misclassified.

More detail about the relationships between the measurements can be gained by comparing the separation distances between the projected data. The average distances between data from two different types of components and the average distance between data from the same type of component can be calculated. For simplicity, the distance between points from two different types of components is called the *class-distance*, and the distance between points belonging to the same type of component is called the *self-distance*. The class-distance is the result of different components having different spectral features. The larger the distance, the greater the difference. The self-distance is caused by differences between spectra that correspond to two instances of the same component. These differences can be the result of minor changes in how the measurements are taken (such as probe type and probe position), changes in program execution over time, and manufacturing variation. By design, the test process described in Sect. 12.4 minimizes the first two factors. However, the influence of manufacturing variation cannot be removed.

These manufacturing variations are the cause of random fluctuation during the manufacturing process of an IC. These fluctuations result in small physical distinctions between individual ICs of the same type and are the basis of physically unclonable functions (PUFs) [2]. These small distinctions can impact the EM emanations generated by the IC, causing slight disparities in the measurements taken on individuals of the same type. While the impact of manufacturing variation cannot be removed, the likelihood of the manufacturing variation causing a misclassification can be evaluated by comparing the class-distances and self-distances of the data. If the self-distance for a component is much smaller than the component's class-distances, it can be concluded that the effect of the manufacturing variation is outweighed by the dissimilarities in the spectral features among the types of components, and manufacturing variation is unlikely to cause misclassification.

The average distances between the measurements for each type of memory are shown below in Table 12.2. The diagonal values (in bold) are the self-distances for each component, while the rest are the average class-distances between different types of components. The values themselves are unitless and their only significance is their size relative to one another.

Reflecting the results in Fig. 12.6, the class-distances between the first four memories and MEM5/MEM6 are significantly higher than the class-distances between the first four memory components only. For example, the class-distance between MEM1 and MEM4 is 9.7 while the class-distance between MEM1 and MEM6 is 59.9.

Furthermore, the table demonstrates that the class-distances between each component are larger than self-distances. This indicates that the impact of the manufacturing variation is outweighed by the differences between different types of components. Therefore, it is unlikely that manufacturing variation would cause misclassification among the tested components.

Table 12.2 Average distances among select memory components [43]

	MEM1	MEM2	MEM3	MEM4	MEM5	MEM6
MEM1	1.8	5.6	7.3	9.7	80.4	59.9
MEM2	5.6	1.1	3.0	4.8	82.1	64.9
MEM3	7.3	3.0	1.2	2.7	85.1	65.0
MEM4	9.7	4.8	2.7	1.9	86.1	67.4
MEM5	80.4	82.1	85.1	86.1	15.4	107.8
MEM6	59.9	64.9	65.0	67.4	107.8	6.1

12.5.4 Recognition of Memory Components

Next the devices with the first four memory components are evaluated. Only one MEM1, MEM3, and MEM4 component is integrated on each A10-MAIN, A13-MICRO, and A20-LIME1, respectively. However, A13-MAIN has two MEM2s, and the A20-LIME2 and A20-MICRO each has two MEM4s. For devices with more than one of the same type of memory component, each component is measured and evaluated separately. The external memory from the A33-MAIN is not included since it uses a spread spectrum memory clock so it is trivial to separate from the others.

Figure 12.7 shows a comparison between the spectra measured from MEM1 on a A10-LIME, MEM2 on a A13-MICRO, MEM3 on a A13-MAIN, and MEM4 on a A20-LIME1. For example, MEM1's carrier has a much stronger and wider spread than the other three components. Furthermore, the relative strength of the

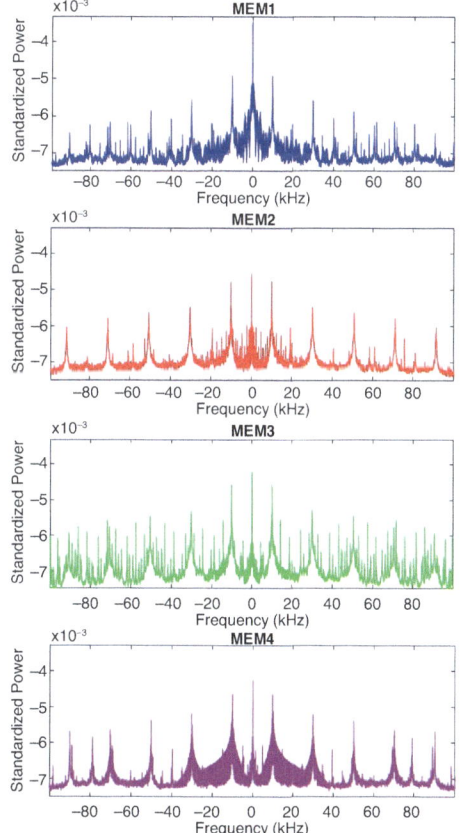

Fig. 12.7 Comparison of the EM signatures from MEM1 (top in blue), MEM2 (second from the top in red), MEM3 (third from the top in green), and MEM4 (bottom in magenta) [43]

Table 12.3 Average distances between memory components [43]

	MEM1	MEM2	MEM3	MEM3
MEM1	1.0	34.7	50.6	32.1
MEM2	34.7	1.9	19.6	13.6
MEM3	50.6	19.6	1.0	27.8
MEM4	32.1	13.6	27.8	4.2

odd sidebands for MEM1 is lower than the other memory components, while the even harmonics are much stronger.

At the same time, the signatures for MEM2 and MEM3 are similar, an unsurprising result given that the components are from the same product line. (The part numbers for MEM2 and MEM3 are H5TQ2G83FFR and H5TQ2G63BFR respectively). However, noticeable differences can be identified between the two components. For example, the harmonics of the MEM3 are more spread out compared to those of MEM2.

The sidebands for MEM4 are the strongest among the memory components. Furthermore, the spread of the sidebands for MEM4 is much larger than the spread of the other components. This spread is especially noticeable at the first harmonics, where it is nearly 10 kHz.

Example measurements from A10-LIME, A13-MAIN, A13-MICRO, and A20-LIME1 are used for training. These devices are selected because they have the cleanest spectra. Since there are two instances of MEM2 on A13-MAIN, only the components designated U1 on the motherboard are used for training. While all the A20 devices use the same memory components, only A20-LIME1 are used for training. The overall classification accuracy for the memory components after cross-validation is 100%. Since there were no classification errors, a confusion matrix for the results is not provided. The algorithm had no difficulty correctly classifying all the memory components, even MEM2 and MEM3, since the distinguishing spectral features are prominent and unique for each type of memory component.

Furthermore, the average distances for the memory components are shown in Table 12.3. The distances are calculated for each iteration of the cross-validation process before being averaged. This table confirms that, the class-distances for all four memory components are significantly larger than the self-distances, so manufacturing variations are unlikely to cause misclassification for the memories.

As a side note, the memory measurements provide an opportunity to examine the impact of other factors on the measured signal. For example, measurements taken on the two MEM4 components on a A20-LIME2 are shown in Fig. 12.8. The only difference between the two components is their physical location on the device; however, the spectra differ noticeably and it can be difficult to visually determine whether these spectra belong to the same family of component. The sidebands generated by the excitation program are much weaker relative to the carrier at U3 compared to U2. Furthermore, the other activity modulating the carrier is stronger at U3. These differences are likely due to differences in PCB traces connected to the components and the relative position of the measurement probe.

12.5 Experimental Validation

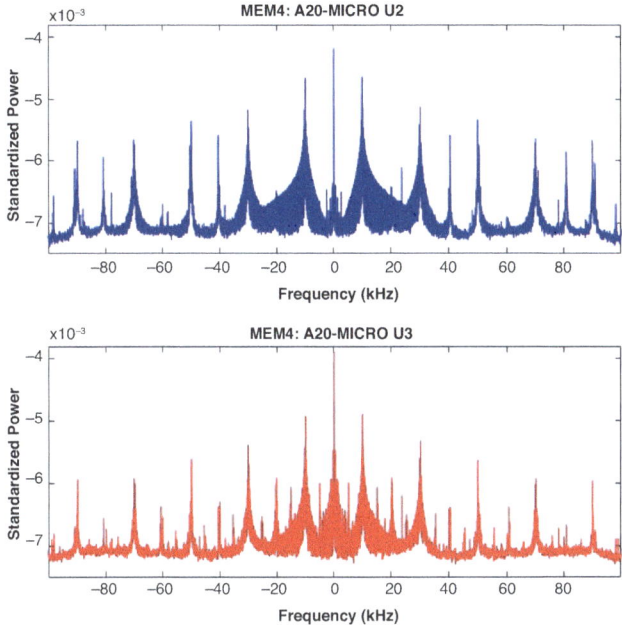

Fig. 12.8 Comparison of the MEM4 EM signatures from A20-MICRO U2 (top in blue) and A20-MICRO U3 (bottom in red) [43]

Despite their differences, both spectra are correctly identified as being from MEM4. This identification is possible because most of the differences in the two spectra are minimized after the measurements are projected into the new feature space and the dimensions are reduced to $K = 4$. Since the strongest variation in the data is represented by the first four singular vectors, smaller variations between measurements are lost when decreasing the number of dimensions.

12.5.5 Recognition of Processor Components

All the IoT devices have their processors integrated into a system-on-chip (SoC) from Allwinner Technology. The specific SoC is identified by the beginning of the device name (Allwinner A10, Allwinner A13, Allwinner A20, and Allwinner A33). Allwinner licensed designs for the processors from ARM Holdings and integrated the design in their SoCs. The A10 and A13 SoCs each have a single Cortex-A8 core [18]. The A20 SoCs each have a dual Cortex-A7 core [17], and the A33 Allwinner has a quad Cortex-A7 core.

While A10 and A13 have the same type of processor and A20 and A33 have the same type of processor, they still need to be classified as separate components.

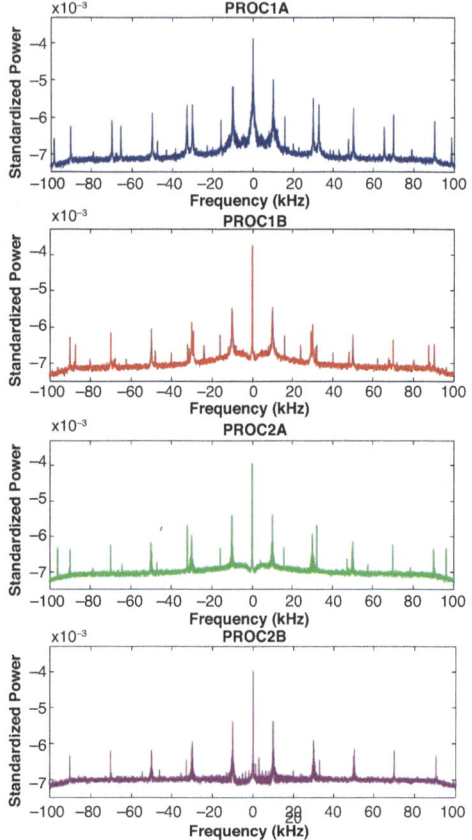

Fig. 12.9 Comparison of the EM signatures from PROC1A (top in blue), PROC1B (second from the top in red), PROC2A (third from the top in green), and PROC2B (bottom in magenta) [43]

In both cases, the processors are based on the same design from ARM; however, the processors are not supplied as discrete components directly from ARM. Instead, Allwinner has to implement the design on each SoC. While the functionality may be the same, there will be slight differences in the processor layout based on the other electronics integrated into the SoC and the layout choices of the designer. From the prospective of classification, the differences between how a processor is implemented on different types of SoC is similar to reverse-engineered or tampered components. Therefore, each type of SoC needs to be classified as a unique group, so we label the Allwinner A10 as PROC1A, the Allwinner A13 as PROC1B, Allwinner A20 as PROC2A, and Allwinner A33 as PROC2B.

Figure 12.9 shows spectra that are measured from an example of each type of processor while they are being excited. The top spectrum is an example of PROC1A from a A10-LIME, the second is an example of PROC1B from an A13-MAIN, the third is example of PROC2A from a A20-LIME1, and the bottom is an example of PROC2B from A33-MAIN.

12.5 Experimental Validation

Table 12.4 Confusion matrix for the processors (in %) [43]

	PROC1A	PROC1B	PROC2A	PROC2B
PROC1A: A10-LIME	100.0	0	0	0
PROC1B: A13-MAIN	0	98.0	2.0	0
PROC1B: A13-MICRO	0	100.0	0	0
PROC2A: A20-LIME1	0	0.2	99.8	0
PROC2A: A20-LIME2	0	0	98.7	1.3
PROC2A: A20-MICRO	0	0	100.0	0
PROC2B: A33-MAIN	0	0	0	100.0

All four spectra have a strong carrier, with harmonics caused by the excitation program at 20 kHz intervals (the even harmonics are too weak to see), giving them the same general shape. Furthermore, the similarities are strongest between SoCs that share the same processor design (i.e., PROC1A and PROC1B are very similar to each other, while PROC2A and PROC2B are very similar to each other but not to PROC1A and PROC1B). However, there are slight differences in properties discussed in Sect. 12.2.1. For example, the sidebands and carrier for PROC1A are stronger and have a larger spread than the others. This and other differences are magnified after projecting the measurements into the new feature space generated from the training data.

During classification, measurements from A10-LIME, A13-MAIN, A20-LIME1, and A33-MAIN are used as training data for the processors. The overall classification accuracy after cross-validation is 99.5%. A breakdown of the classification results for each device type is shown in Table 12.4. In the table, the rows correspond to the measured devices and their correct classification, while the columns correspond to the classification determined by the algorithm. The percentage of correctly classified instances of a device appears in bold font, while the percentage of incorrect classification appears in red.

As the table demonstrates, the individual classification accuracies for all devices are 98% or higher. Importantly, the algorithm is able to accurately classify each type of processor, regardless of the motherboard it is integrated into. Despite using only examples from A13-MAIN and A20-LIME1 for training, the algorithm correctly classified both A13 devices as having a PROC1B and all three A20 devices as having a PROC2A. At the same time, the algorithm is able to correctly distinguish the A10-LIME from the A13 devices and the A20 devices from the A33-MAIN, despite having the similar processor cores. The differences in how the processor is implemented on the SoC are enough to distinguish them. Furthermore, the algorithm correctly differentiated the processors on the A10-LIME and A20-LIME1 despite the A10-LIME's PCB being an older revision of the A20-LIME1 and the A20 Allwinner being pin-to-pin compatible with the A10 Allwinner.

Finally, the average distances for the processors are shown in Table 12.5. As the table demonstrates, the class-distances for all four processors are larger than the self-distances, but not as much as for the memory components.

Table 12.5 Average distances between processor components [43]

	PROC1A	PROC1B	PROC2A	PROC2B
PROC1A	2.7	12.6	17.3	22.5
PROC1B	12.6	2.9	7.5	13.4
PROC2A	17.3	7.5	3.0	6.7
PROC2B	22.5	13.4	6.7	2.4

12.5.6 Recognition of Ethernet Transceiver Components

There are three types of Ethernet transceivers in our experiments: ETH1, ETH2, and ETH3. ETH1 is used on A10-LIME, ETH2 is used on A20-LIME1 and A20-MICRO, and ETH3 is used on A20-LIME2. The A13-MAIN, A13-MICRO, and A33-MAIN do not have Ethernet transceivers.

Spectrum examples for ETH1, ETH2, and ETH3 are shown in Fig. 12.10. For these components, the spectrum is less active than for memory and processor components, indicating that the excitation program is not having as strong of an effect on the behavior of ethernet transceivers. Furthermore, the signatures for all three components share some similar features. However, the differences among them in the spectrum are still significant enough to distinguish the components. For example, the carrier for ETH1 has a larger spread than the others. At the same time, it has more instances of weak activity distributed throughout the spectrum. On the other hand, the spectrum from ETH2 has more activity within the first 10 kHz of the carrier. Finally, the spectrum for ETH3 has less interference than the other components.

Example measurements from A10-LIME, A20-MICRO, and A20-LIME2 are used for training. The classification accuracy for each individual Ethernet transceiver is shown in Table 12.6. The overall classification accuracy for the transceivers is 97.7%. As the table demonstrates, all the transceivers had some classification error, the worst being the A20-LIME1 with a total error of 3.8%. The errors are likely the result of several factors. First, the features of the signature are relatively weak, making them more vulnerable to noise. Second, there are variations in signatures from the same type of component, making it difficult for the algorithm to correctly group all the measurements from the same transceivers together. Third, features shared between signatures from different types of transceivers make it difficult for the algorithm to distinguish the different types of transceivers. The classification accuracy could be potentially improved by changing the measurement settings. Some possibilities include increasing the measurement bandwidth (to increase the number of features for classification), increasing the measurement time (to decrease the influence of noise), or using a different set of instructions for exciting the component.

The average distances for the Ethernet transceivers are shown in Table 12.7. As the table demonstrates, the class-distances are still larger than the self-distances for each transceiver.

12.5 Experimental Validation

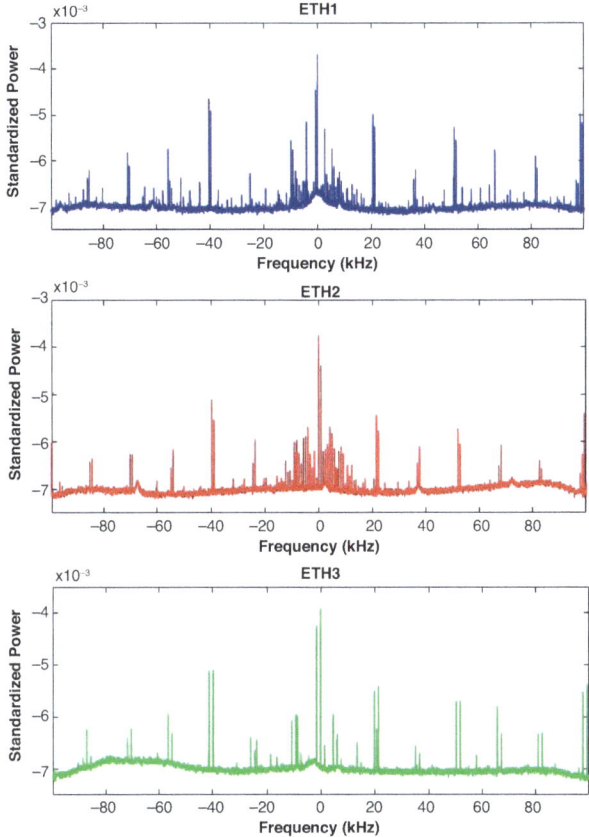

Fig. 12.10 Comparison of the EM signatures from ETH1 (top in blue), ETH2 (middle in red), and ETH3 (bottom in green) [43]

Table 12.6 Confusion matrix for Ethernet transceivers (in %) [43]

	ETH1	ETH2	ETH3
ETH1: A10-LIME	98.2	1.8	0
ETH2: A20-LIME1	2.3	96.2	1.5
ETH3: A20-LIME2	0	3.2	96.8
ETH2: A20-MICRO	0	99.7	0.3

Table 12.7 Average distances between Ethernet transceivers [43]

	ETH1	ETH2	ETH3
ETH1	3.5	11.5	14.4
ETH2	11.5	3.5	5.4
ETH3	14.4	5.4	2.5

References

1. Dakshi Agrawal, Bruce Archambeault, Josyula R Rao, and Pankaj Rohatgi. The EM side–channel (s). In *International workshop on cryptographic hardware and embedded systems*, pages 29–45. Springer, 2002.
2. J. H. Anderson. A PUF design for secure FPGA-based embedded systems. In *2010 15th Asia and South Pacific Design Automation Conference (ASP-DAC)*, pages 1–6, 2010.
3. Claudia Beleites, Richard Baumgartner, Christopher Bowman, Ray Somorjai, Gerald Steiner, Reiner Salzer, and Michael G Sowa. Variance reduction in estimating classification error using sparse datasets. *Chemometrics and intelligent laboratory systems*, 79(1-2):91–100, 2005.
4. Robert Callan, Nina Popovic, Angel Daruna, Eric Pollmann, Alenka Zajic, and Milos Prvulovic. Comparison of electromagnetic side-channel energy available to the attacker from different computer systems. In *2015 IEEE International Symposium on Electromagnetic Compatibility (EMC)*, pages 219–223. IEEE, 2015.
5. Robert Callan, Alenka Zajic, and Milos Prvulovic. A practical methodology for measuring the side-channel signal available to the attacker for instruction-level events. In *2014 47th Annual IEEE ACM International Symposium on Microarchitecture*, pages 242–254. IEEE, 2014.
6. Robert Callan, Alenka Zajic, and Milos Prvulovic. FASE: finding amplitude-modulated side-channel emanations. In *2015 ACM IEEE 42nd Annual International Symposium on Computer Architecture (ISCA)*, pages 592–603. IEEE, 2015.
7. Zentel Electronics Corp. DE0-CV cyclone V board. https://www.intel.com/content/www/us/en/programmable/solutions/partners/partner-profile/terasic-inc-/board/de0-cv-cyclone-v-board.html, OPTurldate = July 15, 2020.
8. Intel Corporation. Altera DE1 board. https://www.intel.com/content/www/us/en/programmable/solutions/partners/partner-profile/terasic-inc-/board/altera-de1-board.html, OPTurldate = July 15, 2020.
9. Zentel Electronics Corporation. A3V64S40GTP/GBF. https://zentel-europe.com/datasheets/A3V64S40GTP_v1.3_Zentel.pdf, OPTurldate = July 15, 2020.
10. ARM Developer. A10-olinuxino-lime. https://www.olimex.com/Products/OLinuXino/A10/A10-OLinuXino-LIME-n4GB/open-source-hardware, OPTurldate = Feb. 9, 2020.
11. ARM Developer. A13-olinuxino. https://www.olimex.com/Products/OLinuXino/A13/A13-OLinuXino/open-source-hardware, OPTurldate = Feb. 9, 2020.
12. ARM Developer. A13-olinuxino-micro. https://www.olimex.com/Products/OLinuXino/A13/A13-OLinuXino-MICRO/open-source-hardware, OPTurldate = Feb. 9, 2020.
13. ARM Developer. A20-olinuxino-lime. https://www.olimex.com/Products/OLinuXino/A20/A20-OLinuXino-LIME/open-source-hardware, OPTurldate = Feb. 9, 2020.
14. ARM Developer. A20-olinuxino-lime2. https://www.olimex.com/Products/OLinuXino/A20/A20-OLinuXino-LIME2/open-source-hardware, OPTurldate = Feb. 9, 2020.
15. ARM Developer. A20-olinuxino-micro. https://www.olimex.com/Products/OLinuXino/A20/A20-OLinuXino-MICRO/open-source-hardware, OPTurldate = Feb. 9, 2020.
16. ARM Developer. A33-olinuxino.
17. ARM Developer. Cortex-a7.
18. ARM Developer. Cortex-a8. https://developer.arm.com/ip-products/processors/cortex-a/cortex-a8, OPTurldate = Feb. 9, 2020.
19. Moumita Dey, Alireza Nazari, Alenka Zajic, and Milos Prvulovic. EMPROF: Memory profiling via EM-emanation in IoT and hand-held devices. In *2018 51st Annual IEEE/ACM International Symposium on Microarchitecture (MICRO)*, pages 881–893. IEEE, 2018.
20. H. Dogan, D. Forte, and M. M. Tehranipoor. Aging analysis for recycled FPGA detection. In *2014 IEEE International Symposium on Defect and Fault Tolerance in VLSI and Nanotechnology Systems (DFT)*, pages 171–176, Oct 2014.
21. Karine Gandolfi, Christophe Mourtel, and Francis Olivier. Electromagnetic analysis: Concrete results. In *International workshop on cryptographic hardware and embedded systems*, pages 251–261. Springer, 2001.

22. U. Guin, K. Huang, D. DiMase, J. M. Carulli, M. Tehranipoor, and Y. Makris. Counterfeit integrated circuits: A rising threat in the global semiconductor supply chain. *Proceedings of the IEEE*, 102(8):1207–1228, Aug 2014.
23. Integrated Silicon Solution Inc. IS42/45R86400D/16320D/32160DIS42/45S86400D/16320D/32160D. http://www.issi.com/WW/pdf/42-45R-S_86400D-16320D-32160Dss.pdf, OPTurldate = July 15, 2020.
24. OLINUXINO is Open Source. Arm developer. https://github.com/OLIMEX/OLINUXINO ,OPTurldate = Feb. 9, 2020.
25. E.J. Jorgensen, F.T. Werner, M. Prvulovic, and Zajic A. Deep learning classification of motherboard components by leveraging em side-channel signals. *J Hardw Syst Security*, 5:114–126, 2021.
26. Nathalie Kae-Nune and Stephanie Pesseguier. Qualification and testing process to implement anti-counterfeiting technologies into IC packages. In *2013 Design, Automation & Test in Europe Conference & Exhibition (DATE)*, pages 1131–1136. IEEE, 2013.
27. Micron. MT41K256M16HA-125 XIT. https://www.micron.com/products/dram/ddr3-sdram/part-catalog/mt41k256m16ha-125-xit, OPTurldate = Feb. 9, 2020.
28. Michael Pecht and Sanjay Tiku. Bogus: electronic manufacturing and consumers confront a rising tide of counterfeit electronics. *IEEE spectrum*, 43(5):37–46, 2006.
29. Milos Prvulovic, Alenka Zajic, Robert L Callan, and Christopher J Wang. A method for finding frequency-modulated and amplitude-modulated electromagnetic emanations in computer systems. *IEEE Transactions on Electromagnetic Compatibility*, 59(1):34–42, 2016.
30. Realtek. RTL8201CP. http://realtek.info/pdf/rtl8201cp.pdf, OPTurldate = Feb. 9, 2020.
31. Riscure. Em probe station. https://getquote.riscure.com/en/quote/2101064/em-probe-station.htm, 2020 (accessed Feb. 9, 2020).
32. Samsung. K4B4G1646E-BYK0. https://www.samsung.com/semiconductor/dram/ddr3/K4B4G1646E-BYK0/, OPTurldate = Feb. 9, 2020.
33. Hynix Semiconductor. H5TQ2G63BFR. https://www.skhynix.com/eolproducts.view.do?pronm=DDR3+SDRAM&srnm=H5TQ2G63BFR&rk=19&rc=computing, OPTurldate = Feb. 9, 2020.
34. Hynix Semiconductor. H5TQ2G83FFR. https://www.skhynix.com/eolproducts.view.do?pronm=DDR3+SDRAM&srnm=H5TQ2G83FFR&rk=19&rc=consumer, OPTurldate = Feb. 9, 2020.
35. Allwinner Technology. Allwinner A10. https://web.archive.org/web/20160729114202/http://www.allwinnertech.com/en/clq/processora/A10.html, OPTurldate = June 10, 2020.
36. Allwinner Technology. Allwinner A13. https://web.archive.org/web/20160811122523/http://www.allwinnertech.com/en/clq/processora/A13.html, OPTurldate = June 10, 2020.
37. Allwinner Technology. Allwinner A20.
38. Allwinner Technology. Allwinner A33. http://www.allwinnertech.com/index.php?c=product&a=index&id=23, OPTurldate = June 10, 2020.
39. Microchip Technology. KSZ9031. https://www.microchip.com/wwwproducts/en/KSZ9031, OPTurldate = Feb. 9, 2020.
40. Microchip Technology. LAN8710A. https://www.microchip.com/wwwproducts/en/LAN8710A, OPTurldate = Feb. 9, 2020.
41. Christopher Wang, Robert Callan, Alenka Zajic, and Milos Prvulovic. An algorithm for finding carriers of amplitude-modulated electromagnetic emanations in computer systems. In *2016 10th European Conference on Antennas and Propagation (EuCAP)*, pages 1–5. IEEE, 2016.
42. Frank Werner, Derrick Albert Chu, Antonije R Djordjević, Dragan I Olćan, Milos Prvulovic, and Alenka Zajic. A method for efficient localization of magnetic field sources excited by execution of instructions in a processor. *IEEE Transactions on Electromagnetic Compatibility*, 60(3):613–622, 2017.
43. Frank T. Werner, Baki Berkay Yilmaz, Milos Prvulovic, and Alenka Zajic. Leveraging em side-channels for recognizing components on a motherboard. *IEEE Transactions on Electromagnetic Compatibility*, 63(2):502–515, 2021.

44. Frank T. Werner, Baki Berkay Yilmaz, Milos Prvulovic, and Alenka Zajic. Leveraging em side-channels for recognizing components on a motherboard. *IEEE Transactions on Electromagnetic Compatibility*, 63(2):502–515, 2021.
45. Baki Berkay Yilmaz, Elvan Mert Ugurlu, Milos Prvulovic, and Alenka Zajic. Detecting cellphone camera status at distance by exploiting electromagnetic emanations. In *MILCOM 2019-2019 IEEE Military Communications Conference (MILCOM)*, pages 1–6. IEEE, 2019.

Chapter 13
Using Analog Side Channels to Attack Cryptographic Implementations

13.1 Introduction

Side channel attacks have historically been the most well-known use for analog side channels. These attacks extract sensitive information, such as cryptographic keys, from analog side channel signals that are collected while the target device is performing computation that uses the sensitive information. The analog signals that have been used for side channel attacks include electromagnetic emanations created by current flows within the device's computational and power-delivery circuitry [2, 3, 19, 26, 40, 55], variation in power consumption [12, 17, 20, 22, 28, 30, 38, 39, 42, 49], and also sound [9, 21, 29, 50], temperature [18, 33], and chasis potential variation [28], all of which can mostly be attributed to variation in power consumption and its interaction with the system's power delivery circuitry. Note that not all side channel attacks use *analog* signals—some rely on probing the state of the processor's caches [11, 51] and/or branch predictors [1], some others use outputs produced in the presence of faults [16], etc.

Most of the research on analog side-channel attacks has focused on relatively simple devices, such as simple embedded systems and smartcards, where side-channel signals can be acquired with bandwidth that is much higher than the clock rates of the target processor and other relevant circuitry (e.g., hardware accelerators for encryption/decryption), and usually with highly intrusive access to the device, e.g., with small probes placed directly onto the chip's package [24, 39]. Recently, attacks on higher-clock-rate devices, such as smartphones and PCs, have been demonstrated [10, 25–27]. These recent results have demonstrated that analog side channel attacks are possible even when signals are acquired with bandwidth that is much lower than the (gigahertz-range) clock rates of the processor, with less-intrusive access to the device, and in the presence of complex behavior caused by advanced performance-oriented features (e.g., multiple instructions per cycle instruction scheduling, etc.), and system software activity (interrupts, multiprocessing, etc.).

To overcome the problems of low signal bandwidth and high signal variability in high-clock-rate systems, attacks on those systems typically identify parts of the computation that are long-lasting (many processor cycles) and have a systematic bias in computational activity that is secret-dependent. As a representative example of this, consider decryption in RSA, which consists of modular exponentiation of the ciphertext using an exponent that is derived from the private key. The attacker's goal in this scenario is to recover enough bits of that secret exponent from the side channel signal, and then compute the remaining parts of the secret key by combining the information that was obtained from side channel signals with other information that is already known to the attacker, such as the public key (usually publicly available) and the ciphertext itself (which may be intercepted by the attacker or even sent by the attacker to the victim). Most of the computational activity in large-integer modular exponentiation is devoted to multiplication and squaring operations, where each squaring (or multiplication) operation operates on large integers and thus takes many processor cycles.

Historically, analog side-channel attacks on RSA have obtained information about a bit in the exponent from a signal snippet that corresponds to an entire large-integer squaring or multiplication operation [15, 25, 28, 29], often relying on a ciphertext that was specifically crafted to produce a large difference in one or more of these signal snippets, e.g., by ensuring that a zero-value bit in the exponent results in having a lot of zero-valued operands during execution of the large-number squaring and/or multiplication function/procedure. This focus on long-lasting operations is understandable, because side channel attacks ultimately recover information by first (1) identifying the relevant snippets in the overall signal timeline and then (2) categorizing the snippet according to which exponent-bit-value it corresponds to. Long signal snippets that correspond to large-number squaring and multiplication are both (1) easier to identify in the overall signal, and (2) provide enough signal samples for successful classification even when using relatively low sampling rates.

However, these large-integer operations are each very regular in terms of the sequence of instructions they perform, and the operands used in those instructions are ciphertext-dependent, so classification of signals according to exponent-related properties is difficult unless either (1) the sequence of square and multiply operations is key-dependent– this happens in some naive implementations, where the entire multiplication operation is skipped if the exponent bit is 0, or (2) the attacker can control the ciphertext that will be exponentiated and can select the ciphertext that will produce systematically different side channel signals for different bit-values in the exponent.

To eliminate the problem of the sequence of operation being secret-dependent, modern cryptographic implementations have introduced the concept of *constant-time code*, i.e., code whose sequence of operations and memory access patterns are the same in every execution of the code. This prevents attackers from extracting secrets from information about *which* parts of the code are executed (or not executed) as the computation progresses, even if they can obtain such information from the side channel.

To eliminate the problem of attacker-controlled ciphertext, some modern cryptographic algorithms, such as the elliptic curve (EC) algorithm Curve25519 [14] are designed to intrinsically tend not to produce significant secret-dependent biases in the operands during computation. Other code, such as modern implementations of RSA, prevents attacker-controlled ciphertext by applying *blinding*, which exploits mathematical properties of exponentiation such that a random number is combined with the ciphertext prior to exponentiation, the exponentiation is then performed on the no-longer-attacker-controlled combined ciphertext, and then the impact of the random number is removed from the result. However, this blinding does come at the cost of lower performance and higher energy consumption for the RSA decryption/signing operation.

Overall, the evolution of implementations for public-key cryptographic operations has mainly been driven by the desire to achieve efficiency while preventing new side channel attacks from easily recovering the private key when it is used, and we next present a historical overview for RSA, a well-known public-key cryptographic approach whose implementations have been subjected to decades of increasingly sophisticated side channel attacks. We then explore the history of Elliptic Curve (EC) cryptographic implementations, which have initially benefited from the lessons learned on the (older) RSA approach.

After considering the history of RSA and of EC cryptographic implementations, we represent two recent analog side channel attacks, one for RSA and one for EC, that target relatively brief computation—the lookup of the exponent bit in RSA and the conditional swap operation in EC—along with mitigation for each attack, to show that even brief computation in constant-time implementations must to be carefully designed to prevent more advanced analog side channel attacks.

13.2 Historical Overview of Analog Side-Channel Attacks on RSA

For clarity, we will focus on RSA's decryption operation, which at its core performs modular exponentiation of the ciphertext c with a secret exponent (d) modulo m or, in more efficient implementations that rely on the Chinese Reminder Theorem (CRT), two such exponentiations, with secret exponents d_p and d_q with modulo p and q, respectively. The side-channel analysis thus seeks to recover either d or, in CRT-based implementations, d_p and d_q, using side-channel measurements obtained while exponentiation is performed. We note, however, that nearly all of the discussion presented here also applies to RSA's signing operation, which also uses modular exponentiation with secret exponents.

The exponentiation is implemented as either left-to-right (starting with the most significant bits) or right-to-left (starting with the least significant bits) traversal of the bits of the exponent, using large-integer modular multiplication to update the result until the full exponentiation is complete. Left-to-right implementations are more common, and without loss of generality our discussion will focus on left-to-right implementations. We use c to denote the ciphertext, d for the secret exponent, and m for the modulus. A simple implementation of exponentiation considers one exponent bit at a time, as shown in Fig. 13.1, which is adapted from OpenSSL's source code.

The BN prefix in Fig. 13.1 stands for "Big Number" (i.e., large integer). Each large integer is represented by an array of *limbs*, where a limb is an ordinary (machine-word-sized) integer. The BN_is_bit_set(d,b) function returns the value (0 or 1) of the b-th bit of large-integer exponent d. This function only requires a few processor instructions to compute the index of the *limb* (array element) that contains the requested bit, load that *limb*, then shift and bit-mask to keep only the requested bit. The instructions that implement the loop, the if statement, and function call/return in Fig. 13.2 are also relatively few in number.

However, the BN_mod_mul operation is much more time-consuming. It requires numerous multiplication instructions that operate on the limbs of the large-integer multiplicands. As we have briefly discussed earlier in this chapter, such long-lasting operations produce numerous signal samples even when the signal is collected with a limited sampling rate. If the timing and/or shape of the signal produced by these long-lasting operations is secret-dependent, the many samples produced by each of the long-lasting operations help attackers accurately categorize each signal snippet, e.g., to determine whether the exponent bit was 0 or 1 for that snippet.

Specifically, large integers c, d, and m (or, in CRT-based implementations the d_q, d_p and the corresponding moduli), all have $O(n)$ bits and thus $O(n)$ limbs, where n is the size of the RSA cryptographic key. A grade-school implementation of BN_mod_mul thus requires $O(n^2)$ limb multiplications, but the Karatsuba multiplication algorithm [35] is typically used to reduce this to $O(n^{log_2 3}) \approx O(n^{1.585})$, In most modern implementations, a significant further performance improvement is achieved by converting the ciphertext to a Montgomery representation, using Montgomery multiplication for BN_mod_mul during exponentiation, and at the end converting the result r back to the standard representation.

```
1   // result r starts out as 1
2   BN_one(r);
3   // For each bit of exponent d
4   for (b=bits-1;b>=0;b--){
5          // r = r*r mod m
6          BN_mod_mul(r,r,r,m);
7          if(BN_is_bit_set(d,b))
8                 // r = r*c mod m
9                 BN_mod_mul(r,r,c,m);
10  }
```

Fig. 13.1 A simple implementation of large-number modular exponentiation

13.2 Historical Overview of Analog Side-Channel Attacks on RSA

```
1   BN_one(r);
2   wstart=bits-1;
3   while(wstart>=0){
4           if(!BN_is_bit_set(d,wstart)){
5                   // Window is 0, square and
6                   // begin a new window
7                   BN_mod_mul(r,r,r,m);
8                   wstart--;
9                   continue;
10          }
11          wval=1;
12          w=1;
13          // Scan up to max window length
14          for(i=1;i<wmax;i++){
15                  // Don't go below exponent's LSB
16                  if(wstart-i<0)
17                          break;
18                  // If 1 extend window to it
19                  if(BN_is_bit_set(d,wstart-i)){
20                          wval=(wval<<(i-w+1))+1;
21                          w=i;
22                  }
23          }
24          // Square result w times
25          for(i=0;i<w;i++)
26                  BN_mod_mul(r,r,r,m);
27          // Multiply window's result
28          // into overall result
29          BN_mod_mul(r,r,ct[wval>>1],m);
30          // Begin a new window
31          wstart-=w;
32  }
```

Fig. 13.2 Sliding-window implementation of large-number modular exponentiation

Even with Montgomery multiplication, however, the vast majority of execution time for large-number exponentiation is spent on large-number multiplications: this is why performance optimizations focus on reducing the number of these multiplications. Likewise, most of the side-channel measurements (e.g., signal samples) collected during large-number exponentiation correspond to large-number multiplication activity, so existing side channel cryptanalysis approaches tend to target this multiplication activity.

One class of attacks focuses on distinguishing between squaring ($r*r$) and multiplication ($r*c$) operations, and recovering information about the secret exponent from the sequence in which they occur. Examples of such attacks include FLUSH+RELOAD [54] (which uses instruction cache behavior) and Percival's attack [47], which uses data cache behavior. In the naive implementation above, an occurrence of squaring tells the attacker that the next bit of the exponent is being used, and an occurrence of multiplication indicates that the value of that bit is 1, so an attack that correctly recovers the square-multiply sequence can trivially obtain all bits of the secret exponent.

To improve performance, most modern implementations of RSA use window-based exponentiation, where squaring is needed for each bit of the exponent, but a multiplication is needed only once per a multi-bit group (called a *window*) of exponent bits. A left-to-right (starting at the most significant bit) *sliding-window* implementation scans the exponent bits and forms windows of varying length. Since a window that contains only zero bits requires no multiplication (and thus cannot benefit from forming multi-bit windows), only windows that begin and end with 1-valued bits are allowed to form multi-bit windows; zero bits in-between these windows are each treated as their own single-bit windows that can omit multiplication. An example of a sliding-window implementation is shown in Fig. 13.2, using code adapted from OpenSSL's source code for sliding-window modular exponentiation. The sliding-window approach chooses a maximum size wmax for the windows it will use, pre-computes a table ct that contains the large-integer value $c^{wval} mod m$ for each possible value $wval$ up to $wmax$ length, and then scans the exponent, forming windows and updating the result for each window.

In this algorithm, a squaring (lines 7 and 26 in Fig. 13.2) is performed for each bit, but the multiplication operation (line 29) is performed only at the (1-valued) LSB of a non-zero window. Thus the square-multiply sequence reveals where some of the 1-valued bits in the exponent are, and additional bits of the exponent have been shown to be recoverable [15] by analyzing the number of squarings between each pair of multiplications. The fraction of bits that can be recovered from the square-multiply sequence depends on the maximum window size wmax, but commonly used values of wmax are relatively small, and prior work [15] has experimentally demonstrated recovery of 49% of the exponent's bits on average from the square-multiply sequence when wmax is 4. Additional techniques [15, 32] have been shown to recover the full RSA private key once enough of the exponent bits are known, and for wmax value of 4 this has allowed full key recovery for 28% of the keys [15]. Finally, recent work has shown that fine-grained control flow tracking through analog side channels can be very accurate [37]. Because this sliding-window implementation uses each bit of the exponent to make at least one control flow decision, highly accurate control flow tracking amounts to discovering the exponent's bits (with a small probability of error).

Concerns about exponent-dependent square-multiply sequences have led to adoption of *fixed window* exponentiation in OpenSSL, which keeps the performance

13.2 Historical Overview of Analog Side-Channel Attacks on RSA

```
1   b=bits-1;
2   while(b>=0){
3       wval=0;
4       // Scan the window,
5       // squaring the result as we go
6       for (i=0;i<w;i++) {
7           BN_mod_mul(r,r,r,m);
8           wval<<=1;
9           wval+=BN_is_bit_set(d,b);
10          b--;
11      }
12      // Multiply window's result
13      // into the overall result
14      BN_mod_mul(r,r,ct[wval],m);
15  }
```

Fig. 13.3 Fixed-window implementation of large-number modular exponentiation

advantages of window-based implementation, but has an exponent-independent square-multiply sequence. This implementation is represented in Fig. 13.3, again adapted from OpenSSL's source code.

All windows now have the same number of bits w, with exactly one multiplication performed for each window—in fact, all of the control flow is now exactly the same regardless of the exponent. Note that the window value (which consists of the bits from the secret exponent) directly determines which elements of ct are accessed. These elements are each a large integer from OpenSSL's "Big Number" BN structure, i.e., an array of ordinary integers. Since each such array is much larger than a cache block, different large integers occupy distinct cache blocks, and thus the cache set that is accessed when reading the elements of the ct array reveals key material. Percival's attack [47], for example, can note the sequence in which the cache sets are accessed by the victim during fixed-window exponentiation, which reveals which window values were used and in what sequence, which in turns yields the bits of the secret exponent. To mitigate such attacks, the implementation in OpenSSL has been changed to store ct such that each of the cache blocks it contains parts from a number of ct elements, and therefore the sequence of memory blocks that are accessed in each ct [wval] lookup leak none or very few bits of that lookup's *wval*.

Another broad class of side channel attacks relies on choosing the ciphertext such that the side-channel behavior of the modular multiplication reveals which of the possible multiplicands is being used. For example, Genkin et al. [28, 29] construct a ciphertext that produces many zero limbs in any value produced by multiplication with the ciphertext, but when squaring such a many-zero-limbed value the result has fewer zero limbs, resulting in an easily-distinguishable side channel signals whenever a squaring operation (BN_mod_mul(r,r,r,m) in our examples) immediately follows a 1-valued window (i.e., when r is equal to $r_{prev} * c \bmod m$). This approach has been extended [26] to construct a (chosen) ciphertext that reveals when a particular window value is used in multiplication in a windowed implementation, allowing full recovery of the exponent by collecting signals that correspond to 2^w chosen ciphertexts (one for each window value). However, chosen-ciphertext attacks can be prevented in the current implementation of OpenSSL by enabling *blinding*, which combines the ciphertext with an encrypted (using the public key) random "ciphertext", performs secret-exponent modular exponentiation on this *blinded* version of the ciphertext, and then "unblinding" the decrypted result.

Overall, because large-integer multiplication is where large-integer exponentiation spends most of its time, most of the side-channel measurements (e.g., signal samples for physical side channels) also correspond to this multiplication activity and thus both attacks and mitigation tend to focus on that part of the signal, leaving the (comparably brief) parts of the signal in-between the multiplications largely unexploited by attacks but also unprotected by countermeasures.

Later in this chapter, we will describe in more detail a recent analog side-channel attack [4] for RSA, which targets the signal that corresponds to computing the value of the window, i.e., the signal *between* the multiplications. Of course, we will also examine the recent changes to RSA implementations that mitigate this attack [4].

13.3 Historical Overview of Analog Side-Channel Attacks on Elliptic Curve

We will use the Elliptic Curve Digital Signature Algorithm (ECDSA) as a representative example of Elliptic Curve cryptography. ECDSA generates a cryptographic signature using a private key, and this signature can be verified using the corresponding public key. First, the participating parties must agree on a choice of curve parameters: the elliptic curve field and equation, denoted here as CURVE, the base point of prime order on the curve, denoted here as G, and an integer order of G, denoted here as n. Given the curve parameters (CURVE,G,n), the signing entity (which we refer to as Alice) secretly chooses as its private key a random positive integer d_A such that $0 < d_A < n$, and computes the curve point $Q_A = d_A \times G$ that will serve as the public key that corresponds to the secret key d_A. The public key is then provided to the verifying entity, which we refer to here as Bob.

To sign a message, Alice first computes e, the cryptographic hash of the message, and then computes z, the value of e truncated to the bit-length of n. Next, Alice randomly chooses an ephemeral secret positive integer (nonce) k such that $0 < k < n$, calculates the curve point $(x, y) = k \times G$ where $\times$ stands for an EC point multiplication by a scalar, then computes $r = x \bmod n$ and $s = k^{-1}(z+r\dot{d}_A) \bmod n$.

If either r or s is zero, a new k is chosen and the signature is recomputed, otherwise the signature consists of the pair (r, s).

To verify the signature, Bob also computes the cryptographic hash of the message and the corresponding value of z. Bob then computes $w = s^{-1} \bmod n$, $u_1 = zw \bmod n$, and $u_2 = rw \bmod n$, then computes the curve point $(x, y) = u_1 \times G + u_2 \times Q$. The signature is valid when r and s are both within the interval $[1, n - 1]$, $(x, y)! = O$, and $r \equiv x \pmod{n}$.

The attacker (which we refer to as Eve) already knows the curve parameters (CURVE,G,n), and can typically obtain Alice's public key Q_A and a large number of messages signed by Alice. Under these conditions, Eve's attempt to calculate the private key d_A requires solving a complex discrete logarithm problem. However, if Eve can obtain the secret nonce k for even one signature, the secret key d_A can be trivially recovered as $d_A = (sk - z)/r$.

One way for Eve to obtain the value of k is to mount a side channel attack during Alice's computation of the signature. Since the value of k is ephemeral—it is generated, used to compute a signature for one message, and then discarded, Eve can try to extract the value of k when it is used in EC point multiplication $(k \times G)$, or when its inverse is used as a multiplicand to compute s. Of the two, the point multiplication is typically the more promising target for the side channel attacks because it is more time-consuming and because its implementation performs operations on curve points depending on the bits of k, which has a tendency to leak information about k unless the implementation is very carefully constructed to avoid doing so.

13.3.1 Point-by-Scalar Multiplication

The naive implementation of EC point multiplication is shown in Fig. 13.4.

In this implementation of point multiplication, for each bit of the scalar k, the previous result r is doubled and, if the bit in k is non-zero, the point p is added to that result. The point-double and point-add operations differ in both the code that

```
1  EC_point_zero(r);  // r=0
2  // For each bit of the scalar k
3  for(b=bits-1;b>=0;b--){
4          EC_point_double(r,r);    // r=2*r
5          if(BN_is_bit_set(k,b))
6                  EC_point_add(r,r,p);  // r = r+p
7
8  }
```

Fig. 13.4 A naive double-and-add implementation of EC point multiplication by a scalar

is executed and the data that is accessed, so various side channels can be used to recover the sequence of these operations and, from that sequence, recover the bits of the scalar k. For example, instruction cache accesses can be used to determine when each point-double and point-add function call occurs. As another example, the timing of data cache accesses to p can be used to determine which point-double operations are followed by point-add operations (note that p is only accessed during a point-add). Also, various analog side channel signals can be used to determine when each point-double and point-add is executed.

For performance reasons, instead of a binary double-and-add implementation, both libgcrypt and OpenSSL have until recently used an implementation based on the non-adjacent form (NAF) of the scalar. In the NAF representation, the scalar is still represented as a sequence of digits, but each digit now represents multiple bits of the scalar. Note the analogy between RSA's exponentiation and EC's point-by-scalar multiplication. In this analogy, RSA's exponent corresponds's to EC's nonce, RSA's big-number squaring operation corresponds to EC's point-double operation, and RSA's big-number multiplication corresponds to EC's point-doubling. Following this analogy, the digits in the NAF representation of the nonce in EC correspond to windows of the RSA's exponent. After pre-computing the value of the point multiplied by each of the possible value of a digit, the NAF-based implementation requires only one point-add for each non-zero digit in the NAF representation of the scalar, and this significantly reduces the number of point-add operations that are needed for the overall point multiplication. However, the pattern of accesses to the table of pre-computed point values now directly corresponds to the NAF representation of k, which allows cache-based attacks to easily recover k [7, 13, 52, 53]. Furthermore, the point-add is still skipped for a zero-valued NAF digit in k, so the sequence of point-double and point-add operations still leaks partial information about k, just like the squaring-multiplication sequence leaked partial information about the exponent in a sliding-window implementation of RSA. This has been exploited by analog side channel attacks, where partial information from multiple signing operations (that use the same private key d_A, but different scalars k) was combined to eventually recover d_A.

To mitigate the side-channel problems of the NAF-based implementation, both libgcrypt and OpenSSL have recently switched to using constant-time implementations of point multiplication during EC signing. Figure 13.5 shows the constant-time implementation adapted from libgcrypt's source code (function _gcry_mpi_ec_mul_point() in mpi/ec.c). This is a double-and-add implementation but, unlike the naive implementation, it executes a point-add for every bit in k, so the sequence of point-double and point-add operations (and the data access pattern) no longer leaks information about k. The output of the point-add operation (tmp in our code example) is then conditionally swapped with the result r, using the bit of k as a condition that dictates whether the swap occurs or not. When a zero-valued bit is encountered in k, the values of tmp and r are left unchanged, which effectively discards the result of the point-add operation. Conversely, when a one-valued bit is encountered in k, the swap results in using the output of the point-add operation as the new value of r.

13.3 Historical Overview of Analog Side-Channel Attacks on Elliptic Curve

```
1  EC_point_zero(r);  // r=0
2  // For each bit of the scalar k
3  for(b=bits-1;b>=0;b--){
4          EC_point_double(r,r);      // r=2*r
5          EC_point_add(tmp,r,p);     // tmp=r+p
6          // Swap r and tmp if b-th bit of k is 1
7          curr_bit=BN_is_bit_set(k,b);
8          EC_point_swap_cond(r,tmp,curr_bit);
9  }
```

Fig. 13.5 libgcrypt's constant-time EC point multiplication

```
1   EC_point_ladder_prep(r,s,p);
2   prev_bit=1;
3   // For each bit of the scalar k
4   for(b=bits-1;b>=0;b--){
5           curr_bit=BN_is_bit_set(k,b);
6           EC_point_swap_cond(r,s,curr_bit^prev_bit);
7           EC_point_ladder_step(r,s,p);
8           prev_bit=curr_bit;
9   }
10  EC_point_swap_cond(r,s,prev_bit);
11  EC_point_ladder_post(r,s,p);
```

Fig. 13.6 OpenSSL's constant-time EC point multiplication

Figure 13.6 shows the constant-time implementation adapted from OpenSSL's source code (function ec_scalar_mul_ladder() in crypto/ec/ec_mult.c). It differs from libgcrypt's implementation in that the doubling and add operations are integrated into a single Montgomery ladder [34, 43, 44] step, which also changes the condition used for the swap—rather than performing the swap depending on each individual bit of the scalar k, the swap is now performed when the i-th bit of k differs from the previous one.

In both libgcrypt's and OpenSSL's implementation of EC point multiplication, the conditional swap is used to avoid having the control flow of the point multiplication depend on individual bits of the scalar k. The implementations of the conditional swap itself in libgcrypt and in OpenSSL are very similar, and Fig. 13.7 shows the implementation of conditional swap adapted from libgcrypt's function _grcy_mpi_swap_cond) in source code file mpi/mpiutil.c. An EC point is stored as one large number for each coordinate, and each number is stored as a sequence of machine words (ordinary word-sized integers). For each machine word in the EC point representation, a bitwise exclusive-or (XOR) is used to compute the bit-wise difference between the word in EC point *a* and the corresponding word in EC point b. The mask is then applied to

```
1  EC_point_swap_cond(a,b,cond){
2          // When cond is 0, set mask to all-zeros
3          // When cond is 1, set mask to all-ones
4          mask=0-cond;
5          // For each machine word in the
6          // EC point representation
7          for(i=0;i<nwords;i++){
8                  delta = (a->w[i] ^ b->w[i]) & mask;
9                  a->w[i] = a->w[i] ^ delta;
10                 b->w[i] = b->w[i] ^ delta;
11         }
12 }
```

Fig. 13.7 Constant-time conditional swap. EC points a and b are swapped if cond is 1, but remain unchanged if cond is 0

this difference, such that the difference is either kept as-is or zeroed out, depending on the condition cond. This masked difference is then applied (via XOR operations) to the two words. The end result is that, when cond is true, the values of the two EC points are swapped, whereas when cond is false the two points are left unchanged. The key property of this conditional swap is that the same instruction sequence is executed, and the same sequence of data accesses is performed, regardless of the value of the condition, which prevents cache-based and many analog-signal side channel attacks from obtaining information about the value of the swap's condition, and thus about the bits of the scalar k in the EC point multiplication.

Because constant-time point multiplication implementations in both libgcrypt and OpenSSL result in executing exactly the same sequence of instructions, and exactly the same sequence of data accesses, regardless of the value of scalar k, they prevent side channel attacks that exploit either instruction or data cache behavior, as well as analog side channel attacks that rely on detecting signal differences caused by executing different program code depending on the bits of the scalar k.

The latest (as of this writing) analog side-channel attack on EC implementation is based on the observation that the conditional swap implementation creates systematic condition-dependent differences in the values of operands used by XOR instructions, and in the bit-toggling activity among words in the internal representation of the two EC points [5]. Specifically, when cond is true, the mask is all-ones, which results in values of delta that have about the same number of 0-valued and 1-valued bits, and this results in toggling about half of the bits in each word of EC points a and b. In contrast, when cond is false, the mask is all-zeroes, so the delta for each pair of words is zero, and no bits are toggled in any of the words that represent EC points a and b. A detailed description of this attack (and proposed defenses) will be presented later in this chapter.

13.4 One and Done: An EM-Based Attack on RSA

13.4.1 Threat Model

The attack model for this attack [4] assumes that there is an adversary who wishes to obtain the secret key used for RSA-based public-key encryption or authentication. It further assumes that the adversary can bring a relatively compact receiver into close proximity of the system performing these RSA secret-key operation. For example, the system to be attacked may be smart-infrastructure or smart-city device, which is located in a public location and uses public key infrastructure (PKI) to authenticate itself and secure its communication over the Internet. Another example is when the adversary can hide a relatively compact receiver in a location where systems can be placed in close proximity to it, e.g., under a cellphone charging station at a public location, under the tabletop surface in a coffee shop, etc.

This attack model also assumes that the adversary can access another device of the same type as the one being attacked and perform RSA decryption/authentication with known keys in preparation for the attack. This is a highly realistic assumption in most attack scenarios described above. Unlike many prior attacks on RSA, ours does **not** assume that the adversary can choose (or even know) the message (ciphertext for RSA decryption) to which the private key will be applied, and it further assumes that the RSA implementation under attack **may utilize blinding** to prevent such chosen-ciphertext attacks. Finally, it assumes that it is highly desirable for the attacker to recover the secret key after only very few uses (ideally only one use) of that key on the target device. This is a very realistic assumption because PKI is typically used only to set up a secure connection, typically to establish the authenticity of the communication parties and establish a symmetric-encryption session key, so in scenarios where the attacker's receiver can only be in close proximity to the target device for a limited time, very few uses of the private RSA key may be observed.

Targeted Software
The software target for this attack is OpenSSL version 1.1.0g [46], the latest version of OpenSSL at the time the attack was presented [4]. RSA decryption in OpenSSL version 1.1.0g used constant-time fixed-window large-number modular exponentiation to mitigate both timing-based attacks and attacks that exploit the exponent-dependent variation in the square-multiply sequence. The lookup tables used to update the result at the end of each window are stored in scattered form to mitigate attacks that examine the cache and memory behavior when reading these tables, and the RSA implementation supports blinding (which we turn on in our experiments) to mitigate chosen-ciphertext attacks.

Targeted Hardware
The targeted hardware are two modern Android-based smartphones and a Linux-based embedded system board, all with ARM processor clocked at frequencies around 1 GHz. In the experiments, probes are placed very close, but without physical

contact with the (unopened) case of the phone, while for the embedded system board we position the probes 20 cm away from the board, so we consider the demonstrated attacks close-proximity but non-intrusive.

13.4.2 Attack Method

In both fixed- and sliding-window implementations, the attack approach focuses on the relatively brief periods of computation that considers each bit of the exponent and forms the window value $wval$.

13.4.3 Receiving the Signal

The targeted computation is brief, and the different values of exponent bits produce relatively small variation in the side-channel signal, so the signals subjected to our analysis need to have sufficient bandwidth and signal-to-noise ratio for our analysis to succeed. To maximize the signal-to-noise ratio while minimizing intrusion, we position EM probes just outside the targeted device's enclosure. We then run RSA decryption in OpenSSL on the target device while recording the signal in a 40 MHz band around the clock frequency. The 40 MHz bandwidth was chosen as a compromise between on one hand, the recovery rate for the bits of the secret exponent and, on the other hand, the availability and cost of receivers capable of capturing the desired bandwidth. Specifically, the 40 MHz bandwidth is well within the capabilities of Ettus USRP B200-mini receiver, which is very compact, costs less than $1,000, and can receive up to 56 MHz of bandwidth around a center frequency that can be set between 70 MHz and 6 GHz. We found that 40 MHz of bandwidth is sufficient to recover nearly all bits of the secret exponent from a single instance of exponentiation that uses that exponent.

We then apply AM demodulation to the received signal, and then upsample it by a factor of 4. The upsampling consists of interpolating through the signal's existing sample points and placing additional points along the interpolated curve. This is needed because our receiver's sampling is not synchronized in any way to the computation of interest, so two signal snippets collected for the same computation may be misaligned by up to half of the sample period. Upsampling allows us to re-align these signals with higher precision, and we found that four-fold upsampling yields sufficient precision for our purposes.

13.4.4 Identifying Relevant Parts of the Signal

Figure 13.8 shows a brief portion of the signal that begins during fixed-window exponentiation in OpenSSL. It includes part of one large-number multiplication

13.4 One and Done: An EM-Based Attack on RSA

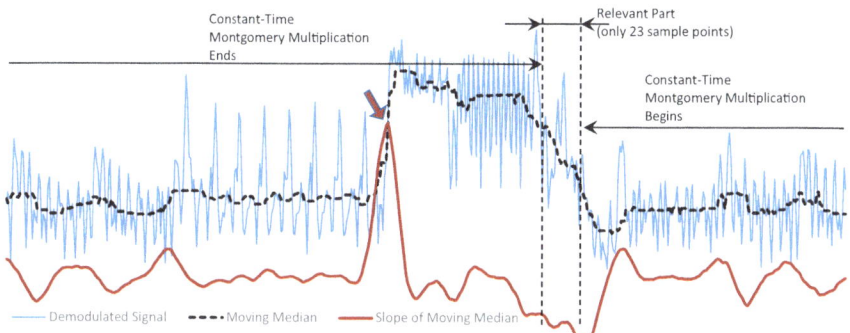

Fig. 13.8 Signal that includes the end of one Montgomery multiplication, then the part relevant to our analysis, and then the beginning of another Montgomery multiplication. The horizontal axis is time (from left to right) and the vertical axis is the magnitude of the AM-demodulated signal [4]

(Line 7 in Fig. 13.3), which in OpenSSL uses the Montgomery algorithm, with a constant-time implementation designed to avoid multiplicand-dependent timing variation that was exploited by prior side-channel attacks. The point in time where Montgomery multiplication returns and the relevant part of the signal begins is indicated by a dashed vertical line in Fig. 13.8. In this particular portion of the signal, the execution proceeds to lines 8 and 9 Fig. 13.3, where a bit of the exponent is obtained and added to $wval$, then lines 10 and 6, and then 7 where, at the point indicated by the second dashed vertical line, it enters another Montgomery multiplication, whose signal continues well past the right edge of Fig. 13.8. As indicated in the figure, the relevant part of the signal is very brief relative to the duration of the Montgomery multiplication.

A naive approach to identifying the relevant snippets in the overall signal would be to obtain reference signal snippets during training and then, during the attack, match against these reference snippets at each position in the signal and use the best-matching parts of the signal. Such signal matching works best when looking for a snippet that has prominent features (that are unlikely to be obscured by the noise), and whose prominent features occur in a pattern which is unlikely to exist elsewhere in the signal. Unfortunately, the signal snippets relevant for our analysis have little signal variation (relative to other parts of the signal), and they have a signal shape (just a few up-and-downs) that resembles many other parts of the signal. In contrast, the signal that corresponds to the Montgomery multiplication has stronger features, and they occur in a very distinct pattern.

Therefore, instead of finding instances of relevant snippets by matching them against their reference signals from training, we use as a reference the signal that corresponds to the most prominent change in the signal during Mongtomery multiplication, where the signal abruptly changes from a period with a relatively low signal level to a period with a relatively high signal level. We identify this point in the signal using a very efficient algorithm. We first compute the signal's moving median (thick dashed black curve in Fig. 13.8) to improve resilience to noise. We then examine the derivative (slope) of this moving median (thick red curve in Fig. 13.8) to identify peaks that significantly exceed its statistically expected variation. In Fig. 13.8 the thick red arrow indicates such a peak, which corresponds to the most prominent change in the Montgomery multiplication that precedes the relevant part of the signal. Because the implementation of the Montgomery multiplication was designed to have almost no timing variation, the signal snippet we actually need for analysis is at a fixed time offset from the point of this match.

Because this method of identifying the relevant snippets of the signal is based on the signal that corresponds to the Montgomery multiplication that precedes each relevant snippet, the same method can be used for extracting relevant signal snippets for both fixed-window and sliding-window exponentiation—in both cases the relevant snippet is at the (same) fixed offset from the point at which a prominent-enough peak is detected in the derivative of the signal's moving median. Essentially, we do not extract exponent bits from the large-number multiplication's signal, but we use the features of that signal to locate the signal snippets that we do use to extract exponent bits.

13.4.5 Recovering Exponent Bits in the Fixed-Window Implementation

In the fixed-window implementation, large-number multiplication is used for squaring (Line 7 in Fig. 13.3) and for updating the result after each window (Line 14). Thus there are four control-flow possibilities for activity between Montgomery multiplications.

The first two control flow possibilities begin when the Montgomery multiplication in line 7 completes. Both control flow possibilities involve updating the window value to include another bit from the exponent (lines 8, 9, and 10), and at line 6 incrementing i and checking it against w, the maximum size of the window. The first control flow possibility is the more common one—the window does not end and the execution proceeds to line 7 when another multiplication at line 7. We label this control flow possibility S-S (from a squaring to a squaring). The second control flow possibility occurs after the last bit of the window is examined and added to $wval$, and in that case the loop at line 6 is exited, the parameters for the result update at line 14 are prepared, and the Montgomery multiplication at line 14 begins. The parameter preparation in our code example would involve computing the address of $ct[wval]$ to create a pointer that would be passed to the Montgomery multiplication as its second multiplicand. In OpenSSL's implementation the ct is kept in a scattered format to minimize leakage of $wval$ through the cache side channel while computing the Montgomery multiplication, so instead the value of $wval$ is used to gather the scattered parts of $ct[wval]$ into a pre-allocated array that is passed to Montgomery multiplication. Since this pre-allocated array is used for all result-update multiplications, memory and cache behavior during the Montgomery multiplication no longer depend on $wval$. This means that this

13.4 One and Done: An EM-Based Attack on RSA

second control-flow possibility involves significant activity to gather the parts of the multiplicand and place them into the pre-allocated array, and only then the Montgomery multiplication at line 14 begins. We label this control flow possibility S-U (from a squaring to an update).

The last two control flow possibilities occur after the result update in line 14 completes its Montgomery multiplication. The loop condition at line 2 is checked, and then one control flow possibility (third of the four) is that the entire exponentiation loop exits. We label this control flow possibility U-X (from an update to an exit). The last control-flow possibility, which occurs for all windows except the last one, is that after line 2 we execute line 3, enter the window-scanning loop at line 6, and begin the next large-number Montgomery multiplication at line 7. We label this control flow possibility U-S (from an update to a squaring).

The sequence in which these four control flow possibilities are encountered in each window is always the same: $w - 1$ occurrences of S-S, then one occurrence of S-U, then either U-S or U-X, where U-X is only possible for the last window of the exponent.

The first part of our analysis involves distinguishing among these four control flow possibilities. The reason for doing so is that noise bursts, interrupts, and activity on other cores can temporarily interfere with our signal and prevent detection of Montgomery multiplication. In such cases, sole reliance on the known sequence of control flow possibilities would cause a "slip" between the observed sequence and the expected one, causing us to use incorrect reference signals to recover bits of the exponent and to put the recovered bits at incorrect positions within the recovered exponent.

The classification into the four possibilities is much more reliable than recovery of exponent's bits. Compared to the other three possibilities, S-U spends significantly more time between Montgomery multiplications (because of the multiplicand-gathering activity), so it can be recognized with high accuracy and we use it to confirm that the exponentiation has just completed a window. The U-X possibility is also highly recognizable because, instead of executing Montgomery multiplication after it, it leads to executing code that converts from Montgomery to standard large-number format, and it serves to confirm that the entire exponentiation has ended. The S-S and U-S snippets both involve only a few instructions between Montgomery multiplications so they are harder to tell apart, but our signal matching still has a very high accuracy in distinguishing between them.

After individual snippets are matched to the four possibilities, that matching is used to find the most likely mapping of the sequence of snippets onto the known valid sequence. For example, if for $w = 5$ we observe S-U, U-S, S-S, S-S, S-S, S-U, all with high-confidence matches, we know that one S-S is missing for that window. We then additionally use timing between these snippets to determine the position of the missing S-S. Even if that determination is erroneous, we will correctly begin the matching for the next window after the S-U, so a missing snippet is unlikely to cause any slips, but even when it does cause a slip, such a slip is very likely to be "contained" within one exponentiation window. Note that a missing S-U or S-S snippet prevents our attack from using its signal matching to recover the value of the corresponding bit. A naive solution would be to assign a random value to that bit (with a 50% error rate among missing bits). However, for full RSA key recovery, known-to-be-missing bits (erasures, where the value of the bit is known to be unknown) are much less problematic than errors (the value of the bit is incorrect but not known a priori to be incorrect), thus label these missing bits as erasures.

Finally, for S-S and S-U snippets we perform additional analysis to recover the bit of the exponent that snippet corresponds to. Recall that, in both S-S and S-U control flow possibilities, in line 9 a new bit is read from the exponent and is added to $wval$, and that bit is the one we will recover from the snippet. For ease of discussion, we will refer to the value of this bit as $bval$. To recover $bval$, in training we obtain examples of these snippets for each value of $bval$. To suppress the noise in our reference snippets and thus make later matching more accurate, these reference snippets are averages of many "identical" examples from training. Clearly, there should be separate references for $bval = 0$ (where only $bval = 0$ examples are averaged) and for $bval = 1$ (where only $bval = 1$ examples are averaged). However, $bval$ is not the only value that affects the signal in a systematic way—the signal in this part of the computation is also affected by previous value of $wval$, loop counter i, etc. The problem is that these variations occur in the same part of the signal where variations due to $bval$ occur, so averaging of these different variants may result in attenuating the impact of $bval$. We alleviate this problem by forming separate references for different bit-positions within the window, e.g., for window size $w = 5$ each value of $bval$ would have 4 sets of S-S snippets and one set of S-U snippets, because the first 4 bits in the window correspond to S-S snippets and the last bit in the window to an S-U snippet. To account for other value-dependent variation in the signal, in each such set of snippets we cluster similar signals together and use the centroid of each cluster as the reference signal. We use the K-Means clustering algorithm, and the distance metric used for clustering is Euclidean distance (sum of squared differences among same-position samples in the two snippets). We found that having at least 6–10 clusters for each set of snippets discussed above improves accuracy significantly. Beyond 6–10 clusters our recovery of secret exponent's bits improves only slightly but requires more training examples to compensate for having fewer examples per cluster (and thus less noise suppression in the cluster's centroid). Thus we use 10 clusters for each window-bit-position for each of the two possible values of $bval$. Overall, the number of S-S reference snippets for $bval$ recovery is $2 * (w - 1) * 10$—two possible values of $bval$, $w - 1$ bit-positions, 10 reference signals (cluster centroids) for each, while for S-U snippets we only have 20 reference snippets because S-U only happens for one (last) bit-position in the window. For commonly used window sizes, this results in a relatively small overall number of reference snippets, e.g., for $w = 5$ there are only 100 reference snippets. To illustrate the difference in the signals created by the value of the exponent's bit, Fig. 13.9 shows two reference S-S snippets (cluster centroids) for each value of the exponent's bit, with the most significant differences between 0-value and 1-value signals indicated by thick arrows.

Recall that, before attempting recovery of an unknown bit of the secret exponent, we will have already identified which control-flow possibility (S-S or S-U) the

13.4 One and Done: An EM-Based Attack on RSA

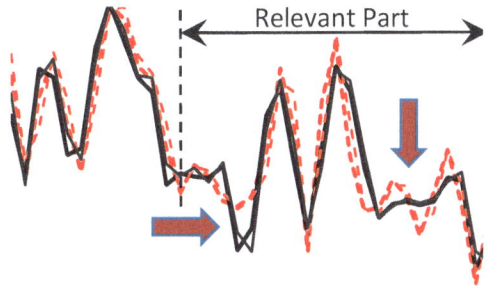

Fig. 13.9 Example signal references (cluster centroid) for S-S snippets. Two references are shown for each value of the exponent's bit that corresponds to the snippet [4]

snippet under consideration belongs to, and for S-S which bit-position it belongs to, so there are 20 reference snippets that each snippet-under-consideration is compared to (10 clusters for $bval = 0$ and 10 clusters for $bval = 1$). Thus the final step of our analysis involves finding the closest match (using Euclidean distance as a metric) among these 20 reference snippets and taking the $bval$ associated with that reference snippet.

13.4.6 Recovering Exponent Bits in the Sliding-Window Implementation

The sliding-window implementation of large-number exponentiation (Fig. 13.2) has three sites where Montgomery multiplication is called: the squaring within a window at line 26, which we label S, the multiplication to update the result at line 29, which we label U, and the squaring for a zero-valued window at line 7, which we label Z. The control flow possibilities between these include going from a squaring to another squaring (which we label as S-S). This transition is very brief (it only involves for instructions to stay in the loop at line 25). The other transitions are S-U, which consumes more time because it performs the $ct[wval]$ computation; U-Z, which involves executing line 31, line 3, line 4 (where a bit of the exponent is examined), and finally entering Montgomery multiplication at line 7; U-S, which involves executing line 31, line 3, line 4, lines 11 and 12, and the entire window-scanning loop at lines 14–23, then line 25 and finally entering Montgomery multiplication at line 26; Z-Z where after line 7 the execution proceeds to line 8, line 9, line 3, line 4, and line 7 again; Z-S where after line 7 the execution proceeds to lines 8, 9, 3, 4, and then to lines 11 and 12, the loop at line 14–23, then line 25 and finally line 26; U-X where after the Montgomery multiplication at line 29 the execution proceeds to line 31 and then exits the loop at line 3; and finally S-X, where after Montgomery multiplication at line 7 the execution proceeds to lines 8 and 9 and then exits the loop at line 3.

Just like in fixed-window implementations, our recovery of the secret exponent begins with determining which snippet belongs to which of these control-flow possibilities. While in Sect. 13.4.5 this was needed only to correct for missing snippets, in the sliding-window implementation the window size varies depending on which bit-values are encountered in the exponent, so distinguishing among the control-flow possibilities is crucial for correctly assigning recovered bits to bit-positions in the exponent even if no snippets are missing. Furthermore, many of the exponent's bits can be recovered purely based on the sequence of these control-flow possibilities.

Our overall approach for distinguishing among control flow possibilities is similar to that in Sect. 13.4.5, except that here there are more control-flow possibilities, and the U-S and Z-S coarse-grained possibilities each have multiple control flow possibilities within the snippet: for each bit considered for the window, line 19 determines whether or not to execute lines 20 and 21. However, at the point in the sequence where U-S can occur, the only other possibility is U-Z, which is much shorter and thus they are easy to tell apart. Similarly, the only alternative to Z-S is the much shorter Z-Z, so they are also easy to tell apart.

By classifying snippets according to which control-flow possibility they exercise (where U-S and U-Z are each treated as one possibility), and by knowing the rules the sequence of these must follow, we can recover from missing snippets and, more importantly, use rules similar to those in [15] to recover many of the bits in the secret exponent. However, in contrast to work in [15] that could only distinguish between a squaring (line 7 or line 26, i.e., S or Z in our sequence notation) and an update (line 29, U in our sequence notation) using memory access patterns within each Montgomery multiplication (which implements both squaring and updates), our method uses the signal snippets between these Montgomery multiplications to recover more detailed information, e.g., for each squaring our recovered sequence indicates whether it is an S or a Z, and this simplifies the rules for recovery of exponent's bits and allows us to extract more of them. Specifically, after a U-S or Z-S, which compute the window value $wval$, the number of bits in the window can be obtained by counting the S-S occurrences that follow before an S-U is encountered. For example, consider the sequence U-S, S-S, S-S, S-U, U-Z, Z-Z, Z-Z, Z-S. The first U-S indicates that a new window has been identified and a squaring for one of its bits is executed. Then the two occurrences of S-S indicate two additional squaring for this window, and S-U indicates that only these three squaring are executed, so the window has only 3 bits. Because the window begins and ends with 1-valued bits, it is trivial to deduce the values of two of these 3 bits. If we also know that $wmax = 5$, the fact that the window only has 3 bits indicates that the two bits after this window are both 0-valued (because a 1-valued bit would have expanded the window to include it). Then, after S-U, we observe U-Z, which indicates that the bit after the window is 0-valued (which we have already deduced), then two occurrences of Z-Z indicate two more 0-valued bits (one of which we have already deduced), and finally Z-S indicates that a new non-zero window begins, i.e., the next bit is 1. Overall, out of the seven bits examined during this sequence, six were recovered solely based on the sequence. Note that two of the bits (the two zeroes after the window) were redundantly recovered, and this redundancy helps us correct mistakes such as missing snippets or miss-categorized snippets.

In general, this sequence-based analysis recovers all zeroes between windows and two bits from each window. In our experiments, when using $wmax = 5$ this analysis alone on average recovers 68% of the secret exponent's bits, and with using $wmax = 6$, another commonly used value for $wmax$, this analysis alone on average recovers 55% of the exponent's bits. These recovery rates are somewhat higher than what square-update sequences alone enable [15], but recall that in our approach sequence recovery is only the preparation for our analysis of exponent-bit-dependent variation within individual signal snippets.

Since the only bits not already recovered are the "inner" (not the first and not the last) bits of each window, and since U-S and Z-S snippets are the only ones that examine these inner bits, our further analysis only focuses on these. To simplify discussion, we will use U-S to describe our analysis because the analysis for Z-S snippets is virtually identical.

Unlike fixed-window implementations, where the bits of the exponent are individually examined in separate snippets, in sliding-window implementations a single U-S or Z-S snippet contains the activity (line 4) for examining the first bit of the window and the execution of the entire loop (lines 14–23) that constructs the $wval$ by examining the next $wmax - 1$. Since these bits are examined in rapid succession without intervening highly-recognizable Montgomery multiplication activity, it would be difficult to further divide the snippet's signal into pieces that each correspond to consideration of only one bit. Instead, we note that $wmax$ is relatively small (typically 5 or 6), and that there are only 2^{wmax} possibilities for the control flow and most of the operands in the entire window-scanning loop. Therefore, in training we form separate reference snippets for each of these possibilities, and then during the attack we compare the signal snippet under consideration to each of the references, identify the best-matching reference snippet (smallest Euclidean distance), and use the bits that correspond to that reference as the recovered bit values.

13.4.7 Full Recovery of RSA Private Key Using Recovered Exponent Bits

Our RSA key recovery algorithm is a variant of the algorithm described by Henecka et al. [31], which is based on Heninger and Shacham's branch-and-prune algorithm [32]. Like Bernstein et al. [15], we recover from the side channel signal only the bits of the private exponents d_p and d_q, and the recovery of the full private key relies on exploiting the numerical relationships (Equations (1) in Bernstein et al. [15]) between these private exponents (d_p and d_q), the public modulus N and exponent e, and p, and q, the private factors of N:

$ed_p = 1 + k_p(p - 1) \ mod \ 2^i,$
$ed_q = 1 + k_q(q - 1) \ mod \ 2^i,$
$pq = N \ mod \ 2^i,$

where k_p and k_q are positive integers smaller than the public exponent e and satisfy $(k_p - 1)(k_q - 1) \equiv k_p k_q N \mod e$. The public exponent practically never exceeds 32 bits [32] and in most cases $e = 65537$, so a key recovery algorithm needs to try at most e pairs of k_p, k_q.

We could not simply apply Bernstein's algorithm [15] to the exponents recovered by our signal analysis because, like the original branch-and-prune algorithm, such recovery requires certain knowledge of the bit values at some fraction of bit-positions in d_p and d_q, while the remaining bits are unknown but *known to be unknown*, i.e., they are *erasures* rather than errors. Such branch-and-prune search has been shown to be efficient when up to 50% of the bit-positions (chosen uniformly at random) in d_p and d_q are erasures, while its running time grows exponentially when the erasures significantly exceed 50% of the bit positions.

Henecka's algorithm [31] can be applied with the above pruning equations to recover the private key when some of the bits are in error. However, its pruning is based on a key assumption that errors are uniformly distributed, and it does not explicitly consider erasures. Recall, however, that for some of the bit positions our analysis cannot identify the relevant signal snippet for matching against training signals (see Sect. 13.4.4), which results in an erasure. A naive approach for handling erasures would be to randomly assign a bit value for each erasure (resulting in a 50% error rate among erasures) and then apply Henecka's algorithm. Unfortunately, the erasures during our recovery are a product of disturbances in the signal that are very large in magnitude, and such a disturbance also tends to last long enough to affect multiple bits. With random values assigned to erasures, this produces 50%-error-rate bursts that are highly unlikely to be produced by uniformly distributed errors, causing Henecka's algorithm to either prune the correct partial candidate key or become inefficient (depending on the choice of the ϵ parameter).

Instead, we modify Henecka's algorithm to handle erasures by branching at a bit position when it encounters an erasure, but ignoring that bit position for the purposes of making a pruning decision. We further extend Henecka's algorithm to not do a "hard" pruning of a candidate key when its error count is too high. Instead, we save such a candidate key so that, if no candidate keys remain but the search for the correct private key is not completed, we can "un-prune" the lowest-error-count candidate keys that were previously pruned due to having too high of an error count. This is similar to adjusting the value of ϵ in Henecka's algorithm and retrying, except that the work of previous tries is not repeated, and this low cost of relaxing the error tolerance allows us to start with a low error tolerance (large ϵ in Henecka et al.) and adjust it gradually until the solution is found.

We further modify Henecka's algorithm to, rather than expand a partial key by multiple bits (parameter t in Henecka et al.) at a time, expand by one bit at a time and, among the newly created partial keys, only further expand the lowest-recent error-count ones until the desired expansion count (t) is reached. In Henecka's algorithm, full expansion by t bits at a time creates 2^t new candidate keys, while our approach discovers the same set of t-times-expanded non-pruned candidates without performing all t expansions on those candidates that encounter too many

13.4 One and Done: An EM-Based Attack on RSA

errors even after fewer than t single-bit expansions. For a constant t, this reduces the number of partial keys that are examined by a constant factor, but when the actual error rate is low this constant factor is close to 2^t.

Overall, our actual implementation of this modified algorithm is very efficient—it considers (expands by one bit) about 300,000 partial keys per second using a single core on then-recent mobile hardware (4th generation Surface Pro with a Core i7 processor), and for low actual error rates typically finds a solution after only a few thousand partial keys are considered. We evaluate its ability to reconstruct private RSA keys using d_p and d_q bits that contain errors and/or erasures by taking 1000 RSA keys, introducing random errors, random erasures, and a half-and-half mix of errors and erasures, at different error/erasure rates, and counting how many partial keys had to be considered (expanded by a bit) before the correct private key was reconstructed. The median number of steps for each error/erasure rate is shown in Fig. 13.10. We only show results for error/erasure rates up to 10% because those are the most relevant to our actual signal-based recovery of the exponent's bits.

We observe that our implementation of reconstruction quickly becomes inefficient when only errors are present and the error rate approaches 7%, which agrees with the theoretical results of Henecka et al.—since d_p and d_q are used, the m factor in Henecka et al. is 2, and the upper bound for efficient reconstruction is at 8.4% error rate. In contrast, when only erasures are present, our implementation of reconstruction remains very efficient even as the erasure rate exceeds 10%, which agrees with Bernstein et al.'s finding that reconstruction should be efficient with up to 50% erasure rates. Finally, when equal numbers of errors and erasures are injected, the efficiency for each injection rate is close to (only slightly worse than) the efficiency for error-only injection at half that rate, i.e., with a mix of errors and erasures, the efficiency of reconstruction is largely governed by the errors.

Figure 13.11 shows the percentage of experiments in which the correct RSA key was recovered in fewer than 5,000,000 steps (about 17 seconds on the Surface 4 Pro). When only errors are present, > 90% of the reconstructions take fewer than 5,000,000 steps until the error rate exceeds 5.4%, at which point the percentage of

Fig. 13.10 Single-bit expansion steps needed to reconstruct the private RSA key (vertical axis, note the logarithmic scale) as a function of the rate at which errors and/or erasures are injected (horizontal axis) [4]

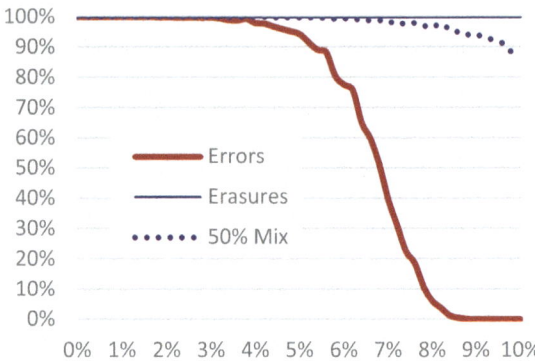

Fig. 13.11 Percentage of keys recovered in fewer than 5,000,000 single-bit expansion steps (vertical axis) as a function of the rate at which errors and/or erasures are injected (horizontal axis) [4]

under-five-million-steps reconstructions rapidly declines and drops below 10% at the 7.9% error rate. In contrast, all erasure-only reconstructions are under 5,000,000 steps even at the 10% erasure rate. Finally, when erasures and errors are both present in equal measure, the percentage of under-5,000,000-step reconstructions remains above 90% until the injection rate reaches 9.8% (4.9% of the bits are in error and another 4.9% are erased). As we will see soon, our actual error and erasure rates are well below these limits.

13.4.8 Measurement Setup

Now, we describe our measurement setup and our results for recovering keys from blinded RSA encryption runs on three different devices.

We run the OpenSSL RSA application on Android smart phones Samsung Galaxy Centura SCH-S738C [48] and Alcatel Ideal [6], and on an embedded device (A13-OLinuXino board [45])). The Alcatel Ideal cellphone has quad-core 1.1 GHz Qualcomm Snapdragon processor with Android OS(version 6) and the Samsung phone has a single-core 800 MHz Qualcomm MSM7625A Chipset with Android OS(version 5). The A13- OLinuXino board is a single-board computer that has an in order, 2-issue Cortex A8 ARM processor [8] and runs Debian Linux operating system.

Signals are received using small magnetic probe, which is placed close to the monitored system as shown in Fig. 13.12. The signals collected by the probe are recorded with Keysight N9020A MXA spectrum analyzer [36]. Our decision to use spectrum analyzer was mainly driven by its existing features such as built-in support for automating measurements, saving and analyzing measured results, visualizing the signals when debugging code, etc. We have observed very similar signals when using less expensive equipment, such as Ettus USRP B200-mini receiver [23]. The analysis was implemented in MATLAB, and on a personal computer the analysis completes in under one minute per recovered 1024-bit exponent.

13.4 One and Done: An EM-Based Attack on RSA

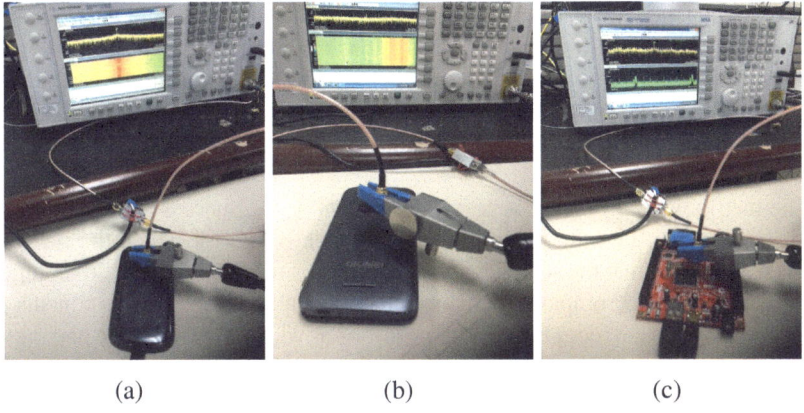

Fig. 13.12 The measurement setup for each of the three devices: (**a**) Samsung galaxy centura sch-S738C smart phone, (**b**) Alcatel Ideal smart phone, and (**c**) the A13-OLinuXino board [4]

13.4.9 Results for OpenSSL's Constant-Time Fixed-Window Implementation

Our first set of experiments evaluates the attack's ability to recover bits of the 1024-bit secret exponent d_p used during RSA-2048 decryption. OpenSSL uses a fixed window size $w = 5$ for exponentiation of this size. Note that RSA decryption involves another exponentiation, with d_q, and uses the Chinese Remainder Theorem to combine their results. However, the two exponentiations use exactly the same code and d_p and d_q are of the same size, so results for recovering d_q are statistically the same to those shown here for recovering d_p.

For each device, our training uses signals that correspond to 15 decryption instances, one for each of 15 randomly generated but known keys, and with ciphertext that is randomly generated for decryption. Note that these 15 decryptions provide around 12,000 examples of S-S signal snippets, 3,000 S-U, 3,000 U-S, and 15 U-X snippets. This is more than enough examples of each control flow possibility to distinguish between these control flow possibilities accurately. More importantly, this provides on average 1,500 snippet examples for each of the 100 ($2 * 5 * w$) clusters whose centroids are used as reference snippets when recovering the bits of the unknown secret exponents. Note that larger RSA keys would proportionally increase the number of snippets produced by each decryption, while w changes little or not at all. Thus for larger RSA keys we expect that even fewer decryptions would be needed for training.

After training, we perform the actual attack. We randomly generate 135 RSA-2048 keys, and for each of these keys we record, demodulate, and upsample the signal that corresponds to *only one decryption* with that key, using a ciphertext that is randomly generated for each decryption. Next, the signal that corresponds to each decryption is processed to extract the relevant snippets from it. Then, each of these snippets is matched against reference snippets (from training) to identify which of the control-flow possibilities each snippet belongs to and, for S-S and S-U snippets, which bit-position in the exponent (and the window) the snippet corresponds to. Finally, S-S and S-U snippets are matched against the 20 clusters that correspond to its position in the window to recover the value of the bit at that position in the secret exponent.

The metric we use for the success of this attack is the success rate for recovery of exponent's bits, i.e., the fraction of the exponent's bits for which the recovery produces the value that the secret exponent at that position actually had. To compute this success rate, we compare the recovered exponents to the actual exponents d_p and d_q that were used, counting the bit positions at which the two agree and, at the end, dividing that count with the total number of bits in the two exponents.

The maximum, median, and minimum success rate for each of the three targeted devices is shown in Fig. 13.13. We observe that the success rate of the attack is extremely high—among all decryptions on all three devices, the lowest recovery rate is 95.7% of the bits. For the OLinuXino board, most decryption instances (>85% of them) had all bits of the exponent recovered correctly, except for the most significant 4 bits. These 4 bits are processed before entering the code in Fig. 13.3 to leave a whole number of 5-bit windows for that code, so we do not attempt to recover them—we treat them as erasures. Among the OLinuXino decryption instances that had any other reconstruction errors, nearly all had only one additional incorrectly recovered bit (error, not erasure), and a few had two.

The results for the Samsung phone were slightly worse—in addition to the 4 most significant bits, several decryption instances had one additional bit that was left unknown (erasure) because of an interrupt that occurs between the derivative-of-moving-median peak and the end of the snippet that follows it, which either obliterates the peak or prevents the snippet from correctly being categorized according to its control flow. In addition to these unknown (but known-to-be-unknown) bits, for the Samsung phone the reconstruction also produced between 0 and 4 incorrectly recovered (error) bits.

Finally, for the Alcatel Ideal phone most instances of the encryption had between 13 and 16 unknown bits in each of the two exponents, mostly because activity on the other three cores interferes with the activity on the core doing the RSA decryption), and a similar number of incorrectly recovered bits (errors).

Fig. 13.13 Success rate for recovery of secret exponent d_p's bits during only one instance of RSA-2048 decryption that uses that exponent. For each device, the maximum, median, and minimum success rate among decryption instances (each with a different randomly generated key) is shown [4]

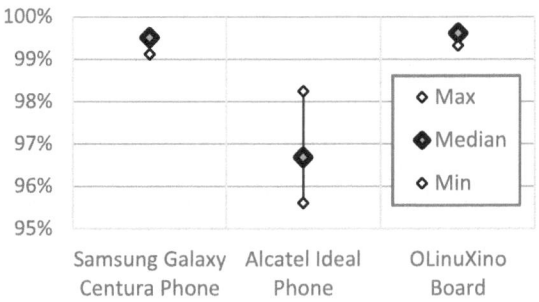

13.4 One and Done: An EM-Based Attack on RSA

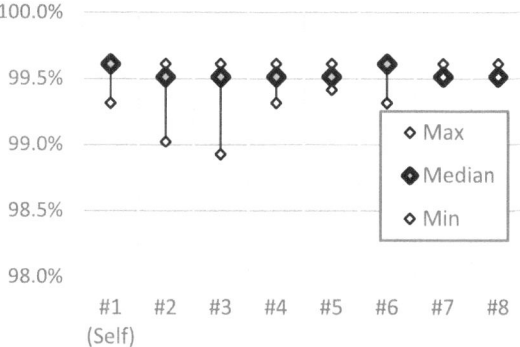

Fig. 13.14 Success rate for recovery of secret exponent d_p's bits during only one instance of RSA-2048 decryption that uses that exponent, when training on OLinuXino board #1 and then using that training data for unknown exponent recovery on the same board and on seven other boards. For each device, the maximum, median, and minimum success rate among decryption instances (each with a different randomly generated key) is shown [4]

To examine how the results would be affected when training using signals collected on one device and then recovering exponent bits using signals obtained from another device of the same kind, we use eight OLinuXino boards,[1] which we label #1 through #8. Our training uses signals obtained only from board #1, and then the unknown keys are used on each of the eight boards and subjected to analysis using the same training data (from board #1). The results of this experiment are shown in Fig. 13.14, where the leftmost data points correspond to training and recovery on the same device, while the remaining seven sets of data points correspond to training on one board and recovery on another.

These results indicate that training on a different device of the same kind does not substantially affect the accuracy of recovery.

Finally, for each RSA decryption instance, the recovered exponent bits, using both the recovered d_p and the recovered d_q, were supplied to our implementation of the full-key reconstruction algorithm. For each instance, the correct full RSA private key was reconstructed within one second on the Core i7-based Surface Pro 4 tablet, including the time needed to find the k_p and k_q coefficients that were not known a priori. This is an expected result, given that even the worst bit recovery rates (for the Alcatel phone) correspond to a an error rate of about 1.5%, combined with an erasure rate of typically 1.5% but sometimes as high as 3% (depending on how much system activity occurs while RSA encryption is execution on the phone), which is well within the range for which our full-key reconstruction is extremely efficient.

---

[1] The OLinuXino boards are much less expensive than the phones, so we could easily obtain a number of OLinuXino boards.

13.4.10 Results for the Sliding-Window Implementation

To improve our understanding of the implications for this new attack approach, we also apply it to RSA-2048 whose implementation uses OpenSSL's sliding-window exponentiation—recall that this was the default implementation used in OpenSSL until it switched to a fixed-window implementation in response to attacks that exploit sliding-window's exponent-dependent square-multiply sequence.

In these experiments, we use 160 MHz of bandwidth and target the OLinuXino board. Recall that, in a sliding-window implementation, our method can categorize the snippets according to their beginning/ending point to recover the sequence of zero-squaring (Z), window-squaring (S), and result update (M) occurrences. The fraction of the exponent's bits recovered by this sequence reconstruction (shown as "S-M-Z Sequence" in Fig. 13.15) is in our experiments between 51.2 and 57.7% with a median of 54.5%. This sequence-based recovery has produces no errors in most cases (keys), and among the few encryptions that had any errors, none had more than one.

In our attack approach, after this sequence-based reconstruction, the U-S and Z-S snippets are subjected to further analysis to recover the remaining bits of the window computed in each U-S and Z-S snippet. At the end of this analysis, the fraction of the exponent's bits that are correctly recovered ("Overall" in Fig. 13.15) is between 97.7 and 99.6%, with a median of 98.7%.

This rate of recovery for exponent bits provides for very rapid reconstruction of the full RSA key. However, we note that it is somewhat inferior to our results on fixed-window exponentiation on the same device (OLinuXino board), in spite of using more bandwidth for attacks on sliding-window (160 MHz bandwidth) than on fixed-window (40 MHz bandwidth) implementation. The primary reason for this is that, in the fixed-window implementation, each analyzed snippet corresponds

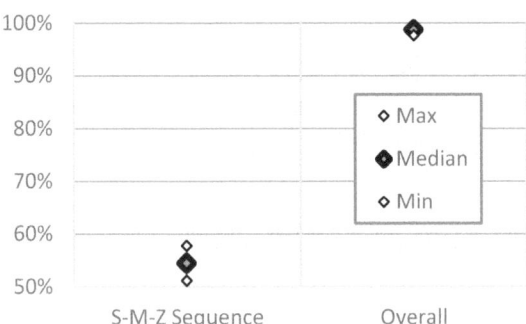

Fig. 13.15 Success rate for recovery of secret exponent d_p's bits during only one instance of RSA-2048 decryption that uses that exponent for sliding-window exponentiation. The maximum, median, and minimum success rate among decryption instances (each with a different randomly generated key) is shown for recovery that only uses the snippet-type sequence (S-M-Z Sequence), and for recovery that also recovers window bits from U-S and Z-S snippets (Overall) [4]

13.4 One and Done: An EM-Based Attack on RSA

to examining only one bit of the exponent. In the sliding-window implementation, however, `wmax` is 6. i.e., 6 bits of the exponent are examined in a single U-S or Z-S snippet, while the exponent-dependent variation in the snippet is not much larger. Since sliding-window recovery tries to extract several times more information from about the same amount of signal change, its recovery is more affected by noise and thus slightly less accurate.

13.4.11 Mitigation

We focus our mitigation efforts on the fixed-window implementation, which is still the implementation of choice in the current version of OpenSSL as of this writing. We identify three key enablers for this attack approach. Successful mitigation requires removing at least one of these enablers, so we now discuss each of the attack enablers along with potential mitigation approaches focused on that enabler.

The first enabler of the specific attack demonstrated in this paper is the existence of computational-activity-modulated EM signals around the processor's clock frequency, and the attacker's ability to obtain these signals with sufficient bandwidth and signal-to-noise ratio. Potential mitigation thus include circuit-level approaches that reduce the effect the differences in computation have the signal, additional shielding that attenuates these signals to reduce their signal-to-noise ratio outside the device, deliberate creation of RF noise and/or interference that also reduces the signal-to-noise ratio, etc. We do not focus on these mitigation because all of them increase the device's overall cost, weight, and/or power consumption, all of them are difficult to apply to devices that are already in use, and all of them may not provide protection against attacks that use this attack approach but through a different physical side channel (e.g., power).

The second enabler of our attack approach is the attacker's ability to precisely locate, in the overall signal during an exponentiation operation, those brief snippets of signal that correspond to examining the bits of the exponent and constructing the value of the window. A simple mitigation approach would thus insert random additional amounts of computation before, during, and/or after window computation. However, additional computation that has significant variation in duration would also have a significant mean of that duration, i.e., it would slow down the window computation. Furthermore, it is possible (and indeed likely) that our attack can be adapted to identify and ignore the signal that corresponds to this additional activity.

The final (third) enabler of our attack approach is the attacker's ability to distinguish between the signals whose computation has the same control flow but uses different values for a bit in the exponent. In this regard, the attack benefits significantly from 1) the limited space of possibilities for value returned by `BN_is_bit_set`—there are only two possibilities: 0 or 1, and from 2) the fact that the computation that considers each such bit is surrounded by computation that operates on highly predictable values—this causes any signal variation caused by the return value of BN_is_bit_set to stand out in a signal that otherwise exhibits very little variation.

Based on these observations, our mitigation relies on obtaining all the bits that belong to one window at once, rather than extracting the bits one at a time. We accomplish this by using the bn_get_bits function (defined in bn_exp.c in OpenSSL's source code), which uses shifts and masking to extract and return a BN_ULONG-sized group of bits aligned to the requested bit-position—in our case, the LSB of the window. The BN_ULONG is typically 32 or 64 bits in size, so there are billions of possibilities for the value it returns, while the total execution time of bn_get_bits is only slightly more than the time that was needed to append a single bit to the window (call to BN_is_bit_set, shifting the wval, and then or-ing to update wval with the new bit). For the attacker, this means that there are now billions of possibilities for the value to be extracted from the signal, while the number of signal samples available for this recovery is similar to what was originally used for making a binary (single-bit) decision. Intuitively, the signal still contains the same amount of information as the signal from which one bit used to be recovered, but the attacker must now attempt to extract tens of bits from that signal.

This mitigation results in a slight *improvement* in execution time of the exponentiation and, as shown in Fig. 13.16, with the mitigation the recovery rate for the exponent's bits is no better than randomly guessing each bit (50% recovery rate). In fact, the recovery rate with the mitigation is lower than 50% because, as in our pre-mitigation results, the bits whose signal snippets could not be located are counted as incorrectly recovered. However, these bits can be treated as erasures, i.e., for each such bit the attacker knows that the value of the bit is unknown, as opposed to a bits whose value is incorrect but the attacker has no a priori knowledge of that, so our recovery rate can be trivially improved by randomly guessing (with 50% accuracy) the value of each erasure, rather than having 0% accuracy on them. With this, the post-mitigation recovery rate indeed becomes centered around 50%, i.e., equivalent to random guessing for all of the bits. This mitigation has been implemented in recent version of OpenSSL.

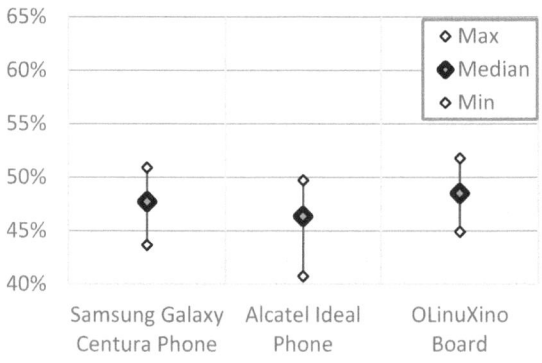

Fig. 13.16 Success rate for recovery of secret exponent d_p's bits after the initial implementation of our window value randomization mitigation is applied [4]

13.5 The Nonce@Once Attack

13.5.1 Attack Setting

The attacked devices are mobile phones, which are equipped with 1.1–1.4 GHz CPU and run full-fledged Android. We do not modify the phones' hardware in any way, and in particular leave the phones' cases closed with all internal EM shielding intact. Moreover, we do not apply any modifications that might help with side-channel analysis to the software stack. This includes running an unmodified Android operating system, with all CPU cores enabled, and with power management settings and background applications in their factory defaults. We also avoid the use of triggers for side-channel acquisition and signal location, leaving all side-channel countermeasures in the targeted software in place. Finally, the phones' WiFi is enabled and connected to our university network.

We do assume, however, that the attacker has a brief and non-invasive access to the target device and can thus place an inexpensive EM probe within a few centimeters of the device. Prior work has shown that such probes can be easily hidden in public locations, such as under a table in a coffee shop or inside the charging surface at a public phone charging station [27]. As such access is inherently limited, we constrain ourselves to short attack duration, using small and easily concealable attack equipment. For equipment, we avoid expensive and bulky top-of-the-line spectrum analyzers used in prior works [4]. Instead, we use a compact, inexpensive software defined radio (SDR) device (around $800), limiting our bandwidth to 40 MHz, which is well below the target devices' clock speeds. We further avoid hardware-based cycle-accurate artificial triggering, running the SDR in continuous sampling mode. Moreover, unlike prior works, which require thousands of traces of emanations from the target device [27], we perform key recovery using only the signal collected during one instance of cryptographic signing that uses the key.

13.5.2 Attack Overview

The attack consists of two main phases. In the training phase, the attacker collects a small number (up to 10) of EM traces from the device, while using known scalars. The attacker uses these traces to train classifiers that first identify the part of the signal that corresponds to the point-by-scalar multiplication operation within the signal, and then identify the condition of each constant-time swap operation executed during that point-by-scalar multiplication.

In the attack phase, the attacker places a probe next to the target device and continuously monitors its EM emanations, waiting for the device to perform a cryptographic signing operation. Once a single EM trace corresponding to such an operation is captured, our attack automatically identifies the point-by-scalar multiplication in the signal and analyzes the condition of each constant-time swap operation. Finally, using these conditions, the attacker recovers the nonce used during the signing operation and combines it with the (public) message and signature to recover the target's secret signing key.

13.5.3 Leveraging Constant-Time Swap

To achieve constant-time coding, the swap operation proposed by RFC 7748 uses a bit mask to swap two machine words, thereby avoiding secret-dependent control flow. The value of this mask is drawn from a random distribution if the swap takes place. If the swap does not take place, the mask is all-zero. Adopting this methodology, newer designs of cryptographic implementations tend to process the secret key one-bit-at-a-time, using the constant-time swap operation to swap two internal values based on the key. Unfortunately, the power consumption of mobile devices typically correlates with the number of bit-flips. Hence, the difference between all-zero and random masks causes a small bias in the CPU's power consumption and with it in the device's electromagnetic emanations. Due to the relatively low bandwidth of the attack, we cannot distinguish between an all-zero and a random-looking value that is used in a single instruction. However, the repeated swapping of multiple machine words, required for swapping large numbers that represent an ECpoint, amplifies the leakage via physical side channels allowing us to distinguish the values. Thus, the using constant-time swap operation, while protecting against microarchitectural side-channel attacks, makes newer cryptographic implementations more vulnerable to low-bandwidth physical side-channel analysis compared to older-generation algorithms that process several key bits at once.

13.5.4 Attack Overview for Constant-Time Swap Operation

In this section we describe our attack on the constant-time swap operation as standardized in RFC 7748 [41] and presented in Fig. 13.7. Note that the constant-time swap operation in Fig. 13.7 hides the value of the condition using secret-independent control flow and memory access pattern. The key observation that enables our attack is that the number of 1 and 0 bits in `delta` used in the XOR instructions in Lines 9 and 10 of Fig. 13.7 changes significantly depending on the value of `cond`.

More specifically, for practical purposes, we can consider a and b to be random. When $cond = 1$, the value of $mask$ is all-ones, resulting in δ being the exclusive-or of random values, so δ tends to have about as many 1-valued bits as it does 0-valued bits. Conversely, when $cond = 0$, $mask$ is all-zero, and so are the bits of δ. Next, we note that the value of δ is used during the main loop of Line 3. Thus, if $cond = 1$

about half of the bits in every word of a and b are toggled, whereas no such toggling occurs when $cond = 0$, (and $\delta = 0$). This $cond$-dependent bias in the amount of toggling causes $cond$-dependent bias in emanated EM signals.

13.5.5 Leakage Amplification

The condition-dependent bias occurs for each word of a and b. As the phones we attack use a 32-bit architecture while every elliptic curve point contains two coordinates of 256 bits each, the bias repeats 16 times for each point. Furthermore, in addition to point coordinates, implementations also store several (up to four) machine words as implementation-specific metadata and context, which are also swapped using program code similar to that shown in Fig. 13.7. Overall, the bias created by a single XOR is multiplied by a factor of about $20 \cdot 2 = 40$ for both points, creating a larger overall difference between the $cond = 0$ case and the $cond = 1$ case.

13.5.6 A Physical Side Channel Attack

Our attack records the unintentional electromagnetic emanations that the target device radiates during a *single* ECDSA signing operation. We then process this signal to identify snippets that correspond to individual EC point swap operations. Each such snippet is then analyzed to recover the value of $cond$. Because most EC algorithms use the nonce bits as condition values, our attack can recover almost all of the nonce bits from a single ECDSA leakage trace. We can then use these nonce bits, along with the signature, to recover the secret signing key that was used.

13.5.7 EM Signal Acquisition and Signature Location

We use an Ettus B200-mini software defined radio (SDR) connected to a custom electromagnetic probe to capture the target's electromagnetic emanations. The SDR is set to capture a 40 MHz-wide frequency band around the CPU frequency of each target (1.4 GHz for ZTE ZFIVE, 1.1 GHz for Alcatel Ideal, and 1 GHz for A13-OLinuXino). The recorded signal is then digitally demodulated and up-sampled before passing it through the custom signal analysis that implements our attack.

For each target, we manually located a position which exhibited the strongest electromagnetic leakage and positioned the probe accordingly. In all experiments, the probe does not make physical contact with the device's chassis, and the attack remains effective up to about 20 millimeters away from the device. While this range does imply physical proximity to the target, we note that this is sufficient for many realistic attack scenarios, such as placing the probe underneath a tablecloth or inside a desk surface [27]. Finally, as we did not modify the phone's chassis, the EM side-channel signal had to penetrate the phone's PCB shielding, battery, and outer case to get to our electromagnetic probe.

The first step in our signal analysis is to identify the part of the signal that corresponds to the overall scalar-by-point multiplication. While past work achieved this by changing the source code of the target library to generate a trigger signal, such modifications are outside our attack model—the attacker is not assumed to be capable of installing malicious software on the target device. Instead, we collect the signal continuously, and use signal analysis to locate the part of the signal that corresponds to the point-by-scalar multiplication.

Because we do not use artificial triggers to identify the signals of the point-by-scalar multiplication operation, we need to scan the continuously captured signal and recognize the patterns that correspond to point-by-scalar multiplication. Moreover, the implementations we consider use unrelated and significantly different code bases (including in terms of count, order, or type of instructions), so the signals they produce vary widely (see Fig. 13.17). However, we observe that most of the execution time of scalar-by-point multiplications is spent on repeated point

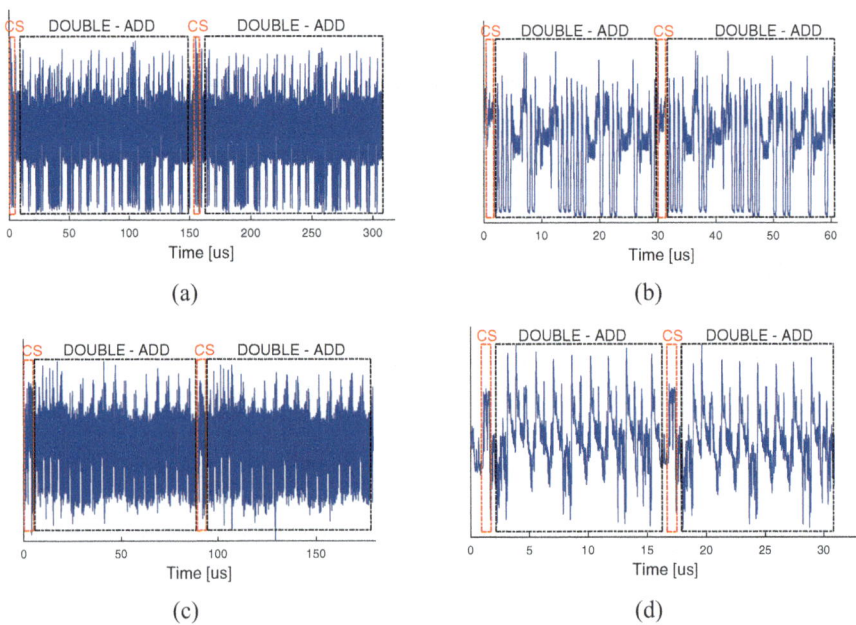

Fig. 13.17 Signal examples covering two scalar-by-point loop iterations each for (**a**) Libgcrypt, (**b**) OpenSSL, (**c**) HACL*, and (**d**) curve25519-donna recorded on the OLinuXino board. The signal for a conditional swap (CS) is indicated by dashed red rectangles, whereas the signal for the point arithmetic (double and add) is indicated by dashed black rectangles. While the signals differ between implementations, the iterations within each implementation are very similar to each other [5]

13.5 The Nonce@Once Attack

addition and point doubling operations. These point operations are designed to eliminate variation in their execution time, control flow, and data access patterns. Consequently, as Fig. 13.17 shows, loop iterations within each implementation display very little variations in their EM leakage patterns.

To identify the point-by-scalar multiplication operation, we exploit this repetition of similar patterns. More specifically, in the profiling phase of the attack, we collect a training trace that includes point-by-scalar operations using randomly generated nonces. We then average the samples of loop iterations in the training set, creating a template for a single loop iteration. In the attack phase we use a moving correlation between the template and the signal, and identify a segment of the signal that contains the expected number of matches to the template. In our experiments we found that the loop iterations are sufficiently long and have enough prominent signal features to reliably locate the point-by-scalar multiplication in the collected trace using a training trace of the signal corresponding to just one point-by-scalar multiplication operation.

13.5.8 Identifying Conditional Swap Signals

Our attack aims at extracting the *cond* argument of the conditional swap operations used during the point-by-scalar multiplication. After identifying the signal segment that corresponds to the point-by-scalar multiplication operation, we proceed to identify the snippets corresponding to the individual conditional swap operations. Here, we again rely on the signal similarity between the individual iterations of the point-by-scalar multiplication loop, using the template matching from Sect. 13.5.7 to identify individual loop iterations. We then identify the signal segments that correspond to the conditional swap, as these are located between two longer sequences of point doubling and adding.

In all four implementations considered in this work (Libgcrypt, OpenSSL, HACL*, and curve25519-donna), the signal that corresponds to the conditional swap (shown as CS in Fig. 13.17) is much shorter than the signal that corresponding to the elliptic curve double and add operations (shown as DOUBLE-ADD in Fig. 13.17). Furthermore, the signal corresponds to the double and add operations has a number of sharply defined spikes that can help precisely locating and aligning it, which helps locate conditional-swap snippets. This is analogous how the signal features in large-number multiplication (squaring) were used to locate is_bit_set signal snippets in the One&Done attack.

13.5.9 Recovering the Value of the Swap Condition from a Snippet

After identifying the signal snippets that correspond to individual conditional swaps, we need to classify the snippets based to the value of the conditionthat has been used. The conditional swap constructs a mask, which is either all-zeros or all-ones depending on the swap condition, which it then applies to an exclusive-or-based swap for each machine word in the internal representation of two elliptic curve points. Next, the internal representation of an elliptic curve point includes the big number representation of each coordinate (stored as an array of several machine words) as well as several machine words that contain point metadata. Overall, the swap operation is responsible for conditionally swapping several tens of machine words.

Notice that when the swap condition is false, there are about 80 exclusive-or operations whose one operand is always zero, and there are 80 machine words that are written to memory without actually changing their values (Lines 9 and 10 in Fig. 13.7). However, when the swap condition is true, the operands in those 80 exclusive-or operations have about the same number zeros and ones, and the values written to those 80 machine words result in toggling about half of the bits in those words. This repetition amplifies the difference in the signal corresponding to the conditional swap operation, resulting in a noticeable bias in the captured signal. Indeed, Fig. 13.18 shows signal snippets that correspond to the relevant part of the conditional swap operation in the four EC implementations considered in this work. In each case, different values of the swap condition produce clearly observable differences in the signal, thanks to the $\times 80$ amplification factor created by the code of the conditional swap.

To recover the swap condition from its signal snippet, we first perform a training phase with a known scalar. During this phase, for each value of the swap condition we use a k-means clustering algorithm, with a Euclidean distance metric, to form c clusters of snippets collected during training. After removing anomalous clusters that contain only one-snippet, we take the centroid of each remaining cluster as a reference signal. During the attack phase, we compute the distance between the new conditional-swap snippet and each of the reference signals (up to c cluster centroids for false and for true swap conditions), and use the swap value of the closest reference signal as the recovered value for the new snippet. We empirically find that, when c is above an implementation-specific threshold, (between 5 and 8), the swap conditions are recovered with high accuracy (practically error-free in

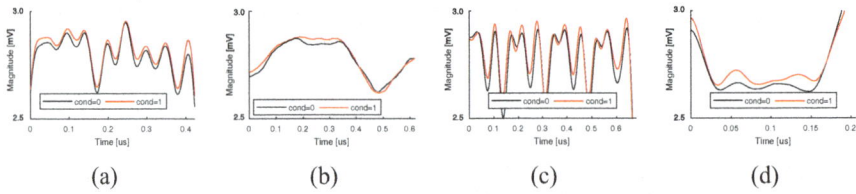

Fig. 13.18 Signal snippets corresponding to the conditional swap operation in (**a**) Libgcrypt, (**b**) OpenSSL, (**c**) HACL*, and (**d**) curve25519-donna when the swap condition is true (black) and when the swap condition is false (red). The recordings where done using the Olimex target and up-sampled for visual clarity [5]

our experiments). We further find that, increasing c beyond the threshold does not degrade accuracy and it remains practically perfect. Increasing c primarily increases the number of anomalous clusters that are removed, without significantly impacting the number of clusters that are actually used for swap-condition recovery. Since the analysis is not very sensitive to the value of c, instead of optimizing c for each combination of device and cryptographic implementation, in our experiments we simply use $c = 10$ for all attacks.

13.5.10 Nonce and Key Recovery

Using the signal analysis described above, we obtain the reconstructed values for the swap conditions of each swap-related signal snippet. Applying this method to the signal that corresponds to the entire scalar-by-point multiplication operation, it is also recovering the sequence of swap conditions, thereby recovering the scalar k.

However, while this approach is capable of recovering most of the bits of k, some bits are inadvertently missing due to signal noise, algorithmic corner cases, and measurement errors. A brute-force approach can recover the missing bits, by generating candidates that match the known bits of k. For each such candidate k_c, the known values of z, r, and s are used to compute a candidate private key d_c, using the equation $d_c = (s \cdot k_c - z)/r \mod n$. Then d_c is used to compute a public key candidate $Q_c = d_c G$, where G is the elliptic curve group generator, and compare it with the actual public key Q used to produce the signature (r, s). If the public keys match, the analysis completes successfully. Otherwise, the procedure is repeated with the next candidate. Because the positions of the missing bits are known, each missing bit roughly doubles the search space. Fortunately, the typical number of missing bits is small, and the search is tractable.

In the experimental evaluation, our signal analysis was tested on 100 ECDSA signing operations, each with a different private key. We found that our signal analysis approach resulted in at most 15 missing bits, thus requiring a brute-force test of up to 32768 candidates for each signing operation. Full key recovery using the above procedure was completed for each of the 100 signing operations, with no signature requiring more than a few minutes of computation time on a laptop system to recover the correct private key [5].

13.5.11 Experimental Evaluation

In this section we describe our measurement setup and the results of recovering private keys from Libgcrypt, OpenSSL, HACL* and curve25519-donna during ECDSA signature computation on three different devices.

13.5.12 Experimental Setup

We run the targeted applications on two Android mobile phones, ZTE ZFIVE LTE2 [56] and Alcatel IDeal [6], and on an Olimex A13-OLinuXino IoT development board [45]. The ZTE ZFIVE has a quad-core 1.4 GHz Qualcomm Snapdragon processor, while the Alcatel Ideal has a quad-core 1.1 GHz Qualcomm Snapdragon processor. The A13-OLinuXino is a single-board computer with an ARM Cortex A8 processor [8] made by Allwinner with Debian Linux.

As mentioned above, we aim for a realistic scenario and avoid modifications that assist with side-channel analysis. In particular, the devices are running unmodified Android / Linux operating systems, with all CPU cores enabled, and with Android's background applications and services running using their factory defaults. Finally, the phones' WiFi is enabled and connected to the university network.

We attack Libgcrypt 1.8.4 and OpenSSL 1.1.1a. For HACL* and curve25519-donna implementations, the latest version from their respective git repositories was used. All four cryptographic implementations use constant-time code for scalar-by-point multiplication, relying on conditional swap operations to avoid control flow that depends on the bits of the scalar. Unless stated otherwise, we use Curve25519, whose maximum order n is a 253-bit value, implying a 253-bit nonce value. We note, however, that the choice of curve has no significant impact on our attack. Any ECDSA implementation that processes the nonce bits using the constant-time operation outlined above is likely to be similarly vulnerable. Finally, we use randomly generated nonces for both the profiling and for the key extraction phases.

Figure 13.19 illustrates our results, which will be discussed in more detail in Sect. 13.5.13, while Figure 13.20 shows our setup for capturing the EM emanations from our target devices. It consists of a small custom-made magnetic probe to receive the EM signals, an Ettus B200-mini software defined radio (SDR) to digitize the EM signal in the desired frequency band, and a personal computer to process the digitized signals. As shown in Fig. 13.20, the probe is placed in close physical proximity to the target device, but without touching it and without opening its enclosure. A mechanical arm is used to hold the probe in the desired position during the experiments. The probe is connected to the SDR which digitizes the signal and sends it, through a USB cable, to a personal computer (not shown in Fig. 13.20) where the signal analysis and ECDSA key recovery is implemented in MATLAB. Note that MATLAB is used mainly for convenience, and that signal analysis and key recovery would likely be significantly faster if they were implemented in the Field Programmable Gate Array (FPGA) that is available within the B-200mini SDR itself.

13.5.13 Attack Results

Our experimental results are based on repeating the attack 100 times for each of the four target implementations on each of the three devices. In each attack, we first use

13.5 The Nonce@Once Attack

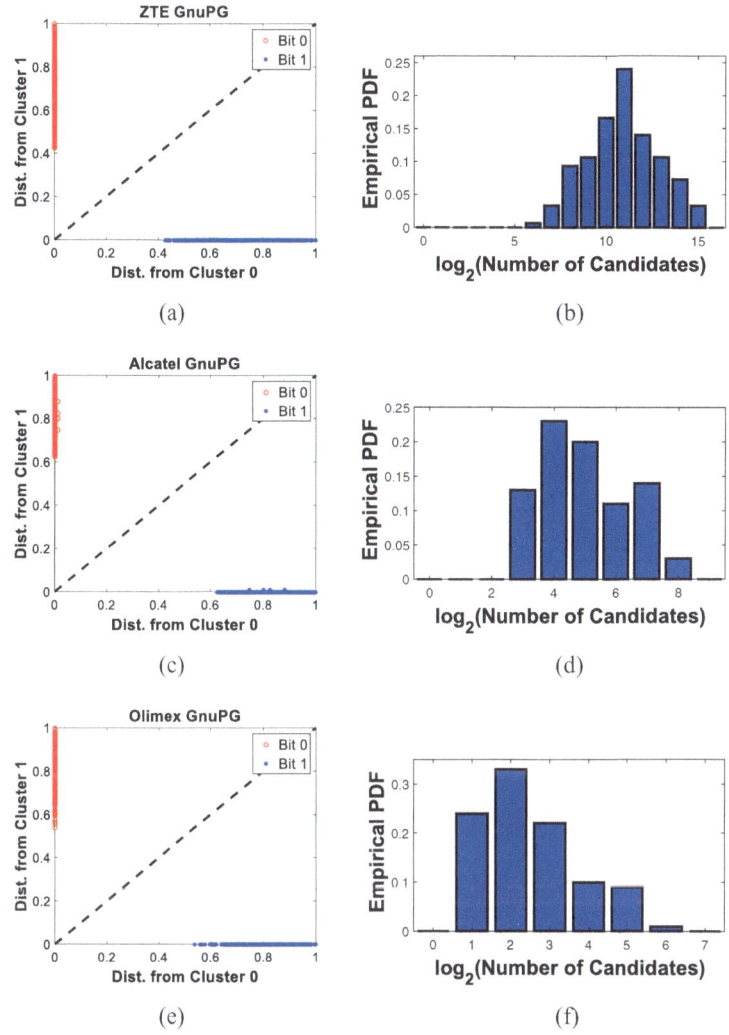

Fig. 13.19 (a) Recovery of swap condition, (b) Number of key candidates in full-key recovery for Libgcrypt on ZTE phone, (c) Recovery of swap condition, (d) Number of key candidates in full-key recovery for Libgcrypt on Alcatel phone, (e) Recovery of swap condition, (f) Number of key candidates in full-key recovery for Libgcrypt on Olimex board [5]

the target's API in order to randomly generate an ECDSA private key and a message. Then, we initiate signal collection while signing the message with the ECDSA key. The signal from this single ECDSA signing operation is then analyzed to recover the nonce k and the ECDSA private key d. Finally, the recovered d is compared to the ground-truth private key to determine if the attack was successful.

Fig. 13.20 Experimental setup for capturing EM emanations from ZTE ZFIVE (left) and Alcatel Ideal (right) phones. In each setup, our custom probe (the flat, beige, circular object at the end of the silver-colored cable) is positioned close to but not touching the phone. The cable is held in position by a mechanical arm, which is visible in the photo on the right. An Ettus B200-mini SDR (white box) digitizes the signal and sends it through a USB cable to a personal computer (not shown) for analysis [5]

Our first set of experimental results consists of attacking a single instance of Libgcrypt's ECDSA signing operation, repeating this attack 100 times on each of the three target devices considered in this work. All 300 of these attack instances successfully recovered the ECDSA private signing key. Figure 13.19 shows further details on the results of our attack. For each device, we show the clustering of the signal snippets according to their distance from the closest 0-value cluster and from the closest 1-value cluster. We also show the histogram for the size of the search space for brute-forcing the nonce k. We can observe that, for all three devices, the signal snippets that correspond to two possible values of the swap condition are well-separated, resulting in correct recovery of the swap condition for all snippets that were identified. We also observe that the number of key candidates to consider during full-key recovery is highest on the ZTE phone, where the number of key candidates is typically 2^{11}, with a maximum of 2^{16}. For the Alcatel phone and the OLinuXino development board, full key recovery requires significantly fewer candidates.

Overall, we observed large signal quality variations between the devices we tested. First, on the two Android-based devices interrupts occur significantly more often than on the Debian-based OLinuXino. Additionally, we found that each of the phones has bursts of activity on its other cores, which creates interference that occasionally prevents identification of swap-related signal snippet (especially on the ZTE phone). Finally, on the ZTE phone the signal for Libgcrypt's point multiplication is unusually "choppy", resulting in sporadic failures to identify the swap snippets even in the absence of interrupts and other interference.

13.5 The Nonce@Once Attack

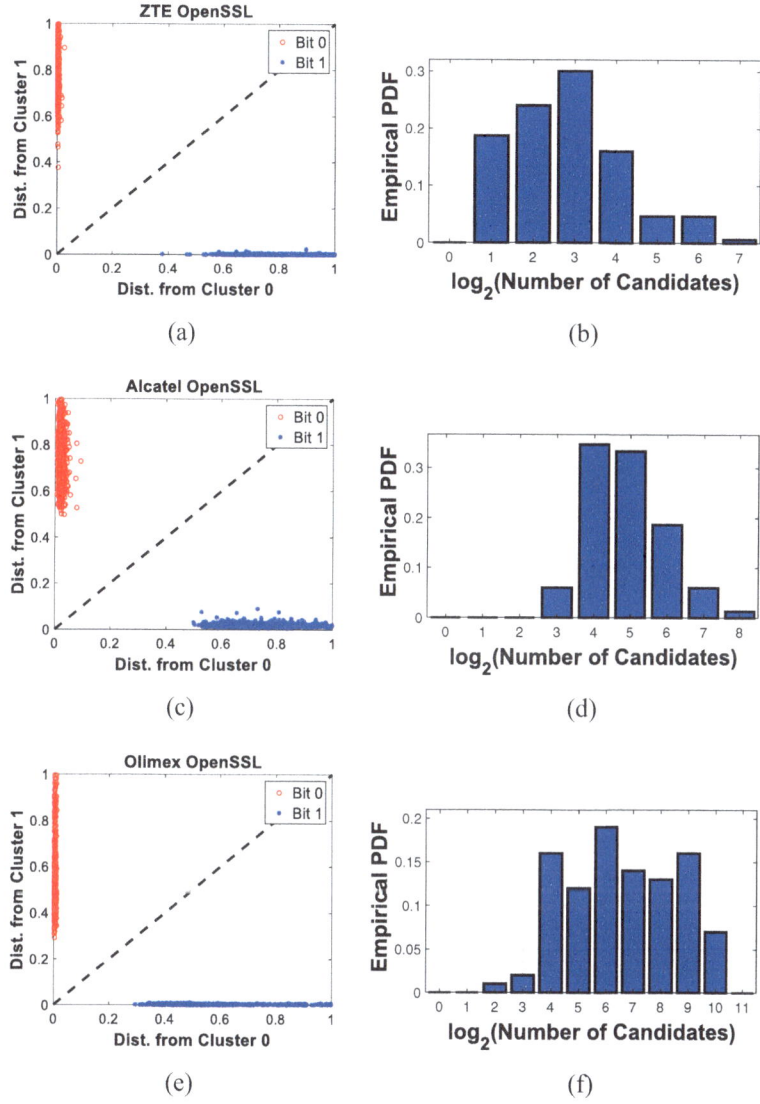

Fig. 13.21 (a) Recovery of swap condition, (b) Number of key candidates in full-key recovery for OpenSSL on ZTE phone, (c) Recovery of swap condition, (d) Number of key candidates in full-key recovery for OpenSSL on Alcatel phone, (e) Recovery of swap condition, (f) Number of key candidates in full-key recovery for OpenSSL on Olimex board [5]

Our second set of experiments targets OpenSSL, repeating the attack 100 times for each device. Figure 13.21 shows the clustering of signal snippets, and the number of candidate values for the nonce, for this set of experiments. As in Libgcrypt, the snippets that correspond to the two swap conditions are clearly separable, and all

300 instances of the attack are successful in recovering the key. However, compared to Libgcrypt, the identification of snippets in OpenSSL performs significantly better on the ZTE phone, with snippets missing only for interrupt and interference related reasons. Overall, the analysis time for an attack for OpenSSL was less than 2 seconds in all attack instances.

Finally, our experiments for HACL* (Fig. 13.22) and curve25519-donna (Fig. 13.23) show results that are mostly similar to those for OpenSSL, with success on all 300 attack instances for HACL* and all 300 attack instances for curve25519-donna, with each attack instance using less than 2 seconds of analysis time.

13.5.14 Other Elliptic Curves and Primitives

We performed a few additional attacks where we used other EC curves in Libgcrypt and OpenSSL, as well as attacks on Libgcrypt's Elliptic Curve Diffie-Hellman (ECDH) implementation. As the scalar-by-point multiplication in all these instances still uses the same conditional swap code, our attack performs similarly to the Curve25519 case.

13.5.15 Mitigation

The main cause enabling our attack is that the conditional swap operation computes, at Line 8 in Fig. 13.7, the value `delta` which, depending on the value of a nonce bit, is either random-looking or zero. For a given `cond` bit, the code in Fig. 13.7 XORs such a `cond`-dependent value of `delta` with each of tens of machine words that make up the internal representation of an elliptic curve point. While we may not be able to recover the value of `cond` (i.e. the current nonce bit) from a single computation at Lines 9 and 10 in Fig. 13.7, the repeated XOR of either all-zeros or random-looking `delta` values with tens of machine words amplifies the leakage caused by the nonce-bit-dependant bias in the Hamming weight (how many bits are 1) of `delta`.

Avoiding Leakage Amplification
The main idea behind our mitigation, which successfully prevents the attack (Fig. 13.24), is to modify the constant-time swap operation to avoid repeated use of values whose Hamming weight (how many bits are 1) directly reveals the current nonce bit. One example of such a modification is shown in Fig. 13.25. For each pair of words, `delta` is computed at line 10, in exactly the same way as it was computed in the original conditional swap function (Fig. 13.7, Line 8). In this modified version, however, the value of `delta` is then masked (Fig. 13.25, Line 11) using the random word `rand` that we have previously (Fig. 13.25, Line 6). This decouples the Hamming weight of `delta` from the current nonce bit, i.e. in this new masked value of `delta` the number of 0-valued and the number of 1-valued

13.5 The Nonce@Once Attack

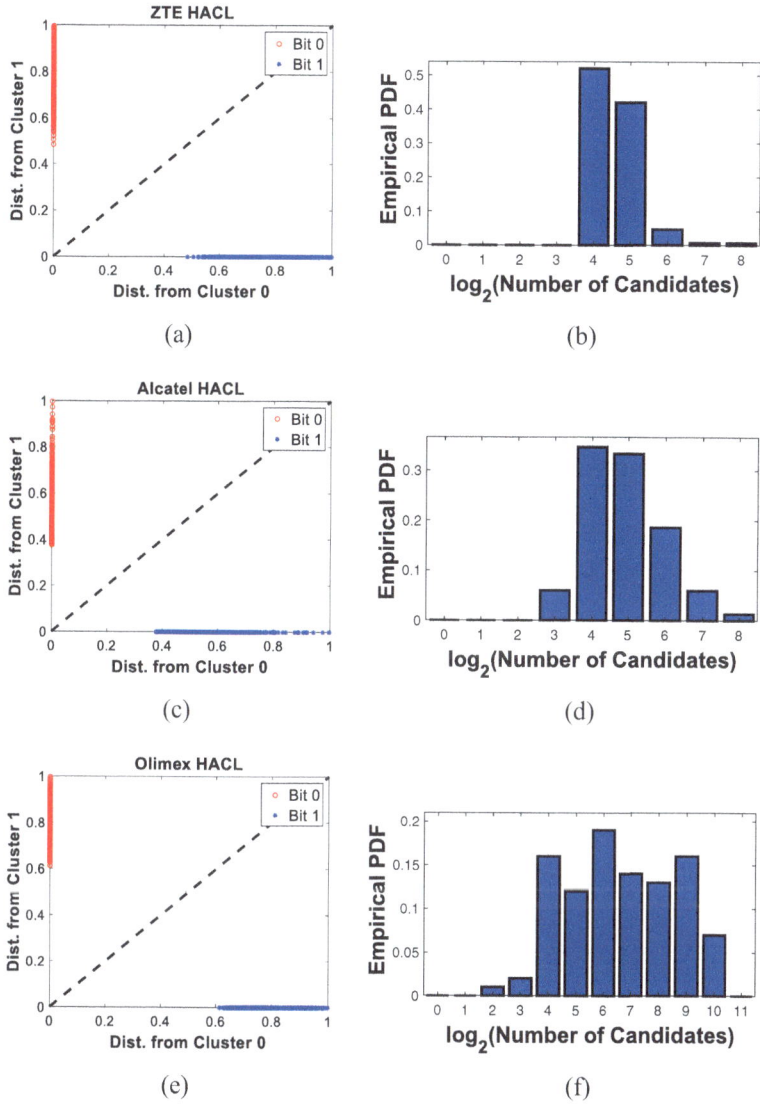

Fig. 13.22 (**a**) Recovery of swap condition, (**b**) Number of key candidates in full-key recovery for HACL* on ZTE phone, (**c**) Recovery of swap condition, (**d**) Number of key candidates in full-key recovery for HACL* on Alcatel phone, (**e**) Recovery of swap condition, (**f**) Number of key candidates in full-key recovery for HACL* on Olimex board [5]

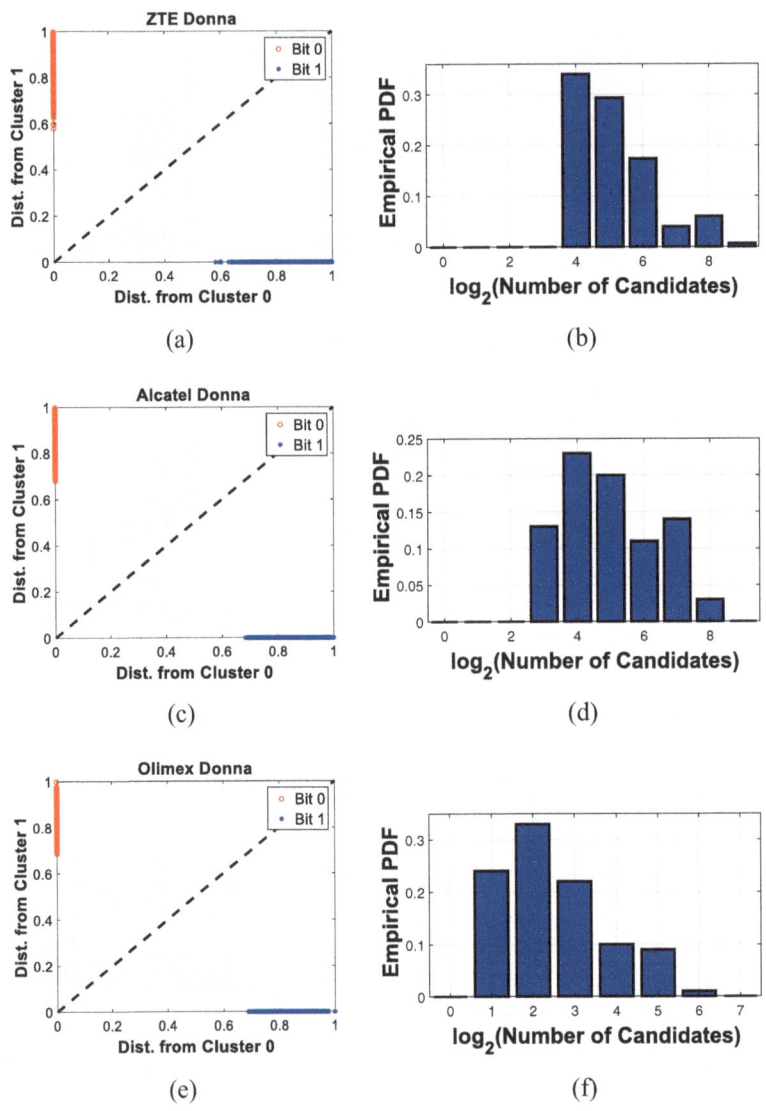

Fig. 13.23 (a) Recovery of swap condition, (b) Number of key candidates in full-key recovery for curve25519-donna on ZTE phone, (c) Recovery of swap condition, (d) Number of key candidates in full-key recovery for curve25519-donna on Alcatel phone, (e) Recovery of swap condition, (f) Number of key candidates in full-key recovery for curve25519-donna on Olimex board [5]

13.5 The Nonce@Once Attack

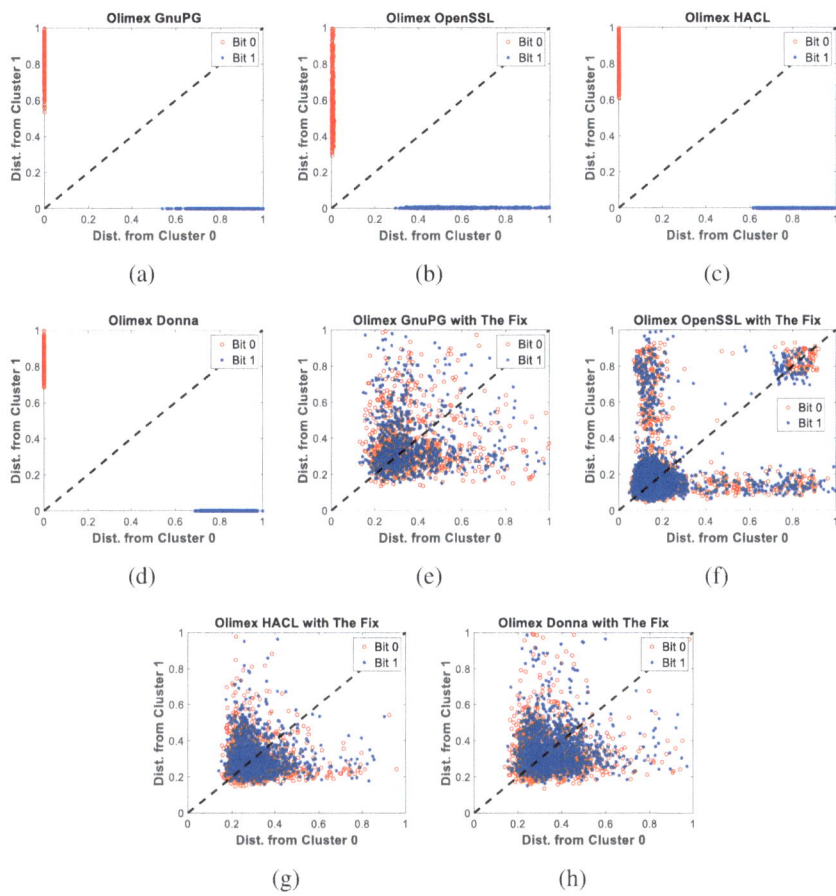

Fig. 13.24 (a) Correlation-based clustering of signal snippets without mitigation for Libgcrypt on Olimex board, (b) Correlation-based clustering of signal snippets without mitigation for OpenSSL on Olimex board, (c) Correlation-based clustering of signal snippets without mitigation for HACL* on Olimex board, (d) Correlation-based clustering of signal snippets without mitigation for curve25519-donna on Olimex board, (e) Correlation-based clustering of signal snippets with mitigation for Libgcrypt on Olimex board, (f) Correlation-based clustering of signal snippets with mitigation for OpenSSL on Olimex board, (g) Correlation-based clustering of signal snippets with mitigation for HACL* on Olimex board, (h) Correlation-based clustering of signal snippets with mitigation for curve25519-donna on Olimex board, with mitigation applied (e)–(h), signal snippets for $cond = 0$ and $cond = 1$ can no longer be separated [5]

bits are statistically the same, regardless of the value of cond. Then,[2] at Lines 18 and 19 we apply this masked delta to the machine words that make up the elliptic

---

[2] For now, we ignore Lines 12–17 in Fig. 13.25, which perfrom no operations on any of the values, but are neded for reasons that will be discussed later in this section.

```
1   EC_point_swap_cond(a,b,cond){
2           // When cond is 0, set mask to all-zeros
3           // When cond is 1, set mask to all-ones
4           mask=0-cond;
5           // Random value for mitigation
6           rand=random_word();
7           // For each machine word in the
8           // EC point representation
9           for(i=0;i<nwords;i++){
10                  delta = (a->w[i] ^ b->w[i]) & mask;
11                  delta = delta ^ rand;
12                  asm volatile (
13                      // No actual assembler code
14                      "";
15                      // But specify that rand is changed
16                      : "+r" (rand)
17                      : :);
18                  a->w[i] = (a->w[i] ^ delta) ^ rand;
19                  b->w[i] = (b->w[i] ^ delta) ^ rand;
20          }
21  }
```

Fig. 13.25 Conditional swap with our mitigation. The value of delta is masked to avoid systematic cond-dependent differences in exclusive-or operands

curve points a and b, just like the original conditional swap function did (Fig. 13.7, Lines 9 and 10), but now that is followed by unmasking each of these machine words (by XOR-ing with rand as the last operation in Line 18 and in Line 19). This unmasking effectively undoes the effects that masking of delta would have on these values, which ensures that this new implementation of the conditional swap function still implements the correct behavior—either swapping a and b or leaving them unchanged, depending on the value of cond.

Empirical Evaluation

To empirically evaluate this countermeasure, we manually checked the signal snippets corresponding to the conditional swap operation. We found that they no longer exhibit condition-dependent differences. Next, Fig. 13.24 shows the results of applying the clustering algorithm from Sect. 13.5.9 on these signal snippets. As can be seen, after applying this countermeasure the clusters that correspond to the two values of the swap condition are no longer separable. Similar results hold even for a larger amount of clusters (e.g., $c = 100, 1000$). Finally, we further verified that, when employing this countermeasure, the accuracy of detecting the swap condition using our attack, is statistically equivalent to a random guess.

Attack Mitigation with Compiler Optimizations

Interestingly, while implementing the countermeasure shown in Fig. 13.25 before we added Lines 12–17, we found that the EM signal from the resulting binary is very similar to the EM signal from the unmodified OpenSSL code, keeping the modified version still vulnerable to our key extraction attack. Further investigation revealed that, when used with high optimization levels ("-O2" and above), the assembler-level code produced by the compiler did not actually contain the countermeasure that we have added to the C source code. We experimented with three versions of the gcc compiler (4.6.3, 5.4.0, and 7.3.0), and also checked the results when compiling for an x86-64 platform.

It appears that, in its effort to improve the performance of the code, the compiler was effectively undoing our mitigation. Intuitively, the compiler finds that, after the value of `delta` is computed at Line 10 in Fig. 13.25, the expression for the new value of `a->w[i]` can be written as `a->w[i] = a->w[i]` $\oplus$ `delta` $\oplus$ `rand` $\oplus$ `rand`. Next, using the properties of the bitwise-XOR operator, the compiler rewrites the above expression as `a->w[i] = a->w[i]` $\oplus$ `delta`. After applying the same optimization to the value of `b->w[i]`, the compiler further realizes that the new value of `delta` computed at Line 11 is not needed, which implies that the value of `rand` is also no longer needed. Thus the compiler eliminates Lines 6 and 11 and removes the XOR with `rand` from lines 18 and 19, which effectively reverts our mitigated code (Fig. 13.25) back to its pre-mitigation state (Fig. 13.7).

While the constant-time swap operation could be written from scratch in assembler, the result might be complicated and hard to verify due to the complex structure of the elliptic curve point representation. Instead, we used gcc's extended inline assembly syntax to tell the compiler (Lines 12–17 in Fig. 13.25) that the value of variable `rand` might change between lines 11 and 18. This forces the compiler to treat the value of `rand` in Lines 18 and 19 as if it were different from the value in Line 11, which prevents the problematic optimization and allows our mitigation to behave as required, regardless of the level of compiler optimization that is applied when this function is compiled.

Performance Overheads

Our mitigation approach increases the duration of the ECDSA signing operation by less than 0.1%. While this demonstrates the efficiency of our countermeasure, we note that this low overhead is not surprising, as ECDSA is dominated by the mathematical operations over elliptic curves, which are orders of magnitude longer than XORs.

References

1. Onur Aciiçmez, Çetin Kaya Koç, and Jean-Pierre Seifert. On the power of simple branch prediction analysis. In *Proceedings of the 2nd ACM Symposium on Information, Computer and Communications security (ASIACCS)*, pages 312–320. ACM Press, March 2007.

2. D Agrawal, B Archambeult, J R Rao, and P Rohatgi. The EM side-channel(s): attacks and assessment methodologies. In http://www.research.ibm.com/intsec/emf-paper.ps, 2002.
3. Dakshi Agrawal, Bruce Archambeault, Josyula R. Rao, and Pankaj Rohatgi. The em side-channel(s). In *Revised Papers from the 4th International Workshop on Cryptographic Hardware and Embedded Systems*, CHES '02, pages 29–45, London, UK, UK, 2003. Springer-Verlag.
4. Monjur Alam, Haider Adnan Khan, Moumita Dey, Nishith Sinha, Robert Callan, Alenka Zajic, and Milos Prvulovic. One&Done: A Single-Decryption EM-Based Attack on OpenSSL's Constant-Time Blinded {RSA}. In *27th {USENIX} Security Symposium ({USENIX} Security 18)*, pages 585–602, 2018.
5. Monjur Alam, Baki Yilmaz, Frank Werner, Niels Samwel, Alenka Zajic, Daniel Genkin, Yuval Yarom, and Milos Prvulovic. Nonce@once: A single-trace em side channel attack on several constant-time elliptic curve implementations in mobile platforms. In *2021 IEEE European Symposium on Security and Privacy (EuroS P)*, pages 507–522, 2021.
6. Alcatel. Alcatel Ideal / Streak Specifications, 2016.
7. Thomas Allan, Billy Bob Brumley, Katrina E. Falkner, Joop van de Pol, and Yuval Yarom. Amplifying side channels through performance degradation. In *ACSAC*, pages 422–435, 2016.
8. ARM. ARM Cortex A8 Processor Manual. https://www.arm.com/products/processors/cortex-a/cortex-a8.php, accessed April 3, 2016.
9. M Backes, M Durmuth, S Gerling, M Pinkal, and C Sporleder. Acoustic side-channel attacks on printers. In *Proceedings of the USENIX Security Symposium*, 2010.
10. Josep Balasch, Benedikt Gierlichs, Oscar Reparaz, and Ingrid Verbauwhede. DPA, Bitslicing and Masking at 1 GHz. In Tim Güneysu and Helena Handschuh, editors, *Cryptographic Hardware and Embedded Systems (CHES)*, pages 599–619. Springer Berlin Heidelberg, 2015.
11. E Bangerter, D Gullasch, and S Krenn. Cache games - bringing access-based cache attacks on AES to practice. In *Proceedings of IEEE Symposium on Security and Privacy*, 2011.
12. A G Bayrak, F Regazzoni, P Brisk, F.-X. Standaert, and P Ienne. A first step towards automatic application of power analysis countermeasures. In *Proceedings of the 48th Design Automation Conference (DAC)*, 2011.
13. Naomi Benger, Joop van de Pol, Nigel P. Smart, and Yuval Yarom. "Ooh aah... just a little bit" : A small amount of side channel can go a long way. In *CHES*, pages 75–92, 2014.
14. Daniel J. Bernstein. Curve25519: New Diffie-Hellman speed records. In *PKC*, pages 207–228, 2006.
15. Daniel J. Bernstein, Joachim Breitner, Daniel Genkin, Leon Groot Bruinderink, Nadia Heninger, Tanja Lange, Christine van Vredendaal, and Yuval Yarom. Sliding right into disaster: Left-to-right sliding windows leak. Conference on Cryptographic Hardware and Embedded Systems (CHES) 2017, 2017.
16. E Biham and A Shamir. Differntial Cryptanalysis of the Data Encryption Standard. In *Proceedings of the 17th Annual International Cryptology Conference*, 1997.
17. Dan Boneh and David Brumley. Remote Timing Attacks are Practical. In *Proceedings of the USENIX Security Symposium*, 2003.
18. J. Brouchier, T. Kean, C. Marsh, and D. Naccache. Temperature attacks. *Security Privacy, IEEE*, 7(2):79–82, March 2009.
19. Robert Callan, Alenka Zajic, and Milos Prvulovic. A practical methodology for measuring the side-channel signal available to the attacker for instruction-level events. In *47th Annual IEEE/ACM International Symposium on Microarchitecture, MICRO 2014, Cambridge, United Kingdom, December 13–17, 2014*, pages 242–254, 2014.
20. S Chari, C S Jutla, J R Rao, and P Rohatgi. Towards sound countermeasures to counteract power-analysis attacks. In *Proceedings of CRYPTO'99, Springer, Lecture Notes in computer science*, pages 398–412, 1999.
21. S Chari, J R Rao, and P Rohatgi. Template attacks. In *Proceedings of Cryptographic Hardware and Embedded Systems - CHES 2002*, pages 13–28, 2002.
22. B Coppens, I Verbauwhede, K De Bosschere, and B De Sutter. Practical Mitigations for Timing-Based Side-Channel Attacks on Modern x86 Processors. In *Proceedings of the 30th IEEE Symposium on Security and Privacy*, pages 45–60, 2009.

23. Ettus. USRP-B200mini. https://www.ettus.com/product/details/USRP-B200mini-i, accessed February 4, 2018.
24. Karine Gandolfi, Christophe Mourtel, and Francis Olivier. Electromagnetic analysis: Concrete results. In *Proceedings of the Third International Workshop on Cryptographic Hardware and Embedded Systems*, CHES '01, pages 251–261, London, UK, UK, 2001. Springer-Verlag.
25. Daniel Genkin, Lev Pachmanov, Itamar Pipman, Adi Shamir, and Eran Tromer. Physical key extraction attacks on pcs. *Commun. ACM*, 59(6):70–79, May 2016.
26. Daniel Genkin, Lev Pachmanov, Itamar Pipman, and Eran Tromer. Stealing keys from PCs using a radio: cheap electromagnetic attacks on windowed exponentiation. In *Conference on Cryptographic Hardware and Embedded Systems (CHES)*, 2015.
27. Daniel Genkin, Lev Pachmanov, Itamar Pipman, Eran Tromer, and Yuval Yarom. ECDSA key extraction from mobile devices via nonintrusive physical side channels. In *CCS*, pages 1626–1638, 2016.
28. Daniel Genkin, Itamar Pipman, and Eran Tromer. Get your hands off my laptop: physical side-channel key-extraction attacks on PCs. In *Conference on Cryptographic Hardware and Embedded Systems (CHES)*, 2014.
29. Daniel Genkin, Adi Shamir, and Eran Tromer. RSA key extraction via low-bandwidth acoustic cryptanalysis. In *International Cryptology Conference (CRYPTO)*, 2014.
30. L Goubin and J Patarin. DES and Differential power analysis (the "duplication" method). In *Proceedings of Cryptographic Hardware and Embedded Systems - CHES 1999*, pages 158–172, 1999.
31. Wilko Henecka, Alexander May, and Alexander Meurer. Correcting Errors in RSA Private Keys. In *Proceedings of CRYPTO*, 2010.
32. Nadia Heninger and Hovav Shacham. Reconstructing rsa private keys from random key bits. In *International Cryptology Conference (CRYPTO)*, 2009.
33. Michael Hutter and Jorn-Marc Schmidt. The temperature side channel and heating fault attacks. In A. Francillon and P. Rohatgi, editors, *Smart Card Research and Advanced Applications*, volume 8419 of *Lecture Notes in Computer Science*, pages 219–235. Springer International Publishing, 2014.
34. Marc Joye and Sung-Ming Yen. The Montgomery powering ladder. In *CHES*, pages 291–302, 2002.
35. A. Karatsuba and Yu. Ofman. Multiplication of many-digital numbers by automatic computers. *Proceedings of the USSR Academy of Sciences*, 145(293–294), 1962.
36. Keysight. N9020A MXA Spectrum Analyzer. https://www.keysight.com/en/pdx-x202266-pn-N9020A/mxa-signal-analyzer-10-hz-to-265-ghz?cc=US&lc=eng, accessed February 4, 2018.
37. Haider Adnan Khan, Monjur Alam, Alenka Zajic, and Milos Prvulovic. Detailed tracking of program control flow using analog side-channel signals: a promise for iot malware detection and a threat for many cryptographic implementations. In *SPIE Defense+Security - Cyber Sensing*, 2018.
38. P Kocher. Timing attacks on implementations of Diffie-Hellman, RSA, DSS, and other systems. In *Proceedings of CRYPTO'96, Springer, Lecture notes in computer science*, pages 104–113, 1996.
39. P Kocher, J Jaffe, and B Jun. Differential power analysis: leaking secrets. In *Proceedings of CRYPTO'99, Springer, Lecture notes in computer science*, pages 388–397, 1999.
40. Markus Guenther Kuhn. Compromising emanations: eavesdropping risks of computer displays. dec 2003.
41. A. Langley, M. Hamburg, and S. Turner. Elliptic curves for security. RFC 7748, 2016.
42. T S Messerges, E A Dabbish, and R H Sloan. Power analysis attacks of modular exponentiation in smart cards. In *Proceedings of Cryptographic Hardware and Embedded Systems - CHES 1999*, pages 144–157, 1999.
43. Peter L. Montgomery. Speeding the Pollard and elliptic curve methods of factorization. In *Mathematics of Computation*, volume 13, pages 243–264, 1987.

44. Katsuyuki Okeya, Hiroyuki Kurumatani, and Kouichi Sakurai. Elliptic curves with the Montgomery-form and their cryptographic applications. In *PKC*, pages 238–257, 2000.
45. Olimex. A13-OLinuXino-MICRO User Manual. https://www.olimex.com/Products/OLinuXino/A13/A13-OLinuXino-MICRO/open-source-hardware, accessed April 3, 2016.
46. OpenSSL Software Foundation. OpenSSL Cryptography and SSL/TLS Toolkit. https://www.openssl.org.
47. Colin Percival. Cache missing for fun and profit. In *Proc. of BSDCan*, 2005.
48. Samsung. Samsung Galaxy Centura SCH-S738C User Manual with Specs. http://www.boeboer.com/samsung-galaxy-centura-sch-s738c-user-manual-guide-straight-talk/, June 7, 2013.
49. W Schindler. A timing attack against RSA with Chinese remainder theorem. In *Proceedings of Cryptographic Hardware and Embedded Systems - CHES 2000*, pages 109–124, 2000.
50. Adi Shamir and Eran Tromer. Acoustic cryptanalysis (On nosy people and noisy machines). http://tau.ac.il/~tromer/acoustic/.
51. Yukiyasu Tsunoo, Etsuko Tsujihara, Kazuhiko Minematsu, and Hiroshi Miyauchi. Cryptanalysis of block ciphers implemented on computers with cache. In *Proceedings of the International Symposium on Information Theory and its Applications*, pages 803–806, 2002.
52. Joop Van de Pol, Nigel P. Smart, and Yuval Yarom. Just a little bit more. In *CT-RSA*, pages 3–21, 2015.
53. Yuval Yarom and Naomi Benger. Recovering OpenSSL ECDSA nonces using the FLUSH+RELOAD cache side-channel attack. IACR Cryptology ePrint Archive 2014/140, 2014.
54. Yuval Yarom and Katrina Falkner. FLUSH+RELOAD: A High Resolution, Low Noise, L3 Cache Side-Channel Attack. In *23rd USENIX Security Symposium (USENIX Security 14)*, pages 719–732, San Diego, CA, 2014. USENIX Association.
55. Alenka Zajic and Milos Prvulovic. Experimental demonstration of electromagnetic information leakage from modern processor-memory systems. *IEEE Transactions on Electromagnetic Compatibility*, 56(4):885–893, 2014.
56. ZTE. Zte zfive 2 lte, 2019.

Chapter 14
Using Analog Side Channels for Hardware Trojan Detection

14.1 Introduction

Integrated circuits (IC) have become an integral aspect of our lives—they control most of our "things", ranging from cellphones and washing machines to airplanes and rockets. Thus, the problem of ensuring authenticity and trust for ICs is already critically important, especially for sensitive fields such as military, finance, and governmental infrastructure, and is gaining in importance as an increasing number of "things" become "smart" and connected into the Internet-of-Things (IoT). However, cost and time-to-market considerations have led IC vendors to outsource some, and in most cases many, steps in the IC supply chain. The sheer number and diversity of entities involved in modern IC supply chain, each with its own set of potentially malicious actors that can insert malicious modifications (HTs) into an IC [40], makes it difficult to trust the resulting ICs, especially when potentially adversarial foreign governments are among these potentially malicious actors in the IC supply chain.

The potential existence of HTs significantly undermines the trust in any system that uses tan IC, because the hardware usually provides the base layer of security and trust that all software layers depend and build on [15, 36, 37]. Specifically, all software protections, correctness analysis, or even proofs rely on the hardware executing instructions as specified; by violating this assumption, HTs can defeat the best software protections and/or subvert even software functionality that is otherwise completely correct and vulnerability-free.

14.2 What Is a Hardware Trojan?

A Hardware Trojan (HT) is a malicious addition or modification to existing circuit elements of an integrated circuit (IC). An HT can change functionality of the IC, reduce its reliability (e.g., reduce battery life), or leak sensitive information. Likely targets of HT-based attacks include military systems, aerospace systems, civilian security-critical applications, financial applications, transportation systems, etc.

Typically, an HT is designed to be stealthy, i.e., to be difficult to discover. To avoid detection during routine testing and normal use of the IC, and HT allows the IC to function normally (i.e., produce correct outputs) until specific conditions, which are highly unlikely to occur during routine testing and/or use, are met. Thus the design of an HT typically has two key components as illustrated in Fig. 14.1: the *payload*, which implements the modification of the original circuit's behavior and the *trigger*, which detects when the conditions for activating the payload have been met. The conditions that activate an HT occur very rarely, and the payload is usually highly inert until activated; while inert, the HT allows the IC to follow its original input/output behavior. The malicious functionality of the payload can be functional, e.g., when HTs modify the outputs of the IC to cause harm or to leak sensitive information, and/or non-functional, e.g., when the payload increases power consumption, causes excessive wear-out to reduce the lifetime of the IC, leaks sensitive information through a side channel, etc. This makes HTs extremely challenging to detect by traditional functional verification and testing—test inputs are unlikely to activate the HT, and until it is activated the HT has no effect on the functional behavior of the IC.

14.2.1 Hardware Trojans: Characteristics and Taxonomy

Conventionally, the hardware has been seen as the root of trust, and the only untrusted parts were assumed to be the software or firmware running on top of the hardware. However, several studies on HTs have shown that even hardware cannot be trusted anymore [34]. Over the past several years, numerous papers have been published on the topic of understanding the intent and behavior [11, 14], implementation [33–43], and taxonomy of hardware Trojans [22–33].

In general, HTs are undesired and unknown malicious modifications to a hardware circuit that have three common characteristics: rarity of activation, malicious purpose, and avoiding detection [11].

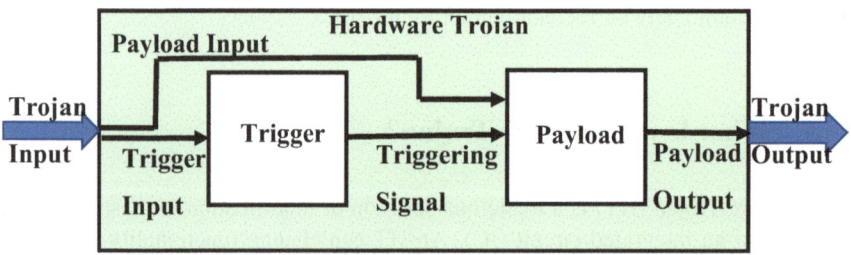

Fig. 14.1 Simplified block diagram of an HT

14.2 What Is a Hardware Trojan?

As discussed earlier in this chapter, a typical HT consists of two components: *trigger* and *payload*. The trigger circuit gets input from the host circuit to constantly check for the right conditions to activate the payload. In these very rare conditions, the payload is activated by the triggering signal from the trigger circuit to perform malicious activities. They could be leaking sensitive information, allowing the attackers to gain access to the hardware, shortening the operational lifetime of the hardware, etc.

As the number and sophistication of HTs has increased dramatically, several studies on the topic of characterizing and classifying HTs have been published over the last few years [33–39]. The most comprehensive characterization/categrization published to date is in [33]. HTs can be classified according to their activation mechanism, according to their functionality, or according to when they are inserted (which phase of the IC design and manufacturing process).

Activation Mechanisms
Some HTs are designed to be always-on, while others remain dormant until they are triggered. A dormant-until-triggered Trojan needs an internal or external event to be activated. An internally triggered HT would, for example, have a counter in the design to trigger the HT after a predetermined time (a "time bomb" HT). Another example of an internally triggered HT would be one that is triggered when the chip temperature exceeds 55°C. An externally triggered HT, on the other hand, monitors the inputs of the IC, waiting for a specific value or a sequence of values. Once activated, an HT can remain active "forever", or it can return to a dormant state after a specified time or under specific conditions.

An externally triggered Trojan requires input from outside the HT to activate. The external trigger can be user input or component output. User-input triggers include push-buttons, switches, keyboards, or keywords and phrases in the input data stream. Component-output triggers might be from any of the components that interact with the target device. For example, data coming through external interfaces such as RS-232 can trigger a Trojan.

Functionality
Functionality of HTs can be classified in many ways. One way is to consider the abstraction level, i.e., does it impact whole system or just one gate. Another way to categorize HTs is according to how they are implemented in the layout. They can also be classified as either combinatorial or sequential. Combinatorial HT circuits consider only the current values of the inputs to check if condition for payload activation is met. On the other hand, sequential HT circuits take fewer inputs and look for a specific *sequence* of values.

Insertion Mechanisms
Based on the phase in the IC design flow when HTs can be inserted into the chip, they can be classified as HTs inserted in specification, design, fabrication, testing and assembly phase.

Specification Phase - In this phase, chip designers define the system's characteristics: the target environment, expected function, size, power, and delay. While the IC's development is in this phase, functional specifications or other design constraints can be altered. For example, a Trojan at the specification phase might change the hardware's timing requirements.

Design Phase - Developers consider functional, logical, timing, and physical constraints as they map the design onto the target technology. Trojans might be in any of the components that aid the design. For example, a standard cell library can be infested with Trojans.

Fabrication Phase - During this phase, developers create a mask set and use wafers to produce the masks. Subtle mask changes can be used to add an HT or, in an extreme case, an adversary can replace the entire mask set.

Testing Phase - The IC testing phase is important for hardware trust, not because it is a likely phase for Trojan insertion, but because it provides an opportunity for Trojan detection. Testing is only useful for such detection if trustworthy. For example, an adversary who has inserted a Trojan in the fabrication phase may want to have control over the test vectors to ensure that the Trojan is not detected during testing. Trustworthy testing ensures that the test vectors will be kept secret and faithfully applied, and that the specified actions—accept/reject, binning—will be faithfully followed. However, a well-designed HT will be highly unlikely to activate during functional testing; in this case, subversion of the testing phase is optional.

Assembly Phase - The tested chip and other hardware components are used to create a printed circuit board (PCB). Every interface in a system where two or more components interact is a potential Trojan insertion site. Even if all the ICs in a system are trustworthy, malicious assembly can introduce security flaws in the system. For example, an unshielded wire connected to a node on the PCB can introduce unintended electromagnetic coupling between the signal on the board and its electromagnetic surroundings. An adversary can exploit this for information leakage, or fault injection, or to reduce reliability.

Effects of Hardware Trojans

The severity of how the HT affects the target hardware or system can range from subtle disturbances to catastrophic system failures. A Trojan can change the functionality of the target device and cause subtle errors that might be difficult to detect. For example, a Trojan might cause an error detection module to accept inputs that should be rejected. A Trojan can downgrade performance by intentionally changing device parameters. These include functional, interface, or parametric characteristics such as power and delay. For example, a Trojan might insert more

14.2 What Is a Hardware Trojan?

buffers for the chip's interconnections and hence consume more power, which in turn could drain the battery quickly. A Trojan can also leak information through both covert and overt channels. Sensitive data can be leaked via radio frequency, optical, thermal, power, or timing side channels, and via interfaces such as RS-232 and JTAG (Joint Test Action Group). For example, a Trojan might leak a cryptographic algorithm's secret key through supposedly unused RS-232 ports. Denial-of-service (DoS) Trojans prevent operation of a function or resource. A DoS Trojan can cause the target module to exhaust scarce resources like bandwidth, computation, and battery power. It could also physically destroy, disable, or alter the device's configuration—for example, causing the processor to ignore the interrupt from a specific peripheral. The DoS created by a DoS HT can be either temporary or permanent.

Locations

A hardware Trojan can be inserted in a single component or spread across multiple components—the processor, memory, input/output, power supply, or clock grid. Trojans distributed across multiple components can act independently of one another, or they can work together as a group to accomplish their attack objectives.

Processor - Any Trojan embedded into the logic units that are part of the processor is under this category. A Trojan in the processor might, for example, change the instructions' execution order, or it may add unwanted behavior to an instruction.

Memory - Trojans in memory blocks or in their interface units fall in this category. These Trojans might alter values stored in the memory, or they can block read or write access to certain memory locations. For example, an HT can change the contents of a programmable read-only memory in an IC, effectively installing malicious firmware.

I/O - Trojans can reside in a chip's peripherals or within the PCB. These peripherals interface with the external components and can give the Trojan control over data communication between the processor and the system's external components. For example, a Trojan might alter the data coming through USB port.

Power Supply - Trojans can alter the voltage and current supplied to the chip, causing failures, extra power consumption, lifetime reduction, etc.

Clock Grid - Trojans in the clock grid can change the clock's frequency, insert glitches in the clock supplied to the chip, and launch fault attacks. These Trojans can also freeze the clock signal supplied to the rest of the chip's functional modules. For example, a Trojan might increase the skew of the clock signal supplied to specific parts of a chip.

14.2.2 Overview of Methods for HT Detection

Some techniques focus on making the IC resilient to the presence of HTs, i.e., on preventing the HT's payload from modifying the behavior of the IC. Such techniques mostly rely on fault-tolerance-inspired approaches to operate correctly even when a HT has been able to modify some of the internal signals. However, these techniques can protect only specific parts of the system, such as a bus [24] or on-chip interconnect [41], and they require redundant activity during normal operation [26], and/or rely on reconfigurable logic [23].

Most counter-HT techniques, however, focus on detecting the presence of HTs. Some HT detection approaches are *destructive*, e.g., they rely on physical removal of the IC's layers to scan the actual layout of the IC, reverse-engineer its GDSII and/or netlist-level design [35], and compare it to a trusted design. However, all the ICs that are found to be HT-free through such analysis are also destroyed by the scan, and the reverse-engineering is extremely expensive and time-consuming, so destructive techniques can only be applied to a small sample of the larger population of IC.

Non-destructive HT detection approaches can be categorized according to whether they are applied to the design of the yet-to-be-fabricated IC (pre-silicon approaches), or to the fabricated IC (post-silicon approaches). Pre-silicon approaches use functional validation, and code and gate-level netlist analysis [32, 38], but they cannot detect HTs that are inserted after the design stage, e.g., by editing the physical layout of the IC at the foundry. Such concerns we addressed by post-silicon methods which attempt to identify HTs in ICs received from the foundry.

Non-destructive post-silicon approaches detect HTs either through testing the functional properties of the IC, or by measuring non-functional (side channel) behavior of the IC as it operates. Functional testing involves finding inputs that are likely to trigger unknown HTs that may exist in the IC, causing the payload of the HT to propagate the effects of the payload to the outputs of the IC, where they can be found to differ from expected outputs [42]. However, trigger conditions for HTs are designed to be difficult to reach accidentally, so the probability of detecting HTs is extremely low for conventional functional testing techniques. Additionally, functional testing techniques are likely to fail in detecting HTs whose payload does not change the input/output behavior or the IC, but rather causes increased power consumption, side channel leakage of sensitive information, etc.

Among post-silicon approaches, HT detection through side channel analysis appears to be the most effective and widely used approach [14, 34]. These methods measure one or more non-functional properties of the IC as it operates, and compare these measurements to reference signals obtained through either simulation or measurement on a device that is known to be genuine (this is often called a "golden" sample of an IC). Side channels that have been used by HT detection techniques include power consumption [6, 8, 9, 19], leakage current [27], temperature [10, 17], and electromagnetic emanations (EM) [7, 20, 28], and some approaches even combine measurements from multiple side channels [21, 31].

Among side-channel-based HT detection approaches, some add the side channel measurement capability to the chip itself, while others rely on measurements that are external to the chip itself. With on-chip measurement, the measurement circuitry is added to the design [12, 13, 25], which allows the specific chosen signals to be measured close to the signal's source. However, the additional circuitry for measurement, and for routing the desired signals to the measurement circuitry, impacts chip size, manufacturing cost, performance, and power, and this impact increases as the set of individually measurable signals increases.

In contrast, external-measurement side channel techniques require no modification to the IC itself. Instead, they rely on externally observable side-effects of the IC's normal activity. Since an HT is typically much smaller than the original circuit, an ideal side channel signal would have little noise and interference, so that the HT's small contribution to the signal is not obscured by the noise. Additionally, the HT's payload is largely inert until activated, and activation during measurement is highly unlikely, so ideally the side channel signal would be affected by the *presence* of the payload circuitry, even when it is inert. Finally, before activation, what little switching activity the HT does create is in its trigger component, which usually has only brief bursts of switching when the inputs it is monitoring change. Therefore, an ideal side channel signal would have high bandwidth, to allow these brief bursts of current fluctuation due to switching activity in the HT can be identified. Unfortunately, existing externally-measurable side channel signals, such as temperature, voltage and power supply current, and electromagnetic emanations [20], tend to vary mostly in response to current variation due to switching activity, so they are largely unaffected by an inert payload. Moreover, temperature changes slowly and has very limited bandwidth; voltage and supply current have low bandwidth [28] because on-chip capacitances that help limit supply voltage fluctuation act as a low-pass filter with respect to both current and voltage as seen from outside the chip. Finally, electromagnetic emanations can have high bandwidth, but their signal-to-noise ratio is affected by noise and interference.

14.3 Hardware Trojan Detection Using the Backscattering Side Channel

As discussed in Chap. 3, switching in digital circuits causes internal impedances to vary, which causes changes in the circuit's radar cross-section (RCS), and thus modulates the carrier wave that is backscattered by the circuit. This new side channel is impedance-based, so it can be beneficial to detection of HTs because the HTs added circuitry, and also the additional connections attached to existing circuitry, result in modifications to the chip's RCS and in how that RCS changes as the on-chip circuits switch. Note that, although the HT's trigger tends to be small, it exhibits switching activity as its logic reacts to inputs from the original circuitry, and it adds connections to the chip's original circuitry to obtain those inputs. If the payload is also connected to the IC's original circuitry e.g., functionally, when activated then those connections also change the IC's RCS, even while the payload is inert.

Most digital logic circuits are synchronous, so the overall switching pattern follows the clock cycle. Furthermore, the clock cycle usually accommodates switching delays along entire paths of logic gates, which means that the impedance change of an individual gate occurs abruptly at some point in the clock cycle, i.e., they have a square-wave-like waveform. This implies that the backscattered signal will contain side-band components for several harmonics of the circuit's clock frequency f_C. These side-band components will be at $f_{carrier} \pm f_C$, $f_{carrier} \pm 2f_C$, $f_{carrier} \pm 3f_C$, etc. The components at $f_{carrier} \pm f_C$ (that correspond to the first harmonic of the clock frequency) will mostly follow the overall RCS change during a cycle, while the components for the remaining harmonics will be influenced by the rapidity (rise/fall times) and timing of the impedance changes within the clock cycle.

Therefore, our detection of HTs using the backscattering side channel will rely on measuring the amplitude of the backscattered signal at $f_{carrier} \pm f_C$, $f_{carrier} \pm 2 * f_C$, ..., $f_{carrier} \pm m * f_C$, i.e., the side-bands for the first m harmonics of the clock frequency. We use only the amplitude (i.e., we ignore the signal's phase and other properties), mainly because the amplitude at some desired frequency is relatively easy to measure, whereas the phase and other properties require much more sophisticated tuning, phase tracking, etc. Furthermore, we note that each of the clock harmonics produces two side-band components that have the same amplitude, so the measurement can be made more efficient by only measuring m points to the left, or m points to the right, of $f_{carrier}$. The results that will be presented here involve measurements of points to the right of the carrier, i.e., $f_{carrier} + f_C$, $f_{carrier} + 2f_C$, etc.

We call the m amplitudes measured for a given circuit a *trace*, and each trace characterizes the circuit's overall amount, timing, and duration of impedance-change activity during a clock cycle. Intuitively, HTs can then be detected by first collecting training traces, using one or more "golden" ICs (that are known to be HT-free), and then HT detection on other ICs would consist of collecting their traces and checking if they are too different from the traces learned in training.

However, the amplitude of a received signal declines rapidly with distance. Our measurements are performed close to the chip, so even small variations in positioning of the probes create significant amplitude changes, and would result in numerous false positives when training and detection are not using identical probe positioning (which is very hard to achieve in practice).

Fortunately, the distance affects all of the points in a trace similarly, i.e., distance attenuates all amplitudes in the trace by the same multiplicative factor. Therefore, rather than using amplitudes for trace comparisons, we use amplitude ratios, i.e., amplitude of a harmonic divided by the amplitude of the previous harmonic,[1] which

---

[1] Measurement of signal amplitude are often expressed in decibels, i.e., on a logarithmic scale, and for these measurements subtraction of logarithmic-scale amplitude values yields the logarithmic-scale value for the amplitude ratio.

14.3 Hardware Trojan Detection Using the Backscattering Side Channel

cancels out the trace's distance-dependent attenuation factor. The resulting $m-1$ amplitude ratios are then used for comparing traces.

To illustrate amplitude ratios and how they are affected by differences in the tests circuit, Fig. 14.2 shows the statistics (mean and standard-deviation error bars) of each amplitude-ratio point, for a genuine AES circuit [33], and for the same AES circuit to which the T1800 Trojan from TrustHub [5] has been added (but remains inactive throughout the measurement). In this experiment, the carrier frequency is $f_{carrier}$=3.031 GHz, the AES circuit is clocked at f_C=20 MHz, and amplitudes for $m=35$ right-side-band harmonics are measured to obtain the 34 amplitude ratios shown in Fig. 14.2.

We observe that different amplitude-ratio points for the same trace vary significantly, from -30 to 35 dB in Fig. 14.2, and that different measurements for the same amplitude-ratio point tend to vary much less than that, making these differences difficult to see in Fig. 14.2, except for the very large differences between the HT-free and HT-afflicted design at the 18th and 19th amplitude ratio.

To more clearly show the differences at other harmonic-ratio points, Fig. 14.3 shows amplitude-ratio points that have been normalized to the mean amplitude ratio for the specifically, genuine AES circuit, for each amplitude ratio the logarithmic-scale points are shifted such that the genuine AES circuit's mean amplitude ratio becomes zero. It can now be observed that, in addition to the 18th and 19th point, the two circuits differ significantly in a number of other points, e.g., measurements for the two circuits are fully separable using the 14th point or the 20th point, and numerous other points have very little overlap between the HT-free and the HT-afflicted sets of measurements.

From Fig. 14.3, it can also be observed that the variance among measurements for the same design tends to increase with the index of the amplitude-ratio point, i.e., for points that correspond to higher harmonics. The primary cause of this increased variance is that higher harmonics of the signal tend to have lower amplitude, which makes their measurement less resilient to noise. Another factor that helps explain this increase in variance among higher harmonics is that they are affected by very

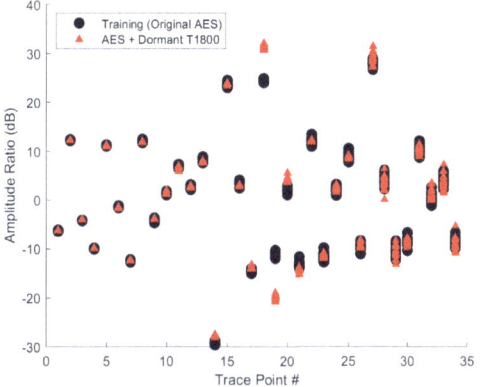

Fig. 14.2 Amplitude ratios for HT-free and HT-afflicted AES [29]

Fig. 14.3 Amplitude ratios for HT-free and HT-afflicted AES, with each point normalized to the mean of its HT-free measurements [29]

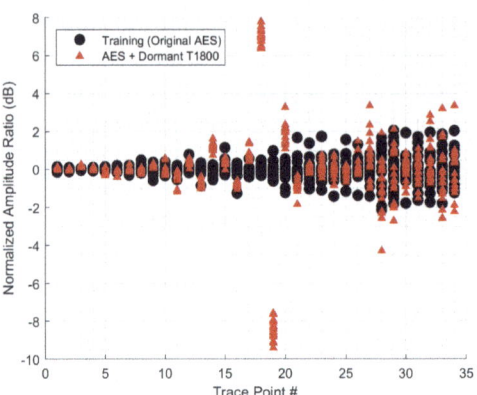

small differences in timing of impedance changes during the clock cycle, and factors such as temperature and power supply voltage fluctuation can create small changes in the switching speed of the gates, and thus in the timing of the resulting impedance changes.

Regardless of the reason for the increasing variance among measurements of higher harmonics, the fact that the variance does increase is an important motivation for using an impedance-based side channel rather than one created by bursts of current. Specifically, for each gate that switches, the impedance change persists for the rest of the cycle, while the burst of current is very brief in duration. This means that the contribution of impedance-change to lower frequencies is more significant than for the current-burst signal. When activity from cycle to cycle is repetitive, the spectrum of the signal's within-a-cycle waveform is projected onto the harmonics of the clock frequency, so gate-switching activity tends to affect lower harmonics of the clock frequency in impedance-based more than it does in current-burst-based side channels. As lower harmonics tend to have less variance from measurement to measurement, impedance-based side channels can be expected to perform better for HT detection than current-burst-based side channels.

14.4 HT Detection Algorithm

Our HT detection algorithm has two phases: *training*, where a circuit that is known to be HT-free is characterized, and *detection*, where an unknown circuit is classified into one of the two categories—HT-free or HT-afflicted, according to how much its measurements deviate from the statistics learned in training.

14.4.1 Training

Figure 14.4 details the training for the prototype implementation of backscattering-based HT detection. This training consists of measuring K times the signal

14.4 HT Detection Algorithm

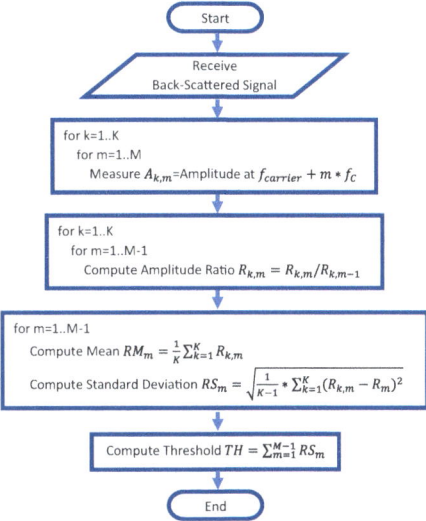

Fig. 14.4 Training algorithm [29]

backscattered from an IC known to be HT-free, each time collecting the m amplitudes at frequencies that correspond to the lowest m harmonics of the IC's clock frequency in the side-band of the received backscattered signal. The $m - 1$ amplitude ratios are then computed from these amplitudes.

Next, for each of the $m - 1$ amplitude ratios, the mean and standard deviation across the K measurements is computed, and the detection threshold for HT detection is computed as the sum of the $m - 1$ standard deviations.

14.4.2 Detection

Figure 14.5 details how the prototype implementation of backscattering detection decides whether to classify an IC as HT-free of HT-afflicted. First, a single measurement is obtained of the m amplitudes that correspond to the lowest m harmonics of the IC's clock frequency in the side-band of the signal that is backscattered from the IC under test, and $m - 1$ amplitude ratios are computed from these amplitudes.

Next, for each of the $m - 1$ amplitude ratios, we compute how much it deviates from the corresponding mean computed during training. This deviation is computed as the absolute value of the difference, and intuitively it measures how much that amplitude ratio differs from what would be expected from an HT-free IC. Finally, this sum of these deviations is compared to the sum of standard deviations from training. Intuitively, the sum of the differences for the IC under test is a measure of how much its overall backscattering "signature" differs from what would be expected from an HT-free IC, and the sum of standard deviations from training corresponds to how much an individual measurement of an HT-free IC can be expected to differ from the average of HT-free measurements. The IC under test

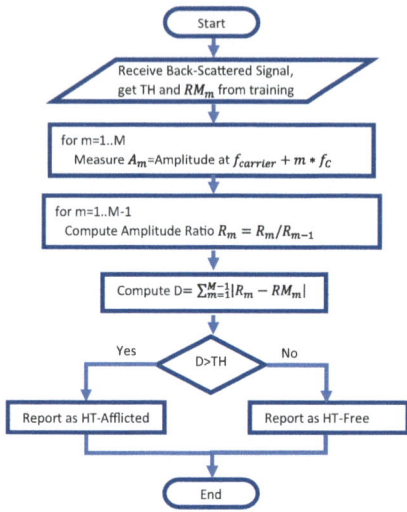

Fig. 14.5 Detection algorithm [29]

is labeled as HT-free if its sum of amplitude-ratio deviations is lower than this detection threshold (sum of standard deviations from training).

14.5 Experimental Setup

Figure 14.6 shows the measurement setup that we use to evaluate the performance of the proposed prototype backscattering-based HT-detection. The measurement setup to detect HTs is shown in Fig. 14.6. The carrier signal is a sinusoid at $f_{carrier}$=3.031 GHz produced by an Agilent MXG N5183A signal generator and transmitted toward the FPGA chip using an Aaronia E1 electric-field near-field probe. To select $f_{carrier}$, we have measured signal strength at the frequency of the reflected carrier signal (i.e., the frequency of the signal we were sending toward the board), the first several harmonics of the modulated FPGA board clock (e.g., 50 MHz away from the carrier), and of the noise floor of the instrument using AARONIA Near Field Probes (0 to 10 GHz). We have found that the side-band signal for the first harmonic of the board's clock is strongest when $f_{carrier}$ is around 3 GHz, but we also found that traditional EM emanations create interference at frequencies that are multiples of the board's clock frequency (50 MHz). Thus we choose $f_{carrier}$=3.031 GHz, a frequency close to 3GHz that avoids interference from the board's traditional EM emanation.

The device-under-test (DuT) is the FPGA chip on the Altera DE0-CV board, and it is positioned using a right-angle ruler so that different DE0-CV boards can be tested using approximately the same position of probes. The backscattered signal is received with an Aaronia H2 magnetic field near-field probe, and this signal preamplified using an EMC PBS2 low-noise amplifier and then the signal amplitudes

14.5 Experimental Setup

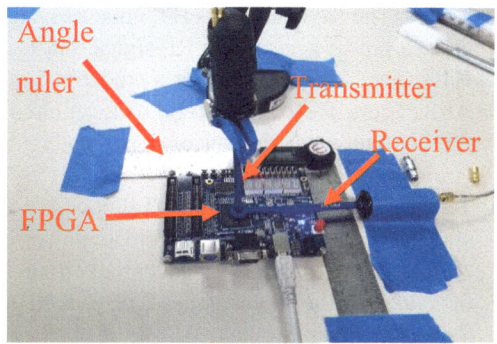

Fig. 14.6 Measurement setup for hardware Trojan detection using IP side channel [29]

at desired frequencies are measured using an Agilent MXA N9020A Vector Signal Analyzer.

14.5.1 Training and Testing Subject Circuit Designs

All circuits used in our experiments are implemented on a Field Programmable Gate Array (FPGA), which allows rapid experimentation by changing the circuit and/or its physical placement and routing, unlike hard-wired ASIC designs that would require fabrication for each layout variant of each circuit. The specific FPGA board we use is the Altera DE0-CV board, and within it the IC on which our backscattering measurement setup focuses is the Altera 5CEBA4F23C7N, a FPGA in Altera's Cyclone V device family.

For our HT detection experiments, we use AES-T1800, AES-T1600, and AES-T1100 hardware Trojan benchmarks from TrustHub [5]. For all three of these HTs, the original HT-free design is an AES-128 cryptographic processor, which uses an 11-stage pipeline to perform 10 stages of AES encryption on 128-bit blocks. Since numerous HTs in the TrustHub repository are similar to each other, we selected these three HT benchmarks because they exhibit different approaches for their triggers and payloads:

- T1800: The payload in this HT is a cyclic shift register that, upon activation, continuously shifts to increase power consumption, which would be a serious problem for small battery-powered or energy-harvesting devices in e.g., medical implants. The HT's trigger circuit consists of combinatorial logic that monitors the 128-bit input of the AES circuit, looking for a specific 128-bit plaintext value: the occurrence of that 128-bit value at the input activates the payload. The size of T1800's trigger circuit is 0.27% of the original AES circuit, and the size of its payload is 1.51% of the size of the AES circuit. Because this HT's trigger and payload can be resized easily, we use this HT to study how our HT detection is affected by HT size and physical location.

- T1600: The payload in this HT, upon activation, creates activity on an otherwise-unused pin to generate an RF signal that leaks the key of the AES circuit. The HT's trigger circuit looks for a predefined *sequence* of values at the input of the AES circuit. The size of T1600's trigger circuit is 0.28% of the size of the original AES circuit, while the size of its payload is 1.76% of the size of the original AES circuit.
- T1100: The payload of this HT, upon activation, modulates its activity using a spread-spectrum technique to create power consumption patterns that leaks the AES key. The trigger is a (sequential) circuit that looks for a predefined sequence of values at the input of the AES circuit to activate the payload. The size of T1800's trigger circuit is 0.28% of the size of the original AES circuit, while the size of its payload is 1.61% of the size of the AES circuit.

A key challenge we faced when implementing the HT-afflicted circuits was that these HTs are specified at the register-transfer level, as modifications to the original AES circuit's Verilog HDL source code. If the modified source code is subjected to the normal compilation, placement, and routing, we found that the addition of the HT causes the EDA tool to change the placement and routing of most logic elements in the overall circuit, and this extensive change makes the modification very easy to detect, regardless of the HT's actual size and activity. The next approach we tried was to compile the AES circuit using the normal compilation, placement, and routing, and then for each HT-afflicted design we used the ECO (Engineering Change Order) tool in Altera's Quartus II suite to add the HT's circuitry, while preserving the placement of logic elements (and the routing of their connections) that belong to the original AES circuit. However, we found that this approach makes it very hard to place the HT's logic elements close to the inputs of the original AES circuit. As will be demonstrated in Sect. 14.6.5, the HT is easier to detect when its trigger is placed away from where it is connected to the original circuit. To make the HTs more stealthy, we instead compile, place, and route the *HT-afflicted* circuit, then create the HT-free circuit by removing (using the ECO tool) the HT's logic elements and their connections. This models the HT "dream scenario" for the malicious entity that wishes to insert the HT, as there is just enough space in the HT-free layout to insert the HT in just the right place to have very short connections to the original circuit. To illustrate this, the placement of the HT-free circuit and the T1800-afflicted circuit are shown in Fig. 14.7, with a zoom-in to show the details where the HT's logic elements are placed.

Finally, for HT detection, the circuit must be supplied with inputs during the evaluation. Since we evaluate our HT detection approach in the dormant-HT scenario, we can use (as long as it avoids triggering the HT), any input sequence that causes logic gates in the original AES circuit to change state, so each cycle we simply flip all of the AES circuit's input bits, as shown in Fig. 14.8.[2]

---

[2] Note that hexadecimal 3 and C correspond to binary 0011 and 1100, while hexadecimal A and 5 correspond to 1010 and 0101, respectively. Thus the inputs we feed to the AES circuit simply toggle each of the input bits, while avoiding all-ones and all-zeros patterns.

Fig. 14.7 Genuine AES circuit (left) and the hardware trojan infected AES circuit (right) [29]

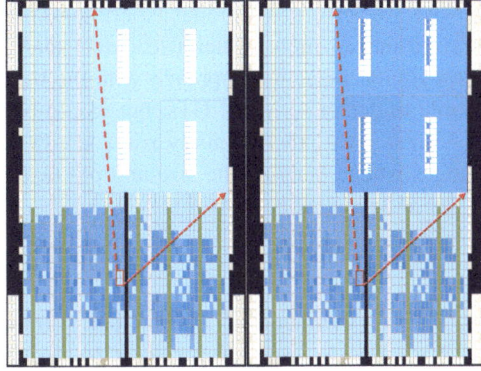

Fig. 14.8 Feeding inputs to the AES circuit [29]

```
always @ (posedge clk or posedge rst)
begin
    if (rst == 1'b1) begin
        cnt = 1'b0 ;
    end else begin
        if (cnt == 1'b1) begin
            cnt = 1'b0 ;
        end else begin
            cnt = cnt + 1'b1 ;
        end
    end
end
always @ (posedge clk or posedge rst)
begin
    if (rst == 1'b1) begin
        r_state <= 128'h55555555_55555555_55555555_55555555 ;
    end
    else begin
        case (cnt)
            1'b0: r_state = 128'h55555555_55555555_55555555_55555555 ;
            1'b1: r_state = 128'hAAAAAAAA_AAAAAAAA_AAAAAAAA_AAAAAAAA ;
        endcase
    end
end
```

14.6 Evaluation

Because it is very difficult to activate an HT without a priori knowledge of its trigger conditions, it is highly desirable for an HT detection scheme to provide accurate detection of *dormant* HTs, i.e., to detect HTs whose payload is never activated while it is characterized by the HT detection scheme. However, a dormant HT is typically more difficult to detect compared to an activated HT. For side channel-based detection methods, in particular, the switching activity in the activated payload, and/or the changes it creates in the switching activity of the original circuit, have more impact on the side channel signal than an inert payload (no switching activity in the payload and no changes to the original circuit's functionality).

Another important practical concern for HT detection is robustness to manufacturing variations and other differences between different physical instances of the same hardware design.

Thus our evaluation focuses on detection of *dormant* HTs with *cross-training*, i.e., training for HT detection is performed on one hardware instance, and then HT detection is performed on others.

The experimental results show that our prototype backscattering-based HT detection, after training with an HT-free design on one DE0-CV board, accurately reports the presence of dormant HTs for each of three different HT designs, on nine other DE0-CV boards, while having no false positives when the HT-free design is used on those nine other DE0-CV boards.

Next, we perform additional experiments to experimentally confirm that dormant HTs are indeed more difficult to detect than activated ones, and also to confirm that a similar detection approach with the traditional EM side channel would still be able to detect activated HTs, but would be unreliable for detection of dormant HTs. Finally, we experimentally evaluate how the accuracy of dormant-HT detection changes when changing the size and physical placement of the hardware Trojan's trigger and payload components.

14.6.1 Dormant-HT Detection with Cross-Training Using the Backscattering Side Channel Signal

We evaluate the effectiveness of our HT detection prototype by training it on one DE0-CV FPGA board with an HT-free AES circuit, then applying HT detection to several test subject circuits implemented on nine *other* DE0-CV FPGA boards.

The test subject designs are:

- *Original AES*. This is the same HT-free AES circuit that was used in training, and we use it to measure the false positive rate of our HT detection,
- *AES + Dormant T1800*. This is the same AES circuit, with the same placement and routing, that was used for training, but with additional logic elements and connections that implement the AES-T1800 Trojan from TrustHub. The size of this HT's trigger (in FPGA logic elements) is 0.27% of the original AES circuit, and we use a payload that was reduced to only 0.03% of the original AES circuit. The reduced payload size helps fit this HT closer to where its input signals can be connected to the original AES circuit, making the HT significantly more difficult to detect.
- *AES + Dormant T1600*. This is the same AES circuit, with the same placement and routing, that was used for training, but with additional logic elements and connections that implement the AES-T1600 Trojan from TrustHub. The size of this HT's trigger is 0.28% of the original AES circuit, while its payload's size is 1.76% of the original AES circuit.
- *AES + Dormant T1100*. This is the same AES circuit, with the same placement and routing, that was used for training, but with additional logic elements and connections that implement the AES-T1100 Trojan from TrustHub. The size of this HT's trigger is 0.28% of the original AES circuit, while its payload's size is 1.61% of the original AES circuit.

For each measurement, the previously measured FPGA board is removed from the measurement setup, and then a different board is positioned using an angle ruler to model a realistic measurement scenario when each measurement uses a very similar but not identical relative position of the chip and the probes. Each test subject design is measured 20 times on each board, and each measurement is used for HT detection in isolation, i.e., for each test subject the detection makes 20 classification

14.6 Evaluation

decisions (HT-free or HT-afflicted) on each of the 9 boards, resulting in a total of 720 decisions. Among these decisions, 180 were on the *Original AES* test subject, and in all 180 of them our prototype has correctly classified the design as HT-free, i.e., the HT detection prototype had no false-positive detections. In the remaining 3 sets of 180 decisions, each test subjects design was HT-afflicted (180 decisions with T1800, 180 decisions with T1600, and 180 with T1100), and in all of them our detection prototype has correctly classified the design as HT-afflicted, i.e., the HT detection prototype has detected the presence of an HT in each measurement in which an HT was present.

Since our HT detection prototype using the backscattering side channel achieves 100% detection of three kinds of dormant HTs, with 0% positives, in the cross-training measurement scenario, we focus the rest of our experimental evaluation on getting more insight into why our HT detection performs so well and how sensitive it is to changes in the position and size of the HT.

14.6.2 HT Detection of Dormant vs. Active HTs Using the Backscattering Side Channel

Figure 14.9 compares the normalized amplitude ratios for an HT-free AES design and for the same AES design (and layout) to which the AES-T1800 Trojan has been added. Two separate sets of 20 measurements are shown for the HT-free design, one that is used for training and one that is used to detect false positives when evaluating HT detection (on another DE-0CV board). For the HT-afflicted design, one set of 20 measurements is collected when the HT is dormant (its payload has not been activated), and another set of 20 measurements is collected with the same HT after its payload is activated. We can observe that there are a number of trace points where both sets of HT-afflicted measurements deviate significantly from HT-free measurements, and that this deviation tends to be larger

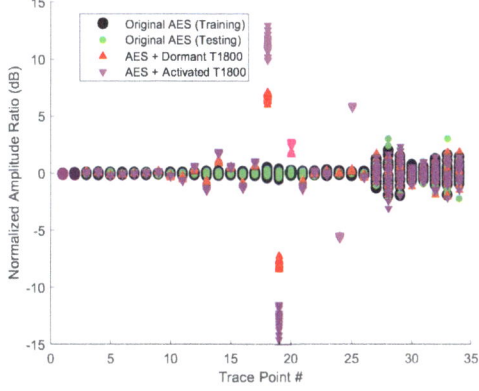

Fig. 14.9 Normalized amplitude ratios for backscattering side channel measurements [29]

for measurements in which the HT has been activated. The higher deviation from HT-free measurements seen for active-HT measurements agrees with the intuitive reasoning that an HT is easier to detect when active then when it is dormant. Even so, our backscattering-based HT detection prototype successfully reports the existence in each dormant-HT experiment (100% detection rate), while correctly reporting all 20 HT-free measurements as HT-free (no false positives).

14.6.3 Comparison to EM-Based HT Detection

The impedance-based backscattering side channel should be more effective for HT detection than existing current-burst-based (e.g., traditional EM) side channels, because backscattering detects change of impedance in a circuit and does not require currents to flow through a part of the circuit in order to detect the change to that part of the circuit. To confirm this, we repeat the same experiment, but this time use amplitudes of EM emanations at the clock frequency and its harmonics, instead of using the clock-frequency harmonics in the side-bands of the backscattered signal. The normalized amplitude ratios from these measurements are shown in Fig. 14.10. We can observe that the HT-afflicted measurements are much less separated from HT-free ones than they were with backscattering—for most trace points even active-HT measurements are all within ±1 dB from the HT-free ones, although for several trace points there is still some separation between the active-HT and HT-free measurements. More importantly, nearly all dormant-HT measurements have a lot of overlap with HT-free measurements, which makes the dormant-HT measurements difficult to distinguish from HT-free ones.

This is confirmed by the results of applying our HT detection analysis to these EM measurements. The ROC (Receiver Operating Characteristic) curves for HT detection using backscattering and EM side channels are shown in Fig. 14.11. Backscattering-based detection correctly identifies the presence of an HT in each

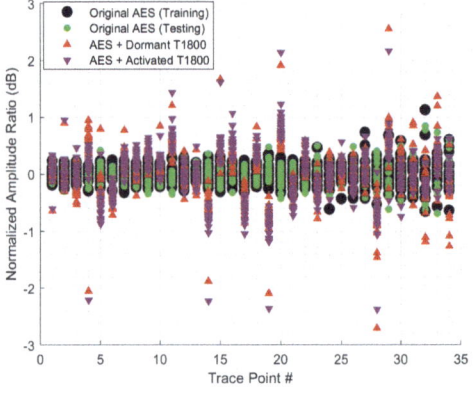

Fig. 14.10 Normalized amplitude ratios for traditional electromagnetic side channel measurements [29]

Fig. 14.11 Detection performance (ROC curve) comparison of backscattering-based and EM-based detection in active-HT and dormant-HT scenarios [29]

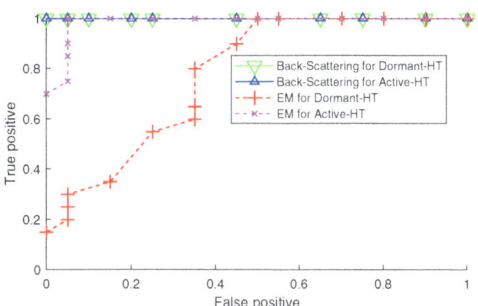

HT-afflicted measurement, without false positives in HT-free measurements, in both active-HT and dormant-HT scenarios. In contrast, detection based on the EM side channel performs less well in the active-HT case, reporting only 70% of the active-HT measurements as HT-afflicted using the default threshold (which produces no false positives). More importantly, EM-based detection in the dormant-HT case performs poorly—in the absence of false positives, only 15% of the dormant-HT measurements are correctly reported as HT-afflicted, and when the detection threshold is reduced to a point where all dormant-HT measurements are reported as HT-afflicted, 50% of the HT-free measurements are also reported as HT-afflicted (a 50% false-positive rate).

In conclusion, these experiments indicate that our HT detection technique's ability to detect dormant HTs comes, at least in large part, from using the backscattering (impedance-based) side channel instead of traditional current-burst-based (EM and power) side channels.

14.6.4 Impact of Hardware Trojan Trigger and Payload Size

To provide more insight into which factors influence our HT detection prototype's ability to detect dormant HTs, we perform experiments in which we reduce the size of the T1800 hardware Trojan's trigger and payload. The T1800 was chosen because it has the smallest trigger among the HTs we used in our experiments, and because both its payload and its trigger can be meaningfully resized.

The T1800 monitors the 128-bit data input of the AES-128 circuit, comparing it to a specific hard-wired 128-bit value, and it activates the payload when that 128-bit value is detected. In terms of logic elements (gates), the size of this 128-bit trigger is only 0.27% of the size of the original AES circuit, i.e., even this full-size trigger is much smaller than the AES circuit to which the HT has been added, and its activity (while the HT is dormant) is difficult to detect using current-based side channels (EM, power, etc.). We implement reduced-trigger variants of this HT that monitor only the 64 least significant bits (the "1/2 Trigger Size" variant, where the trigger circuit size is only 0.15% of the original AES circuit's size), and then only the 32 least significant bits (the "1/4 Trigger Size" variant, where the trigger circuit size is only 0.08% of the original AES circuit size). The normalized harmonic ratio traces for 20 measurements of each design, along with 40 HT-free measurements (20 for training and 20 for false-positives testing) are shown in Fig. 14.12. We observe that smaller trigger sizes result in trace points that are closer to HT-free ones, i.e., that trigger size directly impacts the side-channel-based separation between dormant-HT and HT-free circuits. These results match the intuition that the HT's influence on impedance changes should increase as more input bits are monitored by the HT's trigger, both because of the increased number of connections to the original circuit (which can change impedances "seen" by gates that belong to the original circuit) and because of the increased number of gates whose values can change (switching activity) within a cycle in the HT's trigger circuit itself.

The ROC curves for HT detection with different trigger sizes (Fig. 14.13) confirm that, while the HT with the original-size and even 1/2-size trigger can be detected in each measurement with no false positive, the detection accuracy suffers significantly as the HT's trigger is further reduced to 1/4 of the original size.

We perform additional experiments in which we keep the trigger at full size, but reduce the size of the payload to 50% and then 25%. Our dormant-HT measurement results for these variants are not noticeably different from each other (Fig. 14.14), which implies that the payload size has little impact on our HT detection. This agrees with our theoretical and intuitive expectations: the payload in T1800 has little impact on the impedance changes during a clock cycle, as it has no switching

Fig. 14.12 Normalized amplitude ratios for different sizes of T1800's trigger input [29]

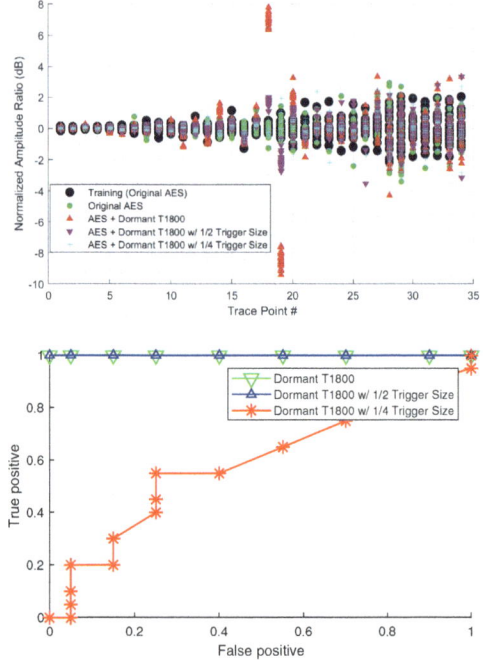

Fig. 14.13 ROC curves for HT detection for different sizes of the HT's trigger circuit [29]

14.6 Evaluation

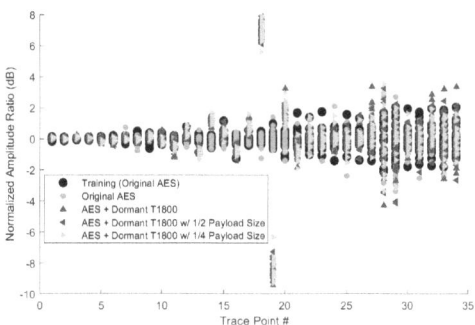

Fig. 14.14 Normalized amplitude ratios for different sizes of T1800's (dormant) payload [29]

activity (until activated), and has no connections to the gates in the original AES circuit (T1800's payload is designed to produce a lot of power-draining switching activity upon activation, not to change the functionality of the AES circuit).

Since the measurements of the full-trigger-and-reduced-payload variants of T1800 HT are very similar to the full-size T1800 HT, they provide the same ROC curves (complete detection without false positives) as the full-size T1800 HT, as shown in Fig. 14.13.

14.6.5 Impact of HT Trigger and Payload Position

We next investigate how backscattering-based HT detection is influenced by the physical location and routing of the HT's connection to the original circuit. For this, we start with the AES circuit with the T1800 HT, whose trigger logic was placed at Position 1 shown in Fig. 14.15 by the placement and routing tool—very close to where its 128-bit input can be connected to the original AES circuit.

We then create a variant of this HT by moving the HT's trigger logic to Position 2, keeping the logic elements and the connections between them in the same position relative to each other, but making the trigger's 128 connections to the original AES circuit much longer. Another variant is similarly created by moving the HT's trigger logic to Position 3.

The dormant-HT measurement results for these three positions are shown in Fig. 14.16. We observe that, at many trace points, in terms of separation of HT-afflicted measurements HT-free ones, Position 2 is significantly more separated than Position 1, and Position 3 provides an additional small increase in separation. This means that HTs placed close to their connection points in the original circuit are more difficult to detect than HTs that require long connections. All of our prior experiments used HTs that were placed by the placement and routing tool in a way that attempts to minimize overall cost (which tends to minimize the total length of the HT's connections to the original circuit), so we can expect the Position 2 and Position 3 variants to also be detected correctly in each dormant-HT measurement

Fig. 14.15 Changing the physical position off the HT's trigger logic [29]

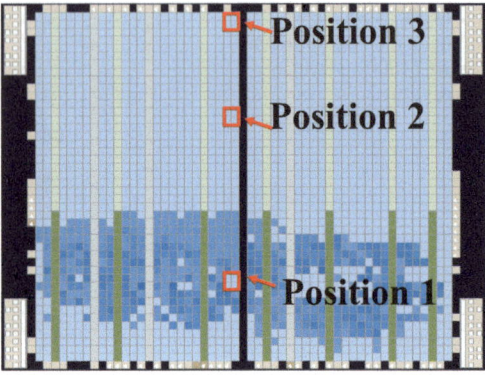

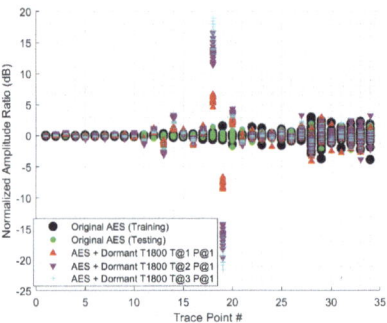

Fig. 14.16 Normalized amplitude ratios for different locations of T1800's trigger logic [29]

(with no false positives in HT-free measurements), and our HT detection results confirm this.

We also performed experiments in which the trigger part of the HT is kept in Position 1, while its payload was moved to Position 2 and then Position 3. Our results show that the payload position has little impact on the measurements, which is as expected given that, in our dormant-HT experiments the payload has no switching activity and the 1-bit "activate" signal between the trigger and the payload never changes its value (it stays at 0, i.e., inactive), and that (Fig. 14.17).

14.6.6 Further Evaluation of HT Detection Using More ICs and HTs

To further evaluate the effectiveness of our HT detection prototype, we implement two different circuits, RS232 and PIC16F84, each with three HTs, from TrustHub [5]. We use the same HT detection prototype described in Sect. 14.4 and the setup described in Sect. 14.5.

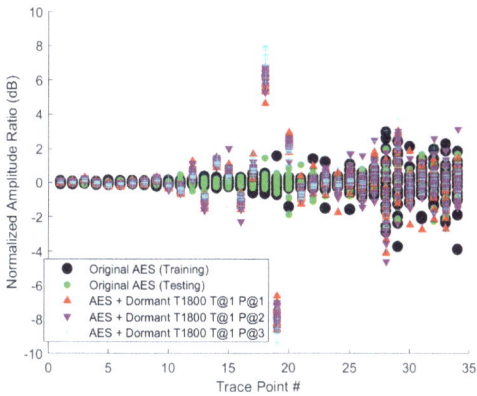

Fig. 14.17 Normalized amplitude ratios for different locations of T1800's (dormant) payload [29]

RS232 circuit

We use RS232-T500, RS232-T600, and RS232-T700 HT benchmarks from TrustHub [5]. For all three of these HTs, the original HT-free design is a RS232 micro-UART core, consisting of a transmitter and a receiver. The transmitter takes input words (128-bit length) and serially outputs each word according to the RS232 standard, while the receiver takes a serial input and outputs 128-bit words.

- RS232-T500: The payload in this HT is a circuit that, upon activation, causes the transmission to fail. The trigger is a sequential circuit that increments its counter every clock cycle, and activates the payload when this counter reaches a certain value. The size of the trigger circuit is 1.67%, and the size of the payload circuit is 1.48%, of the size of the RS232 circuit.
- RS232-T600: The payload in this HT is a circuit that, upon activation, makes the transmitter's "ready" signal become stuck-at-1, and changes specific bits in the transmitted data. The trigger is a sequential circuit that looks for a specific sequence of UART states to activate the payload. The size of the trigger circuit is 1.54%, and the size of the payload circuit is 1.52%, of the size of the RS232 circuit.
- RS232-T700: The payload of this HT is a circuit that, upon activation, makes the transmitter's "finished" signal become stuck-at-0. The trigger is a sequential circuit that looks for a predefined sequence of UART states to activate. The size of the trigger circuit is 1.54%, and the size of the payload circuit is 1.48%, of the size of the RS232 circuit.

The results in Figs. 14.18 and 14.19 show the ratios of harmonics and the ROC curve, respectively. The results show that each of these three Trojans is detected, with 100% accuracy and 0% false positives.

PIC16F84 Circuit

Next, we use PIC16F84-T100, PIC16F84-T200, and PIC16F84-T400 hardware Trojan benchmarks from TrustHub [5]. For all three of these HTs, the original HT-

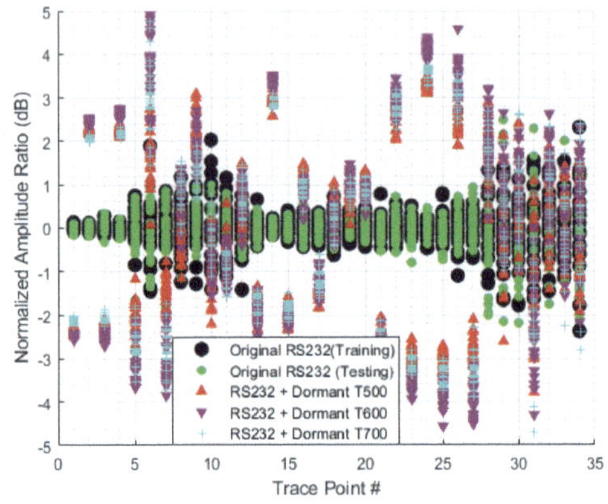

Fig. 14.18 Normalized amplitude ratios for different HTs in the RS232 circuit [29]

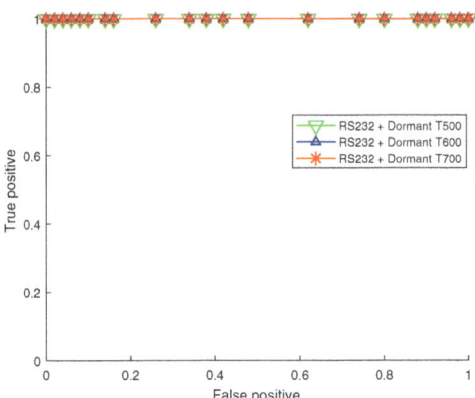

Fig. 14.19 ROC curves for detection of HTs in the RS232 circuit [29]

free design is the PIC16F84 circuit, a RISC micro-controller whose functions and instruction set are very similar to those of the Microchip 16F84 chip.

- PIC16F84-T100: Once activated by its (sequential) trigger circuit, the payload changes the address to PIC16F84's program memory (causing denial of service), or possible injection of malware. The size of the trigger circuit is 1.34%, while the size of the payload circuit is 1.81% of the size of the PIC16F84 circuit.
- PIC16F84-T200: Once activated by its (sequential) trigger circuit, the payload in this HT replaces the instruction register with a sleep command (causing denial of service). The size of the trigger circuit is 1.35%, and the size of the payload circuit is 1.93%, of the size of the PIC16F84 circuit.
- PIC16F84-T400: Once activated by its (sequential) trigger circuit, the payload of this HT changes the address lines to the external EEPROM to 0 (causing denial of service). The size of the trigger circuit is 1.35%, while the size of the payload circuit is 1.75% of the size of the PIC16F84 circuit.

14.6 Evaluation

The results in Figs. 14.20 and 14.21 show the ratios of harmonics and the ROC curve, respectively. The results show that each of these three Trojans is detected, with 100% accuracy and 0% false positives.

Trigger Size Experiment

As discussed in Sect. 14.6.4, trigger size has a significant effect on dormant-HT detectability. We chose RS232-T500 for this experiment because its trigger consists of monitoring the executed instruction stream, counting occurrences of a specific instruction until a threshold value is reached (and then activating the payload). The counter's size can be changed without affecting/changing the overall functionality of the HT.[3] Our reduced-trigger variants of this Trojan by reducing the number of bit of the counter. We have the following design:

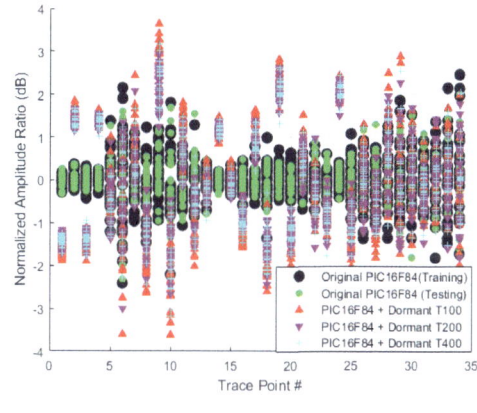

Fig. 14.20 Normalized amplitude ratios for different Trojans on PIC16F84 circuit [29]

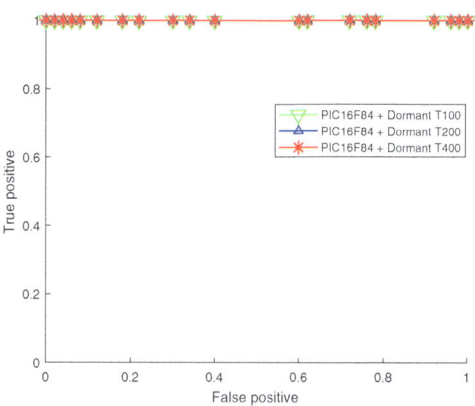

Fig. 14.21 ROC curves for different Trojans on PIC16F84 circuit [29]

---

[3] However, the reduced counter size requires the threshold to be reduced, thus activating the payload sooner and risking detection of the HT during functional and burn-in tests.

- RS232 + Dormant T500: The size of the trigger is 1.67% of the size of the original RS232 circuit.
- RS232 + Dormant T500 w/ 1/2 Trigger: The size of the trigger circuit is 1% of the size of the original RS232 circuit.
- RS232 + Dormant T500 w/ 1/4 Trigger: The size of the trigger circuit is 0.67% of the size of the original RS232 circuit.
- RS232 + Dormant T500 w/ 1/8 Trigger: The size of the trigger circuit is 0.33%, of the size of the original RS232 circuit.

For all four of these variants the payload, circuit remains unchanged, and its size is 1.48% of the original RS232 circuit's size.

The results in Figs. 14.22 and 14.23 show that, the smaller the trigger is, the harder it is to detect the Trojan, which agrees with our previous results for AES-based HTs.

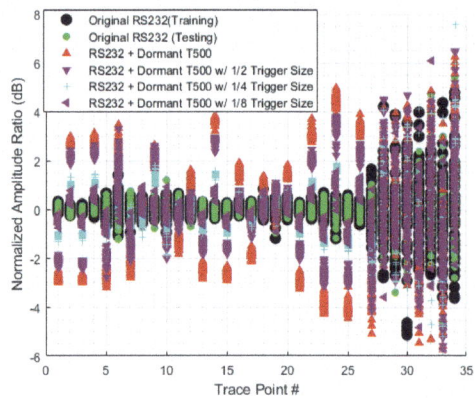

Fig. 14.22 Normalized amplitude ratios for different trigger size [29]

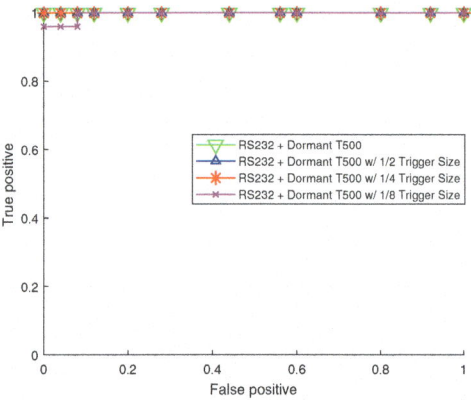

Fig. 14.23 ROC curves for different trigger size [29]

14.7 Detection of Counterfeit IC Using Backscattering Side Channel

Counterfeit ICs are ICs that are misrepresented in terms of their origin or quality. They have been a major concern in IC supply chain because IC counterfeiting infringes on the legitimate producer's intellectual property rights and, more importantly for the end user, counterfeit ICs often have inferior specifications and quality, and may thus represent a hazard if incorporated into critical systems such as aircraft systems, life support, military equipment, or space vehicles [18]. In this section, we evaluate the new backscattering side channel's suitability for detection of IC counterfeits that have a design which is not identical to the legitimate IC, i.e., we do not consider counterfeits where a legitimate IC is relabeled to misrepresent its speed grade, new-vs-recycled status, etc. Specifically, for our experiments, we implement two different kinds of counterfeit IC: 1) Counterfeit ICs with the same functionality as the original but different physical implementation (position) of the circuit, and 2) Counterfeit ICs with the same functionality and position as the original but different physical layout (routing and placement) of the circuit.

14.7.1 Counterfeit ICs with Different Layout

We have implemented several counterfeit IC examples by re-compiling the same design but allowing the EDA tool to change the placement and routing of the circuit. We have four different test subject designs: Original-layout AES IC, the 1st layout AES counterfeit IC, the 2nd layout AES counterfeit IC, and the 3rd layout AES counterfeit IC. The results in Figs. 14.24 and 14.25 show the ratios of harmonics and ROC curve, respectively. The results show that the counterfeit ICs can be easily detected (with no false positives).

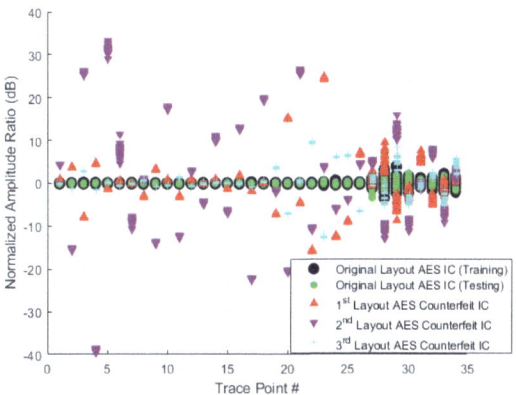

Fig. 14.24 Normalized amplitude ratios for different counterfeit IC layouts [29]

Fig. 14.25 ROC curves for detection of different counterfeit IC layouts [29]

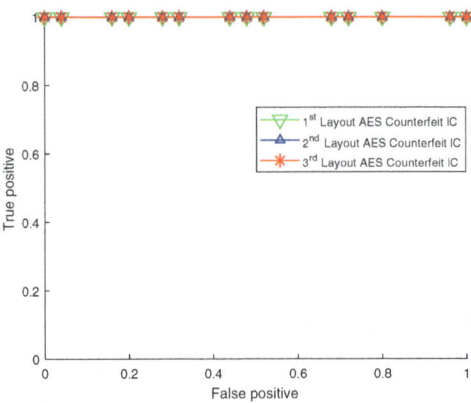

14.7.2 Counterfeit ICs with Changed Position

We have also implemented several counterfeit IC examples by moving the placement of the AES circuit from its original placement, while keeping the relative placement and routing the same, i.e., the entire layout of the IC is just moved to a different position of the chip. We have four different test subject designs: Original-position AES IC, the 1st position counterfeit IC, the 2nd position counterfeit IC, and the 3rd position counterfeit IC. We use the same technique to detect these counterfeit ICs. The results in Figs. 14.26 and 14.27 show the ratios of harmonics and ROC curve, respectively. The results show that we can easily detect all of these counterfeit ICs designs (with no false positives).

14.8 A Golden-Chip-Free Clustering for Hardware Trojan Detection Using the Backscattering Side Channel

As discussed earlier in this Chapter, there are two methods for detection of HTs inserted anywhere in the entire (long) IC manufacturing supply chain: reverse engineering and side-channel analysis. Side-channel analysis techniques have advantages of being non-destructive and relatively fast, which is suitable for testing a large number of ICs. However, side-channel techniques need either (1) a "golden" chip (a chip that is a priori known to be HT-free), which is not a practical assumption if HTs were inserted during manufacturing, or (2) simulation, which only works for the specific circuits that are modeled. These difficulties prevent side-channel techniques from being used without other assisting techniques for HT detection in practice. In contrast, reverse engineering techniques are highly accurate and need neither simulation nor a "golden" chip, which allows them to be used for the detection of HTs without any assisting techniques. However, reverse engineering is extremely

14.8 A Golden-Chip-Free Clustering for Hardware Trojan Detection Using the...

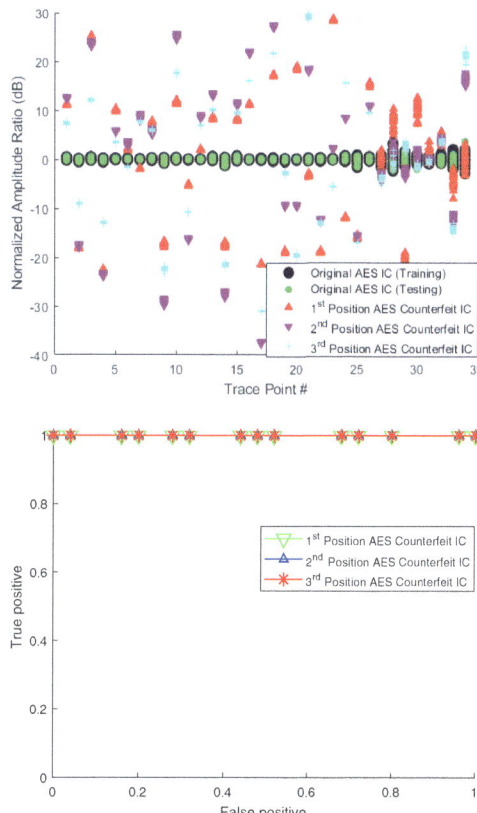

Fig. 14.26 Normalized amplitude ratios for counterfeit IC whose local layout is the same, but position on the chip is changed, relative to the original IC [29]

Fig. 14.27 ROC curves for detection of counterfeit ICs whose local layout is the same, but position on the chip is changed, relative to he original IC [29]

expensive, time-consuming, and destructive. Hence, these techniques could not be deployed for a large population of ICs.

To circumvent the drawbacks of the previous methods, we propose a new clustering method that uses measurements of the backscattering side-channel to enable deployment of reverse engineering techniques to a large population of ICs. Specifically, let there be M fabricated ICs, denoted as IC_1, IC_2,...,IC_{M-1}, IC_M. Utilizing each IC, a trace of features is extracted from its backscattering side-channel signals while it operates. Each IC then can be represented as a point in a high dimensional space. These ICs can be divided into clusters based on how a hardware Trojan (if it is present) affects their backscattering side-channel signals. The objective of the proposed clustering algorithm is to divide all tested ICs into correct clusters, so that every IC in a cluster should belong to the same type in terms of whether they are affected by HTs or not. This helps to tremendously reduce the workload for reverse engineering techniques—only one IC from each cluster needs to be reverse-engineered and more importantly, the conclusions of reverse engineering now apply to not only the IC that has been destroyed in the process, but rather to the entire cluster the reverses-engineered IC came from.

14.8.1 The Impact of HTs on Backscattering Side-Channel Signal

As described in Sect. 14.3 and [29], HTs can be detected by analyzing impedance changes during a clock cycle, where the changes caused by HTs happen and can be observed on the clock signal. Figure 14.28 illustrates a theoretical example of a clock signal modeled as a square wave with added Gaussian noise, and Fig. 14.29 shows a theoretical example of how that clock signal might be affected by a HT. As shown in the figures, if we can capture the backscattered signal of sub-clock samples where the changes caused by HT can be observed, we can detect the presence of HTs. However, the problem with the time-domain signal is that they are often very noisy, therefore, difficult to extract and synchronize measurements to get samples where changes caused by HTs happen.

In contrast, the changes caused by HTs occurring abruptly at some point in the clock cycle can be observed in frequency domain by performing short Time Fourier transformation (STFT) on time-domain signal and observe which frequency components of the time domain signal are affected when dormant HT is present. Figure 14.30 shows Trojan-free and Trojan-affected clock signals in the frequency domain by taking FFT of signals from Figs. 14.28 and 14.29. The signals in the frequency domain are much easier to measure, and the noise power is very small because of focusing a single frequency bin at a time. As a result, instead of measuring the time domain signal, we measure multiple harmonics of the clock in the frequency domain to observe changes in sub-clock samples for HT detection.

14.8.2 Graph Model for Clustering Results

The method for grouping ICs into clusters, based on how HTs (if present) affect their backscattering side-channel signals [30], starts with a vector containing the amplitude ratios of harmonics for the ith board

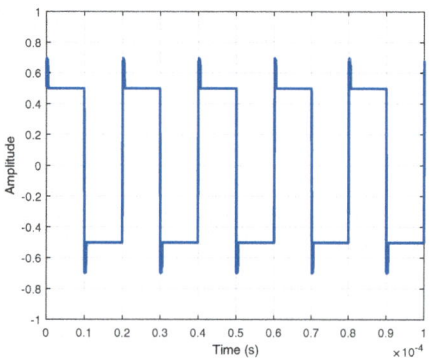

Fig. 14.28 An example of a clock signal with noise [30]

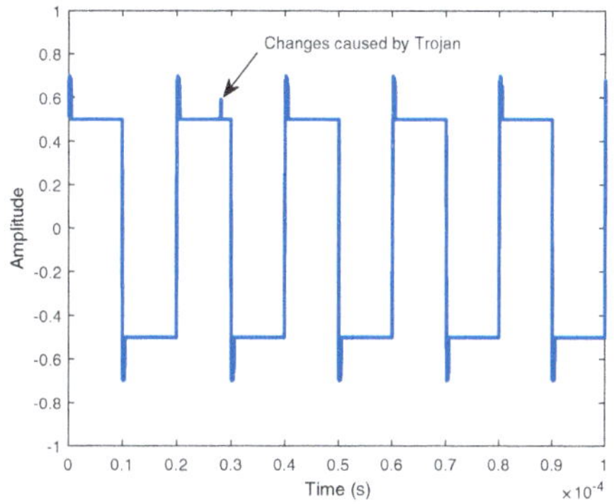

14.8 A Golden-Chip-Free Clustering for Hardware Trojan Detection Using the...

Fig. 14.29 An example of a clock signal affected by hardware Trojan [30]

Fig. 14.30 Trojan-free and Trojan-affected clock signals in frequency domain generated by fast Fourier transforming time domain signals in Figs. 14.28 and 14.29, respectively [30]

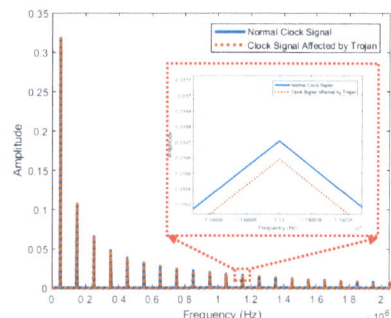

$$\mathbf{y}_i = \begin{bmatrix} y_{i1} & y_{i2} & \cdots & y_{i(N-1)} \end{bmatrix}$$

where

$$y_{ij} = 10 * \log_{10}(\mathbf{h}_i(j+1)/\mathbf{h}_i(j)), \quad (14.1)$$

and $\mathbf{h}_i \in \Re^N$ is a vector containing the harmonic amplitudes for the ith board. Note that these amplitude ratios are the same as those described in Sect. 14.3 and we use the amplitude ratio instead of the amplitude itself to cancel out the attenuation caused by the distance that affects all harmonics. We convert harmonic ratios from linear-domain to dB-domain to prevent the magnitude dominance of the top ratios, and to increase the effect of small harmonic ratios. For clustering, we form matrix $\mathbf{Y}$ that contains he matrix containing the harmonic ratios of all boards:

$$\mathbf{Y} = \begin{bmatrix} - & \mathbf{y}_1 & - \\ - & \mathbf{y}_2 & - \\ & \vdots & \\ - & \mathbf{y}_M & - \end{bmatrix}, \quad (14.2)$$

where M is the number of boards. Our first objective is to reveal hidden information that could be crucial to identifying Trojans in the data, while removing redundant information. A popular technique to reduce the dimensionality of the problem is to keep the significant information by applying Principal Components Analysis (PCA). These methods are especially practical for classification when the data exhibits linear characteristics. Therefore, our first step is to obtain the singular value decomposition (SVD) of $\mathbf{Y}$, which can be written as

$$\mathbf{Y} = \mathbf{U}\mathbf{\Sigma}\mathbf{V}^T. \tag{14.3}$$

We sort the singular values such that the first m singular values are the largest m singular values of the matrix $\mathbf{Y}$. Let then $\mathbf{V}_m$ be a sub-matrix with the first m columns of $\mathbf{V}$ corresponding to these m singular values. We reduce the dimensions of the problem by projecting $\mathbf{Y}$ onto the column space of $\mathbf{V}_m$ as

$$\mathbf{Y}_P = \mathbf{Y}\mathbf{V}_m. \tag{14.4}$$

Here, the value of m is selected so that the power of the projected data is very close to the power of $\mathbf{Y}$, i.e.,

$$\|\mathbf{Y}_P\|_F / \|\mathbf{Y}\|_F \approx 1, \tag{14.5}$$

where $\|\bullet\|_F$ is the Frobenius norm of its argument. For example, in Fig. 14.31, we plot the projected data when $m = 3$, where $\mathbf{Y}_P$ captures 99 % of the power of $\mathbf{Y}$, for a data set in which half of the boards are infected with a Trojan. Here, $\mathbf{s}_j$ denotes the singular value direction corresponding to j^{th} largest singular value.

The next step is to find the clusters in this reduced-size data. The expectation is that each cluster corresponds to a different board group due to either large differences in production, or due to existence of a Trojan. To find the clusters and corresponding centroid points, we use the k-means algorithm. The algorithm requires the number of expected clusters, N_C, and their initial locations, $\mathbf{L}_C \in \Re^{N_C \times m}$ (each row represents the location of the corresponding cluster), as inputs.

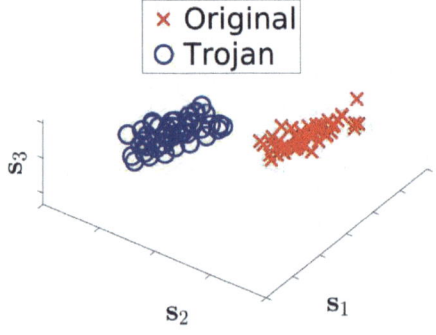

Fig. 14.31 Ground truth information when half of the boards are randomly injected with a Trojan [30]

14.8 A Golden-Chip-Free Clustering for Hardware Trojan Detection Using the...

A careful selection of the initial cluster locations is important to avoid algorithm to converge to a local optimum. To accomplish that, we apply the following procedure to initiate the k-means algorithm:

1. Choose a random sample from the projected data as the location of the first cluster.
2. Find a sample whose total distance is the furthest away from the previously chosen clusters.
3. Repeat until all centroids are initialized.

This procedure ensures wide separation of the centroids. We need to note here that N_C is assumed to be larger than the actual number of clusters in the data, i.e., larger than the number of Trojan types. The assumption follows the fact that we have no information on how many types of Trojan may exist in the testing devices in a realistic scenario. However, having more than the number of actual clusters can be misleading because it can raise suspicion even when there is no Trojan-affected board in the sample space. For example, in Fig. 14.32, we plot the results of the algorithm when $N_C = 6$ for the data given in Fig. 14.31. Comparing the actual labels given in Fig. 14.31, we observe that, while there is no cluster that contains both the original and Trojan-affected circuits, there are several clusters with Original ICs, all of which are actually the same (except for normal manufacturing variation), and several clusters with Trojan ICs, all of which only differ in normal manufacturing variations. Therefore, we need a method that combines redundant clusters to reveal the existence of Trojan-affected circuits more reliably.

To combine redundant clusters, we use the graph method and the shortest path algorithm. To accomplish that, we create a graph where two centroids belong to the same group if they are at the edges of the same arc. Please note that "group" indicates the Trojan type or whether the board is Trojan-affected. Our grouping concept here is that the group of two closest clusters is the same if the distance of these clusters are below some threshold. In other words, the constraint on arcs is that an arc is valid only if the distance between the cluster centroids at the edges is

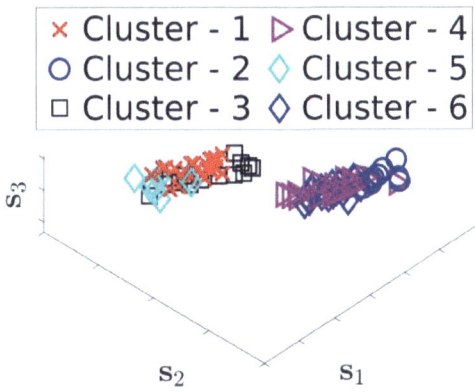

Fig. 14.32 K-means clustering of the boards when the number of center points is chosen to be six [30]

smaller than a given threshold. In that respect, the first step is to obtain a threshold automatically. We can summarize the process of choosing the threshold as follows:

1. Calculate the distance among centroids.
2. Choose the closest two clusters for each cluster, and keep the distances in a list.
3. Assign threshold as the mean distance of this list.

To illustrate how this algorithm works, the graph created by the algorithm is shown in Fig. 14.33a for the clusters in Fig. 14.33b. The nodes corresponding to the same classes are connected. After generating the graph and identifying the valid arcs, the final step is to check whether a node is reachable from other nodes. If there exists a path between any two nodes, we label these as the same type, otherwise, we decide that the sample space contains at least two clusters, therefore, some boards are Trojan-affected. To obtain the connected nodes automatically, we exploit the shortest path algorithm [16] to check whether a node, i.e., a cluster, is reachable from another node. The algorithm returns null if there is no path between two given nodes, and a path if these two nodes are reachable. Based on the outcome of the shortest path, we relabel the sample space indicating whether the connected nodes belong to the same kind. An example of the process is given in Fig. 14.33b. We observe that, although the exact identity of these classes is not known, it is possible to divide data into two groups, and therefore, to determine that the batch contains two circuit designs that are not identical. The implication is that some of the boards are Trojan-affected, or that there are other differences between the two groups that go beyond what ordinary manufacturing variation can explain, i.e., that an IC from each group should be examined in detail (.e.g., reverse-engineered).

14.8.3 Experimental Setup and Testing Scheme Formulation

The measurement setup to evaluate the performance of the clustering algorithm is shown in Fig. 14.34. The setup includes a transmitter Aaronia E1 electric-field

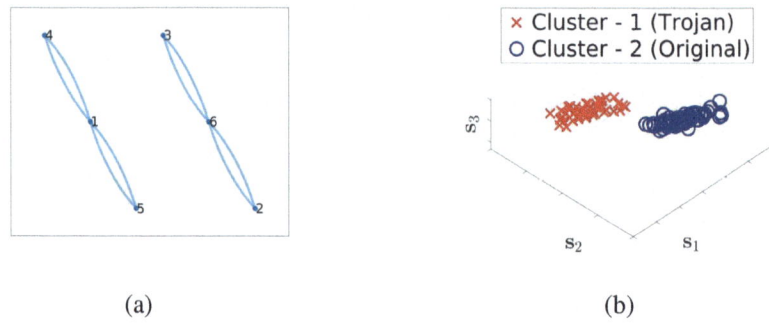

Fig. 14.33 (a) Generation of the graph based on the distances between the centroids of the clusters, (b) Clustering the data into two groups as Trojan injected vs. no-Trojan-free boards. Labels inside the parenthesis indicate the ground truth [30]

14.8 A Golden-Chip-Free Clustering for Hardware Trojan Detection Using the... 453

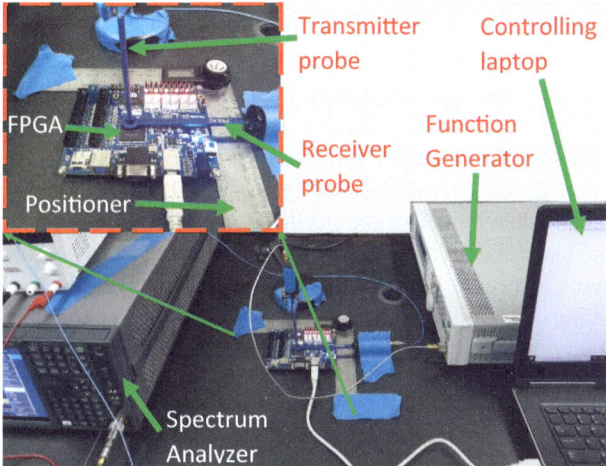

Fig. 14.34 Measurement setup for IC clustering using backscattering side-channel collection for HT detection [30]

near-field probe [1] connected to an Agilent MXG N5183A signal generator [3], and a receiver Aaronia H2 magnetic-field near-field probe [1] connected to an Agilent MXA N9020A spectrum analyzer [4]. The devices-under-test (DuT) are Altera DE0 Cyclone V FPGA boards [2]. An angle ruler is used as a positioner so that different DE0-CV boards can be tested using approximately the same probe positions. A laptop is used to control the devices and automate the measurements. A 3 GHz continuous sinusoid signal is generated by the signal generator, and backscattered signals are recorded by the spectrum analyzer. The measurements are carried in an open environment setup at the room temperature. The effect of environmental conditions such as temperature and voltage source, if existed, should be the same for all clock harmonics, and our technique is based on the ratio between clock harmonics. So that environmental conditions do not significantly affect our accuracy.

As before, we use FPGAs instead of manufacturing ASICs for evaluation because it is much more flexible, time, and cost effective. Although we only evaluate this on FPGAs, there is no reason to believe that the results can not be generalized to ASICs. For example, although the same gate-level design will be smaller in an ASIC, the backscattered signal corresponds to the relative change in impedances, and the relative change of impedances tends to be larger for smaller circuits; if the overall circuit gets smaller in the FPGA-to-ASIC transition and the HT's trigger gets proportionally smaller, a backscattering-based approach may work as well, or possibly even better.

14.8.4 Hardware Trojan Benchmark Implementation

To evaluate our technique, we implement three different benchmark circuits AES, RS232, and PIC16F84 from the TrustHUB Trojan repository [5]. There are total of 21 Trojan designs for AES circuit, 4 Trojan designs for PIC16F84 circuit, and 21 Trojan designs for RS232 circuit. Because numerous HTs in the TrustHub repository are similar to each other, we select circuits that exhibit different approaches for their triggers and payloads. Each of these Trojans has a different triggering mechanism, such as observing a specific sequence of the input, counting number of encryption rounds, observing the number of execution of a specific instruction, etc.. Their different payload functionality also varies, such as shortening the hardware lifetime, leaking private keys, changing the address to program memory, etc. Table 14.1 summarizes the HT benchmarks we use.

The Trojan-affected and Trojan-free designs are carefully mapped to the FPGA by using ECO (Engineering Change Order) tools, so that they have the same layout except for the Trojan part, thus allowing a fair comparison. It is extremely hard to activate an HT without a priori knowledge of its triggering circuit, so it is highly desirable for an HT detection technique to be able to detect HT when it is dormant. As a result, we evaluate our algorithm for dormant HTs. In other words, none of the Trojans is ever activated in any of our experiments.

14.8.5 Testing Scheme Formulation

All HT benchmarks are implemented on Altera DE0 Cyclone V FPGA boards. We use 100 of these boards to prototype a real testing environment. For each HT benchmark, we randomly program each of the 100 boards with HT-free or HT-infected designs, and then record backscattering side-channel signals while each

Table 14.1 Hardware trojan benchmarks and detection results [30]

Benchmark	Size of Trojan (percentage of HT-free circuit)		
	Trigger	Payload	Total
AES-T1200	0.32%	1.61%	1.93%
AES-T500	0.28%	1.51%	1.79%
AES-T700	0.27%	1.76%	2.03%
PIC16F84-T100	1.34%	1.81%	3.15%
PIC16F84-T300	1.37%	1.96%	3.33%
PIC16F84-T400	1.35%	1.75%	3.10%
RS232-T300	1.47%	1.58%	3.05%
RS232-T600	1.50%	1.48%	2.98%
RS232-T901	1.53%	1.61%	3.11%

14.8 A Golden-Chip-Free Clustering for Hardware Trojan Detection Using the... 455

board is running. For each board, we extract the amplitude of the first 40 harmonics of the clock from its backscattering side-channel signal. We only use 40 harmonics because the higher harmonics are very weak (below the noise level). As a result, for each hardware Trojan benchmark, we will have a set of 100 traces (one for each board), and each trace contains 40 points. More formally, the trace for i^{th} board is $\mathbf{h}_i = [h_{i1}, h_{i1}, ..., h_{iN-1}, h_{iN}]$, where $N = 40$, and $1 \leq i \leq 40$. Our clustering algorithm takes these traces as inputs to cluster the ICs.

14.8.6 Evaluation of Existing HT Benchmarks

Overall, our process can be summarized as follows:

→ Collect the data from all 100 boards with the setup given in Fig. 14.34.
→ Compute amplitude ratios for consecutive harmonics, convert them into db-domain,
→ Use them in a matrix to generate $\mathbf{Y}$ as described in Sect. 14.8.2.
→ Obtain SVD of $\mathbf{Y}$, and project it into the space defined by the right-singular vectors corresponding to the largest m singular values to generate $\mathbf{Y}_P$ as described in Sect. 14.8.2. Here, m is chosen such that it is the smallest number of singular values satisfying the following equation:

$$\|\mathbf{Y}_P\|_F / \|\mathbf{Y}\|_F \approx 0.999. \tag{14.6}$$

→ Apply the k-means algorithm by ensuring N_C is larger than the number of possible Trojan types. The initialization of the centroids are done based on the procedure given in Sect. 14.8.2.
→ Generate the graph of similarity with respect to the threshold calculated in Sect. 14.8.2.
→ Apply the shortest path algorithm to reveal possible classes in the sample space. If the algorithm returns more than one cluster, then the batch of boards that are not of the same kind, i.e., some are Trojan-affected.

Since our goal is to separate the Trojan-free designs from all other Trojan-affected designs, we define the accuracy of the measurements as

$$\texttt{accuracy}\ (\%) = \frac{\#\text{of correct labeling}}{\#\text{ of measurements}} \times 100.$$

Please note that the actual labels of the circuits are only required to calculate the accuracy of the proposed method. Therefore, after having the outcome of the procedure given above, we assume that one board from each group was analyzed further to determine its correct labeling, and then apply that label to its entire group.

Thus here one group is labeled as Trojan-free and the other as Trojan-affected. Finally, we compare our labels with the actual labels of each board to calculate the accuracy. If the method classifies all the original designs in a cluster, and if this cluster does not contain any samples from Trojan-affected designs, the accuracy of the algorithm will be equivalent to 100 %.

The tested designs are given in Table 14.1. We first apply our method to the PIC16F84 circuit with 3 different Trojan designs. The results are plotted according to the singular vectors for the largest three singular values. The outcome of the procedure is given in Fig. 14.35a–d. The figures in Fig. 14.35a–c correspond to the scenarios when the batch contains only one Trojan type. However, Fig. 14.35d includes samples from all Trojan designs. The number of singular values that satisfies the condition given in (14.6) is 10, and $N_C = 6$. We also plot for the samples given in Fig. 14.35a the sample distances to each cluster centroid in Fig. 14.36a and their distribution in Fig. 14.36b. The mean distances of "Cluster 1" samples from each of the centroids are 4.96 and 22.27 with standard deviations 3.47 and 5.03, whereas mean distances of "Cluster 2" samples are 23.39 and 6.08 with standard deviations 5.46 and 2.95, respectively. In other words, samples in each cluster are much closer to their own centroid than to the other cluster, so the clustering is very robust. Consequently, we achieve 100 % accuracy for all of the experiments. We need to note here that the legends of the figures do not give any information whether the group is Trojan-affected or original. They only provide the information that the sample space contains two different groups, one of these groups represents the designs with Trojan. However, we provide the actual labels of the classes in parentheses for clarity.

In other experiments, with AES and RS232 circuits, the results are as shown in Fig. 14.35e–h and Fig. 14.35i–l for AES and RS232, respectively. The plots in Fig. 14.35e–h and in Fig. 14.35i–l correspond the experiments when the board batch contains only one Trojan design type for AES and RS232, respectively. The experiments with all considered Trojan designs are shown in Fig. 14.35h, l. We keep the number of clusters, N_C, the same for the PIC16F84 circuit. This time, the number of singular-values satisfying the equation given in (14.6) is 12 for each circuit. We again obtain 100 % accuracy for all these experiments, i.e., all the original circuits are correctly separated from designs that are Trojan-affected.

From the results, we can make the following observations:

(I) The backscattering side channel is a powerful mechanism to separate the Trojan-free from Trojan-affected circuits.
(II) The proposed methodology (backscattered signal plus PCA and k-means algorithm) enables correct clustering of the Trojan-free and Trojan-affected circuits, with no classification errors in our experiments.
(III) When multiple design Trojans exist in IC population, our clustering approach can still separate HT-free from HT-affected ICs. However, if the HTs have similar triggers, they may all be in one cluster together, as is the case here for these HTs.

14.8 A Golden-Chip-Free Clustering for Hardware Trojan Detection Using the... 457

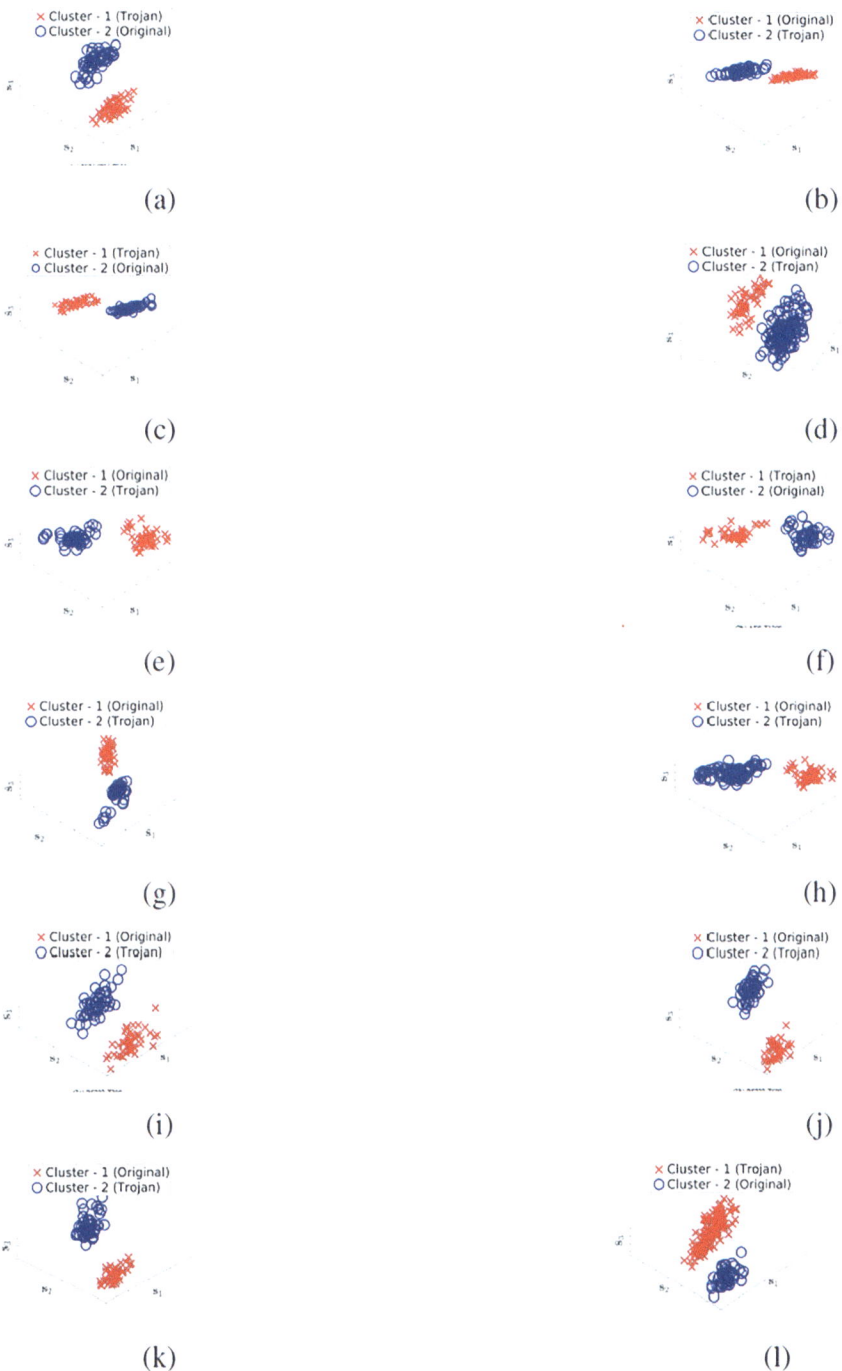

Fig. 14.35 Separation of the Trojan-free and the Trojan-affected circuits. First three columns contain the plots when only one Trojan exists, and the last column of figures are when all considered Trojans exist in the sample space [30]

Fig. 14.36 (a) Distances of each circuit to the cluster centroids. (b) Distribution of distances of each circuit to each cluster centroid [30]

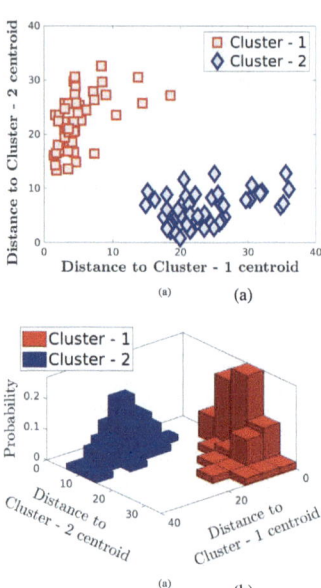

Table 14.2 Hardware trojan benchmarks and detection results [30]

Benchmark	Size of Trojan's trigger (percentage of HT-free circuit)
RS232-T300 w/ 1/2 Trigger Size	0.76%
RS232-T300 w/ 1/4 Trigger Size	0.39%
RS232-T301 w/ 1/8 Trigger Size	0.19%

14.8.7 Evaluation of Changing Size of Hardware Trojan Triggers

Because the algorithm performs so well on the existing HT designs from Table 14.1, this section focuses on testing the limit of our algorithm by reducing the size of HTs. Recall that only the trigger is active, while the payload stays inert, when hardware Trojans are dormant. If the trigger is big enough, Trojan can be detected regardless of its payload size. Therefore, we will focus on changing the size of the trigger to test the limits of the clustering approach. The RS232-T300 Trojan is chosen for this experiment because its trigger can be meaningfully resized. As we change the size of the trigger, we keep the payload the same, to create test designs that are summarized in Table 14.2.

We first investigate whether the clustering method still works when only one HT benchmark exists in the board batch. The same parameters from Sect. 14.8.6 are used for the number of clusters and singular vectors. The clustering results are given in Fig. 14.37, with 100 % accuracy in terms of separating the original circuits from the Trojan-affected ones. Here, one important observation is that, as the size

14.8 A Golden-Chip-Free Clustering for Hardware Trojan Detection Using the... 459

Fig. 14.37 Separation of the Trojan-free and the Trojan-affected circuits when the size of RS232-T300 varies [30]

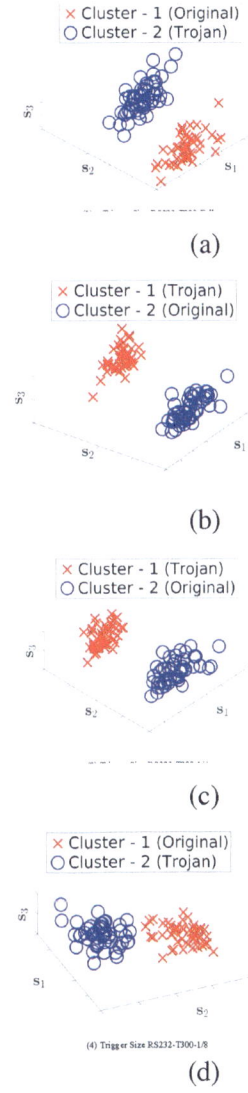

of the Trojan trigger decreases, the distance between centroids of the two classes decreases, i.e., HT-afflicted IC does become more similar to the original HT-free IC when the HT has only 1/8 trigger size compared to its full-size trigger. To illustrate this more directly, we show clustering results where the measurements from all trigger sizes were included, i.e., there are five different designs, one HT-free and four variants of an HT-infected design (with different trigger sizes), are subjected to our clustering technique. The results are shown in Fig. 14.38, with actual (ground-truth) labels (left) and with clustering-produced labels (right). In terms of separating HT-free from HT-infected designs, the accuracy of this clustering is still 100 % (all

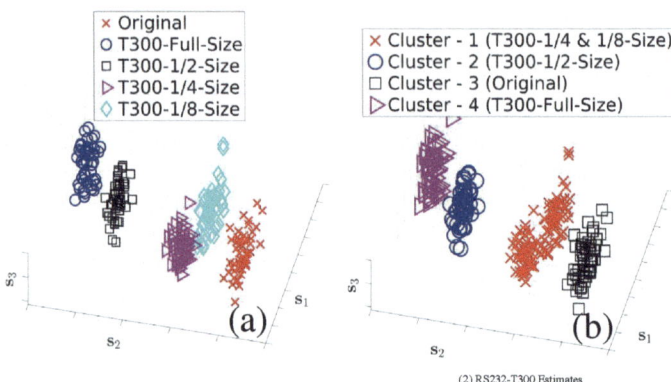

Fig. 14.38 Separation of original and Trojan-affected circuits when the size of RS232-T300 varies. The experiments are performed with original, full-Trigger-size, 1/2-Trigger-size, 1/4-Trigger-size, and 1/8-Trigger-size circuits [30]

HT-free instances are in one cluster while all HT-infected instances are in other clusters). Furthermore, the technique is able to distinguish (put in separate clusters) different variants of the HT, except for the variants with 1/4 and 1/8 triggers, which are in the same cluster. We emphasize that the technique is able to distinguish the 1/8-trigger variant from an HT-free design with the difference between them being 1/8 of the full trigger, even though it did not distinguish 1/4- from the 1/8-trigger variant (the difference among them is also 1/8 of the full trigger). This is because the additional trigger activity in the 1/4 variant is similar to the trigger activity in the 1/8 variant, i.e., it is only a matter of *how much* trigger activity the design has. In contrast, the HT-free design has no trigger activity at all, so the presence of trigger activity in the 1/8 design allows it to be well-separated from the HT-free design. This implies that HTs whose circuitry and activity mimics that of the original design would be more difficult to detect, but only up to a point—even such activity-mimicking HTs would be detected if they are sufficiently large (in this particular experiment, larger than 0.19 % of the original circuit).

Based on the results given in this section (and Sect. 14.8.6), our main observation is that our technique successfully separates HT-free from HT-infected designs, even for very small HTs (0.19 % of the original circuit, in our experiments). Additionally, the technique successfully separates different HT designs from each other, except when the HTs only differ in size (but not nature) of their trigger circuitry, and that difference in size is very small (0.19 % of the original circuit, in our experiments).

References

1. Em probe. http://www.aaronia.com/products/antennas/Near-Field-Probe-Set-PBS2.
2. Fpga. https://www.terasic.com.tw/cgi-bin/page/archive.pl?Language=English&CategoryNo=167&No=921&PartNo=2.

References 461

3. Signal generator. https://www.keysight.com/en/pdx-x201724-pn-N5183A/mxg-microwave-analog-signal-generator-100-khz-to-40-ghz?pm=spc&nid=-32490.1150253&cc=US&lc=eng.
4. Spectrum analyzer. https://www.keysight.com/en/pdx-x202266-pn-N9020A/mxa-signal-analyzer-10-hz-to-265-ghz?pm=spc&nid=-32508.1150426&cc=US&lc=eng.
5. Trusthub. http://www.trust-hub.org/benchmarks/trojan.
6. Dakshi Agrawal, Selcuk Baktir, Deniz Karakoyunlu, Pankaj Rohatgi, and Berk Sunar. Trojan detection using ic fingerprinting. In *Security and Privacy, 2007. SP'07. IEEE Symposium on*, pages 296–310. IEEE, 2007.
7. Josep Balasch, Benedikt Gierlichs, and Ingrid Verbauwhede. Electromagnetic circuit fingerprints for hardware trojan detection. In *Electromagnetic Compatibility (EMC), 2015 IEEE International Symposium on*, pages 246–251. IEEE, 2015.
8. Mainak Banga and Michael S Hsiao. A region based approach for the identification of hardware trojans. In *Hardware-Oriented Security and Trust, 2008. HOST 2008. IEEE International Workshop on*, pages 40–47. IEEE, 2008.
9. Mainak Banga and Michael S Hsiao. Vitamin: Voltage inversion technique to ascertain malicious insertions in ics. 2009.
10. Chongxi Bao, Domenic Forte, and Ankur Srivastava. Temperature tracking: Toward robust run-time detection of hardware trojans. *IEEE Transactions on Computer-Aided Design of Integrated Circuits and Systems*, 34(10):1577–1585, 2015.
11. Swarup Bhunia, Michael S Hsiao, Mainak Banga, and Seetharam Narasimhan. Hardware trojan attacks: threat analysis and countermeasures. *Proceedings of the IEEE*, 102(8):1229–1247, 2014.
12. Byeongju Cha and Sandeep K Gupta. Efficient trojan detection via calibration of process variations. In *Test Symposium (ATS), 2012 IEEE 21st Asian*, pages 355–361. IEEE, 2012.
13. Byeongju Cha and Sandeep K Gupta. Trojan detection via delay measurements: A new approach to select paths and vectors to maximize effectiveness and minimize cost. In *Proceedings of the conference on design, automation and test in Europe*, pages 1265–1270. EDA Consortium, 2013.
14. Rajat Subhra Chakraborty, Seetharam Narasimhan, and Swarup Bhunia. Hardware trojan: Threats and emerging solutions. In *High Level Design Validation and Test Workshop, 2009. HLDVT 2009. IEEE International*, pages 166–171. IEEE, 2009.
15. Wesley K Clark and Peter L Levin. Securing the information highway. *Foreign Aff.*, 88:2, 2009.
16. Thomas H Cormen, Charles E Leiserson, Ronald L Rivest, and Clifford Stein. *Introduction to algorithms*. 2009.
17. Domenic Forte, Chongxi Bao, and Ankur Srivastava. Temperature tracking: An innovative run-time approach for hardware trojan detection. In *Proceedings of the International Conference on Computer-Aided Design*, pages 532–539. IEEE Press, 2013.
18. Ujjwal Guin, Ke Huang, Daniel DiMase, John M Carulli, Mohammad Tehranipoor, and Yiorgos Makris. Counterfeit integrated circuits: A rising threat in the global semiconductor supply chain. *Proceedings of the IEEE*, 102(8):1207–1228, 2014.
19. Chunhua He, Bo Hou, Liwei Wang, Yunfei En, and Shaofeng Xie. A failure physics model for hardware trojan detection based on frequency spectrum analysis. In *Reliability Physics Symposium (IRPS), 2015 IEEE International*, pages PR-1. IEEE, 2015.
20. Jiaji He, Yiqiang Zhao, Xiaolong Guo, and Yier Jin. Hardware trojan detection through chip-free electromagnetic side-channel statistical analysis. *IEEE Transactions on Very Large Scale Integration (VLSI) Systems*, 25(10):2939–2948, 2017.
21. Kangqiao Hu, Abdullah Nazma Nowroz, Sherief Reda, and Farinaz Koushanfar. High-sensitivity hardware trojan detection using multimodal characterization. In *Proceedings of the Conference on Design, Automation and Test in Europe*, pages 1271–1276. EDA Consortium, 2013.

22. Ramesh Karri, Jeyavijayan Rajendran, Kurt Rosenfeld, and Mohammad Tehranipoor. Trustworthy hardware: Identifying and classifying hardware trojans. *Computer*, 43(10):39–46, 2010.
23. Lok-Won Kim and John D Villasenor. Dynamic function replacement for system-on-chip security in the presence of hardware-based attacks. *IEEE Transactions on Reliability*, 63(2):661–675, 2014.
24. Lok-Won Kim, John D Villasenor, et al. A trojan-resistant system-on-chip bus architecture. In *Military Communications Conference, 2009. MILCOM 2009. IEEE*, pages 1–6. IEEE, 2009.
25. Maxime Lecomte, Jacques Fournier, and Philippe Maurine. An on-chip technique to detect hardware trojans and assist counterfeit identification. *IEEE Transactions on Very Large Scale Integration (VLSI) Systems*, 25(12):3317–3330, 2017.
26. D McIntyre, F Wolff, C Papachristou, Swarup Bhunia, and D Weyer. Dynamic evaluation of hardware trust. In *Hardware-Oriented Security and Trust, 2009. HOST'09. IEEE International Workshop on*, pages 108–111. IEEE, 2009.
27. Seetharam Narasimhan, Dongdong Du, Rajat Subhra Chakraborty, Somnath Paul, Francis Wolff, Christos Papachristou, Kaushik Roy, and Swarup Bhunia. Multiple-parameter side-channel analysis: A non-invasive hardware trojan detection approach. In *Hardware-Oriented Security and Trust (HOST), 2010 IEEE International Symposium on*, pages 13–18. IEEE, 2010.
28. Xuan Thuy Ngo, Zakaria Najm, Shivam Bhasin, Sylvain Guilley, and Jean-Luc Danger. Method taking into account process dispersion to detect hardware trojan horse by side-channel analysis. *Journal of Cryptographic Engineering*, 6(3):239–247, 2016.
29. Luong N. Nguyen, Chia-Lin Cheng, Milos Prvulovic, and Alenka Zajic. Creating a backscattering side channel to enable detection of dormant hardware trojans. *IEEE Transactions on Very Large Scale Integration (VLSI) Systems*, 27(7):1561–1574, 2019.
30. Luong N. Nguyen, Baki Berkay Yilmaz, Milos Prvulovic, and Alenka Zajic. A novel golden-chip-free clustering technique using backscattering side channel for hardware trojan detection. In *2020 IEEE International Symposium on Hardware Oriented Security and Trust (HOST)*, pages 1–12, 2020.
31. Abdullah Nazma Nowroz, Kangqiao Hu, Farinaz Koushanfar, and Sherief Reda. Novel techniques for high-sensitivity hardware trojan detection using thermal and power maps. *IEEE Transactions on Computer-Aided Design of Integrated Circuits and Systems*, 33(12):1792–1805, 2014.
32. Hassan Salmani. Cotd: reference-free hardware trojan detection and recovery based on controllability and observability in gate-level netlist. *IEEE Transactions on Information Forensics and Security*, 12(2):338–350, 2017.
33. Bicky Shakya, Tony He, Hassan Salmani, Domenic Forte, Swarup Bhunia, and Mark Tehranipoor. Benchmarking of hardware trojans and maliciously affected circuits. *Journal of Hardware and Systems Security*, 1(1):85–102, 2017.
34. Mohammad Tehranipoor and Farinaz Koushanfar. A survey of hardware trojan taxonomy and detection. *IEEE design & test of computers*, 27(1), 2010.
35. Randy Torrance and Dick James. The state-of-the-art in ic reverse engineering. In *Cryptographic Hardware and Embedded Systems-CHES 2009*, pages 363–381. Springer, 2009.
36. John Villasenor. The hacker in your hardware. *Scientific American*, 303(2):82–87, 2010.
37. John Villasenor. *Compromised by design?: Securing the defense electronics supply chain*. Center for Technology Innovation at Brookings, 2013.
38. Adam Waksman, Matthew Suozzo, and Simha Sethumadhavan. Fanci: identification of stealthy malicious logic using boolean functional analysis. In *Proceedings of the 2013 ACM SIGSAC conference on Computer & communications security*, pages 697–708. ACM, 2013.
39. Xiaoxiao Wang, Mohammad Tehranipoor, and Jim Plusquellic. Detecting malicious inclusions in secure hardware: Challenges and solutions. In *Hardware-Oriented Security and Trust, 2008. HOST 2008. IEEE International Workshop on*, pages 15–19. IEEE, 2008.
40. Kan Xiao, Domenic Forte, Yier Jin, Ramesh Karri, Swarup Bhunia, and M Tehranipoor. Hardware trojans: Lessons learned after one decade of research. *ACM Transactions on Design Automation of Electronic Systems (TODAES)*, 22(1):6, 2016.

41. Qiaoyan Yu and Jonathan Frey. Exploiting error control approaches for hardware trojans on network-on-chip links. In *Defect and Fault Tolerance in VLSI and Nanotechnology Systems (DFT), 2013 IEEE International Symposium on*, pages 266–271. IEEE, 2013.
42. Jie Zhang, Feng Yuan, Linxiao Wei, Yannan Liu, and Qiang Xu. Veritrust: Verification for hardware trust. *IEEE Transactions on Computer-Aided Design of Integrated Circuits and Systems*, 34(7):1148–1161, 2015.
43. Jie Zhang, Feng Yuan, and Qiang Xu. Detrust: Defeating hardware trust verification with stealthy implicitly-triggered hardware trojans. In *Proceedings of the 2014 ACM SIGSAC Conference on Computer and Communications Security*, pages 153–166. ACM, 2014.

Index

A
Acoustic side channel, 7, 13
Active analog side channel, 10, 31
Advanced-persistent-threat attack, 165
AES circuit, 427
Alternation frequency, 95, 115
Amplitude modulated signal, 77
Amplitude Modulation, 38
Analog side channel, 7, 13
Angle Modulation, 38
Attestation, 323
Autocorrelation function, 118

B
Backscattering side-channel, 8, 25, 109, 430
Basic block program tracking, 240
Bit error rate (BER), 122
Bit resolution, 81
Bounds on the capacity of channels, 115
Branch prediction, 115

C
Cache hits and misses, 115
Cache miss, 324
Capacity, 84
Capacity of side channel created by execution of series of instructions, 131
Center frequency, 77, 81
Challenge-response paradigm, 323
Checksum computation, 324
Clock frequency, 156
Code reuse attack, 162

Computer hardware, 57
Convolutional neural network (CNN), 191
Counterfeits, 345
Coupling noise, 84
Covert channel communication, 113, 114
curve25519-donna, 403
Cyber-physical system, 151, 225

D
Data rate, 84
DDoS bot, 167
Deletion, 115
Dictionary, 176
Dictionary reduction, 176
Differences in instruction execution, 74
Differential analysis, 11
Digital circuits, 13
Digital communications, 43
Distance-dependent pathloss, 99
Distance-dependent path loss model, 101
Dynamic root of trust, 325
Dynamic scheduling of instructions, 115

E
Electromagnetic emanations, 74, 115, 164, 227
Electromagnetic side channel, 7, 13, 298, 325, 345
Electromagnetic side channel energy (ESE), 57
Electromagnetic side channel energy of sequence of instructions, 67
Elliptic curve digital signature algorithm (ESDCA), 376

EM-monitoring attestation (EMMA), 326
Emanated signal power (ESP), 68
Embedded device, 323
Environment-dependent shadowing gain, 102
Euclidean distance, 386

F
Far-field, 94
Fixed-window implementation, 87
FLUSH+RELOAD, 374
Fourier transform, 77
Frequency, 77
Frequency-domain signal, 77, 152

H
HACL, 403
Henecka's algorithm, 390
Horizontal polarization, 99

I
Input/output-observable side channel, 2
Insertion, 115
Integrated circuits, 345, 419
Internet of Things (IoT), 151
Inverse Fourier transform, 78

J
Jitter, 122

K
K-Means clustering algorithm, 386

L
Large-number multiplication, 373
Libgcrypt, 403
Lower sideband, 95
Low-frequency sinusoids, 78
Low-latency attacks, 325
Low-pass filter, 123

M
Magnitude, 77
Markovian convolutional neural network for malware detection, 191
Markov model based program profiling, 193
Markov source model, 132
Markov state model, 191
Measured pairwise side channel signal power, 67

Memory-copy attack, 335
Memory profiling using analog emanations, 280
Memory-shadow attack, 337
Mibench applications, 167, 218
Micro-architectural events, 324
Model for covert channel communication system, 117
Modeling information leakage from computer program, 132
Modular exponentiation, 88
Montgomery multiplication, 372

N
Near-field, 94

O
OpenSSL, 87, 403
Orientation-dependent shadowing gain, 102
Oscilloscope, 17

P
Pairwise side-channel signal power available to an attacker, 57
Passive analog side channel, 10, 31
Path loss, 94
Payload, 420
Percival's attack, 374
Permutations-based instruction tracking, 255
Phase, 77
Photonic side channel, 8, 13, 31
Physically-observable side channel, 2, 114
Pipeline stall, 115
Power side channel, 7, 13
Power spectral density (PSD), 122
Printed circuit boards (PCB), 345
Probability of detection, 114
Profiling network interrupts, 314
Profiling page faults, 310
Programmable logic controllers, 151
Program profiling, 211
Program profiling for multiple cores, 203
Propagation of side channel signals, 96
Proxy attack, 339
Pseudo-Random Number Generator (PRNG), 327

R
Radar cross-section, 425
Random erasures, 391

Ransomwmare, 167
Real-time oscilloscope, 17
Recognition/authentication of electronic components, 346
Relationship between emanated energy and program execution, 74
Request-to-response time, 324
Resolution bandwidth, 77, 82
Return-oriented programming, 338
Rootkit-based attack, 338

S

Sampling oscilloscope, 17
Sampling rate, 80, 156
Screaming side channel, 42
Shadow gain model, 99
Shadowing gain, 99
Shannon's information theory, 115
Shannon's Sampling Theorem, 81
Shellcode attack, 166
Shielded loop probes, 107
Side channel, 1
Side channel attacks, 1
Side-channel measurements, 373
Signal bandwidth, 77
Signal compression and processing, 349
Signal-to-noise ratio (SNR), 84, 255
Simple analysis, 11
Single instruction ESE, 58
Sliding-window implementation of exponentiation, 87, 374
Software-defined radio, 18
Software-observable side channel, 2, 113
Span, 77, 82
Spectral features, 347

Spectral profiling, 212
Spectral sample, 156
Spectrogram, 42
Spectrum analyzer, 17, 77, 218
Stuxnet, 165
Sum of sinusoidal functions, 77
Sweep time, 77, 82
Synchronization errors, 115

T

Temperature side channel, 7, 13
Thermal noise, 83
Time-domain analysis, 77, 224
Time-domain signal, 77, 152
Time resolution, 156
Trigger, 420

U

Unintentional carrier, 95
Unintentionally modulated emanation, 57
Unintentionally modulated side channel, 95
Upper sideband, 95

V

Vertical polarization, 99

W

Window-based exponentiation, 374

Z

Zero-overhead profiling (ZoP), 225

SPRINGER NATURE

GPSR Compliance

The European Union's (EU) General Product Safety Regulation (GPSR) is a set of rules that requires consumer products to be safe and our obligations to ensure this.

If you have any concerns about our products, you can contact us on ProductSafety@springernature.com

In case Publisher is established outside the EU, the EU authorized representative is:

Springer Nature Customer Service Center GmbH
Europaplatz 3
69115 Heidelberg, Germany

The manufacturer's authorised representative in the EU is Springer Nature Customer Service Centre GmbH, Europaplatz 3, 69115 Heidelberg, Germany. If you have any concerns regarding our products, please contact ProductSafety@springernature.com

Printed and bound by CPI Group (UK) Ltd, Croydon, CR0 4YY

26/03/2026

02078983-0001